Perinatal and Developmental Epigenetics

Translational Epigenetics Series

Trygve Tollefsbol, Ph.D., D.O., Distinguished Professor - Series Editor

Professor of Biology, University of Alabama at Birmingham, and Senior Scientist, Comprehensive Cancer Center, Comprehensive Center for Healthy, Birmingham, AL, United States
Aging, Comprehensive Diabetes Center and Nutrition Obesity Research Center, Birmingham, AL, United States
Director, Cell Senescence Culture Facility, Birmingham, AL, United States

Transgenerational Epigenetics
Edited by Trygve O. Tollefsbol, 2014

Personalized Epigenetics
Edited by Trygve O. Tollefsbol, 2015

Epigenetic Technological Applications
Edited by Y. George Zheng, 2015

Epigenetic Cancer Therapy
Edited by Steven G. Gray, 2015

DNA Methylation and Complex Human Disease
By Michel Neidhart, 2015

Epigenomics in Health and Disease
Edited by Mario F. Fraga and Agustin F. F Fernández, 2015

Epigenetic Gene Expression and Regulation
Edited by Suming Huang, Michael Litt and C. Ann Blakey, 2015

Epigenetic Biomarkers and Diagnostics
Edited by Jose Luis García-Giménez, 2015

Drug Discovery in Cancer Epigenetics
Edited by Gerda Egger and Paola Barbara Arimondo, 2015

Medical Epigenetics
Edited by Trygve O. Tollefsbol, 2016

Chromatin Signaling and Diseases
Edited by Olivier Binda and Martin Fernandez-Zapico, 2016

Genome Stability
Edited by Igor Kovalchuk and Olga Kovalchuk, 2016

Chromatin Regulation and Dynamics
Edited by Anita Göndör, 2016

Neuropsychiatric Disorders and Epigenetics
Edited by Dag H. Yasui, Jacob Peedicayil and Dennis R. Grayson, 2016

Polycomb Group Proteins
Edited by Vincenzo Pirrotta, 2016

Epigenetics and Systems Biology
Edited by Leonie Ringrose, 2017

Cancer and Noncoding RNAs
Edited by Jayprokas Chakrabarti and Sanga Mitra, 2017

Nuclear Architecture and Dynamics
Edited by Christophe Lavelle and Jean-Marc Victor, 2017

Epigenetic Mechanisms in Cancer
Edited by Sabita Saldanha, 2017

Epigenetics of Aging and Longevity
Edited by Alexey Moskalev and Alexander M. Vaiserman, 2017

The Epigenetics of Autoimmunity
Edited by Rongxin Zhang, 2018

Epigenetics in Human Disease, Second Edition
Edited by Trygve O. Tollefsbol, 2018

Epigenetics of Chronic Pain
Edited by Guang Bai and Ke Ren, 2018

Epigenetics of Cancer Prevention
Edited by Anupam Bishayee and Deepak Bhatia, 2018

Computational Epigenetics and Diseases
Edited by Loo Keat Wei, 2019

Pharmacoepigenetics
Edited by Ramón Cacabelos, 2019

Epigenetics and Regeneration
Edited by Daniela Palacios, 2019

Chromatin Signaling and Neurological Disorders
Edited by Olivier Binda, 2019

Transgenerational Epigenetics, Second Edition
Edited by Trygve Tollefsbol, 2019

Nutritional Epigenomics
Edited by Bradley Ferguson, 2019

Prognostic Epigenetics
Edited by Shilpy Sharma, 2019

Epigenetics of the Immune System
Edited by Dieter Kabelitz, 2020

Stem Cell Epigenetics
Edited by Eran Meshorer and Giuseppe Testa, 2020

Epigenetics Methods
Edited by Trygve Tollefsbol, 2020

Histone Modifications in Therapy
Edited by Pedro Castelo-Branco and Carmen Jeronimo, 2020

Environmental Epigenetics in Toxicology and Public Health
Edited by Rebecca Fry, 2020

Genome Stability
Edited by Igor Kovalchuk and Olga Kovalchuk, 2021

Twin and Family Studies of Epigenetics
Edited by Shuai Li and John Hopper, 2021

Epigenetics and Metabolomics
Edited by Paban K. Agrawala and Poonam Rana, 2021

Medical Epigenetics, Second Edition
Edited by Trygve Tollefsbol, 2021

Epigenetics in Precision Medicine
Edited by José Luis García-Giménez, 2022

Epigenetics of Stress and Stress Disorders
Edited by Nagy A. Youssef, 2022

Post-Transcriptional Gene Regulation in Human Disease
Edited by Buddhi Prakash Jain, Shyamal K. Goswami and Tapan Sharma, 2022

Epigenetics in Organ Specific Disorders
Edited by Chandra S. Boosani and Ritobrata Goswami, 2022

Translational Epigenetics

Perinatal and Developmental Epigenetics

Volume 35

Edited by

Garima Singh

Assistant Professor, Department of Pathology, Anatomy and Cell Biology, Thomas Jefferson University, Philadelphia, PA, USA

ELSEVIER

ACADEMIC PRESS

An imprint of Elsevier

Academic Press is an imprint of Elsevier
125 London Wall, London EC2Y 5AS, United Kingdom
525 B Street, Suite 1650, San Diego, CA 92101, United States
50 Hampshire Street, 5th Floor, Cambridge, MA 02139, United States
The Boulevard, Langford Lane, Kidlington, Oxford OX5 1GB, United Kingdom

ISBN: 978-0-12-821785-6

For information on all Academic Press publications visit our website at https://www.elsevier.com/books-and-journals

Publisher: Stacy Masucci
Acquisitions Editor: Peter B. Linsley
Editorial Project Manager: Matthew Mapes
Production Project Manager: Stalin Viswanathan
Cover Designer: Matthew Limbert

Typeset by TNQ Technologies

Contents

SECTION 4 Epigenetic programming by adverse perinatal influences

SECTION 5 Pharmacological and nonpharmacological approaches to reprogram the developmental programming to improve the health outcomes

Contributors

Taqveema Ali
Postgraduate Institute of Medical Education and Research, Chandigarh, Punjab, India

Nafisa H. Balasinor
National Institute for Research in Reproductive and Child Health, Department of Neuroendocrinology, Mumbai, Maharashtra, India

Taylor S. Campbell
University of Delaware, Psychological and Brain Sciences, Newark, DE, United States

Akin Cayir
Vocational Health College, Canakkale Onsekiz Mart University, Canakkale, Turkey

Keran Chen
Section of Cell Biology and Functional Genomics, Department of Metabolism, Digestion and Reproduction, Faculty of Medicine, Imperial College London, London, United Kingdom

Laura R. Cortes
Neuroscience Institute, Georgia State University, Atlanta, GA, United States

Chloe M. Farrell
Institute of Psychiatry, Psychology & Neuroscience (IoPPN), King's College London, London, United Kingdom

Nancy G. Forger
Neuroscience Institute, Georgia State University, Atlanta, GA, United States

Zhengying He
Section of Cell Biology and Functional Genomics, Department of Metabolism, Digestion and Reproduction, Faculty of Medicine, Imperial College London, London, United Kingdom

Lauren E. Hood
Department of Psychology, Arizona State University, Tempe, AZ, United States

Jiachun Jiang
Section of Cell Biology and Functional Genomics, Department of Metabolism, Digestion and Reproduction, Faculty of Medicine, Imperial College London, London, United Kingdom

Kirtan Kaur
Department of Medicine, Hematology, and Medical Oncology, Icahn School of Medicine at Mount Sinai, New York, NY, United States

Jyotdeep Kaur
Postgraduate Institute of Medical Education and Research, Chandigarh, Punjab, India

Shagufta Khan
CSIR - Centre for Cellular and Molecular Biology, Hyderabad, Telangana, India

Jonna M. Leyrer-Jackson
Department of Medical Education, School of Medicine, Creighton University, Phoenix, AZ, United States

Simeiyun Liu
Department of Molecular, Cell and Developmental Biology, University of California, Santa Cruz, CA, United States

X. Shawn Liu
Department of Physiology & Cellular Biophysics, Columbia Stem Cell Initiative, Herbert Irving Comprehensive Cancer Center, Columbia University Medical Center, Columbia University, New York, NY, United States

Steven J. Millership
Section of Cell Biology and Functional Genomics, Department of Metabolism, Digestion and Reproduction, Faculty of Medicine, Imperial College London, London, United Kingdom

Rakesh K. Mishra
CSIR - Centre for Cellular and Molecular Biology, Hyderabad, Telangana, India; Tata Institute for Genetics and Society, Bangalore, Karnataka, India

Kumari Nishi
National Institute for Research in Reproductive and Child Health, Department of Neuroendocrinology, Mumbai, Maharashtra, India

M. Foster Olive
Department of Psychology, Arizona State University, Tempe, AZ, United States

Beenish Rahat
National Institute of Child Health and Human Development, National Institutes of Health, Bethesda, MD, United States

Valerie Jane Robert
Laboratory of Biology and Modeling of the Cell (LBMC), Ecole Normale Supérieure de Lyon, CNRS, UMR 5239, Inserm, U1293, Université Claude Bernard Lyon 1, Lyon, France

Tania L. Roth
University of Delaware, Psychological and Brain Sciences, Newark, DE, United States

Guy A. Rutter
Section of Cell Biology and Functional Genomics, Department of Metabolism, Digestion and Reproduction, Faculty of Medicine, Imperial College London, London, United Kingdom; CHUM Research Center and University of Montréal, Montréal, QC, Canada; Lee Kong Chian School of Medicine, Nanyang Technological University, Singapore, Singapore

Joshua F. Shaffer
Department of Molecular, Cell and Developmental Biology, University of California, Santa Cruz, CA, United States

Renuka Sharma
Postgraduate Institute of Medical Education and Research, Chandigarh, Punjab, India

Upasna Sharma
Department of Molecular, Cell and Developmental Biology, University of California, Santa Cruz, CA, United States

Surabhi Srivastava
CSIR - Centre for Cellular and Molecular Biology, Hyderabad, Telangana, India; Tata Institute for Genetics and Society, Bangalore, Karnataka, India

Zoe Stamoulis
Section of Cell Biology and Functional Genomics, Department of Metabolism, Digestion and Reproduction, Faculty of Medicine, Imperial College London, London, United Kingdom

Minglu Wang
Department of Physiology & Cellular Biophysics, Columbia University Medical Center, Columbia University, New York, NY, United States

Vanessa Yu
Section of Cell Biology and Functional Genomics, Department of Metabolism, Digestion and Reproduction, Faculty of Medicine, Imperial College London, London, United Kingdom

SECTION 1

The ever-growing complexity of epigenetic regulation of gene expression

CHAPTER

1

Epigenetic regulation of gene expression: an overview of classical and recently discovered novel players

Beenish Rahat[1], Renuka Sharma[2], Taqveema Ali[2], Jyotdeep Kaur[2]

[1]*National Institute of Child Health and Human Development, National Institutes of Health, Bethesda, MD, United States;* [2]*Postgraduate Institute of Medical Education and Research, Chandigarh, Punjab, India*

Introduction

The inefficiency of nuclear cloning as a result of associated epigenetic abnormalities highlights the crucial role of epigenetic modifications and their reprogramming during embryonic development.[1] Epigenetic mechanisms are defined as the processes that cause stable heritable alterations in gene expression without altering the deoxyribonucleic acid (DNA) sequence.[2]

The mammalian life cycle goes through a unique sequence of steps that changes a fertilized cell zygote into a complex, multicellular adult organism. It includes critical reprogramming in germ cells and the zygote preparing the genome for a new cycle. This is followed by extended phases of cellular differentiation with the help of transcription factors, which promotes strict lineage restriction—based development toward a particular cell type. Such regulations are both encoded in *cis* and directed by transcription factors as well as in heritable, covalent modifications to DNA and histones known as epigenetic mechanisms that promote cell fate decisions and avoid reversible changes.[3] Various epigenetic mechanisms such as DNA methylation, genomic imprinting, histone modifications, and noncoding RNAs (ncRNAs) work to set up epigenome that propagates gene expression profiles through multiple rounds of meiosis and mitosis.[4]

Although different cells in a multicellular organism have similar genetic material, differential expression of the genes regulated by multiple epigenetic mechanisms makes these cells functionally heterogenous. Such differential expression is established during early development and maintained through mitosis. Epigenetic modifications can also arise under the influence of environmental changes in mature humans and mice.[5]

Perinatal and Developmental Epigenetics. https://doi.org/10.1016/B978-0-12-821785-6.00010-4

DNA methylation plays an essentially important role during embryonic development regulating crucial developmental processes such as genomic imprinting, X-inactivation, and gene regulation. Advances in the detection sequencing technologies have led to better understanding of the role of DNA methylation and its distribution over different genomic regions such as promoters, exons, and imprinted control regions (ICRs). DNA methylation plays a critical role during epigenetic reprogramming in primordial germ cells (PGCs) and preimplantation developmental stages.[4]

Genomic imprinting regulates allele-specific expression of imprinted genes in parent-of-origin-specific manner. Different types of epigenetic marks present at the ICRs of these imprinted genes maintain allele-specific expression of these genes. These epigenetic marks are established on the maternal and paternal allele during gamete formation in germline cells and then maintained after fertilization.[6] Genomic imprinting regulates normal development of both placenta and the fetus. Abnormal genomic imprinting in placenta not only alters the functioning of the placenta but also affects the normal fetal development. Placenta is known to express various imprinted genes, which are more sensitive to early environmental cues.[7]

Histone methylation is related to early developmental process and plays a crucial role in differentiation of trophectoderm (TE) and inner cell mass (ICM). Various types of known histone modifications are linked specifically to both gene activation and gene repression. Recent studies on the role of noncoding RNAs (ncRNAs) have emphasized the importance of these RNAs in various cellular processes such as in gene expression during development.

Alteration in these epigenetic mechanisms results in the loss of cellular identity, cellular transformation, or development of several diseases.[4] Environmental factors can specifically influence epigenetic modifications, leading to defective cellular differentiation as a consequence developing multiple disorder during pregnancy and neonatal period. Various antenatal as well as postnatal factors, such as pollution, maternal nutrition, neonatal diet, and the composition of microbiota, promote epigenetic alterations changing the adaptation of particular individual to the environment and in addition can also enhance the lifelong health risk and making an individual prone to late-onset diseases such as allergies, autoimmunity, obesity, diabetes, and asthma.[2] Some of these epigenetic changes are also inherited transgenerationally.[8]

DNA methylation

DNA methylation is the most studied and essential epigenetic modifications. It is a postreplication modification that is primarily found at C_5 position of cytosine (mC) of the CpG dinucleotides. It undergoes dynamic remodeling including reprogramming and de novo methylation during mammalian development. DNA methylation[9] maintains stable gene expression through mitotic cell division by stably propagating the methylation pattern through mitosis,[10] thus contributing to the stability of gene expression states. In combination with nucleosome-modifying proteins, DNA methylation helps to establish repressed state of chromatin.[11]

Although 80% of CpGs are methylated in nonembryonic cells, CpG islands, which includes short sequence domains, are generally unmethylated irrespective of the gene expression.[12] These CpG islands are associated with active promoters and genes.[13] Transcriptional expression of such genes is regulated by transcription factor Sp1 and various CpG-binding proteins.[14] Methylation of such CpG islands during development results in long-term repression of the corresponding genes.[15]

DNMTs and DNA-binding proteins

DNA methylation is regulated during development by DNMT1 (maintenance methyltransferase) and Dnmt3a and Dnmt3b (de novo methyltransferases). Recent studies based on DNMTs have emphasized the importance of DNA methylation during development. Studies have shown activation of transposable elements and derepression of ectopic gene expression in absence of DNA methylation.[16] Dnmt1[17] is an oocyte-specific isoform that plays a role in maintenance of maternal imprints. Dnmt3a and Dnmt3b are involved in global de novo methylation after implantation[18] and in combination with Dnmt3L, which has no methyltransferases activity; these de novo methyltransferases establish methylation imprints in the female germline.[19]

DNA methylation represses gene expression by forming nonaccessible boundaries and blocking the binding of transcription regulatory proteins.[20] It inhibits binding of CTCF (CCCTC-binding factor) that is a chromatin boundary element–binding protein, which blocks association between an enhancer and its promoter[21] and, thus, allows interaction between a promoter and an enhancer across the inert boundary site. Certain proteins known as methyl-CpG-binding proteins have specific affinity for methylated CpG sites. Such proteins such as MeCP2,[22] MBD1–MBD4 (methyl-CpG-binding domain proteins),[23] and Kaiso[24] act as mediators of transcriptional repression. MeCP2 further interacts with corepressor complex containing HDACs to facilitate gene repression.[25] Defects in the methyl-binding proteins are related to certain disorders. Rett syndrome, a neurological disorder, is caused by a missense mutations, leading to decreased binding of MECP2 to methylated DNA.[26] Null mutation in *Mecp2* also results in abnormal brain structure and function.[27] The mutations associated with the DNMTs and methyl-binding proteins have been related with multiple disorders as listed in Table 1.1.

Importance of epigenetic reprogramming during development

In general, DNA methylation patterns are stable across different tissues throughout life; however, during implantation and in germ cells, paternal and maternal genomes undergoes rapid demethylation.[9] DNA methylation regulates long-term silencing of the gene expression, during development; however, at the "midblastula transition" extensive gene expression is associated with promoter demethylation.[16] During

Table 1.1 Defects in DNMTs and methyl-binding proteins and related phenotypes.

DNMT/MBP	Related abnormalities	References
DNMT1 (maintenance methyltransferase)	Embryonic lethality, delayed development, hereditary sensory autonomic and neuropathy type 1E (HSAN1E), autosomal dominant cerebellar ataxia, deafness, and narcolepsy. Loss of imprinting, Loss of heterozygosity (LOH), X-linked gene expression	28,29
DNMT3A (de novo methyltransferase)	Birth of runted mice with null DNMT3a, chromosomal rearrangements, embryonic lethality, acute myeloid leukemia, myelodysplastic syndrome, body overgrowth	18,30,31
DNMT3B (de novo methyltransferase)	No viable mice with DNMT3B, immunodeficiency, centromere instability, facial anomalies (ICF syndrome)	18
DNMT2	Abnormal RNA methylation with no observed change in genomic DNA methylation patterns	32
DNMT1O (oocyte-specific isoform)	Loss of maternal imprints	17
DNMT3L (no catalytic activity, of its own, colocalizes with DNMT3A and 3B)	Maternal imprinting defects, azoospermia	19
MeCP2	RTT, rapid weight loss, and death between 6 and 12 weeks in mice, hind limb clasping, misaligned jaws, decreased brain weight	33
MBD1	Impaired spatial learning, reduced neurogenesis, impaired genomic stability, prostate cancer	34
MBD2	Behavioral abnormality, maternal nurturing defects with small litter size and reduced weight of pups, breast cancer	35
MBD3	Embryonic lethality, glioblastoma	35
MBD4	Increased frequency of mutations, higher C:G to T:A transitions, colorectal cancer, endometrial cancer,	36
Kaiso	Reduced tumorigenesis, colorectal cancer	37

mammalian development, the amount of DNA methylation changes in a systematic pattern. Immediately after fertilization, the male genome is actively stripped of DNA methylation,[38] while demethylation of maternal genome occurs passively and subsequently in further cleavage divisions.[39] Implantation is generally followed by a wave of genome-wide de novo methylation increasing the amount of DNA methylation in gastrulating embryo, which then decreases with differentiation in specific tissues.[16]

DNMT1 ensures epigenetic inheritance by playing a role in maintaining DNA methylation by recognizing nascent strand of DNA opposite to a previously methylated position. In mammalian genomes, with the exception of CpG islands specifically present at the promoter regions of many housekeeping and developmentally regulated genes, the CpG islands especially the ones at intragenic regions are methylated during development.[40] Aberrant methylation at these sites is prevented by enzymes such as TET and thymidine DNA glycosylase. Exclusion of DNMT1 from promoter region CpG islands and transcription start sites (TSS) is ensured by binding of specific transcription factor to ensure the hypomethylated status. This has been confirmed by observed demethylation upon incorporation of SP1-binding site into an endogenously methylated locus.[41] H3K4 methyltransferases and nucleosome exclusion provides additional protection from DNA methylation. De novo methylation is also prevented by formation of R-loops of single-stranded DNA due to active transcription at these sites. However, certain promoters are stably silenced by repressive transcription, which recruit a variety of proteins working together to silence the promoter. These proteins include de novo DNA methyltransferases, heterochromatin protein 1, and H3K9 methyltransferases.[9]

DNA methyltransferase 3B (DNMT3B) is involved in silencing of germline gene promoters and is mainly mediated by DNA methyltransferase 3B (DNMT3B), which acts downstream of transcriptional repressors (E2F transcription factor 6). DNMT3A and DNMT3B target certain low CpG promoters for their de novo methylation in association with histone deacetylases and methyltransferases. Cooperation between nucleosome remodeler (lymphoid-specific helicase) and epigenetic silencers forms a heterochromatin assembly at these promoters.[42] Furthermore, H3K9 dimethyltransferase G9A is also recruited along with DNMT3A or DNMT3B[43] accelerating repression.[44] H3K9 methylation is believed to trigger the formation of heterochromatin followed by stable silencing by DNA methylation.[45]

DNA methylation plays an important role in repression of transposable elements such as long terminal repeat (LTR), long interspersed nuclear elements (LINEs), and short interspersed nuclear elements (SINEs).[46] The promoter regions of these elements are constitutively hypermethylated to block their function. In embryonic stem cells (ESCs), activities of both DNMT1 and DNMT3 are required to maintain stable DNA methylation at these transposable elements. Tripartite motif-containing protein 28 (TRIM28) and H3K9 methyltransferase SETDB1 play an essential role in silencing of LTRs.[47,48] The expression of transposable elements silencing factors is restricted to early embryogenesis and ESCs. In adults, the repression of transposable elements is mediated by methyl-CpG-binding protein 2 (MECP2) and histone deacetylases (HDACs).[49,50]

Role of DNA methylation in embryonic stem cells

Although DNA methylation has no role establishment and maintenance of pluripotency, it plays an essential role in regulating commitment of ESCs, which is highlighted by loss of DNA methylation. In absence of DNA methyltransferases, ESCs show no observable aneuploidy,[51] while in the absence of DNA methylation, these loose the potential to differentiate and are unable to repress pluripotency factors and upregulate germ layer-associated markers.[52]

In mouse embryonic stem cells, transcriptional repression of trophectodermal transcription factor gene E74-like factor 5 (*Elf5*) by hypermethylation maintains extraembryonic potential and trophectodermal lineage commitment of these cells. Only a small portion of cells have extraembryonic potential. DNA methylation regulates extraembryonic commitment.[52] Absence of DNMT1 leads to upregulation and demethylation of ELF5 in hematopoietic stem cells (HSCs).[53]

DNA methylation determines the lineage specificity toward myeloid vs. lymphoid lineages when they exit from self-renewal to become restricted progenitors. Absence of DNMT1 blocks the self-renewal of HSCs, while the commitment toward myeloid lineage is favored in DNMT1 hypomorphic HSCs. Methylation of *Elf5* in ESCs and that of *Oct4* and *Nanog* in trophectodermal stem cells regulate the extraembryonic or embryonic fate of these cells.[54] Loss of either maintenance or de novo methyltransferase leads to markedly disparate rates of global methylation. DNMT1 loss leads to loss of methylation, which stabilizes to ~20%, while loss of DNMT3A or DNMT3A results in the loss of all the methylation with ongoing divisions.[52]

Pathological impact of altered DNA methylation

Mammalian development cannot tolerate any substantial amount of epigenetic abnormality as evident from the global deregulation of gene expression in the clones that are able to survive. This gene deregulation and the resulting phenotypes are observed in clones of all species.[55] The reprogramming errors impact majorly on the mammalian development. Recent studies have demonstrated the role of epigenetic abnormalities in the development of cancer. Global hypomethylation accompanied with hypermethylation of CpG islands is a general feature of cancerous cells.[56] Epigenetic deregulation leads to several alterations such as aberrant expression of oncogenes and tumor suppressor genes and disables DNA repair machinery and chromosomal instability.[57] Altered promoter methylation of oncogenes such as *c-myc*[58] and tumor suppressor genes[59] leads to the development preeclampsia and choriocarcinoma.

Genomic imprinting

In mammals, a set of genes are expressed specifically from either maternal or paternal copy of the gene in a unique parent-of-origin-specific manner. These are known as imprinted genes. Initial period of the life ranging from conception to the second anniversary is especially important for genomic imprinting. Epigenetic alterations and nutritional factors play a vital role during this time period in determining the developmental progress and future susceptibility to various diseases.[60] In genomic imprinting, one of the two homologous chromosomal sets is marked reversibly during development blocking the expression of the particular set. This phenomenon was discovered in mammals in 1980s.[7] These genes are generally present in clusters of three or four genes spanning 1 MB and encode for both protein-coding and noncoding RNAs.[6] Some imprinted genes are not present within these clusters and are called as microimprinted loci.[61]

Epigenetic mechanisms of gene regulation for imprinted genes

Imprinted genes are regulated via two different mechanisms: (1) the insulator model (e.g., *H19/Igf2* locus), where allele-specific expression is facilitated by insulator protein CTCF, and (2) ncRNA model (e.g., *Igf2r/Airn* and *Kcnq1/Kcnq1ot1*),[62] where the expression of ncRNA that has promoter located within the ICR of imprinting locus regulates the allele-specific expression of imprinted genes.[63] There are approximately 150 known imprinted genes in mice and are highly conserved among mammals.[62] This conservative nature of imprinted genes helps in mouse and human research. Deregulation or specific mutation of parental allele causes rare congenital disorders.[64]

Establishing epigenetic marks on imprinted genes

Various epigenetic marks are present differentially between the two alleles regulating imprinted gene expression. These marks are present especially within the regulatory region known as imprinting control region (ICR). ICRs are associated with different epigenetic marks such as DNA methylation, histone modifications, noncoding RNA, and chromatin organizations, resulting in parent-of-origin-specific expression of imprinted genes.[6,65] These imprints are not only faithfully replicated during the cell division but also are perfectly established on the maternal and paternal allele during gamete formation. The acquittance of these imprinting marks in germline cells and their maintenance after fertilization has been well studied. The epigenetic reprogramming occurs in germ cells during gametogenesis ensuring independent establishment of parental-specific imprints that drives proper embryonic development.[1] Methylation imprints are erased actively to reset imprinting during the development of PGCs[66] and reestablished in allele-specific manner during oogenesis and spermatogenesis.[16] De novo DNA methyltransferase 3a (DNMT3A) establishes DNA methylation with the help of DNMT3L, which has no methyltransferase activity, but acts as an accessory protein in germlines.[19] In *H19/Igf2* locus, paternal-specific methylation is established in male germline in prospermatogonia,[67] while methylation at ICR occurs in growing oocytes in maternal-specific manner.[68]

During PGC reprogramming, various epigenetic remodeling steps occur in addition to DNA demethylation, resulting in the increased expression of DNA-binding factors Stella, TET1, and cell cycle that arrest the G2 phase. Enhanced levels of TET1 are observed at germline gene promoters and ICRs,[69] which play a role in the activation of germline-associated genes and global hydroxymethylcytosine (hmC) during PGC progression.[70] PGCs also have lower expression of DNMT1 destabilizing factor UHRF1 and DNMT3; thus, global demethylation is eased by cell division in the absence of methyltransferase activity.[69,71]

Erasure of DNA methylation and an extensive genome reprogramming occur after fertilization; however, the parental-specific imprints are maintained by DNA methyltransferase DNMT1. In addition, during early embryonic development, proteins such as STELLA (maternal factor PGC7)[72] and zinc finger protein homolog 57 (ZFP57)[73] play a role in protecting and regulating DNA methylation at ICRs. In somatic cells, such imprints are retained throughout lifetime. However, the biparental imprints are finally erased in PGCs recruited from somatic cells.[7] This process is carried out by various active and passive events, such as translocation (Tet) family of methyl cytosine dioxygenases oxidizing 5-methylcytosine to 5-hydroxymethylcytosine[70] and action of DNA repair machinery.[74] In PGCs, DNMT1 is blocked from reaching to replication fork by repression of Uhrf1, hence preventing methylation of newly replicated DNA.[71]

Imprinting loci show differential rates of de novo methylation, suggesting the role of epigenetic memory in reestablishing methylation marks. Proteins zinc finger protein ZFP57 and TRIM28 play a role in maintaining imprints. In embryos, ZFP57 recruits TRIM28 to imprints, while in ESCs, ZFP57 recruits a complex containing TRIM28, the DNMT3s, and UHRF1 to a methylated CpG-containing motif.[9]

In addition to DNA methylation, PGCs also show demethylation of histones. H3K9me2 marks are erased by decreased levels of binding partner of G9A, GLP,[75] and recruitment of histone chaperones.[76] Interestingly, global H3K27 methylation occurs to compensate demethylation at H3K9 and DNA levels.[75]

Role of genomic imprinting in development

Increased evidence of gene imprinting disorders associated with various assisted reproductive techniques (ART) emphasize the importance of genomic imprinting in early embryonic development.[7] Imprinted genes have a vital role in mammalian development regulating fetal and placental growth, pluripotency, differentiation, and behavior.[65] These genes regulate embryogenesis, placental implantation, and fetal growth during fetoplacental development including both fetal growth–promoting and fetal growth–restricting pathways.[77]

Role of genomic imprinting in fetoplacental development

Genomic imprinting is observed only in the placental mammals with various genes showing placenta-specific imprinting.[78] Imprinting genes are conserved between species. Placenta is a vital organ for proper fetal development.[79] Many imprinted genes exhibit splice variants specific to placenta based on their expression from placenta-specific promoter.[78]

Imprinting disorders have been associated with several placental abnormalities such as (1) abnormal placental weights,[80] (2) defective proliferation, apoptosis, and differentiation of trophoblasts,[79] (3) malfunctioning spongiotrophoblast, (4) morphological abnormalities related to disproportionate labyrinth and junctional, (5) the number of glycogen cells, (6) placental vascularity zones, (7) imperfect fetal/placental weight ratio, (8) atypical placental endocrine function,[81] and (9) altered nutrient uptake.[78,79,81]

Altered expression of imprinted gene can also pose a potential threat for proper fetal growth, leading to large-for-gestational-age (LGA) fetus,[82] abnormal birth weight, head circumference of newborn,[77] neurobehavioral development,[83] and altered invasive potential of trophoblasts.[84] Placental growth and functions are more prone to environmental cues such as air pollution, alcohol consumption, and exposure of pregnant women to endocrine disruptors (phthalates, BPA, and phthalates), leading to imprinting defects in placenta and developing fetus.[7]

Genomic imprinting especially has a critical role in the brain and neuronal development. Studies based on gynogenic and androgenetic mouse chimeras have emphasized the developmental regulation of imprinted genes in the brain. These chimeras have revealed the effect of abnormal genomic imprinting on the brain size.[85] Transcriptome sequencing analysis has shown differential expression of imprinted genes in different brain regions, with higher proportion of these genes within hypothalamus and hindbrain. Higher maternally expressed genes are found during early embryonic development, compared with higher expression of paternally expressed genes in adult brain regions.[86]

Imprinted genes such as *Dlk1, Igf2,* and *Ube3a* show diverse transcriptional dosage in brain during neuronal development.[87] In brain development, several genes also show brain-specific imprinting bias related to parental allele expression.[86]

Imprinted genes influence both embryonic neurodevelopmental processes, such as proliferation, differentiation, migration, self-renewal, axonal, and dendritic outgrowth, and functioning of adult brain such as regulating synaptic plasticity, transmission, and action potentials.[88] Table 1.2 shows a list of imprinted genes with altered expression related to fetal/neonatal and placental abnormalities and known disorders.

Genomic imprinting and X-chromosome inactivation

Like genomic imprinting, X-chromosome inactivation is recognized by monoallelic gene expression. Several mechanisms are involved in maintaining the inactivation of X-chromosome, which includes DNA methylation, expression of ncRNA *Xist*, and

Table 1.2 Altered imprinted gene expression related to fetal/neonatal and placental abnormalities and known disorders.

Gene	Fetal–neonatal abnormalities	Placental abnormalities	Known disorders	References
Igf2	Growth restriction/ overgrowth, macroglossia, abdominal wall defects, higher embryonic tumors	Decreased labyrinth size and trophoblast surface area, impaired placental permeability	Beckwith–Wiedemann syndrome, Silver–Russell syndrome	65,89,90
Igf2r	Overgrowth, postaxial polydactyly, heart Abnormalities, organomegaly	Altered placental functioning		91
H19	Growth restriction/ overgrowth	Increased labyrinthine trophoblast, impaired placental efficiency, impaired placental permeability	Beckwith–Wiedemann syndrome, Silver–Russell syndrome	7,65,92
Peg1	Growth restriction, behavioral abnormalities	Growth restriction, impeded angiogenesis		7,93
Peg10	Intrauterine growth restriction	Early placental development, altered invasive potential of trophoblasts		84,94,95
Cdkn1c	Growth retardation	Increased spongiotrophoblast	Beckwith–Wiedemann syndrome	7,96
Dlk1	Growth restriction, increased adiposity, skeletal deformation, early puberty, developmental delay, scoliosis, hyptonia	Impaired maternal–fetal exchange	Uniparental disomy chr14	97,98
Grb10	Disproportionate overgrowth	Enhanced labyrinth size, increased surface area for exchange	Silver–Russell syndrome	99
Snrpn	Feeding difficulties, hypothalamic dysfunction, intellectual delay, impaired speech, poor motor control function	Altered invasive potential of trophoblasts	Prader–Willi syndrome, Angelman syndrome	84

state of deacetylation of histones.[100] DNMT1 plays a crucial role in maintaining X-chromosome inactivation.[101] While in somatic cells, X inactivation is random, in extraembryonic lineages, it is regulated by parental origin. The inactive X chromosome is determined by the expression of *Xist* from X inactivation center (Xic) and its coating of the specific X chromosome.[102] Furthermore, the expression of *Xist* is regulated by a noncoding antisense transcript known as *Tsix*,[103] with the mediation of CTCF that binds *Tsix* ICR,[104] hence, modulating X inactivation.

Histone modifications

After Allfrey's pioneering studies in the 1960s, it is known that histones are post-translationally modified. Histone modification is a covalent posttranslational modification (PTM) to histones. These influence chromatin structure and employ remodeling enzymes for repositioning nucleosomes.[105] These modifications act in various biological processes—transcriptional activation/inactivation, chromosome packaging, and DNA damage/repair.

Types of histone modifications

Numerous modifications mark histone, which alters the gene expression, which is given in Table 1.3. Acetylation and phosphorylation caused transcriptional activation; SUMOylation, deimination, and proline isomerization usually exist in transcriptionally silent regions; methylation and ubiquitination are known to influence both activation and repression.[120] However, whether histones are conjugated to O-GlcNAc in mammals is still unknown.[121] The noncovalent proline isomerization on H3 holds an impact on H3K36 methylation to boost transcription.[122] The deimination of arginine by PADI4 provokes arginine methylation and converts arginine/methylarginine to citrulline.[123,124] Hypercitrullination encourages chromatin decondensation.[125] In general, these newly recognized histone modifications can affect conventional modifications such as methylation etc. via competing for the same site or via cross-talk conferring a conformational transformation, thus altering the downstream signaling and finally gene expression.[126]

Regulation of gene expression by histone modifications during embryonic and perinatal development

Histone methylation is linked to early differentiation processes, i.e., trophectoderm (TE) and inner cell mass (ICM) formation.[127] A few histones directly inherited from the sperm contain increased hyperacetylated H4K8 and K12 and could transmit epigenetic information.[128] ESCs involve hydroxylation of methylated cytosines by the Tet family of enzymes.[129] Embryonic system exhibits distinctive histone methylation patterns of bivalent (activating and repressive) marks at lineage-specific genes that get converted to monovalent marks while differentiation, to define apt gene

Table 1.3 Types of histone modifications and their role in transcriptional regulation.

Histone modification	Group attached/ removed	Enzymes involved	Role
Methylation	Methyl group	Lysine methyltransferases (KMTs), arginine methyltransferases, and histone demethylases (HDMs)	Controls transcription by influencing chromatin structure, recruiting transcription factors (TFs), binding with initiation and elongation factors, and also affecting RNA processing.[106]
Acetylation	Acetyl-CoA group	Histone acetyltransferases (HATs) and histone deacetylases (HDACs)	Facilitates open chromatin by recruiting TFs. HDACs erase acetyl residues and cause repression.[107,108]
Phosphorylation	Phosphate group	Kinases and phosphatases	DNA damage repair, chromatin compaction control associated with mitosis and meiosis, and control of transcriptional activity.[105,109]
Ubiquitination	Ubiquitin group	Ubiquitin ligases and peptidases	Monoubiquitination-critical event in protein translocation, DNA damage signaling, and transcriptional regulation. Monoubiquitination of histone-2A and histone-2B linked with transcriptional repression and induction, respectively. Polyubiquitination symbolizes the protein for either degradation or activation.[110–113]
Crotonylation	Crotonyl group	Histone crotonyltransferases, acyl CoA synthetase, and decrotonylases	Exclusive mark for active sex chromosome-linked genes in male germ cells by associating it with active chromatin,

Table 1.3 Types of histone modifications and their role in transcriptional regulation.—*cont'd*

Histone modification	Group attached/ removed	Enzymes involved	Role
			together with promoters and enhancers.[114] Impact on gene expression is unclear and can activate/repress transcription.
Neddylation	NEDD8	E3 ligase RNF168	Neddylation on K119 of H2A disturbs ubiquitination at this site and decreased response to DNA damage.[115]
Biotinylation	Biotin residue	Holocarboxylase synthetase (HCS)	Biological implication is not well understood.[116]
ADP ribosylation	ADP-ribose moiety	ADP-ribosyl transferases	Biological importance is divergent ranging from DNA repair to replication and transcription.[117]
SUMOylation	SUMO (small ubiquitin-related modifier) proteins	Enzyme cascade (E1 −E2−E3). E1; SUMO-activating enzyme E2; SUMO-conjugating enzyme E3; SUMO ligases	Induce transcriptional repression via recruiting HDACs and heterochromatin protein 1.[118]
N-formylation	Formyl moiety of 3′-formylphosphate	Nonenzymatic reaction	N-formylation of histone lysine originates from oxidative stress. It interferes with methylation/acetylation of the same residue and add on to the pathology of oxidative and nitrosative stress.[119]

expression in tissues.[130] Methylation and demethylation of both activating and repressive marks are essential for establishing embryonic and extraembryonic lineages.[130] Altered methylation throughout embryogenesis can induce faulty development of organs and body patterning.[130] Histone modifications play a role in the segregation between ICM and TE, with a higher H3K27me3 and lower H2A and/or H4 phosphorylation in ICM. These changes, however, cannot be directly correlated to gene expression.[131] In four-celled mouse embryo, one blastomere showed hypomethylated H3H3R26, diminished H3R26me3, and upregulated *Nanog* and *Sox2*, thus promoting ICM fate of these cells.[132] The acetylated lysine and methylated arginine decline globally on egg chromatin.[133] H3K27me3 is vital in development and is reversible, and therefore acts as an on–off switch according to developmental signals.[134] H3K9me3 is a feature of heterochromatin, whereas H3K9me2 is restricted to null or less expressive genes of euchromatin.[135–137] H3K4 methylation commonly occurs at transcription start sites in hESCs (human), zebrafish, and mouse embryos. H3K4me3-marked genes were 80% expressed in hESCs, marginally expressed in early embryos of zebrafish, whereas H3K4 methylation sites were common between the TE and ICM in mouse embryos.[138,139] H3K27me was found in certain discrete genes in mouse embryos.[127] In ESCs, H3K27me3 domains were less broadened when compared with differentiated cells.[140,141] During biological development, H3K9me3 was dynamically altered; lineage-specific genes activate by losing it, while pluripotency genes gain this mark.[142] In early embryogenesis, in zebrafish and mouse, H3K4me3 and H3K27me3 bivalently marked genes were absent.[143,144] Bivalently marked genes were initially considered peculiar to early embryo; they play special roles in later developmental stages; however, some studies explained mechanisms separate from bivalency for gene suppression.[145,146] Mouse embryos showed H3K27 marks in gametes, which disappeared postfertilization and reestablished after implantation.[146] On the contrary, maternal genome of preimplantation embryos has broad, noncanonical tracts of H3K4me3; these were associated with gene suppression in mouse embryos.[146] H3K4me3 and H3K27me3 are elevated in the ova compared with sperm chromatin, which during development progressively attains these marks.[147] Also, immunostaining for histone marks has depicted an unequal distribution of histone modifications among the parental genomes.[131,148,149] H4 hyperacetylated lysine and H3 andand H4 methylated arginine at 17th andand third positions, respectively, are reduced in metaphase II eggs, fertilized eggs, and early embryonic metaphase stage blastomeres, which is pivotal for transition from egg to embryo during genomic reprogramming.[133] Carrozza et al. found that Rpd3 removes histone acetylation marks associated with gene elongation.[150] Out of all histone modifications, methylation is the most studied and has been detailed in Table 1.4.

Nucleosome remodeling

Chromatin remodeling (CR) has a primary role in transcriptional modulation. There are two types of CR: one begins with various histone modifications and the other by

Table 1.4 Histone methylation at specific sites with their effects on gene expression.

Methylation site	Expression	Reference
H3K4me3	Activation	151
Di- and trimethylation at H3K4 (promoters), H3K36, H3K79	Activation	152
Monomethylated H3K27, H3K9, H4K20, H3K79, H2BK5	Activation	152–154
Trimethylated H3K27, H3K9, and H3K79	Repression	152
H3K4me1 (enhancers) with H3K27 acetylation	Activation	155
H3K9 and H3K27	Repression	152,154
H3R8me2s and H4R3me2s	Repression	156

ATP-dependent drafting of nucleosomes; the latter is called nucleosome remodeling (NR). NR mediates transcription in an ATP-dependent approach and influences the embryonic development.[157,158] Classically, remodelers can (1) mediate nucleosome sliding in which the nucleosome location on DNA varies, (2) generate contact with trans-acting factors where histones remain bound, here DNA is more approachable, (3) full eviction of the octameric histone, and (4) histone replacement with another histone.[159] NR enzymes make DNA free by total or partial disassembly of nucleosomes, which modify the nucleosome composition in comparison with histone variants and could affect folding of the nucleosomal fiber.[160] Postreplication and transcription events, it recovers the accurate nucleosome spacing and rectifies the repair of damaged DNA.[161,162] Defective NR affects the integrity of genome along with cellular functions, leading to acceleration of both cellular senescence and transformation.[162] Remodeling complexes are subclassified into four groups on the basis of other functional domains within the ATPases: "SWI/SNF (mating type switching/sucrose nonfermenting), ISWI (imitation switch), CHD (chromodomain helicase DNA binding), and INO80 (inositol autotroph 80)".[163] SWI–SNF remodelers are master regulators implicated in the developmental processes in most cells.[164] These ATP-dependent NR complexes play imperative function in maintaining ESC self-renewal and differentiation.[165] Two other ATP-dependent nucleosome remodelers—Lsh (role in retrotransposon silencing) and NURF—are also vital for correct ESC differentiation. Knockdown of *Lsh* resulted in failure to repress the pluripotency-related promoters.[166] *NURF*-deficient ESCs display defective differentiation of all germ layers.[167] In ESCs, a direct communication of remodeling complexes with pluripotency TFs verifies that TFs recruit remodelers to sites of action, promoting self-renewal.[168] For instance, barrier-to-autointegration factor (*BAF*) class members of SWI–SNF complexes communicate with pluripotency TFs to refrain developmental regulators in ESCs.[168] Opposingly, Tip60–p400 locates to the same pathway as *Nanog* but does not interact with this TF directly.[164] On differentiation, BAF turns off pluripotency genes such as *Nanog* and induces expression of regulators. In mammals, almost 30 SWI–SNF2 family members comprise a larger number of ATP-dependent NR complexes.[165]

Chromatin structure of embryonic stem cells

hESCs firstly were isolated by Thomson et al.[169] In ESCs, chromatin configuration is open, more relaxed with respect to differentiated cells.[170] In ESCs, a condensed chromatin is not often observed, and ample transcriptionally active regions are predominant.[171,172] Concerning morphological differences, chromatin in ESCs is spread out in relaxed regions, diffused, and amorphous.[173] Additionally, nuclear lamina, characterized by a reduced level of A-type lamins in ESCs, becomes densely stained by specific lamin antibody after differentiation.[174] Precise distribution of euchromatin and heterochromatin-related factors was observed within ESC colonies.[175] A main feature associated with the self-renewal as well as pluripotent ESCs is their indefinite replication, which is ensured by telomerase enzyme that prevents shortening of telomere during cell division.[169] In ESCs, self-renewal and pluripotency are interlinked and can be manifested as symmetric or asymmetric cell division.[176] A study revealed that Nanog-positive cells are seen in the very interior regions of pluripotent ESC colonies. However, H3K27me3 marks are seen in the nuclei located on the margin of the mESC (mouse) colony. The nuclei found in the interior show higher positivity for the splicing factor SC-35, but HP1β-positive cells located mainly to the periphery of the mESC colony.[175] Chromatin and chromatin-related proteins occurring in ESCs are extremely dynamic as proved by living cell experiments. For example, hyperdynamic interaction of the structural chromatin protein is functionally essential trait of pluripotent ESCs and is accountable for the undifferentiated phase of ESCs.[173] In ESCs, HP1 and histone-H1 are loosely bound to chromatin unlike brain progenitor cells and primary mouse embryonic fibroblasts where HP1 and H1 bind strongly and form a "closed" chromatin arrangement. This infers that chromatin–protein dynamics are under regulation during cellular processes and not an outcome of the overall biophysical status of chromatin.[177] Furthermore, the open confirmation and also the hyperdynamic plasticity of chromatin connect well with biological properties of ESCs.[173] It is linked to appropriate transcriptional profile, which is associated with the movement of epigenetically significant proteins.[172] However, heterochromatins do change from an active to repressive category during ontogenesis.[178] During differentiation of female ESCs, X-chromosome is inactivated and relocated near nuclear periphery with raised levels of H3K27me3.[179]

Ontogeny of chromatin remodeling

CR of developmentally regulated genes happens in combination with commitment to a lineage. Oocytes, early embryos, and ESCs enclose effective CR activities, which infer that chromatin dynamics is imperative in early lineage decisions.[180] During differentiation, ESCs spontaneously perform developmentally regulated programs of genes induction and silencing, which are closely coordinated with modification in chromatin structure. The best studied instance of this is X chromosome inactivation (XCI), where numerous CR events stabilize the silenced X chromosomes in female ESCs.[181] The succeeding CR processes take place to stabilize the inactive X chromosome, which include a momentary connection of the HMT complex Eed-Ezh2 methylating H3K27 residue.[182,183] Subsequently, histones are hypoacetylated on lysines and methylated at lysine 9.[184,185] Inactivating CR complexes generally cause developmental arrest at blastocyst phase. Likely, the loss of *SNF5*, a constituent of SWI/SNF CR complexes while implanting, might lead to developmental arrest and is obligatory for ESC viability.[186] In F9 embryonal carcinoma cells, deletion of *SNF2β* results in lethality.[187] The polycomb group gene *rae28* is mandatory for renewal in HSC,[188] while *Ezh2* is necessary for postimplantation processes and ESC viability too.[189] The modifications of H3K27ac and H3K9ac signals in Drosophila embryos in enhancers and promoter regions downregulate genes associated with neural stem/progenitor cell pluripotency and upregulate genes linked to neuron projection morphogenesis and differentiation.[190] CR proteins such as CHD1 and esBAF are responsible for the open state of chromatin and preservation of self-renewal ability and pluripotency in ESCs.[164,191] ESC differentiation and CR proteins perturbation can transform heterochromatin abundance and ESCs distribution.[191] Thus, on differentiation, epigenetic modifications do not change globally, but chromatin configuration of stem cells is rather modified.[192]

Higher-order chromatin organization (topologically associating domains)

In classical depiction of chromosome structure, chromatin is depicted to have active (A) and inactive (B) sections, each comprising bulky topologically associating domains (TADs).[193] TADs are divided into smaller nested loops that define the regulatory space within which genes can communicate with regulatory proteins. Loop boundaries are decided by CTCF (CCCTC-binding factor) binding to DNA, provideD transcriptional isolation from adjoining loops.[194,195] Targeting CTCF sites, which are essential for TADs, resulted in new promoter—enhancer binding and abnormal gene expressions, phenocopying human appendages malformation syndromes.[196] Inside TADs, chromatin configuration is revealed by nucleosomal density, position, and composition.[197] The regulatory landscape inside TADs is extremely dynamic through cell lineage specification.[198] The contacts within regulatory areas are shuffled during development; TADs usually maintain the same conformation, although an extensive switching between A (active) and B (not active) chromosomal compartments occurs; this has been confirmed in both mouse and hESCs. Studies have shown that TADs change their regulatory interactions among cell types when naive and pluripotency states were compared. During transcription, signaling-mediated binding of TFs and enhancer within TADs play a key role in determining the transcriptional signature of the differentiating cells.[199,200]

Pathological impact of deregulation in histone modifications

The fault in histone methylation could be lethal at diverse developmental stages or cause disturbed organ functioning in mature animals, depending on cell type specificity of the methylation defect or cause developmental arrest.[201] The mutated histone methylation regulators led to defective skeletal and neurodevelopment.[130] Mutations in the H3K4 methyltransferase genes *SETD1B* and *MLL1–4* have been characterized by intellectual, facial, and skeletal abnormalities, specifically reduced body size and microcephaly.[202,203] Aberrations in H3K4 demethylase genes *KDM5C* and *LSD1* or in the LSD1-binding partner PHD finger protein 21A led to similar syndromes causing intellectual disability and bone deformities.[204–206] Considerable mutations in *MLL4* and its antagonist *LSD1* gene have been linked to Kabuki syndrome conferring that opposite roles at a molecular level can lead to same consequences at an organismal level.[204,207] Alterations in both the H3K9 demethylase-encoding PHF8 and the methyltransferase-encoding GLP resulted in craniofacial deformity and intellectual disability in humans and animal models.[208,209] The overgrowth pathologies of Weaver, Sotos, and Beckwith–Wiedemann syndromes were observed when mutation occurs in *EZH2* and *NSD1*, which encodes H3K27 and H3K36 methyltransferases, respectively.[210,211] Abnormal H3K9 and H3K37 trimethylation at promoter regions of oncogenes (VEGF, EGFR, and c-jun)[212] and matrix metalloproteinases (MMP2 and MMP9)[213] leads to the development of choriocarcinoma. Function loss of PRMT7 caused mild intellectual disability with obesity and poor bone development.[214]

Noncoding RNAs

A noncoding RNA (ncRNA) is a functional RNA molecule that is not translated into a protein. ncRNAs specifically function at the transcriptional and posttranscriptional level to regulate gene expression ncRNAs that are implicated in epigenetic processes can be alienated into two major groups: the short ncRNAs (<30 nts) and the long ncRNAs (>200 nts). microRNAs (miRNAs), short interfering RNAs (siRNAs), and Piwi-interacting RNAs (piRNAs) are the three main classes of short ncRNAs. They have been shown to play an important role in the formation of heterochromatin, modification of histones, gene silencing, and DNA methylation targeting. Long noncoding RNA (lncRNA) are a large and diverse group of regulatory RNAs around 200 nts in length. They play an important role in various cellular activities or diseases.

Regulation of gene expression by noncoding RNAs during embryonic and perinatal development

In cellular processes such as in gene expression regulation, ncRNAs have been observed to play an important role. Various mammalian ncRNAs emerge to be well conserved and articulated in a developmentally synchronized manner with expression patterns that are tissue specific.

Small noncoding RNAs (miRNAs and siRNAs)

miRNAs are small single-stranded molecules (20–24 nts), conserved evolutionarily, and have been assumed to regulate the expression at the posttranscriptional level in ~50% of the genes in a cell. In comparison with other endogenous small RNAs, miRNAs originate from transcripts making distinguishing hairpin arrangements.[215] Drosha and Dicer processes hairpin into the mature miRNA by allowing their interaction with argonaute (Ago) proteins that further lead to the formation of RNA-induced silencing complex (RISC).[216,217] The miRNAs then bind with mRNAs at 3′-untranslated region (UTR) most favorably, to direct their translational repression or deadenylation and degradation.[218,219] Besides classical roles, miRNAs have also been observed to regulate gene expression via targeting promoters as well as via translational activation[220]; the earlier is expected to engage epigenetic mechanisms. siRNAs play roles in both the nucleus and cytoplasm. Initially, they were implicated by annealing to mRNA sequence elements to PTGs (posttranscriptional gene silencing) that further leads to instability of mRNA and its degradation.[221] In addition, siRNAs also take part in heterochromatin formation and TGS (transcriptional gene silencing).

An individual's phenotype is believed to be shaped by the genetic code that is inherited from parents. However, it has been now known that genetic code manifestation can be modified by flexibility in the settings of epigenetic modifications occurred during the formation of embryo. The milieu, in which the embryo, fetus, and neonate grow up, appears to confer changes to epigenetic settings that can cause enduring modifications in homeostasis. The significance of miRNAs during early embryo and gamete development has been broadly investigated. The vivacity of various proteins implicated in biogenesis of miRNA, including DICER, DiGeorge syndrome critical region gene 8 (DGCR8), and DROSHA in both oogenic and spermatogenic pathways, has been assessed through knockout studies.[222,223] Developmental potential of resulting embryos can be affected by various miRNAs which are paternally contributed from the sperms. Nevertheless, miR-34c is considered to start the initial cleavage divisions in the mouse and thus constitutes a typical example of the influence of a single paternally derived miRNA on development of embryo.[224] In various knockout studies, it has been observed that small RNAs, including miRNAs, are important for embryonic development. Both *Dgcr8 and Dicer1* knockouts in mice are lethal embryonically.[225,226] Embryos that have decreased number of stem cells are devoid of *Dicer1*.[225] miRNA dysregulation in embryos can potentially affect cell differentiation, implantation, and early pregnancy maintenance.[227] Implantation is an intricate process that depends on the synchronized expression and secretion of molecular signals in the microenvironment of uterus including growth factors.[228,229] In the uterine fluid, secreted miRNAs from the developing embryo mediate these processes.

Piwi-interacting RNAs

piRNAs are 23- to 36-nucleotide single-stranded RNAs, which act as guide for argonaute proteins, the Piwi proteins. The first piRNAs were derived in 2001 from the testes of *Drosophila melanogaster*.[230] Until 2007, the mechanism leading to piRNA biogenesis remained indefinable, but later on, "piRNA ping-pong cycle" was described by two groups independently.[231] Small RNAs associated with all three *D. melanogaster* Piwi-clade proteins—argonaute 3 (Ago3) Piwi and aubergine (Aub) were sequenced, and it was discovered that each protein binds to specific piRNA populations. Antisense piRNA, complexed with Aub, would recognize and cleave a sense transposon transcript. The cleaved product would then be processed into an Ago3-bound sense piRNA, which could seek out target transcripts. Ago3-directed cleavage triggers generation of the original antisense piRNA, capable of both silencing the target element and further amplifying the response.[232,233] piRNA clusters act as a source of primary antisense precursor RNAs. These "piRNA clusters" are primarily composed of faulty transposon sequences.[234] Many studies have provided exciting insights into the pathway of piRNA and its role to protect genome integrity and fertility. Also, diverse roles of silencing by piRNAs are still unanswered. piRNAs have a particular role among sncRNAs in reproductive biology. Earlier in various studies, it was demonstrated that piRNAs take part in tumorigenesis, gametogenesis, epigenetic regulation, transposon silencing germline development, and regulation of gene expression.[235] In germline, piRNAs have been mainly uncovered, but there is a growing evidence that suggests that their defensive role extends into somatic cells.

Long noncoding RNAs

The majority of the non-protein-coding transcripts belong to the group of lncRNAs, which are arbitrarily considered as 200 nt in length.[236] However, the categorization of this group of ncRNAs is ambiguous, as many of these lncRNAs can also act as primary transcripts for the production of short RNAs. lncRNAs are mainly characterized by nuclear localization, low expression, and low level of sequence conservation. Furthermore, they are composed of both poly A + and poly A2 transcripts.[237,238] lncRNAs can be classified on the basis of their proximity to protein-coding genes dividing them into five categories, which are sense, antisense, bidirectional, intronic, and intergenic. Recently, a subgroup of lncRNAs named large intergenic noncoding RNAs (lincRNAs) was described on the basis of distinctive chromatin signature that marks actively transcribed genes. Trimethylation of lysine 4 of histone H3 (H3K4me3) marks the lincRNAs at their promoter, and trimethylation of lysine 36 of histone H3 (H3K36me3) marks them along the transcribed region. In contrast to most of the lncRNAs, lncRNAs are highly conserved among different species. lncRNAs have been reported to play a role in guiding chromatin-modifying complexes to specific genomic loci, and by doing so, lncRNAs participate in the establishment of cell type–specific epigenetic states.[239] The most important examples which are involved in epigenetic gene silencing include the role of X in active specific transcript (*XIST*) in X-chromosome inactivation and *H19* or

Air in genomic imprinting.[236] The *H19* gene encodes a 2.3-kb ncRNA, which is highly expressed during embryogenesis, but after the birth, it is shut off in most tissues.[240] It has also been seen that environmental factors such as maternal undernutrition also play a role in regulation of expression of *H19* in a sex-specific manner, where maternal low-protein diet was shown to cause abnormalities in male; however, there was no such scenario in female mice blastocysts.[241] Interestingly, this mechanism could potentially contribute to the different susceptibility of cardiovascular diseases between male and female.[242] Apart from embryogenesis, lncRNAs also play some part in tumor development by promoting the expression of genes involved in metastasis and angiogenesis.[240]

Majority of the noncoding genomes including lncRNAs are transcribed at a certain point during development.[240,243,244] The T-box transcription factor, *EOMES* (Eomesodermin), is the first specific marker of the cardiogenic lineage in mammals, and it is expressed within the primitive streak at E5. It is involved in activation of the transcription factor *MESP1* in cardiogenic precursors at the primitive streak stage.[245] *MESP1* in turn activates the CGRN as cardiac precursors migrate laterally to form the heart fields.[246,247] The lncRNA *Bvht* (braveheart) is expressed in mouse pluripotent stem cells, before it gets restricted to the cardiogenic lineage during gastrulation.[248] Expression of *Bvht* is lost in noncardiogenic lineages, suggesting that it plays a permissive rather than instructive role in cardiac specification. Table 1.5 illustrates function of various lnRNAs genes in embryonic development and cancer.

Table 1.5 Functional role of various long non-coding RNAs.

lncRNA genes	Function	Literature
BCAR4 (breast cancer antiestrogen resistance 4)	Preimplantation embryo development	[249]
PTENP1 (PTEN pseudogene 1)	Fetal/adult tissues	[250]
COLCA1 (colorectal cancer associated 1)	Early embryo development	[251]
HULC (hepatocellular carcinoma upregulated long noncoding RNA)	Upregulated in liver cancer	[252]
HOTAIR	Embryo development, breast cancer, gastric cancer, colorectal cancer	[253]
MALAT1 (metastasis-associated lung adenocarcinoma transcript 1)	Oncosuppressive role in liver, pancreas, breast, colon cancers	[254]
MEG3 (maternally expressed gene 3)	Oncosuppressive role in pituitary adenoma and meningioma	[255]
AACSP1 (acetoacetyl-CoA pseudogene 1)	Embryo development, llung, and colon cancers	[251]
LUCAT1 (lung cancer associated transcript 1)	Embryo development, bladder, and lung cancers	[256]

Interaction of lncRNAs with chromatin-modifying complexes

Regulation of transcription by lncRNAs takes place by recruitment of chromatin-remodeling complexes, which in turn arbitrate epigenetic modifications.[236] A well-described transcriptional complex called as repressive polycomb gene complex (PcG) commences and supports epigenetic alterations. PcG has two multiprotein complexes, namely, polycomb repressive complex 1 (PRC1) and 2 (PRC2). PRC2 components establish the silent chromatin state by trimethylating H3K27. Binding of PRC1components to H3K27me3 causes ubiquitination at lysine 119 on histone 2A.[257,258] PRC1 and PRC2 components are also RNA-binding proteins. lncRNAs guide PcG for locus-specific silencing. Interaction of PcG with histone deacetylases leads to transcriptional repression, and interaction with DNA methyltransferase 3 alpha causes silencing by CpG methylation.[259,260] In the developmental stages of *Drosophila melanogaster*, homeotic (Hox) genes were found to be silenced by PcG. HOX antisense intergenic RNA (*HOTAIR*) is 2.2-kb lncRNA that is required for silencing of HOXD cluster and is mediated by PcG.[261] In an earlier study, it was demonstrated that PRC2 binds to the 5′ end of *HOTAIR*, while REST/CoREST complex (lysine-specific demethylase 1) binds to its 3′ end. Removal of dimethyl mark by *LSD1* from H3K4 represses gene expression. *HOTAIR* allows binding of two different repressive complexes to chromatin for combined methylation of H3K27 and demethylation of H3K4 bound by PcG.[262] Polycomb repressive complex 1 component chromobox 7 (CBX7) can bind H3K27me3 and *ANRI* L. Repression of *INK4A/ARF/INK4B* is impaired if there is mutation in either of the binding domains for H3K27me3 or RNA. Interestingly, *ANRIL* is the only transcript associated with cardiovascular disease susceptibility and is observed to be localized at the chromosome region 9q21.[263]

Pathological impact of deregulation in lncRNAs

Gene expression patterns in a cell are mainly controlled by epigenetic modifications. These changes are constant and are heritable somatically.[264] An epigenotype is defined as the assembly of DNA methylation states, histone variant composition, and histone modifications of noncoding RNAs.[265] Non-protein-coding RNAs (ncRNAs) are the most recent class of molecules contributing to the epigenetic signal. For a long time, the category of ncRNA comprised only of transfer RNA, spliceosomal RNA, and ribosomal RNA. With the increased access to the genomic sequences from numerous organisms and the enhanced genetic studies at molecular level in cross-species (from *Escherichia coli* to humans), the record of ncRNAs has expanded.[266] Imprinting in a genomic region describes the phenotypic differences provided that paternally inherited deletions cause Prader–Willi syndrome (PWS), while in Angelman syndrome (AS), there is deletion of maternal origin. PWS, which occurs in about 1/10,000 births, was illustrated about 50 years ago and is distinguished by infantile hypotonia, failure to thrive due to poor feeding, developmental delay, lethargy, short stature, and secondary hypogonadism.[267–269] It has been observed that there is an implication of lncRNAs (derived from the *SNORD116* snoRNA cluster) in PWS pathogenesis.[270] *Ube3a-as* is a candidate gene for AS. It is basically a lncRNA and is suggested to be responsible for the downregulation of paternal *Ube3a* expression.[271] An excellent example that represents the importance of epigenetics is uncovered by a human disorder Beckwith–Wiedemann syndrome (BWS); it makes us understand the role of epigenetics not only in normal development but also in the cell growth regulation and tumorigenesis. This disorder is defined by somatic overgrowth, predisposition to childhood embryonal malignancies, and congenital abnormalities.[272] Development of BWS and various cancer is attributed to imprinting disruption of the *CDKN1C/KCNQ1OT1* domain, changing the maternal epigenotype to the paternal type, and thus resulting in the repression of *CDKN1C* expression.[273] Silver–Russell syndrome (SRS) is another developmental disorder defined by retardation of growth, some dysmorphic facial and cranial features, and short stature often with asymmetry.[274] Most extensively studied lncRNA in IUGR pathogenesis is *H19*. This lncRNA was observed to have a role in SRS.[275] A dominant X-linked disorder Rett syndrome (RTT) is characterized by postnatal neurological disorder, motor abnormalities, seizures ataxia, and replacement of hand use by purposeless hand wringing.[276] The high concordance rate in monozygotic twins collectively with a few incidences of mother-to-child transmission suggested that there is genetic basis of this disease, and it is most likely autosomal dominant inheritance.

Effect of different conditions on epigenetic mechanisms and programming of human health and disease risk

Changes in chromatin conformation occur by epigenetic marks to regulate gene expression. There is implication of epigenetic mechanisms during in utero development and at the cellular level, so exposures to environment may damage the fetus by harming the developing organisms epigenome to adjust risk of disease in later life.[277] Comparatively, protective epigenetic changes may be elicited by bioactive food components all through the life, especially in early life. Afar from genetics, largely health condition of an individual may be considered as an amalgamation of many signals provided by the environment that starts at gestation through epigenetic changes. Our health and epigenome are influenced by our life experiences, lifestyle, and environment and eventually outlines what and who we are. Even though common genome is shared by identical twins and are seemingly similar phenotypically, yet they are distinctive individuals with definable distinctions. These dissimilarities are consequences from different gene expression influenced by various epigenetic factors. The lifestyle factors such as nutrition, behavior, toxin, and pollutants exposure are known to be related with epigenetic alterations. From gestation to death, nutrition remains as an important environmental exposure that influences our health by its impact on epigenetic phenomena. It is also one of the most researched and well-understood epigenetic factors of environment.[278] It has been observed that there is an association between unfavorable prenatal nutritional circumstances, postnatal health, and increased risk of disease. For example, Dutch Famine Birth Cohort of 1944–45 used to observe starvation effect during pregnancy, and consequent physical condition and developmental outcomes were recorded. Maximum effect was observed on disease risk in the initial months of pregnancy.[279,280] It seems in response to a restricted nutrient supply that fetus adapts by itself epigenetically. In humans, prenatal exposure to famine leads to continual epigenetic differences that have been attributed to a minor degree of methylation in insulin metabolism genes compared with unexposed siblings.[281] Health complications were experienced by the grandchildren of famine-exposed women due to dietary changes according to evidence from historical records. In nutrient-restricted offspring, the molecular mechanisms behind transgenerational inheritance and adult diseases of fetal origin have been studied. The results from such studies suggested that gamete methylation by means of the maternal and paternal lineage together with mitochondrial dysfunction and oxidative stress were among the initial events.[282,283] Various studies have observed the dietary effects of epigenetics on phenotype and vulnerability to diseases during lifetime. Folate, a water-soluble B vitamin is a source of one-carbon for AdoMet synthesis, which is important for DNA methylation metabolism and is associated with phenotypic alterations through DNA methylation.[284,285] Another study observed significant epigenetic changes due to maternal dyslipidemia in placenta as well as fetal liver with increased accumulation of triacylglycerols in fetal liver. It has been reported from animal experiments that changes in DNA methylation cause cardiovascular and metabolic diseases mostly in males. Earlier a study demonstrated development of metabolic and cardiovascular diseases and effect on central nervous system plasticity due to epigenetic changes by parental diet in offspring.[286] Therefore, reversible changes in DNA methylation status, chromatin remodeling and histone modifications, and subsequent change in gene expression by nutrients and bioactive components have an impact on overall health. There might be beneficial effects of bioactive food components and various dietary patterns that surmount the harmful impact of negative life behaviors that include smoking or chemical exposure.[287] Nevertheless, nutritional epigenetics is relatively a new subfield of epigenetics, so there is limited knowledge on the exact effect on epigenetics due to bioactive food components and their link with phenotypes. Bioactive food components that trigger epigenetic signatures can be deciphered and might pave the means for personalized nutritional intercession and help in understanding of how our bodies react to diets or nutrients in particular.

Impact of prenatal stress on epigenetic modifications

Stress conditions while pregnant found linked to epigenetic modifications and psychiatric troubles in childhood. Under the conditions of prenatal intimate partner violence (IPV), glucocorticoid receptor (*NR3C1*) and its repressor *FKBP51* depicted differential methylation in prenatally stressed children. These children showed hypermethylation in heterochromatin-like regions, which has been connected to stress/disease resilience earlier.[288] Chronically stressed mothers throughout pregnancy can lead to preterm birth.[289] In addition to *NR3C1* and *FKBP5, BDNF,* which controls neuronal growth, and serotonin transporter *SLC6A4*, which codes for 5-HTT protein, are widely explored in context to stress.[289,290] Promoter methylation of *NR3C1* and DNA methylation of genes such as *FKBP5, BDNF, 11B-HSD2*, and SLC6A4 together affect hypothalamic–pituitary–adrenal (HPA) axis and organism's reaction to stress.[291–293] Turecki and Meaney revealed 9/10 studies reported exon-1F hypermethylation with early-life hardships (reason of anxiety and depression) in humans.[294] Pregnancy-related anxiety illustrated significant association at CpG9 of 1F.[295] Sparrow et al. revealed preterm birth found associated with altered methylome specifically at the sites, which is known to influence neural development.[296] Oberlander et al. were first to confirm the association between prenatal maternal depression and *NR3C1* promoter methylation in infants.[297] The depressive moods in the third trimester were also seen associated with higher methylation of the NGFI-A binding site in *NR3C1*, which further elevated HPA stress reactivity.[295] On the other hand, antidepressants exposure in pregnancy results in the differential methylation of 42 CpG sites including that of *COL7A1* in the cord blood and *TNFRSF21* and *CHRNA2* in the fetus.[298] Infants born to severe depressed women had 2.4% raised methylation at *MEG3-DMR*. However, infants with lower birth weight had 1.6% hypomethylation of *IGF2-DMR*, whereas infants with high birth weight had 5.9% higher methylation of *PLAGL1-DMR*.[299] Prenatal maternal objective adversity was well correlated with DNA methylation of 1675-CpGs.[300] Interpersonal violence that the grandmother suffered throughout pregnancy was linked to differentially methylated 27CpGs in the offspring which maps to 22-annotated genes. The chief differentially methylated genes in this study were *CFTR (cystic fibrosis transmembrane conductance regulator), CORIN (corin, serine peptidase), BARX1 (BARX homeobox1), and SMYD3*.[301] Also the hypermethylated GR promoter at NGFI-A-binding site in males can be attributed to early prenatal stress.[302]

Long-term impacts of epigenetic deregulation during childhood

Stressful conditions in pregnancy may impact increased vulnerability to childhood and adulthood health troubles in the children and grandchildren too.[301] Children had psychiatric issues when they were exposed to maternal IPV after birth. Unexpectedly, these problems were gone when maternal IPV occurred prenatally and antenatally.[288] HPA axis is a key element of human stress reactivity and its vulnerability reasoned as the etiological factor for several chronic diseases in adulthood.[289] Radtke et al. examined *NR3C1* hypermethylation in 10- to 19-year-old children born to mothers who were exposed to IPV pre- and postpregnancy. Such association even after more than a decade proves that transgenerational effects can be sustained lifelong.[303] Maternal socioeconomic challenges in pregnancy are associated with reduced placental *HSD11B2* methylation in males.[304] Reports in animal models verified that DNA methylation modifications are transmitted to further generations.[305] Altered *CFTR* methylation was earlier shown in adolescents from orphanages and also in abandoned children having the history of adversities.[306] This dysregulated CFTR affects release of vitamin D, which might cause depressive symptoms.[307] Stressful prenatal conditions become the most powerful factor affecting mental health in later life via inducing epigenetic changes in fetal brain.[308] Prenatal maternal depression resulted in neonatal methylation of glucocorticoid receptor *NR3C1* and infant cortisol stress-related responses in humans.[297] Certain factors such as chronic stress and war trauma influence HPA axis methylation with noteworthy changes at TF binding loci in all target genes tested. Methylation at several CpGs of *NR3C1* and *CRH* was found associated with birth weight. These observations suggest that prenatal stress impacts developmental processes by modifying HP axis genes epigenetically.[309] Pregnant women's exposure to a natural disaster causes DNA methylation changes in their children even after 13 years of the exposure.[310] Given that prenatal stress might affect the mother, child, and the grandchild, pregnancy is a vital stage to screen for psychosocial risk factors.[301]

Effect of acute intrapartum events on epigenetic mechanisms

To prepare for extrauterine living, fetal genome remodeled epigenetically during intrapartum.[311] The intrapartum period is too small to exert an epigenetic influence; however, researchers address that it could affect postnatal and longer-term health consequences in the neonate.[312] Studies have linked cesarean delivery to increased asthma, eczema, diabetes, bronchiolitis, multiple sclerosis, and obesity.[313–317] The lack of exposure to infectious agents in early childhood provided an explanation for raised atopic diseases.[318] Applying this hypothesis to cesarean deliveries, absence of exposure to vaginal flora could change the immune responses.[319] The epigenetic impact of childbirth (EPIIC) hypothesis suggests that physiological labor and birth exert eustress (positive stress) on the fetus and it modifies the genes related to immunity, weight regulation, and certain tumor suppressors epigenetically. Altered levels of cortisol, adrenalin, and oxytocin produced during labor might remodel fetal epigenome, which further alters gene expression. This reprogramming could manifest noncommunicative diseases and behavioral issues in neonate and adulthood.[311] Maladaptive perinatal stress was found associated with labor interventions, such as cesarean section and altered DNA methylation.[320] Perinatal manipulation of the hormone oxytocin disturbs motherly care in animals; the *oxytocin receptor (OXTR)* is likely a candidate for epigenetic modifications.[321] Hypermethylated *OXTR* in blood and cortex tissue appreciably relate with autism spectrum disease in humans.[322] Csoka and Szyf hypothesized that pharmaceuticals (intrapartum antibiotics) can alter CR and DNA methylation, and such changes persist even after the drug cessation.[323] Cesarean newborns had hypermethylation at birth than those born vaginally. 3–5 days after-birth methylation was not altered via vaginal delivery but extensively reduced in the cesarean group.[320] However, in 2012, it was found that delivery mode was not associated with global methylation at birth, when cord blood DNA of both the modes was compared.[324,325] Also, cesarean deliveries may increase immune disorders in later life and methylation of neutrophil elastase and interferon regulatory factor-1.[324] A study deciphered that mode of delivery differentially methylates 343-loci (enriched for immunoglobulin biosynthetic process and glycolysis and ketone metabolism regulation) in neonatal CD34þ cells.[326] Due to such "adaptive epigenetic tuning," genome is epigenetically remodeled in intrapartum phase to prepare a newborn for postnatal survival.[327]

References

1. Rideout 3rd WM, Eggan K, Jaenisch R. Nuclear cloning and epigenetic reprogramming of the genome. *Science*. 2001;293:1093–1098.
2. Indrio F, Martini S, Francavilla R, et al. Epigenetic matters: the link between early nutrition, microbiome, and long-term health development. *Front Pediatr*. 2017;5:178.
3. Reik W. Stability and flexibility of epigenetic gene regulation in mammalian development. *Nature*. 2007;447:425–432.
4. Hackett JA, Surani MA. DNA methylation dynamics during the mammalian life cycle. *Philos Trans R Soc Lond B Biol Sci*. 2013;368:20110328.
5. Ramsahoye BH, Biniszkiewicz D, Lyko F, Clark V, Bird AP, Jaenisch R. Non-CpG methylation is prevalent in embryonic stem cells and may be mediated by DNA methyltransferase 3a. *Proc Natl Acad Sci U S A*. 2000;97:5237–5242.
6. Barlow DP. Genomic imprinting: a mammalian epigenetic discovery model. *Annu Rev Genet*. 2011;45:379–403.
7. Thamban T, Agarwaal V, Khosla S. Role of genomic imprinting in mammalian development. *J Biosci*. 2020;45.
8. Miller Jr WB. The eukaryotic microbiome: origins and implications for fetal and neonatal life. *Front Pediatr*. 2016;4:96.
9. Smith ZD, Meissner A. DNA methylation: roles in mammalian development. *Nat Rev Genet*. 2013;14:204–220.
10. Riggs AD. X inactivation, differentiation, and DNA methylation. *Cytogenet Cell Genet*. 1975;14:9–25.
11. Wolffe AP, Matzke MA. Epigenetics: regulation through repression. *Science*. 1999;286:481–486.
12. Larsen F, Gundersen G, Lopez R, Prydz H. CpG islands as gene markers in the human genome. *Genomics*. 1992;13:1095–1107.
13. Adachi N, Lieber MR. Bidirectional gene organization: a common architectural feature of the human genome. *Cell*. 2002;109:807–809.
14. Lee JH, Voo KS, Skalnik DG. Identification and characterization of the DNA binding domain of CpG-binding protein. *J Biol Chem*. 2001;276:44669–44676.
15. Riggs AD, Xiong Z, Wang L, Lebon JM. Methylation dynamics, epigenetic fidelity and X chromosome structure. *Novartis Found Symp*. 1998;214:214–225. ; discussion 225-232.
16. Jaenisch R, Bird A. Epigenetic regulation of gene expression: how the genome integrates intrinsic and environmental signals. *Nat Genet*. 2003;33(Suppl):245–254.
17. Howell CY, Bestor TH, Ding F, et al. Genomic imprinting disrupted by a maternal effect mutation in the Dnmt1 gene. *Cell*. 2001;104:829–838.

18. Okano M, Bell DW, Haber DA, Li E. DNA methyltransferases Dnmt3a and Dnmt3b are essential for de novo methylation and mammalian development. *Cell*. 1999;99: 247–257.
19. Hata K, Okano M, Lei H, Li E. Dnmt3L cooperates with the Dnmt3 family of de novo DNA methyltransferases to establish maternal imprints in mice. *Development*. 2002; 129:1983–1993.
20. Watt F, Molloy PL. Cytosine methylation prevents binding to DNA of a HeLa cell transcription factor required for optimal expression of the adenovirus major late promoter. *Genes Dev*. 1988;2:1136–1143.
21. Ohlsson R, Renkawitz R, Lobanenkov V. CTCF is a uniquely versatile transcription regulator linked to epigenetics and disease. *Trends Genet*. 2001;17:520–527.
22. Lewis JD, Meehan RR, Henzel WJ, et al. Purification, sequence, and cellular localization of a novel chromosomal protein that binds to methylated DNA. *Cell*. 1992;69: 905–914.
23. Hendrich B, Bird A. Identification and characterization of a family of mammalian methyl-CpG binding proteins. *Mol Cell Biol*. 1998;18:6538–6547.
24. Prokhortchouk A, Hendrich B, Jorgensen H, et al. The p120 catenin partner Kaiso is a DNA methylation-dependent transcriptional repressor. *Genes Dev*. 2001;15:1613–1618.
25. Nan X, Ng HH, Johnson CA, et al. Transcriptional repression by the methyl-CpG-binding protein MeCP2 involves a histone deacetylase complex. *Nature*. 1998;393: 386–389.
26. Free A, Wakefield RI, Smith BO, Dryden DT, Barlow PN, Bird AP. DNA recognition by the methyl-CpG binding domain of MeCP2. *J Biol Chem*. 2001;276:3353–3360.
27. Shahbazian M, Young J, Yuva-Paylor L, et al. Mice with truncated MeCP2 recapitulate many Rett syndrome features and display hyperacetylation of histone H3. *Neuron*. 2002;35:243–254.
28. Lei H, Oh SP, Okano M, et al. De novo DNA cytosine methyltransferase activities in mouse embryonic stem cells. *Development*. 1996;122:3195–3205.
29. Li E, Bestor TH, Jaenisch R. Targeted mutation of the DNA methyltransferase gene results in embryonic lethality. *Cell*. 1992;69:915–926.
30. Ehrlich M, Tsien F, Herrera D, Blackman V, Roggenbuck J, Tuck-Muller CM. High frequencies of ICF syndrome-like pericentromeric heterochromatin decondensation and breakage in chromosome 1 in a chorionic villus sample. *J Med Genet*. 2001;38: 882–884.
31. Tatton-Brown K, Seal S, Ruark E, et al. Mutations in the DNA methyltransferase gene DNMT3A cause an overgrowth syndrome with intellectual disability. *Nat Genet*. 2014; 46:385–388.
32. Goll MG, Kirpekar F, Maggert KA, et al. Methylation of tRNAAsp by the DNA methyltransferase homolog Dnmt2. *Science*. 2006;311:395–398.
33. Chen RZ, Akbarian S, Tudor M, Jaenisch R. Deficiency of methyl-CpG binding protein-2 in CNS neurons results in a Rett-like phenotype in mice. *Nat Genet*. 2001;27: 327–331.
34. Zhao X, Ueba T, Christie BR, et al. Mice lacking methyl-CpG binding protein 1 have deficits in adult neurogenesis and hippocampal function. *Proc Natl Acad Sci U S A*. 2003;100:6777–6782.
35. Hendrich B, Guy J, Ramsahoye B, Wilson VA, Bird A. Closely related proteins MBD2 and MBD3 play distinctive but interacting roles in mouse development. *Genes Dev*. 2001;15:710–723.

36. Millar CB, Guy J, Sansom OJ, et al. Enhanced CpG mutability and tumorigenesis in MBD4-deficient mice. *Science*. 2002;297:403–405.
37. Prokhortchouk A, Sansom O, Selfridge J, et al. Kaiso-deficient mice show resistance to intestinal cancer. *Mol Cell Biol*. 2006;26:199–208.
38. Oswald J, Engemann S, Lane N, et al. Active demethylation of the paternal genome in the mouse zygote. *Curr Biol*. 2000;10:475–478.
39. Li E. Chromatin modification and epigenetic reprogramming in mammalian development. *Nat Rev Genet*. 2002;3:662–673.
40. Illingworth RS, Gruenewald-Schneider U, Webb S, et al. Orphan CpG islands identify numerous conserved promoters in the mammalian genome. *PLoS Genet*. 2010;6: e1001134.
41. Macleod D, Charlton J, Mullins J, Bird AP. Sp1 sites in the mouse aprt gene promoter are required to prevent methylation of the CpG island. *Genes Dev*. 1994;8: 2282–2292.
42. Myant K, Stancheva I. LSH cooperates with DNA methyltransferases to repress transcription. *Mol Cell Biol*. 2008;28:215–226.
43. Myant K, Termanis A, Sundaram AY, et al. LSH and G9a/GLP complex are required for developmentally programmed DNA methylation. *Genome Res*. 2011;21:83–94.
44. Dong KB, Maksakova IA, Mohn F, et al. DNA methylation in ES cells requires the lysine methyltransferase G9a but not its catalytic activity. *EMBO J*. 2008;27: 2691–2701.
45. Ayyanathan K, Lechner MS, Bell P, et al. Regulated recruitment of HP1 to a euchromatic gene induces mitotically heritable, epigenetic gene silencing: a mammalian cell culture model of gene variegation. *Genes Dev*. 2003;17:1855–1869.
46. Mouse Genome Sequencing C, Waterston RH, Lindblad-Toh K, et al. Initial sequencing and comparative analysis of the mouse genome. *Nature*. 2002;420:520–562.
47. Matsui T, Leung D, Miyashita H, et al. Proviral silencing in embryonic stem cells requires the histone methyltransferase ESET. *Nature*. 2010;464:927–931.
48. Wolf D, Goff SP. TRIM28 mediates primer binding site-targeted silencing of murine leukemia virus in embryonic cells. *Cell*. 2007;131:46–57.
49. Jones PL, Veenstra GJ, Wade PA, et al. Methylated DNA and MeCP2 recruit histone deacetylase to repress transcription. *Nat Genet*. 1998;19:187–191.
50. Muotri AR, Marchetto MC, Coufal NG, et al. L1 retrotransposition in neurons is modulated by MeCP2. *Nature*. 2010;468:443–446.
51. Tsumura A, Hayakawa T, Kumaki Y, et al. Maintenance of self-renewal ability of mouse embryonic stem cells in the absence of DNA methyltransferases Dnmt1, Dnmt3a and Dnmt3b. *Gene Cell*. 2006;11:805–814.
52. Jackson M, Krassowska A, Gilbert N, et al. Severe global DNA hypomethylation blocks differentiation and induces histone hyperacetylation in embryonic stem cells. *Mol Cell Biol*. 2004;24:8862–8871.
53. Ng RK, Dean W, Dawson C, et al. Epigenetic restriction of embryonic cell lineage fate by methylation of Elf5. *Nat Cell Biol*. 2008;10:1280–1290.
54. Nakanishi MO, Hayakawa K, Nakabayashi K, Hata K, Shiota K, Tanaka S. Trophoblast-specific DNA methylation occurs after the segregation of the trophectoderm and inner cell mass in the mouse periimplantation embryo. *Epigenetics*. 2012;7:173–182.
55. Jaenisch R, Wilmut I. Developmental biology. Don't clone humans. *Science*. 2001;291: 2552.
56. Jones PA. DNA methylation and cancer. *Oncogene*. 2002;21:5358–5360.

57. Jones PA, Baylin SB. The fundamental role of epigenetic events in cancer. *Nat Rev Genet*. 2002;3:415–428.
58. Rahat B, Hamid A, Ahmad Najar R, Bagga R, Kaur J. Epigenetic mechanisms regulate placental c-myc and hTERT in normal and pathological pregnancies; c-myc as a novel fetal DNA epigenetic marker for pre-eclampsia. *Mol Hum Reprod*. 2014;20: 1026–1040.
59. Rahat B, Thakur S, Hamid A, Bagga R, Kaur J. Association of aberrant methylation at promoter regions of tumor suppressor genes with placental pathologies. *Epigenomics*. 2016b;8:767–787.
60. Bhutta ZA, Das JK, Rizvi A, Gaffey MF, Walker N, Horton S, Webb P, Lartey A, Black RE, Lancet Nutrition Interventions Review Group TM, Child Nutrition Study G. Evidence-based interventions for improvement of maternal and child nutrition: what can be done and at what cost? *Lancet*. 2013;382:452–477.
61. Mccole RB, Oakey RJ. Unwitting hosts fall victim to imprinting. *Epigenetics*. 2008;3: 258–260.
62. Lee JT, Bartolomei MS. X-inactivation, imprinting, and long noncoding RNAs in health and disease. *Cell*. 2013;152:1308–1323.
63. Latos PA, Pauler FM, Koerner MV, et al. Airn transcriptional overlap, but not its lncRNA products, induces imprinted Igf2r silencing. *Science*. 2012;338:1469–1472.
64. Butler MG. Genomic imprinting disorders in humans: a mini-review. *J Assist Reprod Genet*. 2009;26:477–486.
65. Plasschaert RN, Bartolomei MS. Genomic imprinting in development, growth, behavior and stem cells. *Development*. 2014;141:1805–1813.
66. Hajkova P, Erhardt S, Lane N, et al. Epigenetic reprogramming in mouse primordial germ cells. *Mech Dev*. 2002;117:15–23.
67. Davis TL, Yang GJ, Mccarrey JR, Bartolomei MS. The H19 methylation imprint is erased and re-established differentially on the parental alleles during male germ cell development. *Hum Mol Genet*. 2000;9:2885–2894.
68. Lucifero D, Mann MR, Bartolomei MS, Trasler JM. Gene-specific timing and epigenetic memory in oocyte imprinting. *Hum Mol Genet*. 2004;13:839–849.
69. Seisenberger S, Andrews S, Krueger F, et al. The dynamics of genome-wide DNA methylation reprogramming in mouse primordial germ cells. *Mol Cell*. 2012;48: 849–862.
70. Yamaguchi S, Hong K, Liu R, et al. Tet1 controls meiosis by regulating meiotic gene expression. *Nature*. 2012;492:443–447.
71. Kagiwada S, Kurimoto K, Hirota T, Yamaji M, Saitou M. Replication-coupled passive DNA demethylation for the erasure of genome imprints in mice. *EMBO J*. 2013;32: 340–353.
72. Nakamura T, Liu YJ, Nakashima H, et al. PGC7 binds histone H3K9me2 to protect against conversion of 5mC to 5hmC in early embryos. *Nature*. 2012;486:415–419.
73. Mackay DJ, Callaway JL, Marks SM, et al. Hypomethylation of multiple imprinted loci in individuals with transient neonatal diabetes is associated with mutations in ZFP57. *Nat Genet*. 2008;40:949–951.
74. Pastor WA, Aravind L, Rao A. TETonic shift: biological roles of TET proteins in DNA demethylation and transcription. *Nat Rev Mol Cell Biol*. 2013;14:341–356.
75. Seki Y, Yamaji M, Yabuta Y, et al. Cellular dynamics associated with the genome-wide epigenetic reprogramming in migrating primordial germ cells in mice. *Development*. 2007;134:2627–2638.

76. Hajkova P, Ancelin K, Waldmann T, et al. Chromatin dynamics during epigenetic reprogramming in the mouse germ line. *Nature*. 2008;452:877–881.
77. Lambertini L, Marsit CJ, Sharma P, et al. Imprinted gene expression in fetal growth and development. *Placenta*. 2012;33:480–486.
78. Cassidy FC, Charalambous M. Genomic imprinting, growth and maternal-fetal interactions. *J Exp Biol*. 2018;221.
79. Fowden AL, Coan PM, Angiolini E, Burton GJ, Constancia M. Imprinted genes and the epigenetic regulation of placental phenotype. *Prog Biophys Mol Biol*. 2011;106: 281–288.
80. Ounap K. Silver-russell syndrome and beckwith-Wiedemann syndrome: opposite phenotypes with heterogeneous molecular etiology. *Mol Syndromol*. 2016;7:110–121.
81. John RM. Imprinted genes and the regulation of placental endocrine function: pregnancy and beyond. *Placenta*. 2017;56:86–90.
82. Kappil MA, Green BB, Armstrong DA, et al. Placental expression profile of imprinted genes impacts birth weight. *Epigenetics*. 2015;10:842–849.
83. Green BB, Kappil M, Lambertini L, et al. Expression of imprinted genes in placenta is associated with infant neurobehavioral development. *Epigenetics*. 2015;10:834–841.
84. Rahat B, Mahajan A, Bagga R, Hamid A, Kaur J. Epigenetic modifications at DMRs of placental genes are subjected to variations in normal gestation, pathological conditions and folate supplementation. *Sci Rep*. 2017a;7:40774.
85. Barton SC, Ferguson-Smith AC, Fundele R, Surani MA. Influence of paternally imprinted genes on development. *Development*. 1991;113:679–687.
86. Gregg C, Zhang J, Weissbourd B, et al. High-resolution analysis of parent-of-origin allelic expression in the mouse brain. *Science*. 2010;329:643–648.
87. Albrecht U, Sutcliffe JS, Cattanach BM, et al. Imprinted expression of the murine Angelman syndrome gene, Ube3a, in hippocampal and Purkinje neurons. *Nat Genet*. 1997;17:75–78.
88. Perez JD, Rubinstein ND, Dulac C. New perspectives on genomic imprinting, an essential and multifaceted mode of epigenetic control in the developing and adult brain. *Annu Rev Neurosci*. 2016;39:347–384.
89. Constancia M, Angiolini E, Sandovici I, et al. Adaptation of nutrient supply to fetal demand in the mouse involves interaction between the Igf2 gene and placental transporter systems. *Proc Natl Acad Sci U S A*. 2005;102:19219–19224.
90. Leighton PA, Saam JR, Ingram RS, Stewart CL, Tilghman SM. An enhancer deletion affects both H19 and Igf2 expression. *Genes Dev*. 1995;9:2079–2089.
91. Sferruzzi-Perri AN, Sandovici I, Constancia M, Fowden AL. Placental phenotype and the insulin-like growth factors: resource allocation to fetal growth. *J Physiol*. 2017; 595:5057–5093.
92. Gao WL, Liu M, Yang Y, et al. The imprinted H19 gene regulates human placental trophoblast cell proliferation via encoding miR-675 that targets Nodal Modulator 1 (NOMO1). *RNA Biol*. 2012;9:1002–1010.
93. Mayer W, Hemberger M, Frank HG, et al. Expression of the imprinted genes MEST/Mest in human and murine placenta suggests a role in angiogenesis. *Dev Dynam*. 2000;217:1–10.
94. Diplas AI, Lambertini L, Lee MJ, et al. Differential expression of imprinted genes in normal and IUGR human placentas. *Epigenetics*. 2009;4:235–240.
95. Ono R, Nakamura K, Inoue K, et al. Deletion of Peg10, an imprinted gene acquired from a retrotransposon, causes early embryonic lethality. *Nat Genet*. 2006;38:101–106.

96. Andrews SC, Wood MD, Tunster SJ, Barton SC, Surani MA, John RM. Cdkn1c (p57Kip2) is the major regulator of embryonic growth within its imprinted domain on mouse distal chromosome 7. *BMC Dev Biol*. 2007;7:53.
97. Appelbe OK, Yevtodiyenko A, Muniz-Talavera H, Schmidt JV. Conditional deletions refine the embryonic requirement for Dlk1. *Mech Dev*. 2013;130:143–159.
98. Cleaton MA, Dent CL, Howard M, et al. Fetus-derived DLK1 is required for maternal metabolic adaptations to pregnancy and is associated with fetal growth restriction. *Nat Genet*. 2016;48:1473–1480.
99. Charalambous M, Cowley M, Geoghegan F, et al. Maternally-inherited Grb10 reduces placental size and efficiency. *Dev Biol*. 2010;337:1–8.
100. Csankovszki G, Nagy A, Jaenisch R. Synergism of Xist RNA, DNA methylation, and histone hypoacetylation in maintaining X chromosome inactivation. *J Cell Biol*. 2001; 153:773–784.
101. Beard C, Li E, Jaenisch R. Loss of methylation activates Xist in somatic but not in embryonic cells. *Genes Dev*. 1995;9:2325–2334.
102. Avner P, Heard E. X-chromosome inactivation: counting, choice and initiation. *Nat Rev Genet*. 2001;2:59–67.
103. Cohen DE, Lee JT. X-chromosome inactivation and the search for chromosome-wide silencers. *Curr Opin Genet Dev*. 2002;12:219–224.
104. Chao W, Huynh KD, Spencer RJ, Davidow LS, Lee JT. CTCF, a candidate trans-acting factor for X-inactivation choice. *Science*. 2002;295:345–347.
105. Bannister AJ, Kouzarides T. Regulation of chromatin by histone modifications. *Cell Res*. 2011;21:381–395.
106. Greer EL, Shi Y. Histone methylation: a dynamic mark in health, disease and inheritance. *Nat Rev Genet*. 2012;13:343–357.
107. Josling GA, Selvarajah SA, Petter M, Duffy MF. The role of bromodomain proteins in regulating gene expression. *Genes*. 2012;3:320–343.
108. Swygert SG, Peterson CL. Chromatin dynamics: interplay between remodeling enzymes and histone modifications. *Biochim Biophys Acta*. 2014;1839:728–736.
109. Rossetto D, Avvakumov N, Cote J. Histone phosphorylation: a chromatin modification involved in diverse nuclear events. *Epigenetics*. 2012;7:1098–1108.
110. Cao J, Yan Q. Histone ubiquitination and deubiquitination in transcription, DNA damage response, and cancer. *Front Oncol*. 2012;2:26.
111. Ravid T, Hochstrasser M. Diversity of degradation signals in the ubiquitin-proteasome system. *Nat Rev Mol Cell Biol*. 2008;9:679–690.
112. Schwertman P, Bekker-Jensen S, Mailand N. Regulation of DNA double-strand break repair by ubiquitin and ubiquitin-like modifiers. *Nat Rev Mol Cell Biol*. 2016;17: 379–394.
113. Weake VM, Workman JL. Histone ubiquitination: triggering gene activity. *Mol Cell*. 2008;29:653–663.
114. Borkin D, He S, Miao H, et al. Pharmacologic inhibition of the Menin-MLL interaction blocks progression of MLL leukemia in vivo. *Cancer Cell*. 2015;27:589–602.
115. Li T, Guan J, Huang Z, Hu X, Zheng X. RNF168-mediated H2A neddylation antagonizes ubiquitylation of H2A and regulates DNA damage repair. *J Cell Sci*. 2014;127: 2238–2248.
116. Kothapalli N, Camporeale G, Kueh A, et al. Biological functions of biotinylated histones. *J Nutr Biochem*. 2005;16:446–448.

117. Messner S, Hottiger MO. Histone ADP-ribosylation in DNA repair, replication and transcription. *Trends Cell Biol*. 2011;21:534–542.
118. Shiio Y, Eisenman RN. Histone sumoylation is associated with transcriptional repression. *Proc Natl Acad Sci U S A*. 2003;100:13225–13230.
119. Jiang T, Zhou X, Taghizadeh K, Dong M, Dedon PC. N-formylation of lysine in histone proteins as a secondary modification arising from oxidative DNA damage. *Proc Natl Acad Sci U S A*. 2007;104:60–65.
120. Kouzarides T. Chromatin modifications and their function. *Cell*. 2007;128:693–705.
121. Gagnon J, Daou S, Zamorano N, et al. Undetectable histone O-GlcNAcylation in mammalian cells. *Epigenetics*. 2015;10:677–691.
122. Nelson CJ, Santos-Rosa H, Kouzarides T. Proline isomerization of histone H3 regulates lysine methylation and gene expression. *Cell*. 2006;126:905–916.
123. Cuthbert GL, Daujat S, Snowden AW, et al. Histone deimination antagonizes arginine methylation. *Cell*. 2004;118:545–553.
124. Wang Y, Wysocka J, Sayegh J, et al. Human PAD4 regulates histone arginine methylation levels via demethylimination. *Science*. 2004;306:279–283.
125. Wang Y, Li M, Stadler S, et al. Histone hypercitrullination mediates chromatin decondensation and neutrophil extracellular trap formation. *J Cell Biol*. 2009;184:205–213.
126. Zhao Z, Shilatifard A. Epigenetic modifications of histones in cancer. *Genome Biol*. 2019;20:245.
127. Dahl JA, Reiner AH, Klungland A, Wakayama T, Collas P. Histone H3 lysine 27 methylation asymmetry on developmentally-regulated promoters distinguish the first two lineages in mouse preimplantation embryos. *PLoS One*. 2010;5:e9150.
128. Van De Werken C, Van Der Heijden GW, Eleveld C, et al. Paternal heterochromatin formation in human embryos is H3K9/HP1 directed and primed by sperm-derived histone modifications. *Nat Commun*. 2014;5:5868.
129. Ficz G, Branco MR, Seisenberger S, et al. Dynamic regulation of 5-hydroxymethylcytosine in mouse ES cells and during differentiation. *Nature*. 2011; 473:398–402.
130. Jambhekar A, Dhall A, Shi Y. Roles and regulation of histone methylation in animal development. *Nat Rev Mol Cell Biol*. 2019;20:625–641.
131. Burton A, Torres-Padilla ME. Chromatin dynamics in the regulation of cell fate allocation during early embryogenesis. *Nat Rev Mol Cell Biol*. 2014;15:723–734.
132. Torres-Padilla ME, Parfitt DE, Kouzarides T, Zernicka-Goetz M. Histone arginine methylation regulates pluripotency in the early mouse embryo. *Nature*. 2007;445: 214–218.
133. Sarmento OF, Digilio LC, Wang Y, et al. Dynamic alterations of specific histone modifications during early murine development. *J Cell Sci*. 2004;117:4449–4459.
134. Trojer P, Reinberg D. Facultative heterochromatin: is there a distinctive molecular signature? *Mol Cell*. 2007;28:1–13.
135. Peters AH, Kubicek S, Mechtler K, et al. Partitioning and plasticity of repressive histone methylation states in mammalian chromatin. *Mol Cell*. 2003;12:1577–1589.
136. Rea S, Eisenhaber F, O'carroll D, et al. Regulation of chromatin structure by site-specific histone H3 methyltransferases. *Nature*. 2000;406:593–599.
137. Rice JC, Briggs SD, Ueberheide B, et al. Histone methyltransferases direct different degrees of methylation to define distinct chromatin domains. *Mol Cell*. 2003;12: 1591–1598.

138. Lindeman LC, Andersen IS, Reiner AH, et al. Prepatterning of developmental gene expression by modified histones before zygotic genome activation. *Dev Cell.* 2011; 21:993–1004.
139. Zhao XD, Han X, Chew JL, et al. Whole-genome mapping of histone H3 Lys4 and 27 trimethylations reveals distinct genomic compartments in human embryonic stem cells. *Cell Stem Cell.* 2007;1:286–298.
140. Hawkins RD, Hon GC, Lee LK, et al. Distinct epigenomic landscapes of pluripotent and lineage-committed human cells. *Cell Stem Cell.* 2010;6:479–491.
141. Zhu J, Adli M, Zou JY, et al. Genome-wide chromatin state transitions associated with developmental and environmental cues. *Cell.* 2013;152:642–654.
142. Nicetto D, Donahue G, Jain T, et al. H3K9me3-heterochromatin loss at protein-coding genes enables developmental lineage specification. *Science.* 2019;363:294–297.
143. Zhang B, Wu X, Zhang W, et al. Widespread enhancer dememorization and promoter priming during parental-to-zygotic transition. *Mol Cell.* 2018;72:673–686 e676.
144. Zheng H, Huang B, Zhang B, et al. Resetting epigenetic memory by reprogramming of histone modifications in mammals. *Mol Cell.* 2016;63:1066–1079.
145. Wu J, Xu J, Liu B, et al. Chromatin analysis in human early development reveals epigenetic transition during ZGA. *Nature.* 2018;557:256–260.
146. Zhang B, Zheng H, Huang B, et al. Allelic reprogramming of the histone modification H3K4me3 in early mammalian development. *Nature.* 2016;537:553–557.
147. Santos F, Peters AH, Otte AP, Reik W, Dean W. Dynamic chromatin modifications characterise the first cell cycle in mouse embryos. *Dev Biol.* 2005;280:225–236.
148. Beaujean N. Histone post-translational modifications in preimplantation mouse embryos and their role in nuclear architecture. *Mol Reprod Dev.* 2014;81:100–112.
149. Burton A, Torres-Padilla ME. Epigenetic reprogramming and development: a unique heterochromatin organization in the preimplantation mouse embryo. *Brief Funct Genomics.* 2010;9:444–454.
150. Carrozza MJ, Li B, Florens L, et al. Histone H3 methylation by Set2 directs deacetylation of coding regions by Rpd3S to suppress spurious intragenic transcription. *Cell.* 2005;123:581–592.
151. Andreu-Vieyra CV, Chen R, Agno JE, et al. MLL2 is required in oocytes for bulk histone 3 lysine 4 trimethylation and transcriptional silencing. *PLoS Biol.* 2010;8.
152. Barski A, Cuddapah S, Cui K, et al. High-resolution profiling of histone methylations in the human genome. *Cell.* 2007;129:823–837.
153. Bernstein BE, Humphrey EL, Erlich RL, et al. Methylation of histone H3 Lys 4 in coding regions of active genes. *Proc Natl Acad Sci U S A.* 2002;99:8695–8700.
154. Bernstein BE, Kamal M, Lindblad-Toh K, et al. Genomic maps and comparative analysis of histone modifications in human and mouse. *Cell.* 2005;120:169–181.
155. Calo E, Wysocka J. Modification of enhancer chromatin: what, how, and why? *Mol Cell.* 2013;49:825–837.
156. Majumder S, Alinari L, Roy S, et al. Methylation of histone H3 and H4 by PRMT5 regulates ribosomal RNA gene transcription. *J Cell Biochem.* 2010;109:553–563.
157. Becker PB, Workman JL. Nucleosome remodeling and epigenetics. *Cold Spring Harbor Perspect Biol.* 2013;5.
158. Berger ND, Stanley FKT, Moore S, Goodarzi AA. ATM-dependent pathways of chromatin remodelling and oxidative DNA damage responses. *Philos Trans R Soc Lond B Biol Sci.* 2017;372.

159. Clapier CR, Iwasa J, Cairns BR, Peterson CL. Mechanisms of action and regulation of ATP-dependent chromatin-remodelling complexes. *Nat Rev Mol Cell Biol.* 2017;18: 407–422.
160. Hargreaves DC, Crabtree GR. ATP-dependent chromatin remodeling: genetics, genomics and mechanisms. *Cell Res.* 2011;21:396–420.
161. Van Attikum H, Gasser SM. Crosstalk between histone modifications during the DNA damage response. *Trends Cell Biol.* 2009;19:207–217.
162. Watanabe R, Kanno SI, Mohammadi Roushandeh A, Ui A, Yasui A. Nucleosome remodelling, DNA repair and transcriptional regulation build negative feedback loops in cancer and cellular ageing. *Philos Trans R Soc Lond B Biol Sci.* 2017;372.
163. Curtis CD, Davis RB, Ingram KG, Griffin CT. Chromatin-remodeling complex specificity and embryonic vascular development. *Cell Mol Life Sci.* 2012;69:3921–3931.
164. Ho L, Jothi R, Ronan JL, Cui K, Zhao K, Crabtree GR. An embryonic stem cell chromatin remodeling complex, esBAF, is an essential component of the core pluripotency transcriptional network. *Proc Natl Acad Sci U S A.* 2009;106:5187–5191.
165. Fazzio TG, Panning B. Control of embryonic stem cell identity by nucleosome remodeling enzymes. *Curr Opin Genet Dev.* 2010;20:500–504.
166. Xi S, Zhu H, Xu H, Schmidtmann A, Geiman TM, Muegge K. Lsh controls Hox gene silencing during development. *Proc Natl Acad Sci U S A.* 2007;104:14366–14371.
167. Landry J, Sharov AA, Piao Y, et al. Essential role of chromatin remodeling protein Bptf in early mouse embryos and embryonic stem cells. *PLoS Genet.* 2008;4:e1000241.
168. Liang J, Wan M, Zhang Y, et al. Nanog and Oct4 associate with unique transcriptional repression complexes in embryonic stem cells. *Nat Cell Biol.* 2008;10:731–739.
169. Thomson JA, Itskovitz-Eldor J, Shapiro SS, et al. Embryonic stem cell lines derived from human blastocysts. *Science.* 1998;282:1145–1147.
170. Bhattacharya D, Talwar S, Mazumder A, Shivashankar GV. Spatio-temporal plasticity in chromatin organization in mouse cell differentiation and during Drosophila embryogenesis. *Biophys J.* 2009;96:3832–3839.
171. Cremer T, Cremer M. Chromosome territories. *Cold Spring Harbor Perspect Biol.* 2010;2:a003889.
172. Efroni S, Duttagupta R, Cheng J, et al. Global transcription in pluripotent embryonic stem cells. *Cell Stem Cell.* 2008;2:437–447.
173. Meshorer E, Yellajoshula D, George E, Scambler PJ, Brown DT, Misteli T. Hyperdynamic plasticity of chromatin proteins in pluripotent embryonic stem cells. *Dev Cell.* 2006;10:105–116.
174. Constantinescu D, Gray HL, Sammak PJ, Schatten GP, Csoka AB. Lamin A/C expression is a marker of mouse and human embryonic stem cell differentiation. *Stem Cell.* 2006;24:177–185.
175. Sustackova G, Legartova S, Kozubek S, Stixova L, Pachernik J, Bartova E. Differentiation-independent fluctuation of pluripotency-related transcription factors and other epigenetic markers in embryonic stem cell colonies. *Stem Cell Dev.* 2012; 21:710–720.
176. Walczak P, Kedziorek DA, Gilad AA, Barnett BP, Bulte JW. Applicability and limitations of MR tracking of neural stem cells with asymmetric cell division and rapid turnover: the case of the shiverer dysmyelinated mouse brain. *Magn Reson Med.* 2007;58: 261–269.
177. Dundr M. Nuclear bodies: multifunctional companions of the genome. *Curr Opin Cell Biol.* 2012;24:415–422.

178. Arney KL, Fisher AG. Epigenetic aspects of differentiation. *J Cell Sci*. 2004;117: 4355–4363.
179. Bartova E, Galiova G, Krejci J, Harnicarova A, Strasak L, Kozubek S. Epigenome and chromatin structure in human embryonic stem cells undergoing differentiation. *Dev Dynam*. 2008;237:3690–3702.
180. Rasmussen TP. Embryonic stem cell differentiation: a chromatin perspective. *Reprod Biol Endocrinol*. 2003;1:100.
181. Lee JT. Disruption of imprinted X inactivation by parent-of-origin effects at Tsix. *Cell*. 2000;103:17–27.
182. Plath K, Fang J, Mlynarczyk-Evans SK, et al. Role of histone H3 lysine 27 methylation in X inactivation. *Science*. 2003;300:131–135.
183. Silva J, Mak W, Zvetkova I, et al. Establishment of histone h3 methylation on the inactive X chromosome requires transient recruitment of Eed-Enx1 polycomb group complexes. *Dev Cell*. 2003;4:481–495.
184. Keohane AM, Lavender JS, O'neill LP, Turner BM. Histone acetylation and X inactivation. *Dev Genet*. 1998;22:65–73.
185. Mermoud JE, Popova B, Peters AH, Jenuwein T, Brockdorff N. Histone H3 lysine 9 methylation occurs rapidly at the onset of random X chromosome inactivation. *Curr Biol*. 2002;12:247–251.
186. Klochendler-Yeivin A, Fiette L, Barra J, Muchardt C, Babinet C, Yaniv M. The murine SNF5/INI1 chromatin remodeling factor is essential for embryonic development and tumor suppression. *EMBO Rep*. 2000;1:500–506.
187. Sumi-Ichinose C, Ichinose H, Metzger D, Chambon P. SNF2beta-BRG1 is essential for the viability of F9 murine embryonal carcinoma cells. *Mol Cell Biol*. 1997;17: 5976–5986.
188. Ohta H, Sawada A, Kim JY, et al. Polycomb group gene rae28 is required for sustaining activity of hematopoietic stem cells. *J Exp Med*. 2002;195:759–770.
189. O'carroll D, Erhardt S, Pagani M, Barton SC, Surani MA, Jenuwein T. The polycomb-group gene Ezh2 is required for early mouse development. *Mol Cell Biol*. 2001;21: 4330–4336.
190. Ye Y, Li M, Gu L, et al. Chromatin remodeling during in vivo neural stem cells differentiating to neurons in early Drosophila embryos. *Cell Death Differ*. 2017;24: 409–420.
191. Gaspar-Maia A, Alajem A, Polesso F, et al. Chd1 regulates open chromatin and pluripotency of embryonic stem cells. *Nature*. 2009;460:863–868.
192. Ugarte F, Sousae R, Cinquin B, et al. Progressive chromatin condensation and H3K9 methylation regulate the differentiation of embryonic and hematopoietic stem cells. *Stem Cell Rep*. 2015;5:728–740.
193. Ernst J, Kellis M. ChromHMM: automating chromatin-state discovery and characterization. *Nat Methods*. 2012;9:215–216.
194. Ea V, Baudement MO, Lesne A, Forne T. Contribution of topological domains and loop formation to 3D chromatin organization. *Genes*. 2015;6:734–750.
195. Pombo A, Dillon N. Three-dimensional genome architecture: players and mechanisms. *Nat Rev Mol Cell Biol*. 2015;16:245–257.
196. Lupianez DG, Kraft K, Heinrich V, et al. Disruptions of topological chromatin domains cause pathogenic rewiring of gene-enhancer interactions. *Cell*. 2015;161:1012–1025.
197. Lawrence M, Daujat S, Schneider R. Lateral thinking: how histone modifications regulate gene expression. *Trends Genet*. 2016;32:42–56.

198. Bertero A, Madrigal P, Galli A, et al. Activin/nodal signaling and NANOG orchestrate human embryonic stem cell fate decisions by controlling the H3K4me3 chromatin mark. *Genes Dev.* 2015;29:702–717.
199. Dowen JM, Fan ZP, Hnisz D, et al. Control of cell identity genes occurs in insulated neighborhoods in mammalian chromosomes. *Cell.* 2014;159:374–387.
200. Ji X, Dadon DB, Powell BE, et al. 3D chromosome regulatory landscape of human pluripotent cells. *Cell Stem Cell.* 2016;18:262–275.
201. Kim JD, Kim E, Koun S, et al. Proper activity of histone H3 lysine 4 (H3K4) methyltransferase is required for morphogenesis during zebrafish cardiogenesis. *Mol Cell.* 2015;38:580–586.
202. Hiraide T, Nakashima M, Yamoto K, et al. De novo variants in SETD1B are associated with intellectual disability, epilepsy and autism. *Hum Genet.* 2018;137:95–104.
203. Jones WD, Dafou D, Mcentagart M, et al. De novo mutations in MLL cause Wiedemann-Steiner syndrome. *Am J Hum Genet.* 2012;91:358–364.
204. Chong JX, Yu JH, Lorentzen P, et al. Gene discovery for Mendelian conditions via social networking: de novo variants in KDM1A cause developmental delay and distinctive facial features. *Genet Med.* 2016;18:788–795.
205. Goncalves TF, Goncalves AP, Fintelman Rodrigues N, Dos Santos JM, Pimentel MM, Santos-Reboucas CB. KDM5C mutational screening among males with intellectual disability suggestive of X-Linked inheritance and review of the literature. *Eur J Med Genet.* 2014;57:138–144.
206. Kim HG, Kim HT, Leach NT, et al. Translocations disrupting PHF21A in the Potocki-Shaffer-syndrome region are associated with intellectual disability and craniofacial anomalies. *Am J Hum Genet.* 2012;91:56–72.
207. Hannibal MC, Buckingham KJ, Ng SB, et al. Spectrum of MLL2 (ALR) mutations in 110 cases of Kabuki syndrome. *Am J Med Genet A.* 2011;155A:1511–1516.
208. Abidi F, Miano M, Murray J, Schwartz C. A novel mutation in the PHF8 gene is associated with X-linked mental retardation with cleft lip/cleft palate. *Clin Genet.* 2007;72: 19–22.
209. Iacono G, Dubos A, Meziane H, et al. Increased H3K9 methylation and impaired expression of Protocadherins are associated with the cognitive dysfunctions of the Kleefstra syndrome. *Nucleic Acids Res.* 2018;46:4950–4965.
210. Baujat G, Rio M, Rossignol S, et al. Paradoxical NSD1 mutations in Beckwith-Wiedemann syndrome and 11p15 anomalies in Sotos syndrome. *Am J Hum Genet.* 2004;74:715–720.
211. Gibson WT, Hood RL, Zhan SH, et al. Mutations in EZH2 cause Weaver syndrome. *Am J Hum Genet.* 2012;90:110–118.
212. Rahat B, Najar RA, Hamid A, Bagga R, Kaur J. The role of aberrant methylation of trophoblastic stem cell origin in the pathogenesis and diagnosis of placental disorders. *Prenat Diagn.* 2017b;37:133–143.
213. Rahat B, Sharma R, Bagga R, Hamid A, Kaur J. Imbalance between matrix metalloproteinases and their tissue inhibitors in preeclampsia and gestational trophoblastic diseases. *Reproduction.* 2016a;152:11–22.
214. Akawi N, Mcrae J, Ansari M, et al. Discovery of four recessive developmental disorders using probabilistic genotype and phenotype matching among 4,125 families. *Nat Genet.* 2015;47:1363–1369.
215. Ha M, Kim VN. Regulation of microRNA biogenesis. *Nat Rev Mol Cell Biol.* 2014;15: 509–524.

216. Babiarz JE, Ruby JG, Wang Y, Bartel DP, Blelloch R. Mouse ES cells express endogenous shRNAs, siRNAs, and other Microprocessor-independent, Dicer-dependent small RNAs. *Genes Dev*. 2008;22:2773–2785.
217. Ruby JG, Jan CH, Bartel DP. Intronic microRNA precursors that bypass Drosha processing. *Nature*. 2007;448:83–86.
218. Huntzinger E, Izaurralde E. Gene silencing by microRNAs: contributions of translational repression and mRNA decay. *Nat Rev Genet*. 2011;12:99–110.
219. Ipsaro JJ, Joshua-Tor L. From guide to target: molecular insights into eukaryotic RNA-interference machinery. *Nat Struct Mol Biol*. 2015;22:20–28.
220. Xu W, San Lucas A, Wang Z, Liu Y. Identifying microRNA targets in different gene regions. *BMC Bioinf*. 2014;15(Suppl 7):S4.
221. Moazed D. Small RNAs in transcriptional gene silencing and genome defence. *Nature*. 2009;457:413–420.
222. Murchison EP, Stein P, Xuan Z, et al. Critical roles for Dicer in the female germline. *Genes Dev*. 2007;21:682–693.
223. Tang F, Kaneda M, O'carroll D, et al. Maternal microRNAs are essential for mouse zygotic development. *Genes Dev*. 2007;21:644–648.
224. Liu WM, Pang RT, Chiu PC, et al. Sperm-borne microRNA-34c is required for the first cleavage division in mouse. *Proc Natl Acad Sci U S A*. 2012a;109:490–494.
225. Bernstein E, Kim SY, Carmell MA, et al. Dicer is essential for mouse development. *Nat Genet*. 2003;35:215–217.
226. Wang Y, Medvid R, Melton C, Jaenisch R, Blelloch R. DGCR8 is essential for microRNA biogenesis and silencing of embryonic stem cell self-renewal. *Nat Genet*. 2007; 39:380–385.
227. Ticconi C, Zicari A, Belmonte A, Realacci M, Rao Ch V, Piccione E. Pregnancy-promoting actions of HCG in human myometrium and fetal membranes. *Placenta*. 2007;28(Suppl A):S137–S143.
228. Armant DR. Blastocysts don't go it alone. Extrinsic signals fine-tune the intrinsic developmental program of trophoblast cells. *Dev Biol*. 2005;280:260–280.
229. Davidson LM, Coward K. Molecular mechanisms of membrane interaction at implantation. *Birth Defects Res C Embryo Today*. 2016;108:19–32.
230. Thomson T, Lin H. The biogenesis and function of PIWI proteins and piRNAs: progress and prospect. *Annu Rev Cell Dev Biol*. 2009;25:355–376.
231. Czech B, Hannon GJ. One loop to rule them all: the ping-pong cycle and piRNA-guided silencing. *Trends Biochem Sci*. 2016;41:324–337.
232. Faehnle CR, Joshua-Tor L. Argonautes confront new small RNAs. *Curr Opin Chem Biol*. 2007;11:569–577.
233. Senti KA, Brennecke J. The piRNA pathway: a fly's perspective on the guardian of the genome. *Trends Genet*. 2010;26:499–509.
234. Toth KF, Pezic D, Stuwe E, Webster A. The piRNA pathway guards the germline genome against transposable elements. *Adv Exp Med Biol*. 2016;886:51–77.
235. Yu Y, Xiao J, Hann SS. The emerging roles of PIWI-interacting RNA in human cancers. *Cancer Manag Res*. 2019;11:5895–5909.
236. Ponting CP, Oliver PL, Reik W. Evolution and functions of long noncoding RNAs. *Cell*. 2009;136:629–641.
237. Kapranov P, Cheng J, Dike S, et al. RNA maps reveal new RNA classes and a possible function for pervasive transcription. *Science*. 2007;316:1484–1488.

238. Wu Q, Kim YC, Lu J, et al. Poly A—transcripts expressed in HeLa cells. *PLoS One.* 2008;3:e2803.
239. Guttman M, Amit I, Garber M, et al. Chromatin signature reveals over a thousand highly conserved large non-coding RNAs in mammals. *Nature.* 2009;458:223–227.
240. Gabory A, Jammes H, Dandolo L. The H19 locus: role of an imprinted non-coding RNA in growth and development. *Bioessays.* 2010;32:473–480.
241. Kwong WY, Miller DJ, Ursell E, et al. Imprinted gene expression in the rat embryo-fetal axis is altered in response to periconceptional maternal low protein diet. *Reproduction.* 2006;132:265–277.
242. Ordovas JM, Smith CE. Epigenetics and cardiovascular disease. *Nat Rev Cardiol.* 2010; 7:510–519.
243. De Hoon M, Shin JW, Carninci P. Paradigm shifts in genomics through the FANTOM projects. *Mamm Genome.* 2015;26:391–402.
244. Djebali S, Davis CA, Merkel A, et al. Landscape of transcription in human cells. *Nature.* 2012;489:101–108.
245. Costello I, Pimeisl IM, Drager S, Bikoff EK, Robertson EJ, Arnold SJ. The T-box transcription factor Eomesodermin acts upstream of Mesp1 to specify cardiac mesoderm during mouse gastrulation. *Nat Cell Biol.* 2011;13:1084–1091.
246. Bondue A, Lapouge G, Paulissen C, et al. Mesp1 acts as a master regulator of multipotent cardiovascular progenitor specification. *Cell Stem Cell.* 2008;3:69–84.
247. Saga Y, Miyagawa-Tomita S, Takagi A, Kitajima S, Miyazaki J, Inoue T. MesP1 is expressed in the heart precursor cells and required for the formation of a single heart tube. *Development.* 1999;126:3437–3447.
248. Klattenhoff CA, Scheuermann JC, Surface LE, et al. Braveheart, a long noncoding RNA required for cardiovascular lineage commitment. *Cell.* 2013;152:570–583.
249. Angulo L, Perreau C, Lakhdari N, et al. Breast-cancer anti-estrogen resistance 4 (BCAR4) encodes a novel maternal-effect protein in bovine and is expressed in the oocyte of humans and other non-rodent mammals. *Hum Reprod.* 2013;28: 430–441.
250. Kovalenko TF, Morozova KV, Ozolinya LA, Lapina IA, Patrushev LI. The PTENP1 pseudogene, unlike the PTEN gene, is methylated in normal endometrium, as well As in endometrial hyperplasias and carcinomas in middle-aged and elderly females. *Acta Naturae.* 2018;10:43–50.
251. Bouckenheimer J, Assou S, Riquier S, et al. Long non-coding RNAs in human early embryonic development and their potential in ART. *Hum Reprod Update.* 2016;23: 19–40.
252. Du Y, Kong G, You X, et al. Elevation of highly up-regulated in liver cancer (HULC) by hepatitis B virus X protein promotes hepatoma cell proliferation via down-regulating p18. *J Biol Chem.* 2012;287:26302–26311.
253. Hajjari M, Salavaty A. HOTAIR: an oncogenic long non-coding RNA in different cancers. *Cancer Biol Med.* 2015;12:1–9.
254. Zhou X, Liu S, Cai G, et al. Long non coding RNA MALAT1 promotes tumor growth and metastasis by inducing epithelial-mesenchymal transition in oral squamous cell carcinoma. *Sci Rep.* 2015;5:15972.
255. Yu W, Shi Q, Wu C, Shen X, Chen L, Xu J. Promoter hypermethylation in fl uences the suppressive role of long non-coding RNA MEG3 in the development of multiple myeloma. *Exp Ther Med.* 2020;20:637–645.

256. Huan L, Guo T, Wu Y, et al. Hypoxia induced LUCAT1/PTBP1 axis modulates cancer cell viability and chemotherapy response. *Mol Cancer.* 2020;19:11.
257. Bernstein E, Duncan EM, Masui O, Gil J, Heard E, Allis CD. Mouse polycomb proteins bind differentially to methylated histone H3 and RNA and are enriched in facultative heterochromatin. *Mol Cell Biol.* 2006;26:2560–2569.
258. Pandey RR, Mondal T, Mohammad F, et al. Kcnq1ot1 antisense noncoding RNA mediates lineage-specific transcriptional silencing through chromatin-level regulation. *Mol Cell.* 2008;32:232–246.
259. Van Der Vlag J, Otte AP. Transcriptional repression mediated by the human polycomb-group protein EED involves histone deacetylation. *Nat Genet.* 1999;23: 474–478.
260. Vire E, Brenner C, Deplus R, et al. The Polycomb group protein EZH2 directly controls DNA methylation. *Nature.* 2006;439:871–874.
261. Gupta RA, Shah N, Wang KC, et al. Long non-coding RNA HOTAIR reprograms chromatin state to promote cancer metastasis. *Nature.* 2010;464:1071–1076.
262. Tsai MC, Manor O, Wan Y, et al. Long noncoding RNA as modular scaffold of histone modification complexes. *Science.* 2010;329:689–693.
263. Broadbent HM, Peden JF, Lorkowski S, et al. Susceptibility to coronary artery disease and diabetes is encoded by distinct, tightly linked SNPs in the ANRIL locus on chromosome 9p. *Hum Mol Genet.* 2008;17:806–814.
264. Berger SL, Kouzarides T, Shiekhattar R, Shilatifard A. An operational definition of epigenetics. *Genes Dev.* 2009;23:781–783.
265. Frias-Lasserre D, Villagra CA. The importance of ncRNAs as epigenetic mechanisms in phenotypic variation and organic evolution. *Front Microbiol.* 2017;8:2483.
266. Perez P, Jang SI, Alevizos I. Emerging landscape of non-coding RNAs in oral health and disease. *Oral Dis.* 2014;20:226–235.
267. Ledbetter DH, Riccardi VM, Airhart SD, Strobel RJ, Keenan BS, Crawford JD. Deletions of chromosome 15 as a cause of the Prader-Willi syndrome. *N Engl J Med.* 1981; 304:325–329.
268. Magenis RE, Brown MG, Lacy DA, Budden S, Lafranchi S. Is Angelman syndrome an alternate result of del(15)(q11q13)? *Am J Med Genet.* 1987;28:829–838.
269. Nicholls RD, Knoll JH, Butler MG, Karam S, Lalande M. Genetic imprinting suggested by maternal heterodisomy in nondeletion Prader-Willi syndrome. *Nature.* 1989;342: 281–285.
270. Mccann KL, Baserga SJ. Long noncoding RNAs as sinks in Prader-Willi syndrome. *Mol Cell.* 2012;48:155–157.
271. Greer PL, Hanayama R, Bloodgood BL, et al. The Angelman Syndrome protein Ube3A regulates synapse development by ubiquitinating arc. *Cell.* 2010;140:704–716.
272. Weksberg R, Nishikawa J, Caluseriu O, et al. Tumor development in the Beckwith-Wiedemann syndrome is associated with a variety of constitutional molecular 11p15 alterations including imprinting defects of KCNQ1OT1. *Hum Mol Genet.* 2001;10: 2989–3000.
273. Higashimoto K, Soejima H, Saito T, Okumura K, Mukai T. Imprinting disruption of the CDKN1C/KCNQ1OT1 domain: the molecular mechanisms causing Beckwith-Wiedemann syndrome and cancer. *Cytogenet Genome Res.* 2006;113:306–312.
274. Eggermann T, Wollmann HA, Kuner R, et al. Molecular studies in 37 Silver-Russell syndrome patients: frequency and etiology of uniparental disomy. *Hum Genet.* 1997; 100:415–419.

275. Gicquel C, Rossignol S, Cabrol S, et al. Epimutation of the telomeric imprinting center region on chromosome 11p15 in Silver-Russell syndrome. *Nat Genet.* 2005;37: 1003–1007.
276. Hagberg B, Aicardi J, Dias K, Ramos O. A progressive syndrome of autism, dementia, ataxia, and loss of purposeful hand use in girls: rett's syndrome: report of 35 cases. *Ann Neurol.* 1983;14:471–479.
277. Tiffon C. The impact of nutrition and environmental epigenetics on human health and disease. *Int J Mol Sci.* 2018;19.
278. Torano EG, Garcia MG, Fernandez-Morera JL, Nino-Garcia P, Fernandez AF. The impact of external factors on the epigenome: in utero and over lifetime. *BioMed Res Int.* 2016;2016:2568635.
279. Argente J, Mehls O, Barrios V. Growth and body composition in very young SGA children. *Pediatr Nephrol.* 2010;25:679–685.
280. Barker DJ, Clark PM. Fetal undernutrition and disease in later life. *Rev Reprod.* 1997;2: 105–112.
281. Heijmans BT, Tobi EW, Stein AD, et al. Persistent epigenetic differences associated with prenatal exposure to famine in humans. *Proc Natl Acad Sci U S A.* 2008;105: 17046–17049.
282. Aiken CE, Tarry-Adkins JL, Ozanne SE. Transgenerational effects of maternal diet on metabolic and reproductive ageing. *Mamm Genome.* 2016;27:430–439.
283. Rando OJ, Simmons RA. I'm eating for two: parental dietary effects on offspring metabolism. *Cell.* 2015;161:93–105.
284. Davis CD, Ross SA. Dietary components impact histone modifications and cancer risk. *Nutr Rev.* 2007;65:88–94.
285. Kim KC, Friso S, Choi SW. DNA methylation, an epigenetic mechanism connecting folate to healthy embryonic development and aging. *J Nutr Biochem.* 2009;20: 917–926.
286. Mukerjee S, Zhu Y, Zsombok A, Mauvais-Jarvis F, Zhao J, Lazartigues E. Perinatal exposure to Western diet programs autonomic dysfunction in the male offspring. *Cell Mol Neurobiol.* 2018;38:233–242.
287. Dolinoy DC, Huang D, Jirtle RL. Maternal nutrient supplementation counteracts bisphenol A-induced DNA hypomethylation in early development. *Proc Natl Acad Sci U S A.* 2007;104:13056–13061.
288. Serpeloni F, Radtke KM, Hecker T, et al. Does prenatal stress shape postnatal resilience?—an epigenome-wide study on violence and mental health in humans. *Front Genet.* 2019;10:269.
289. M R, B P. Epigenetic signature of chronic maternal stress load during pregnancy might be a potential biomarker for spontaneous preterm birth. *Balkan J Med Genet.* 2018;21: 27–33.
290. Braithwaite EC, Kundakovic M, Ramchandani PG, Murphy SE, Champagne FA. Maternal prenatal depressive symptoms predict infant NR3C1 1F and BDNF IV DNA methylation. *Epigenetics.* 2015;10:408–417.
291. Non AL, Binder AM, Barault L, Rancourt RC, Kubzansky LD, Michels KB. DNA methylation of stress-related genes and LINE-1 repetitive elements across the healthy human placenta. *Placenta.* 2012;33:183–187.
292. Paquette AG, Lester BM, Koestler DC, Lesseur C, Armstrong DA, Marsit CJ. Placental FKBP5 genetic and epigenetic variation is associated with infant neurobehavioral outcomes in the RICHS cohort. *PLoS One.* 2014;9:e104913.

293. Provenzi L, Fumagalli M, Sirgiovanni I, et al. Pain-related stress during the Neonatal Intensive Care Unit stay and SLC6A4 methylation in very preterm infants. *Front Behav Neurosci*. 2015;9:99.
294. Turecki G, Meaney MJ. Effects of the social environment and stress on glucocorticoid receptor gene methylation: a systematic review. *Biol Psychiatr*. 2016;79:87–96.
295. Hompes T, Izzi B, Gellens E, et al. Investigating the influence of maternal cortisol and emotional state during pregnancy on the DNA methylation status of the glucocorticoid receptor gene (NR3C1) promoter region in cord blood. *J Psychiatr Res*. 2013;47: 880–891.
296. Sparrow S, Manning JR, Cartier J, et al. Epigenomic profiling of preterm infants reveals DNA methylation differences at sites associated with neural function. *Transl Psychiatry*. 2016;6:e716.
297. Oberlander TF, Weinberg J, Papsdorf M, Grunau R, Misri S, Devlin AM. Prenatal exposure to maternal depression, neonatal methylation of human glucocorticoid receptor gene (NR3C1) and infant cortisol stress responses. *Epigenetics*. 2008;3: 97–106.
298. Schroeder JW, Smith AK, Brennan PA, et al. DNA methylation in neonates born to women receiving psychiatric care. *Epigenetics*. 2012;7:409–414.
299. Liu Y, Murphy SK, Murtha AP, et al. Depression in pregnancy, infant birth weight and DNA methylation of imprint regulatory elements. *Epigenetics*. 2012b;7: 735–746.
300. Cao-Lei L, Massart R, Suderman MJ, et al. DNA methylation signatures triggered by prenatal maternal stress exposure to a natural disaster: project Ice Storm. *PLoS One*. 2014;9:e107653.
301. Serpeloni F, Radtke K, De Assis SG, Henning F, Natt D, Elbert T. Grandmaternal stress during pregnancy and DNA methylation of the third generation: an epigenome-wide association study. *Transl Psychiatry*. 2017;7:e1202.
302. Mueller BR, Bale TL. Sex-specific programming of offspring emotionality after stress early in pregnancy. *J Neurosci*. 2008;28:9055–9065.
303. Radtke KM, Ruf M, Gunter HM, et al. Transgenerational impact of intimate partner violence on methylation in the promoter of the glucocorticoid receptor. *Transl Psychiatry*. 2011;1:e21.
304. Appleton AA, Armstrong DA, Lesseur C, et al. Patterning in placental 11-B hydroxysteroid dehydrogenase methylation according to prenatal socioeconomic adversity. *PLoS One*. 2013;8:e74691.
305. Skinner MK, Guerrero-Bosagna C, Haque M, Nilsson E, Bhandari R, Mccarrey JR. Environmentally induced transgenerational epigenetic reprogramming of primordial germ cells and the subsequent germ line. *PLoS One*. 2013;8:e66318.
306. Esposito EA, Jones MJ, Doom JR, Macisaac JL, Gunnar MR, Kobor MS. Differential DNA methylation in peripheral blood mononuclear cells in adolescents exposed to significant early but not later childhood adversity. *Dev Psychopathol*. 2016;28: 1385–1399.
307. Hoang MT, Defina LF, Willis BL, Leonard DS, Weiner MF, Brown ES. Association between low serum 25-hydroxyvitamin D and depression in a large sample of healthy adults: the Cooper Center longitudinal study. *Mayo Clin Proc*. 2011;86:1050–1055.
308. Babenko O, Kovalchuk I, Metz GA. Stress-induced perinatal and transgenerational epigenetic programming of brain development and mental health. *Neurosci Biobehav Rev*. 2015;48:70–91.

309. Kertes DA, Kamin HS, Hughes DA, Rodney NC, Bhatt S, Mulligan CJ. Prenatal maternal stress predicts methylation of genes regulating the hypothalamic-pituitary-adrenocortical system in mothers and newborns in the democratic republic of Congo. *Child Dev.* 2016;87:61−72.
310. Cao-Lei L, Elgbeili G, Massart R, Laplante DP, Szyf M, King S. Pregnant women's cognitive appraisal of a natural disaster affects DNA methylation in their children 13 years later: project Ice Storm. *Transl Psychiatry.* 2015;5:e515.
311. Dahlen HG, Kennedy HP, Anderson CM, et al. The EPIIC hypothesis: intrapartum effects on the neonatal epigenome and consequent health outcomes. *Med Hypotheses.* 2013;80:656−662.
312. Hyde MJ, Mostyn A, Modi N, Kemp PR. The health implications of birth by Caesarean section. *Biol Rev Camb Phil Soc.* 2012;87:229−243.
313. Cardwell CR, Stene LC, Joner G, et al. Caesarean section is associated with an increased risk of childhood-onset type 1 diabetes mellitus: a meta-analysis of observational studies. *Diabetologia.* 2008;51:726−735.
314. Hakansson S, Kallen K. Caesarean section increases the risk of hospital care in childhood for asthma and gastroenteritis. *Clin Exp Allergy.* 2003;33:757−764.
315. Huh SY, Rifas-Shiman SL, Zera CA, et al. Delivery by caesarean section and risk of obesity in preschool age children: a prospective cohort study. *Arch Dis Child.* 2012; 97:610−616.
316. Maghzi AH, Etemadifar M, Heshmat-Ghahdarijani K, Nonahal S, Minagar A, Moradi V. Cesarean delivery may increase the risk of multiple sclerosis. *Mult Scler.* 2012;18:468−471.
317. Pistiner M, Gold DR, Abdulkerim H, Hoffman E, Celedon JC. Birth by cesarean section, allergic rhinitis, and allergic sensitization among children with a parental history of atopy. *J Allergy Clin Immunol.* 2008;122:274−279.
318. Strachan DP. Hay fever, hygiene, and household size. *BMJ.* 1989;299:1259−1260.
319. Strachan DP. Family size, infection and atopy: the first decade of the "hygiene hypothesis. *Thorax.* 2000;55(Suppl 1):S2−S10.
320. Schlinzig T, Johansson S, Gunnar A, Ekstrom TJ, Norman M. Epigenetic modulation at birth—altered DNA-methylation in white blood cells after Caesarean section. *Acta Paediatr.* 2009;98:1096−1099.
321. Leng G, Meddle SL, Douglas AJ. Oxytocin and the maternal brain. *Curr Opin Pharmacol.* 2008;8:731−734.
322. Gregory SG, Connelly JJ, Towers AJ, et al. Genomic and epigenetic evidence for oxytocin receptor deficiency in autism. *BMC Med.* 2009;7:62.
323. Csoka AB, Szyf M. Epigenetic side-effects of common pharmaceuticals: a potential new field in medicine and pharmacology. *Med Hypotheses.* 2009;73:770−780.
324. Franz MB, Poterauer M, Elhenicky M, et al. Global and single gene DNA methylation in umbilical cord blood cells after elective caesarean: a pilot study. *Eur J Obstet Gynecol Reprod Biol.* 2014;179:121−124.
325. Virani S, Dolinoy DC, Halubai S, et al. Delivery type not associated with global methylation at birth. *Clin Epigenet.* 2012;4:8.
326. Almgren M, Schlinzig T, Gomez-Cabrero D, et al. Cesarean delivery and hematopoietic stem cell epigenetics in the newborn infant: implications for future health? *Am J Obstet Gynecol.* 2014;211:502 e501−508.
327. Vaiserman AM. Epigenetic programming by early-life stress: evidence from human populations. *Dev Dynam.* 2015;244:254−265.

CHAPTER

Histone modifications in germline development and maintenance

2

Valerie Jane Robert

Laboratory of Biology and Modeling of the Cell (LBMC), Ecole Normale Supérieure de Lyon, CNRS, UMR 5239, Inserm, U1293, Université Claude Bernard Lyon 1, Lyon, France

Cécile Bedet, Francesca Palladino, Valérie J. Robert

Germ cells, by opposition to somatic cells, are specifically required for the transmission of genetic information from generation to generation and are thus considered immortal. Germ cells are also totipotent, as they have the potential to produce any kind of cell type after fertilization. These two features, immortality and totipotency, make germ cells crucial for the survival of species. Within the gonad, the specialized reproductive organ of animals, germ cells are organized along a polarized axis with proliferating stem cells and mature gametes found at opposite ends and differentiating cells occupying the space in between. Primordial germ cells (PGCs) are the embryonic precursors of germ cells. Two main modes of PGC specification have been described.[1] The first one is observed in mammals and relies on a coordinated combination of transcriptional and epigenetic mechanisms in response to instructive signaling.[2] The second mode of PGC specification is observed in flies, nematodes, and zebrafish. In these species, germ cells are instead specified by the deposition of maternally provided cytoplasmic material found in specialized cytoplasmic ribonucleoprotein organelles located at the nuclear periphery. These organelles, collectively known as germ granules, are evolutionary conserved and have been described in many species including mice, Drosophila, and *C. elegans*.[3,4]

Over the past two decades, epigenetic mechanisms have been shown to be essential for the establishment and maintenance of the transcriptional programs supporting germline totipotency and immortality. Epigenetic regulation in the germline relies, among others, on chemical modifications of DNA and histones. In mammals and zebrafish, DNA methylation on cytosines of CpG islands in the germline is a major player of epigenetic processes such as parental imprinting and X-inactivation. In *C. elegans*, methylation of CpG islands appears to be absent. Most posttranslational modifications (PTMs) of histones occur at the N-terminus. Histone PTMs are widely conserved and include methylation, acetylation, and phosphorylation, among others. Methylation mostly occurs at lysine (K) and arginine (R) residues, and acetylation at lysine residues.[5] Methylation is catalyzed by conserved histone methyltransferases (HMTase). With the exception of DOT1,[6] all HMTs contain a conserved catalytic

Perinatal and Developmental Epigenetics. https://doi.org/10.1016/B978-0-12-821785-6.00012-8

domain known as SET (Su(var)3-9, E(z) and *Trithorax*) able to mono-, di-, and trimethylate lysine residues and mono- and dimethylate arginine residues on their amino groups. Methylation is a dynamic modification that can be reversed by the activity of histone demethylases (see Refs. 7,8 for recent reviews). The first histone lysine demethylase to be identified was LSD1/KDM1,[9] followed by the identification of a large class of lysine histone demethylases containing the conserved jumonji domain (jmjC). Methylation on histone H3 at Lys4 (H3K4me) or Lys36 (H3K36me) are associated with a more accessible chromatin structure. H3K4me3 is found at active promoters and enhancers, while H3K36me3 is deposited throughout gene bodies.[10] Methylation of lysine 9 (H3K9me) and 27 (H3K27me) of histone H3 are instead most abundant at regions associated with facultative or constitutive heterochromatin, respectively. Methylated histone residues are recognized by specific domains, including chromo, tudor or PHD finger domains present in proteins referred to as "readers" that mediate downstream effects. Acetylation is dynamically regulated by the activity of histone acetyl transferases (HAT) and deacetylases (HDACs) that contribute to chromatin regulation in two distinct ways. First, by removing the positive charge from histones and decreasing their interaction with negatively charged phosphate groups of DNA, acetylation contributes to chromatin relaxation. Second, binding of acetyl groups by bromodomains present in chromatin factors can mediate downstream effects.

In this chapter, we will focus on histone PTMs in germ cells. We will describe our current knowledge on the distribution of histone modifications during germ cells specification and gametogenesis in mammals and the model system *C. elegans*. We will also briefly discuss some of the functions attributed to histone PTMs during these developmental processes.

Dynamic reorganization of histone PTMs supports germ cell development, identity, and functions in mammals

Description of histone PTMs and their dynamics

From PGC specification to birth

Early in mammalian development, the epigenetic landscape of newly formed PGCs is similar to that observed in their adjacent somatic cells,[11,12] and parental genomic imprinting established in embryos is still present. Genomic imprinting, which involves DNA methylation (mostly at cytosines of CpG islands) and histone methylation of autosomal somatic genes in a parent-of-origin-specific manner, is responsible for the differential expression of loci depending of their parental origin. In addition, in somatic cells at this stage of development, the X chromosome is submitted to dosage compensation, a process that takes place in embryos at the beginning of cellular differentiation from the pluripotent state to equalize the expression level of X-linked genes between XX females and XY males. Germline development is accompanied by extensive chromatin reprogramming including the removal and reestablishment of DNA methylation and histone PTMs (see below for details). This process is required to erase parental imprinting and X-chromosome compensation, and to reset the germline genome for the acquisition of totipotency and the establishment of proper transcriptional programs for germline development and function.[13–16]

Whole mount immunohistochemistry analysis performed at different stages of mouse embryonic development has shown that the major changes in histone PTMs that accompany chromatin reprogramming in germ cells occur in two steps, before and after PCGs have reached the gonadal ridges (Fig. 2.1). First, at day 8.5 of embryonic development (E8.5), a specific chromatin signature reminiscent of totipotency is detected in migrating PGCs, but not in surrounding somatic cells. This signature is associated with the loss of H3K9me2 and a strong increase in H3K27me3. At this stage, H3K4me2/me3 and H3K9 acetylation (H3K9ac) levels are comparable in migrating germ cells and surrounding somatic cells. Germ cells upregulate both H3K4me2/3 and H3K9ac when they colonize the genital ridges at day 10.5 of embryonic development. At day 11.5, nuclear reorganization characterized by transient decompaction of chromatin is observed and is associated with loss of both repressive H3K27me3 and H3K9me3 marks that subsequently reappear at stage E12.5. Chromatin immunoprecipitation sequencing (ChIP-seq) was used to establish genome-wide maps of four histone PTMs (H3K4me3, H3K27me3, H3K27ac, and H2BK20ac) in purified male mouse fetal germ cells of stage 13.5, when PGC migration to the genital ridge is completed and just before birth.[17]

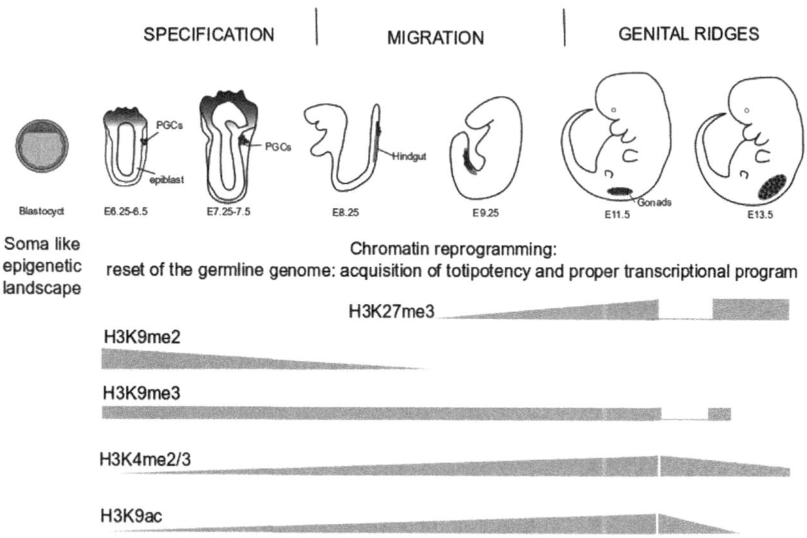

FIGURE 2.1

Posttranslational histone modification dynamics in mouse germ cells during embryonic development.

Most of the detected H3K4me3 peaks mapped to promoters or CpG islands. When coupled with transcriptomic analysis, these data revealed that genes highly expressed in E13.5 germ cells are enriched for H3K4me3 and H3K27ac, while developmental pathway, retrotransposon, and major histocompatibility complex (MHC) genes silenced in germ cells are enriched for H3K27me3.

In humans, genome-wide studies of epigenetic marks are not available for migrating PGCs. However, immunostaining experiments on fetal gonads between week 6 and 13 of gestation, when cells have reached the somatic gonad and are proliferating,[18] showed that both male and female PGCs are characterized by a low level of H3K9me2 in comparison with the surrounding somatic cells of the fetal gonad. In addition, between week 6 and 8, PGCs exhibited higher levels of H3K27me3 than somatic cells, but these did not persist at later stages. Similar patterns of H3K9me2 and H3K27me3 are observed in pigs, suggesting that a chromatin germline signature might be conserved in mammals during the early steps of germline establishment. In human PGCs, high levels of H3K4me1, H3K4me3 and H3K9ac were detected between week 7–13, reminiscent of what is observed in mouse.[16]

During gametogenesis

Following completion of their migration to the somatic gonad, PGCs proliferate and form mitotic diploid cells known as oogonia in females and spermatogonia in males.[19,20] Oogonia undergo either programmed cell death or further maturation in primary oocytes that arrest in prophase-I of meiosis before birth. Spermatogonia instead arrest in mitosis before birth. In both sexes, gametogenesis resumes after puberty and sex-specific patterns of de novo DNA methylation and histone PTMs are established.[21,22]

In female germ cells, DNA methylation imprinting occurs during the diplotene stage of meiotic prophase I, when oocytes are in their growth phase.[23,24] Most histone methylations, including H3K4me2, H3K9me2, and H3K79me2/3, appear to have conserved and stable patterns during oocyte maturation in mouse, porcine, sheep, bovine, and human (Fig. 2.2).[25–27] Recently, H3K23me3, a poorly described histone PTM associated with heterochromatin in *C. elegans* and trypanosomes, was described in porcine.[28] During oocyte growth and at metaphase I, a strong H3K23me3 signal was observed that became weaker on the chromatin of MII-stage oocytes. The function of this modification remains to be established. In mouse, superresolution imaging revealed that the distribution of repressive and activating histone methylation states varies along the pachytene chromosome and defines three distinct nanoscale compartments.[29] First, repressive H3K27me3 is observed in periodic clusters along the chromosome; second, repressive H3K9me3 is present at centromeres; and third, activating H3K4me3 emanates radially from the axis of the chromosome. Although additional studies are necessary to understand the function of these patterns, it suggests that a strong correlation exists between transcriptional competence and higher chromatin organization in which histone methylation might be involved.

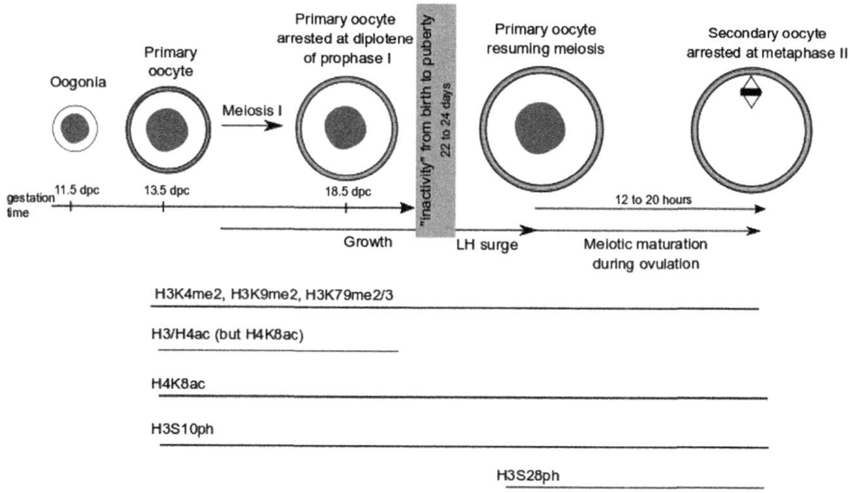

FIGURE 2.2

Posttranslational histone modification dynamics in mouse germ cells during oogenesis. Gestation time is indicated in days post coitum (dpc). Lutein hormone (LH).

In contrast to what is observed for histone methylation, dynamic patterns of acetylation on histone H3 and H4 and phosphorylation on H3Ser10 and H3Ser28 appear to differ depending on the species (see Ref. 30 for a review). For instance, immunofluorescence experiments conducted in mouse following the surge of luteinizing hormone that precedes ovulation revealed a decrease in all acetylation marks with the exception of H4K8ac. In contrast, H4K8ac, H4K12ac, and H3K9ac are all clearly detectable at both meiotic I and II stages in in vitro porcine mature oocytes.[31–33] Experiments using inhibitors of HDAC activity during oogenesis in mouse and pig suggest that histone acetylation plays an important role in chromatin organization both during oocyte growth and oocyte meiosis (see Ref. 30 for a review). In addition, it has been proposed that in mouse, histone deacetylation contributes to the reestablishment of transcriptional programs required for a totipotent zygote at the next generation.[32] Phosphorylation of H3Ser10, but not H3Ser28, is observed earlier in mouse than in porcine oocytes and significantly increases before metaphase I in both species.[27] The functional significance of H3Ser phosphorylation is still largely debated.[30]

In males, spermatogonia first divide and then start differentiation to produce primary spermatocytes (Fig. 2.3). De novo DNA methylation imprints are established during this mitotic step prior to entry into meiosis. After two rounds of meiosis, primary spermatocytes give birth to haploid male maturing germ cells known as round spermatids. In round spermatids, canonical histones are replaced by histone variants that will in turn be replaced by protamines, at a time when round spermatids further differentiate into sperm cells. Protamines are highly basic proteins required for the 10-fold compaction of the sperm genome. Approximately 99% of histones are

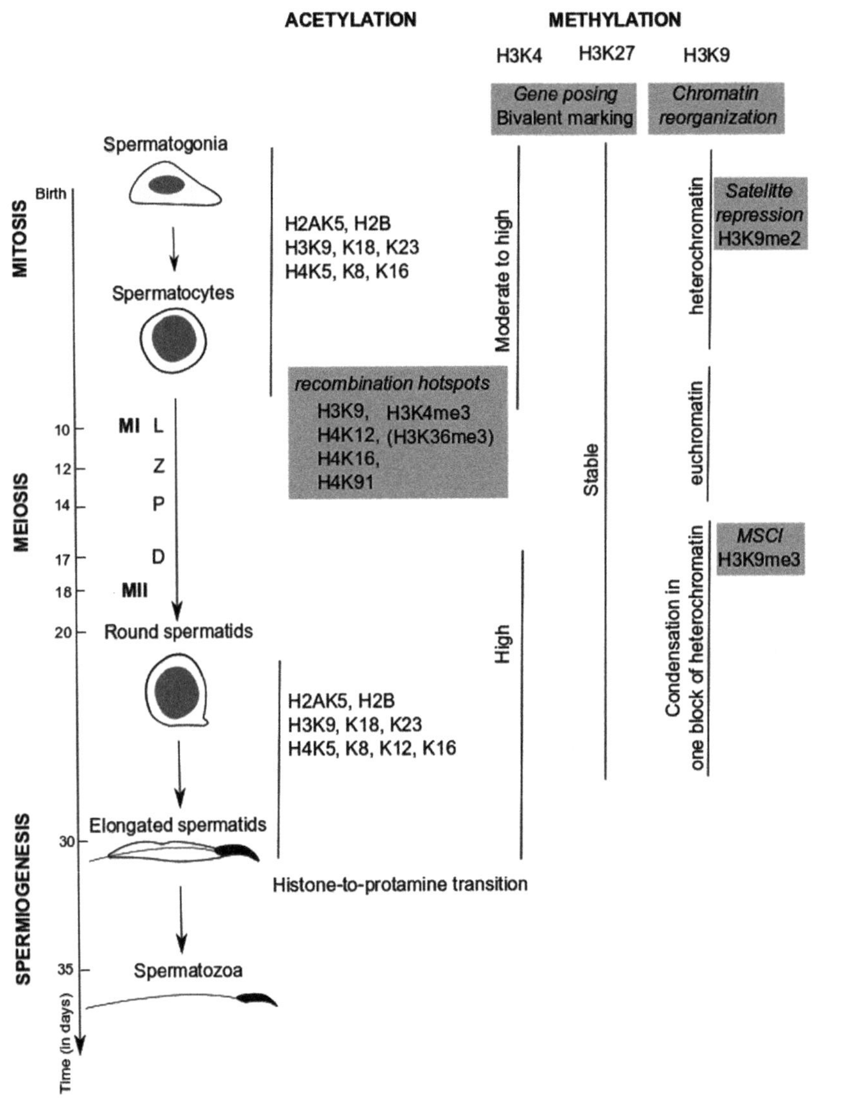

FIGURE 2.3

Post-translational histone modification dynamics in mouse germ cells during spematogenesis.

replaced by protamines in mice mature sperm and 85% in human sperm. In humans, retained histones are found at promoters of genes involved in embryonic development, and the associated modifications are thought to be instructive for gene expression.[34] PTMs are found not only on histones, histones variants, and retained histones, but also on protamines.[35–37]

Initial immunostaining experiments revealed dynamic patterns of histone acetylation during mouse spermatogenesis.[38] It was shown that while histones H2A, H2B, H3, and H4 are acetylated in spermatogonia, no acetylation is detected in spermatocytes and round spermatids. Acetylated forms of H2A, H2B, H3, and H4 reappear in elongating spermatids and finally disappear before final maturation in sperm cells, when histones are replaced by protamines. Subsequent immunochemical analysis on paraffin-embedded sections of adult mouse testis revealed strong H3 acetylation at lysines K9, K18, and K23 in mitotic spermatogonia, which decreased during early steps of meiosis but again increased at diplotene stages.[39] A more recent analysis showed that histone acetyltransferases are dynamically expressed throughout meiosis I and that some H4 and H3 acetylations (including H3K9ac, H4K12ac, H4K16ac, and H4K91ac) can be detected by ChIP at specific loci where meiotic recombination occurs (see the following).[40] In rats, a combination of biochemical techniques was used to examine the modification patterns of TH2B, a male-specific histone variant.[41] TH2B was found to be acetylated at K16 in spermatogonia, spermatocytes, and round spermatids, and combined acetylations were found at K13 and K17 and at K13, K16, and K21 in spermatogonia and spermatocytes.

Initial immunohistochemical analysis showed that H3K4 methylation fluctuates during spermatogenesis.[42] The authors detected moderate H3K4 mono-, di-, and trimethylation levels in spermatogonia that increased in preleptotene and leptotene spermatocytes, disappeared in zygotene and pachytene spermatocytes and reappeared in spermatids.[42] However, this observation is not fully supported by a more recent study that reported a weak H3K4me3 signal until diplotene that became stronger in spermatids.[39] Overexpressing the H3K4 demethylase KDM1 during spermatogenesis results in decreased H3K4me2 levels in sperm[43] and severely impairs progeny development and survival for several generations. H3K4me2 reduction is associated with altered RNA profiles in both sperm and progeny. Initial immunostaining experiments showed that H3K9me patterns are also highly dynamic from spermatogonia to elongating spermatids.[44] The H3K9me signal that is detected at heterochromatic foci in spermatogonia and preleptotene spermatocytes extends into euchromatin at early meiotic prophase and early pachytene to condense into one block of heterochromatin in elongating spermatids. Recent ChIP experiments performed on spermatogonial cells, pachytene spermatocytes, and round spermatids revealed large dynamic changes in H3K9me3 distribution during spermatogenesis, especially during and after meiosis.[45] During mitosis, H3K9me3 domains are preferentially found at LTRs and genes not expressed during spermatogenesis. At pachytene by contrast, a strong H3K9me3 signal that progressively disappears during meiosis is found at gene promoters. Defective H3K9me during spermatogenesis results in an altered survival of spermatogonial cells, increased apoptosis, and meiotic defects.[44,46,47] Contrasting with the dynamic patterns observed for acetylation and methylation of H3K4 and H3K9, H3K27me3 is reported to be stable during spermatogenesis.[39]

More recently, a proteomic approach combining peptide-based bottom-up and intact protein top-down mass spectrometry was used to describe histone and protamine posttranslational modifications in mouse mature sperm cells.[37] This approach identified 26 histone PTMs, including already known lysine and arginine methylation and lysine acetylation marks, as well as four novel histone PTMs not previously described in any tissue or cell type, suggesting they might be sperm specific. This study also identified for the first time 11 PTMs on mouse protamines, including serine and threonine phosphorylation, serine and lysine acetylation, and lysine methylation. The role of these modifications remains unknown.

Histone PTMs cooperate to regulate germline functions and shape the epigenetic landscape

The identification of the enzymes involved in the deposition and removal of histone PTMs, as well as the development of NGS-based comprehensive epigenomic and transcriptomic strategies, has revealed the functional importance of histone PTMs in germline establishment and maintenance. Examples of the role of specific histone PTMs in the germline of mammals are discussed in the next section.

Histone PTMs in meiotic recombination

Meiotic recombination is a major source of genetic diversity and is essential to promote the recognition, pairing, and accurate segregation of homologous chromosomes. Meiotic recombination initiates at the onset of meiosis when the Spo11 endonuclease specifies DNA double-strand breaks (DSB) in discrete genomic regions known as recombination hot spots.[48] Formation of recombination hot spots is facilitated by H4 acetylation that promotes chromatin opening and H3 methylation.[40,49,50] In mice spermatocytes, the position of recombination hot spots is determined by PRDM9, a meiosis-specific HMTase with a PR/SET domain that catalyzes both H3K4me3 and H3K36me3.[51–53] H3K4me3 and H3K36me3 on the same nucleosome create a hot spot–specific signature recognized by the endonuclease Spo11. DSB are repaired at diakinesis when a D-loop structure is formed between the broken chromosome and its intact homolog that is used as a repair template.[48] This D-loop structure is resolved by either crossover or strand displacement. A recent study showed that the binding of PRDM9 on the intact chromosome increases the probability that a DSB is resolved as a crossover.[54] Moreover, H4K16ac catalyzed by the acetyl transferase MOF and H2AX phosphorylation are essential for the recruitment of the DSB repair machinery and correct repair.[38,55] It is not known yet whether PRDM9 functions are conserved in mouse oocytes. However, in these cells, H3K4me3 provides a permissive signal for the G2–M transition.[55]

Histone PTMs and meiotic sex chromosome inactivation

At the pachytene stage of male meiosis, sex chromosomes migrate to the nuclear periphery, resulting in the formation of a condensed structure named the “XY body”.[56] Sex chromosomes in XY bodies undergo partial pairing at their pseudoautosomal regions, while their noncomplementary regions remain asynapsed. Meiotic sex chromosome inactivation (MSCI) is a process triggered by the unsynapsed regions that result in megabase-scale chromatin remodeling to ensure transcriptional silencing of sex chromosome genes. H3 and H4 acetylations increase on the X and Y chromosomes before the start of MSCI to subsequently decrease, along with all histone PTMs associated with more open chromatin states.[57] The only exception is H4K16ac, which contributes to H2AX phosphorylation, a key factor of MSCI.[57] H2AX phosphorylation induces the deposition of histone PTMs involved in heterochromatin formation, including H3K9me2 at the transition between pachytene and diplotene, and H3K9me3 at diplotene. H3K9me3 deposition is catalyzed by SETDB1, requires an intact DNA damage response pathway (DDR), and is necessary for complete XY chromatin remodeling.[58] Interestingly, this correlates with a strong decrease in H3K4me3, but only a partial decrease in H3K4me2, suggesting that MSCI is established without complete removal of active modifications.[59]

Histone PTMs and transcriptional control during gametogenesis

During gametogenesis, specific and distinct transcriptional programs support first proliferation in mitotic self-renewing germ cells and later on, meiosis during differentiation. Specifically, as male germ cells exit mitosis to enter meiosis, a massive transcriptional change is observed that results in the transcriptional repression of genes highly expressed in spermatogenesis progenitor cells and the activation of late spermatogenesis genes.[59] ChIP-seq analysis showed that autosomal genes activated in late spermatogenesis are silenced in germline stem cells, but poised for rapid reactivation through the deposition of active modifications and RNAPII, and reduction in H3K27me3. Later, in round spermatids, these same genes are characterized by H3K4me2 enrichment that spreads out broadly throughout the gene body. Autosomal genes expressed in progenitor cells become silenced at the mitotic-to-meiotic transition and are characterized by the deposition of H3K27me3 in round spermatids. The deposition of H3K27me3 at promoters with H3K4me3 marks genes for activation after fertilization. By contrast, X-linked genes subject to MSCI are poised for activation after fertilization by a H3K9me3 dependent mechanism, independently of the formation of bivalent domains (see previous section). In spermatids, a small number of X-linked genes escape MSCI and are expressed. This specific set of genes is not marked by H3K4me3 in germ stem cells, suggesting that they are not poised for later activation. Transcriptional activation of these genes occurs in the presence of H3K9me2 on their coding sequence, whereas H3K4me2 and lysine crotonylation are present at their transcription start site.[60]

A similar system of bivalent signature consisting of H3K4me3 and H3K27me3 is present at the transcriptional start sites of developmentally important genes that are silenced in both male and female germ cells from fetal stages through completion of meiosis.[61] Bivalent marking silences developmental genes to preserve germ cell identity while keeping genes in a poised state to allow quick reactivation during postembryonic development. Bivalent marking is required for highly specialized and differentiated gametes to retain the ability to activate a variety of distinct transcriptional programs and produce a totipotent zygote. Bivalent marking of promoter regions has also been described in embryonic stem cells and is thought to promote efficient entry into differentiation programs following induction.[62]

Histone PTMs distribution and function in *Caenorhabditis elegans* hermaphrodite germ cells

In *C. elegans*, the early germline blastomeres (known as P0, P1, P2, P3) undergo four successive asymmetric cell divisions that generate P4, the founder cell of the germ line (Fig. 2.4).[63–65] P4, which arises at the 16–24 cell stage, divides symmetrically around the 100 cell stage to give birth to two PGCs, Z2 and Z3. PGCs remain mitotically quiescent until the mid-L1 stage, when larvae start feeding. During divisions of the P-lineage, P blastomeres specifically inherit from their parent cell P-granules, the *C. elegans* germ granules. At the time of hatching, Z2 and Z3 are found in the

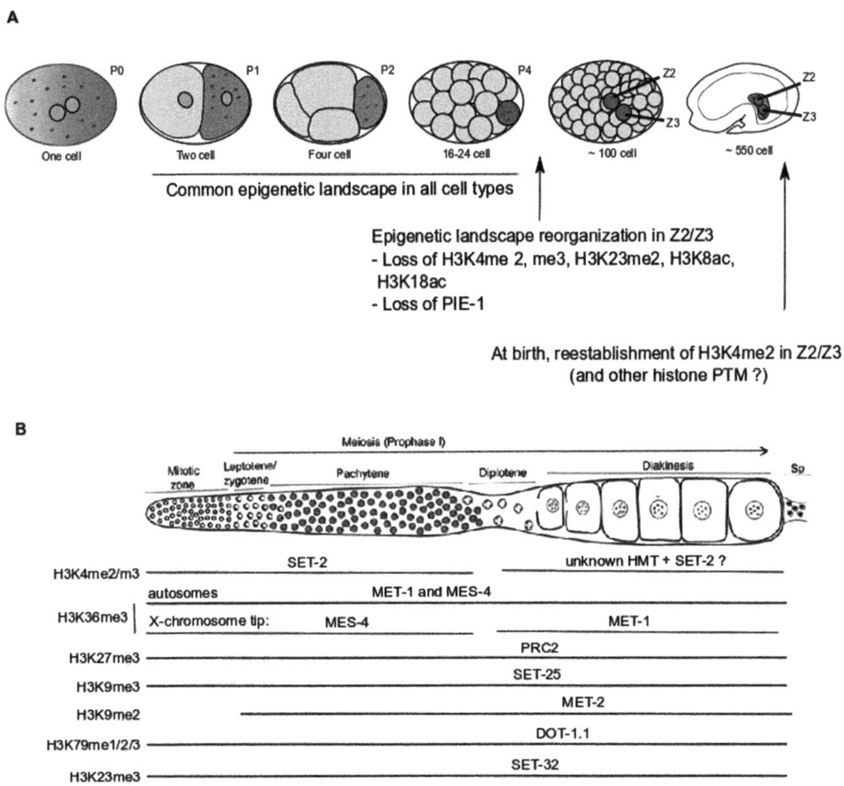

FIGURE 2.4

Posttranslational histone modification dynamics in *Caenorhabditis elegans* germ cells. (A) During embryogenesis. (B) In the adult germline.

ventral midbody of the animal and are tightly associated with somatic gonad precursors. At the mid-L1 stage, Z2 and Z3 start to proliferate, while by the late L2 stage, the somatic gonad precursors Z1 and Z4 proliferate to form the somatic gonad primordium. By L3, germ cells in the proximal-most part of the extending gonad enter meiosis. In hermaphrodites, spermatogenesis occurs during the L4 stage and produces spermatozoa that will be stored in the spermatheca. At the late L4 stage of hermaphrodite development, spermatogenesis is negatively regulated to allow a switch to oogenesis, which then continues throughout the adult life.[66]

In each of the symmetrical U-shaped arms of the adult *C. elegans* hermaphrodite gonad, germ cells are organized in a distal-to-proximal manner, with mitotic undifferentiated cells in the distal-most part and differentiated gametes in the proximal-most part (Fig. 2.4). Between these two stages a gradient of cell stransitioning from mitosis to meiosis I prophase is observed. Germline nuclei reside in a cellular syncytium up to the loop made by the gonadal arm, and become cellularized only as they reach diplotene more proximally. Out of a total of 20,000 genes encoded by the *C. elegans* genome, 6000 were found to be expressed throughout the hermaphrodite germline and show spatiotemporal dynamic expression patterns.[67]

Distribution of histone PTMs in the *Caenorhabditis elegans* germline

Repressive chromatin structure in the P blastomeres and PGC

Maintenance of germline identity in early embryogenesis relies on transcriptional repression mediated by the maternally-inherited PIE-1 transcriptional factor and DNA compaction.[63,65] PIE-1 localizes in nuclei in the early P-lineage blastomeres until the division of P4 into Z2 and Z3, when it is degraded. Immunostaining experiments revealed that at this stage, the activating marks H3K4me2 and me3, H4K8ac, and H3K18ac, as well as the repressive mark H3K23me2, disappear from these cells but are maintained in the somatic blastomeres (Figure 2.4).[68–70] The dynamic pattern of H3K4me2 in PGCs is required to preserve germ cell identity and relies on the maternal NANOS proteins that belong to an evolutionary conserved family of factors involved in germ cell survival and pluripotency. Together, these results suggest that the reorganization of the epigenetic landscape in PGCs is required to establish chromatin-based transcriptional repression in the absence of PIE-1.[69] H3K4me2 is quickly reestablished in the germline after hatching, when PGCs resume proliferation. Interestingly, this reestablishment is sensitive to environmental conditions since it requires presence of food.[68] In the future, ChIP or Cut and Run assays,[71] followed by sequencing, should make it possible to study how the chromatin reorganization observed by microscopy-based approaches correlates with the global chromatin redistribution of histone PTMs in Z2 and Z3. The recent development of a protocol to isolate GFP-expressing PGCs by fluorescent-activated cell sorting (FACS) should facilitate these types of approaches.[72]

Chromatin landscape in the adult hermaphrodite germline

The presence of a number of histone PTMs has been described in both mitotic and meiotic nuclei of hermaphrodite and male adult germlines (see Refs. 73–82 and 83 for a review) (Fig. 2.4). In contrast to mammals, histones are largely retained in mature sperm[82,84] and carry canonical PTMs. Immunostaining experiments performed on hermaphrodite germ cells at pachytene, the first stage in which chromosomes can be easily distinguished, revealed that histone PTMs associated with active transcription and repression are differentially distributed on chromosomes. Autosomes are enriched in marks associated with transcriptionally active chromatin such as histone H3 and H4 acetylation or H3K4 and H3K36 methylation, while the X-chromosome is mostly (with the exception of its left tip) decorated by the repressive mark H3K27me3.

Many of the enzymes involved in the deposition of histone methylations have been identified.[73–77,85–88] SET-2, the *C. elegans* ortholog of yeast Set1, is responsible for both H3K4me2 and H3K4me3 in the distal germline, together with other subunits of the SET1/COMPASS complex.[77,78,89,90] The enzyme responsible for H3K4me in the proximal germline remains to be identified. SET-16, the only other SET1 family member in *C. elegans*, is a likely candidate (see Ref. 89, and V. Robert, personal observations). MES-4 and MET-1 (an ortholog of human NSD) are responsible for H3K36me3 on autosomes at all stages of germline differentiation. In addition, MES-4 deposits H3K36me3 on the X-chromosome tip from mitosis to late pachytene and MET-1 at diakinesis.[78] H3K27me2 and -me3, catalyzed by the polycomb repressive complex 2 (PRC2), are present on chromatin throughout the germline.[73] However, while H3K27me2 is uniformly distributed on both autosomes and X-chromosome, H3K27me3 is enriched on the X chromosome and found organized in domains on autosomes H3K36 and H3K27 methylation together play an essential role in regulating the germline transcriptional program and organize the genome into extended chromatin domains in both embryos and adults.[91] In hermaphrodites, H3K9me3 is present on all chromosomes in all germ cells, while meiotic nuclei display an H3K9me2 gradient starting at early pachytene and progressively increasing in late pachytene and diplotene.[81]

Interestingly, the unpaired X chromosome of males and any unpaired regions (i.e., free duplications and extrachromosomal arrays) in hermaphrodites are highly enriched for H3K9me2 deposited by MET-2, an ortholog of mammalian SETDB1.[79,92] This process is reminiscent of MSCI in mammals.[56] MET-2 is also responsible for global H3K9me2 in both male and hermaphrodite germlines,[81] while SET-25, an ortholog of mammalian EHMT1/G9a, is responsible for global H3K9me3.It was initially reported that H3K9me3 is still detected in the germline of *met-2* mutants, suggesting that SET-25 has a H3K9me3 activity independent of MET-2.[81] However, subsequent studies using validated H3K9me3 antibodies have shown that H3K9me3 is altered in *met-2* mutant germ lines.[93,94] The current model is that MET-2 and SET-25 act together in a step-wise manner with MET-2 depositing H3K9me1 and me2, and SET-25 adding H3K9me3, as described in embryos.[88]

Recent data has shown that H3K79 methylation, deposited by DOT-1.1, plays an important role in coordinating events during meiosis.[80,95] H3K79me1, me2, and me3 are observed from mitotic to late pachytene cells on both autosomes and the X chromosome, and DOT-1.1 plays a role in the regulation of pairing, synapsis, and recombination.[80]

Role of histone PTMs in transcriptional regulation in the *Caenorhabditis elegans* germline

The finding that PRC2, responsible for H3K27me3, and MES-4, an H3K36 HMT, are both essential for germline proliferation provided the first evidence that histone PTMs play an important role in the germline of *C. elegans*.[96,97] This was further supported by studies showing that multicopy repetitive transgenes in the germline are silenced via the deposition of repressive H3K9 methylation.[79,98] In the following section, we provide examples of the role of specific histone PTMs in the germline of *C. elegans*.

X chromosome silencing

The subunits of the *C. elegans* PRC2 complex, responsible for deposition of H3K27me, were identified more than 20 years ago in a genetic screen for mutants with maternal effect sterility.[87,96,97] In hermaphrodite germlines, H3K27me3 is present all along the X-chromosome and participates in its silencing.[73,79] This function is indirectly controlled by MES-4 that deposits H3K36me3 on autosomes. The current model is that H3K36me3 deposited by MES-4 on autosomes repels MES-2/6-dependent H3K27me3, thereby concentrating it on the X. Loss of MES activity results in derepression of X-linked genes and sterility.[99] These observations support the "either H3K36me3 or H3K27me3" rule consistent with an antagonism between PRC2-dependent H3K27me3 and trithorax-dependent H3K36me3 initially described in mammals and flies.[100,101]

Histone PTMs and transcriptional memory

Analysis of progenies from asymmetric crosses in which chromosomes carrying H3K27me and H3K36me were provided by either hermaphrodites or males have elegantly shown that these two histone PTMs are transmitted from both oocytes and sperm to zygotes.[78,102] In zygotes that have inherited differentially marked sets of chromosomes, modified histone remains associated with DNA from the gamete-of-origin through several rounds of replication, even in the absence of a functional HMTase. In wild-type zygotes, PRC2 and MES-4 proteins are maternally inherited. Maternally supplied PRC2 ensures that H3K27me3 marking persists after 16–24 cell stage of embryonic development, while MES-4 is required for H3K36me3 maintenance after the four-cell stage. MES-4 associates with sperm-inherited chromosomes soon after fertilization, and this association depends on their prior marking with H3K36me3.[78] These observations suggest that both H3K27 and H3K36 methylation transmit a memory of transcriptional activity across generations.

The global distribution of H3K4me3, H3K36me3, and H3K27me3 in mature sperm and unfertilized oocytes was recently described by ChIP-seq.[84] This study revealed alternating clusters of H3K36me3 and H3K27me3 on autosomes, while the X chromosome carries H3K27me3 along its length with very few H3K36me3-enriched regions (with the exception of the pseudo-autosomal region).[84,99] Their comparison with transcriptional profiling revealed that genes that are expressed in both spermatogenic and oogenic germlines are generally marked by H3K36me3 and lack H3K27me3, while genes that are not expressed either in spermatogenic or oogenic germlines are marked with H3K27me3. Interestingly, in sperm (but not in oocytes), genes that are exclusively expressed in spermatogenic germlines show a multivalent marking over their gene body consisting of H3K4me3 (that is typically found at transcriptional start sites of transcribed genes), H3K36me3, and H3K27me3 and do not follow the "either H3K36me3 or H3K27me3" rule. Multivalent marking on this specific class of genes is not observed in embryos, where only H3K27me3 is present at genes exclusively expressed in spermatogenic germlines. These results suggest that at these genes, both H3K36me3 and H3K4me3 inherited from sperm are lost or erased in early embryos.[93] Surprisingly, sperm also displays active histone marking on genes known to be expressed during oogenesis, and these genes are expressed in sperm. One plausible explanation is that transcription of these genes is required to maintain them in an open chromatin state in condensed sperm. Altogether, these results suggest that histone PTMs can combine in multiple and diverse ways to assign a transcriptional identity to gene subsets.

Histone PTMs and maintenance of germline identity

Several studies underline the importance of histone PTMs in the maintenance of germline identity in hermaphrodites. The first indication that histone modifications are key players in restricting the ability of a transcription factor to reprogram cellular identity in the germline came from experiments in which ectopic expression of a neuronal transcription factor was able to convert germ cells into neurons following HDAC inhibition.[103] Conversion of germ cells into neurons has also been observed following PRC2 knockdown,[104] and in animals lacking the H3K4 HMT SET-2.[105] In the case of *set-2* mutants, loss of germ cell identity and conversion to somatic cell types was only observed after growing animals multiple generations at the stressful temperature of 25°C.[75] This was accompanied by upregulation of X-linked genes as well as ectopic expression of TGFβ pathway components that could contribute to the loss of germline identity.[106]

Histone PTMs and transgenerational epigenetic inheritance of environmental conditions

Physiological plasticity is crucial for organisms to adapt to changes in their environment. Epigenetic regulators are key factors of physiological plasticity as they can influence gene expression and cellular homeostasis in animals directly exposed to environmental stress. A highly controversial topic is whether environmentally induced epigenetic changes can be inherited across generations in mammals (see Ref. 107 for a recent and complete review of the subject).

While it remains to be established whether histone PTMs are causative agents in the transgenerational transmission of environmental cues to progeny, over the last two decades *C. elegans* has become an organism of choice to study this process, and exciting examples of this type of transmission have been reported (see Refs. 108,109 for case studies and Refs. 110–113 for recent reviews). Transgenerational memory of RNAi-induced silencing is one of the most studied model of epigenetic inheritance in *C. elegans*.[114] In this system, RNAi silencing induced on endogenous genes or single-copy transgenes is maintained for many generations after elimination of the double-strand RNA (dsRNA) trigger through a process dependent on the nuclear RNAi machinery.[115] In the exposed generation, small RNAs produced from the dsRNA trigger through the canonical RNAi machinery are loaded on the argonaute HRDE-1 that shuttles to the nucleus to induce the deposition of H3K9me3, H3K27me3, and H3K23me3 heterochromatin marks at targeted loci.[77,116,117] Transgenerational inheritance of these marks, and their progressive loss over generations as the silencing process is reversed, are observed.

While it is widely accepted that small RNAs are agents of heritability, the roles of histone PTMs in transgenerational epigenetic inheritance are still debated. SET-25 and SET-32, which catalyze H3K9me3 and H3K23me3, respectively, were shown to be required for the establishment, but not the long-term maintenance of silencing.[118,119,120] The MET-2 H3K9 dimethylase is indirectly required to limit the duration of transgenerational inheritance through an effect on the biogenesis of heritable small RNA.[121] The SET-25 H3K9 trimethylase is also required for the biogenesis of a specific class of endogenous small RNA that target newly evolved genes.[122]

H3K9 methylation and genome stability

met-2 set-25 double mutants that lack all H3K9 methylation are sterile at the stressful temperature of 25°C.[94] Transcriptomic analysis performed in both germline and embryos showed that MET-2, responsible for H3K9 mono- and di-methylation, is essential for the repression of satellite sequences, while SET-25, which deposits tri-methylation, represses a subset of DNA and RNA transposons.[123,124] Derepression of repetitive elements in *met-2 set-25* mutants correlates with increased apoptosis in the germline, indicative of DNA damage, R-loop formation, and hypersensitivity to replicative stress. Based on these observations, it was proposed that H3K9 methylation protects the *C. elegans* genome by suppressing transcription-induced replication stress. Recently, it was shown that satellite sequences repressed by MET-2 are enriched in the monoubiquitination of H2A (H2Aub) deposited by the BRCA-1/E3 ubiquitin ligase.[125] BRCA-1 was shown to be involved in satellite repression, and a cross-talk between H3K9me2 and H2Aub may contribute to genome stability.

Concluding remarks

Regardless of the mode used to specify a germline during early development, histone PTMs are essential players in its establishment and maintenance, and crucial to support germline immortality and totipotency. Initial investigations on the role of PTMs in these processes have relied on histological and immunostaining analysis. The recent developments of whole genome-sequencing approaches, targeted genome engineering, and novel strategies to collect biological material of interest now open new perspectives to study the distribution of histone PTMs, identify enzymes required for their writing and reading, and describe how they cooperate to maintain germline function. These questions are of primary interest in public health since histone PTMs largely contribute to the transgenerational memory of cellular response to environmental stress, which, in turn, significantly impact germline maintenance and species survival.

References

1. Extavour CG, Akam M. Mechanisms of germ cell specification across the metazoans: epigenesis and preformation. *Development (09501991)*. 2003;130:5869.
2. Sybirna A, Wong FCK, Surani MA. Genetic basis for primordial germ cells specification in mouse and human: conserved and divergent roles of PRDM and SOX transcription factors. *Curr Top Dev Biol*. 2019;135:35–89. https://doi.org/10.1016/bs.ctdb.2019.04.004.
3. Trcek T, Lehmann R. Germ granules in Drosophila. *Traffic*. 2019;20:650–660. https://doi.org/10.1111/tra.12674.
4. Voronina E, Seydoux G, Sassone-Corsi P, Nagamori I. RNA granules in germ cells. *Cold Spring Harbor Perspect Biol*. 2011;3:a002774. https://doi.org/10.1101/cshperspect.a002774.
5. Kouzarides T. Chromatin modifications and their function. *Cell*. 2007;128:693–705. https://doi.org/10.1016/j.cell.2007.02.005.
6. Feng Q, Wang H, Ng HH, et al. Methylation of H3-lysine 79 is mediated by a new family of HMTases without a SET domain. *Curr Biol*. 2002;12:1052–1058. https://doi.org/10.1016/S0960-9822(02)00901-6.
7. Nottke A, Colaiácovo MP, Shi Y. Developmental roles of the histone lysine demethylases. *Development*. 2009;136:879–889. https://doi.org/10.1242/dev.020966.
8. Zhang J, Jing L, Li M, He L, Guo Z. Regulation of histone arginine methylation/demethylation by methylase and demethylase (review). *Mol Med Rep*. 2019;19:3963–3971. https://doi.org/10.3892/mmr.2019.10111.
9. Shi Y, Lan F, Matson C, et al. Histone demethylation mediated by the nuclear amine oxidase homolog LSD1. *Cell*. 2004;119:941–953. https://doi.org/10.1016/j.cell.2004.12.012.
10. Jambhekar A, Dhall A, Shi Y. Roles and regulation of histone methylation in animal development. *Nat Rev Mol Cell Biol*. 2019;20:625–641. https://doi.org/10.1038/s41580-019-0151-1.

11. Hackett JA, Zylicz JJ, Surani MA. Parallel mechanisms of epigenetic reprogramming in the germline. *Trends Genet.* 2012;28:164–174. https://doi.org/10.1016/j.tig.2012.01.005.
12. Saitou M, Kagiwada S, Kurimoto K. Epigenetic reprogramming in mouse pre-implantation development and primordial germ cells. *Development (Cambridge, England).* 2012;139:15–31. https://doi.org/10.1242/dev.050849.
13. Seki Y, Hayashi K, Itoh K, Mizugaki M, Saitou M, Matsui Y. Extensive and orderly reprogramming of genome-wide chromatin modifications associated with specification and early development of germ cells in mice. *Dev Biol.* 2005;278:440–458. https://doi.org/10.1016/j.ydbio.2004.11.025.
14. Seki Y, Yamaji M, Yabuta Y, et al. Cellular dynamics associated with the genome-wide epigenetic reprogramming in migrating primordial germ cells in mice. *Development.* 2007;134:2627–2638. https://doi.org/10.1242/dev.005611.
15. Hajkova P, Ancelin K, Waldmann T, et al. Chromatin dynamics during epigenetic reprogramming in the mouse germ line. *Nature.* 2008;452:877–881. https://doi.org/10.1038/nature06714.
16. Matsui Y, Mochizuki K. A current view of the epigenome in mouse primordial germ cells. *Mol Reprod Dev.* 2014;81:160–170. https://doi.org/10.1002/mrd.22214.
17. Ng J-H, Kumar V, Muratani M, et al. In vivo epigenomic profiling of germ cells reveals germ cell molecular signatures. *Dev Cell.* 2013;24:324–333. https://doi.org/10.1016/j.devcel.2012.12.011.
18. Eguizabal C, Herrera L, De Onate L, Montserrat N, Hajkova P, Izpisua Belmonte JC. Characterization of the epigenetic changes during human gonadal primordial germ cells reprogramming. *Stem Cell.* 2016:2418. https://doi.org/10.1002/stem.2422.
19. Clarke HJ. Growth and meiotic maturation of mammalian oocytes: an overview. In: Skinner MK, ed. *Encyclopedia of reproduction.* 2nd ed. Oxford: Academic Press; 2018:144–152. https://doi.org/10.1016/B978-0-12-801238-3.64442-2.
20. Goldberg E, Zirkin BR. Spermatogenesis: overview. In: *Encyclopedia of reproduction.* Elsevier; 2018:13–18. https://doi.org/10.1016/B978-0-12-801238-3.64422-7.
21. Sasaki H, Matsui Y. Epigenetic events in mammalian germ-cell development: reprogramming and beyond. *Nat Rev Genet.* 2008;9:129–140. https://doi.org/10.1038/nrg2295.
22. Smallwood SA, Kelsey G. De novo DNA methylation: a germ cell perspective. *Trends Genet.* 2012;28:33–42. https://doi.org/10.1016/j.tig.2011.09.004.
23. Smallwood SA, Tomizawa S, Krueger F, et al. Dynamic CpG island methylation landscape in oocytes and preimplantation embryos. *Nat Genet.* 2011;43:811–814. https://doi.org/10.1038/ng.864.
24. Farthing CR, Ficz G, Ng RK, et al. Global mapping of DNA methylation in mouse promoters reveals epigenetic reprogramming of pluripotency genes. *PLoS Genet.* 2008;4:e1000116. https://doi.org/10.1371/journal.pgen.1000116.
25. Sarmento OF, Digilio LC, Wang Y, et al. Dynamic alterations of specific histone modifications during early murine development. *J Cell Sci.* 2004;117:4449–4459. https://doi.org/10.1242/jcs.01328.
26. Ooga M, Inoue A, Kageyama S, Akiyama T, Nagata M, Aoki F. Changes in H3K79 methylation during preimplantation development in mice. *Biol Reprod.* 2008;78:413–424. https://doi.org/10.1095/biolreprod.107.063453.
27. Wang Q, Wang C-M, Ai J-S, et al. Histone phosphorylation and pericentromeric histone modifications in oocyte meiosis. *Cell Cycle.* 2006;5:1974–1982. https://doi.org/10.4161/cc.5.17.3183.

28. Lin T, Sun L, Lee JE, Lee JB, Kim SY, Jin DI. Changes of histone H3 lysine 23 acetylation and methylation in porcine somatic cells, oocytes and preimplantation embryos. *Theriogenology*. 2020;148:162–173. https://doi.org/10.1016/j.theriogenology.2020.03.006.
29. Prakash K, Fournier D, Redl S, et al. Superresolution imaging reveals structurally distinct periodic patterns of chromatin along pachytene chromosomes. *Proc Natl Acad Sci USA*. 2015;112:14635–14640. https://doi.org/10.1073/pnas.1516928112.
30. Gu L, Wang Q, Sun Q-Y. Histone modifications during mammalian oocyte maturation: dynamics, regulation and functions. *Cell Cycle*. 2010;9:1942–1950. https://doi.org/10.4161/cc.9.10.11599.
31. Akiyama T, Kim J-M, Nagata M, Aoki F. Regulation of histone acetylation during meiotic maturation in mouse oocytes. *Mol Reprod Dev*. 2004;69:222–227. https://doi.org/10.1002/mrd.20121.
32. Kim J-M, Liu H, Tazaki M, Nagata M, Aoki F. Changes in histone acetylation during mouse oocyte meiosis. *J Cell Biol*. 2003;162:37–46. https://doi.org/10.1083/jcb.200303047.
33. Nagashima T, Maruyama T, Furuya M, et al. Histone acetylation and subcellular localization of chromosomal protein BRD4 during mouse oocyte meiosis and mitosis. *Mol Hum Reprod*. 2007;13:141–148. https://doi.org/10.1093/molehr/gal115.
34. Hammoud SS, Nix DA, Zhang H, Purwar J, Carrell DT, Cairns BR. Distinctive chromatin in human sperm packages genes for embryo development. *Nature*. 2009;460:473–478. https://doi.org/10.1038/nature08162.
35. Govin J, Caron C, Lestrat C, Rousseaux S, Khochbin S. The role of histones in chromatin remodelling during mammalian spermiogenesis. *Eur J Biochem*. 2004;271:3459–3469. https://doi.org/10.1111/j.1432-1033.2004.04266.x.
36. Carrell DT. Epigenetics of the male gamete. *Fertil Steril*. 2012;97:267–274. https://doi.org/10.1016/j.fertnstert.2011.12.036.
37. Brunner AM, Nanni P, Mansuy IM. Epigenetic marking of sperm by post-translational modification of histones and protamines. *Epigenet Chromatin*. 2014;7:2. https://doi.org/10.1186/1756-8935-7-2.
38. Hazzouri M, Pivot-Pajot C, Faure AK, et al. Regulated hyperacetylation of core histones during mouse spermatogenesis: involvement of histone deacetylases. *Eur J Cell Biol*. 2000;79:950–960. https://doi.org/10.1078/0171-9335-00123.
39. Song N, Liu J, An S, Nishino T, Hishikawa Y, Koji T. Immunohistochemical analysis of histone H3 modifications in germ cells during mouse spermatogenesis. *Acta Histochem Cytoc*. 2011;44:183–190. https://doi.org/10.1267/ahc.11027.
40. Getun IV, Wu Z, Fallahi M, et al. Functional roles of acetylated histone marks at mouse meiotic recombination hot spots. *Mol Cell Biol*. 2017;37. https://doi.org/10.1128/MCB.00942-15.
41. Lu S, Xie YM, Li X, et al. Mass spectrometry analysis of dynamic post-translational modifications of TH2B during spermatogenesis. *Mol Hum Reprod*. 2009;15:373–378. https://doi.org/10.1093/molehr/gap028.
42. Godmann M, Auger V, Ferraroni-Aguiar V, et al. Dynamic regulation of histone H3 methylation at lysine 4 in mammalian spermatogenesis. *Biol Reprod*. 2007;77:754–764. https://doi.org/10.1095/biolreprod.107.062265.
43. Siklenka K, Erkek S, Godmann M, et al. Disruption of histone methylation in developing sperm impairs offspring health transgenerationally. *Science*. 2015;350. https://doi.org/10.1126/science.aab2006. aab2006–aab2006.

44. Peters AHFM, O'Carroll D, Scherthan H, et al. Loss of the Suv39h histone methyltransferases impairs mammalian heterochromatin and genome stability. *Cell*. 2001;107:323–337. https://doi.org/10.1016/S0092-8674(01)00542-6.
45. Liu Y, Zhang Y, Yin J, et al. Distinct H3K9me3 and DNA methylation modifications during mouse spermatogenesis. *J Biol Chem*. 2019;294:18714–18725. https://doi.org/10.1074/jbc.RA119.010496.
46. An J, Zhang X, Qin J, et al. The histone methyltransferase ESET is required for the survival of spermatogonial stem/progenitor cells in mice. *Cell Death Dis*. 2014;5. https://doi.org/10.1038/cddis.2014.171. e1196–e1196.
47. Au Yeung WK, Brind'Amour J, Hatano Y, et al. Histone H3K9 methyltransferase G9a in oocytes is essential for preimplantation development but dispensable for CG methylation protection. *Cell Rep*. 2019;27:282–293. e4. https://doi.org/10.1016/j.celrep.2019.03.002.
48. Cole F, Keeney S, Jasin M. Preaching about the converted: how meiotic gene conversion influences genomic diversity. *Ann N Y Acad Sci*. 2012;1267:95–102. https://doi.org/10.1111/j.1749-6632.2012.06595.x.
49. Buard J, Barthès P, Grey C, de Massy B. Distinct histone modifications define initiation and repair of meiotic recombination in the mouse. *EMBO J*. 2009;28:2616–2624. https://doi.org/10.1038/emboj.2009.207.
50. Tamburini BA, Tyler JK. Localized histone acetylation and deacetylation triggered by the homologous recombination pathway of double-strand DNA repair. *Mol Cell Biol*. 2005;25:4903–4913. https://doi.org/10.1128/MCB.25.12.4903-4913.2005.
51. Baudat F, Buard J, Grey C, et al. PRDM9 is a major determinant of meiotic recombination hotspots in humans and mice. *Science*. 2010;327:836–840. https://doi.org/10.1126/science.1183439.
52. Diagouraga B, Clément JAJ, Duret L, Kadlec J, de Massy B, Baudat F. PRDM9 methyltransferase activity is essential for meiotic DNA double-strand break formation at its binding sites. *Mol Cell*. 2018;69:853–865. https://doi.org/10.1016/j.molcel.2018.01.033. e6.
53. Powers NR, Parvanov ED, Baker CL, Walker M, Petkov PM, Paigen K. The meiotic recombination activator PRDM9 trimethylates both H3K36 and H3K4 at recombination hotspots in vivo. *PLoS Genet*. 2016;12:e1006146. https://doi.org/10.1371/journal.pgen.1006146.
54. Hinch AG, Zhang G, Becker PW, et al. Factors influencing meiotic recombination revealed by whole-genome sequencing of single sperm. *Science*. 2019;363. https://doi.org/10.1126/science.aau8861.
55. Jiang H, Gao Q, Zheng W, et al. MOF influences meiotic expansion of H2AX phosphorylation and spermatogenesis in mice. *PLoS Genet*. 2018;14:e1007300. https://doi.org/10.1371/journal.pgen.1007300.
56. Baarends WM, Grootegoed JA. Chromatin dynamics in the male meiotic prophase. *Cytogenet Genome Res*. 2003;103:225–234. https://doi.org/10.1159/000076808.
57. Khalil AM, Boyar FZ, Driscoll DJ. Dynamic histone modifications mark sex chromosome inactivation and reactivation during mammalian spermatogenesis. *Proc Natl Acad Sci USA*. 2004;101:16583–16587. https://doi.org/10.1073/pnas.0406325101.
58. Hirota T, Blakeley P, Sangrithi MN, et al. SETDB1 links the meiotic DNA damage response to sex chromosome silencing in mice. *Dev Cell*. 2018;47:645–659. e6. https://doi.org/10.1016/j.devcel.2018.10.004.
59. Sin H-S, Kartashov AV, Hasegawa K, Barski A, Namekawa SH. Poised chromatin and bivalent domains facilitate the mitosis-to-meiosis transition in the male germline. *BMC Biol*. 2015;13:1–15. https://doi.org/10.1186/s12915-015-0159-8.

60. Tan M, Luo H, Lee S, et al. Identification of 67 histone marks and histone lysine crotonylation as a new type of histone modification. *Cell*. 2011;146:1016–1028. https://doi.org/10.1016/j.cell.2011.08.008.
61. Lesch BJ, Dokshin GA, Young RA, McCarrey JR, Page DC. A set of genes critical to development is epigenetically poised in mouse germ cells from fetal stages through completion of meiosis. *Proc Natl Acad Sci USA*. 2013;110:16061–16066. https://doi.org/10.1073/pnas.1315204110.
62. Bernstein BE, Mikkelsen TS, Xie X, et al. A bivalent chromatin structure marks key developmental genes in embryonic stem cells. *Cell*. 2006;125:315–326. https://doi.org/10.1016/j.cell.2006.02.041.
63. Strome S. *Specification of the germ line*. WormBook; 2005. https://doi.org/10.1895/wormbook.1.9.1.
64. Hubbard EJA, Greenstein D. *Introduction to the germ line*. WormBook; 2005. http://www.ncbi.nlm.nih.gov/books/NBK19766/. Accessed May 7, 2021.
65. Wang JT, Seydoux G. Germ cell specification. *Adv Exp Med Biol*. 2013;757:17–39. https://doi.org/10.1007/978-1-4614-4015-4_2.
66. Pazdernik N, Schedl T. Introduction to germ cell development in C. elegans. *Adv Exp Med Biol*. 2013;757:1–16. https://doi.org/10.1007/978-1-4614-4015-4_1.
67. Tzur YB, Winter E, Gao J, Hashimshony T, Yanai I, Colaiácovo MP. Spatiotemporal gene expression analysis of the Caenorhabditis elegans germline uncovers a syncytial expression switch. *Genetics*. 2018;210:587–605. https://doi.org/10.1534/genetics.118.301315.
68. Schaner CE, Deshpande G, Schedl PD, Kelly WG. A conserved chromatin architecture marks and maintains the restricted germ cell lineage in worms and flies. *Dev Cell*. 2003;5:747–757. https://doi.org/10.1016/S1534-5807(03)00327-7.
69. Kelly WG. Transgenerational epigenetics in the germline cycle of Caenorhabditis elegans. *Epigenet Chromatin*. 2014;7:1–17. https://doi.org/10.1186/1756-8935-7-6.
70. Vandamme J, Sidoli S, Mariani L, et al. H3K23me2 is a new heterochromatic mark in *Caenorhabditis elegans*. *Nucleic Acids Res*. 2015:gkv1063. https://doi.org/10.1093/nar/gkv1063.
71. Skene PJ, Henikoff S. An efficient targeted nuclease strategy for high-resolution mapping of DNA binding sites. *ELife*. 2017;6:e21856. https://doi.org/10.7554/eLife.21856.
72. Lee C-YS, Lu T, Seydoux G. Nanos promotes epigenetic reprograming of the germline by down-regulation of the THAP transcription factor LIN-15B. *ELife*. 2017:1–31. https://doi.org/10.7554/eLife.30201.001.
73. Bender LB, Cao R, Zhang Y, Strome S. The MES-2/MES-3/MES-6 complex and regulation of histone H3 methylation in C. elegans. *Curr Biol*. 2004;14:1639–1643. https://doi.org/10.1016/j.cub.2004.08.062.
74. Bender LB, Suh J, Carroll CR, et al. MES-4: an autosome-associated histone methyltransferase that participates in silencing the X chromosomes in the C. elegans germ line. *Development (09501991)*. 2006;133:17.
75. Xiao Y, Bedet C, Robert VJP, et al. Caenorhabditis elegans chromatin-associated proteins SET-2 and ASH-2 are differentially required for histone H3 Lys 4 methylation in embryos and adult germ cells. *Proc Natl Acad Sci U S A*. 2011;108:8305.
76. Li T, Kelly WG. A role for Set1/MLL-related components in epigenetic regulation of the Caenorhabditis elegans germ line. *PLoS Genet*. 2011;7:e1001349.
77. Schwartz-Orbach L, Zhang C, Sidoli S, et al. Caenorhabditis elegans nuclear RNAi factor SET-32 deposits the transgenerational histone modification, H3K23me3. *ELife*. 2020;9:e54309. https://doi.org/10.7554/eLife.54309.

78. Kreher J, Takasaki T, Cockrum C, et al. Distinct roles of two histone methyltransferases in transmitting H3K36me3-based epigenetic memory across generations in Caenorhabditis elegans. *Genetics*. 2018;210:969–982. https://doi.org/10.1534/genetics.118.301353.
79. Kelly WG, Schaner CE, Dernburg AF, et al. X-chromosome silencing in the germline of C. elegans. *Development*. 2002;129:479–492.
80. Lascarez-Lagunas LI, Herruzo E, Grishok A, San-Segundo PA, Colaiácovo MP. DOT-1.1-dependent H3K79 methylation promotes normal meiotic progression and meiotic checkpoint function in C. elegans. *PLoS Genet*. 2020;16:e1009171. https://doi.org/10.1371/journal.pgen.1009171.
81. Bessler JB, Andersen EC, Villeneuve AM. Differential localization and independent acquisition of the H3K9me2 and H3K9me3 chromatin modifications in the Caenorhabditis elegans adult germ line. *PLoS Genet*. 2010;6:e1000830. https://doi.org/10.1371/journal.pgen.1000830.
82. Samson M, Jow MM, Wong CCL, et al. The specification and global reprogramming of histone epigenetic marks during gamete formation and early embryo development in C. elegans. *PLoS Genet*. 2014;10:e1004588. https://doi.org/10.1371/journal.pgen.1004588.
83. Wynsberghe PM, Maine EM. Epigenetic control of germline development. *Germ Cell Dev C Elegans*. 2012:373. https://doi.org/10.1007/978-1-4614-4015-4_13.
84. Tabuchi TM, Rechtsteiner A, Jeffers TE, Egelhofer TA, Murphy CT, Strome S. Caenorhabditis elegans sperm carry a histone-based epigenetic memory of both spermatogenesis and oogenesis. *Nat Commun*. 2018;9. https://doi.org/10.1038/s41467-018-06236-8.
85. Furuhashi H, Takasaki T, Rechtsteiner A, et al. Trans-generational epigenetic regulation of C. elegans primordial germ cells. *Epigenet Chromatin*. 2010;3:1–21. https://doi.org/10.1186/1756-8935-3-15.
86. Beurton F, Stempor P, Caron M, et al. Physical and functional interaction between SET1/COMPASS complex component CFP-1 and a Sin3S HDAC complex in C. elegans. *Nucleic Acids Res*. 2019;47:11164–11180. https://doi.org/10.1093/nar/gkz880.
87. Xu L, Fong Y, Strome S. The Caenorhabditis elegans maternal-effect sterile proteins, MES-2, MES-3, and MES-6, are associated in a complex in embryos. *Proc Natl Acad Sci U S A*. 2001.
88. Towbin BD, González-Aguilera C, Sack R, et al. Step-wise methylation of histone H3K9 positions heterochromatin at the nuclear periphery. *Cell*. 2012;150:934–947. https://doi.org/10.1016/j.cell.2012.06.051.
89. Fisher K, Southall SM, Wilson JR, Poulin GB. Methylation and demethylation activities of a C. elegans MLL-like complex attenuate RAS signalling. *Dev Biol*. 2010;341:142–153. https://doi.org/10.1016/j.ydbio.2010.02.023.
90. Herbette M, Robert V, Bailly A, et al. A role for Caenorhabditis elegans COMPASS in germline chromatin organization. *Cells*. 2020. https://doi.org/10.3390/cells9092049.
91. Evans KJ, Huang N, Stempor P, Chesney MA, Down TA, Ahringer J. Stable Caenorhabditis elegans chromatin domains separate broadly expressed and developmentally regulated genes. *Proc Natl Acad Sci U S A*. 2016;113:E7020–E7029. https://doi.org/10.1073/pnas.1608162113.
92. Bean CJ, Schaner CE, Kelly WG. Meiotic pairing and imprinted X chromatin assembly in Caenorhabditis elegans. *Nat Genet*. 2004;36:100–105. https://doi.org/10.1038/ng1283.
93. Ho JWK, Jung YL, Liu T, et al. Comparative analysis of metazoan chromatin organization. *Nature*. 2014;512:449–452. https://doi.org/10.1038/nature13415.

94. Zeller P, Padeken J, van Schendel R, Kalck V, Tijsterman M, Gasser SM. Histone H3K9 methylation is dispensable for Caenorhabditis elegans development but suppresses RNA:DNA hybrid-associated repeat instability. *Nat Genet*. 2016;48:1385. https://doi.org/10.1038/ng.3672.
95. Cecere G, Hoersch S, Jensen MB, Dixit S, Grishok A. The ZFP-1(AF10)/DOT-1 complex opposes H2B ubiquitination to reduce Pol II transcription. *Mol Cell*. 2013;50: 894–907. https://doi.org/10.1016/j.molcel.2013.06.002.
96. Holdeman R, Nehrt S, Strome S. MES-2, a maternal protein essential for viability of the germline in Caenorhabditis elegans, is homologous to a Drosophila Polycomb group protein. *Development*. 1998;125:2457–2467. https://doi.org/10.1242/dev.125.13.2457.
97. Korf I, Fan Y, Strome S. The Polycomb group in Caenorhabditis elegans and maternal control of germline development. *Development*. 1998;125:2469–2478. https://doi.org/10.1242/dev.125.13.2469.
98. Kelly WG, Fire A. Chromatin silencing and the maintenance of a functional germline in Caenorhabditis elegans. *Development (Cambridge, England)*. 1998;125:2451–2456.
99. Gaydos LJ, Rechtsteiner A, Egelhofer TA, Carroll CR, Strome S. Antagonism between MES-4 and Polycomb repressive complex 2 promotes appropriate gene expression in C. elegans germ cells. *Cell Rep*. 2012;2:1169.
100. Yuan W, Xu M, Huang C, Liu N, Chen S, Zhu B. H3K36 methylation antagonizes PRC2-mediated H3K27 methylation. *J Biol Chem*. 2011;286:7983–7989. https://doi.org/10.1074/jbc.M110.194027.
101. Schmitges FW, Prusty AB, Faty M, et al. Histone methylation by PRC2 is inhibited by active chromatin marks. *Mol Cell*. 2011;42:330–341. https://doi.org/10.1016/j.molcel.2011.03.025.
102. Gaydos LJ, Wang W, Strome S. H3K27me and PRC2 transmit a memory of repression across generations and during development. *Science*. 2014;345:1515–1518. https://doi.org/10.1126/science.1255023.
103. Tursun B, Patel T, Kratsios P, Hobert O. Direct conversion of C. elegans germ cells into specific neuron types. *Science*. 2011;331:304–308. https://doi.org/10.1126/science.1199082.
104. Patel T, Tursun B, Rahe DP, Hobert O. Removal of Polycomb repressive complex 2 makes C. elegans germ cells susceptible to direct conversion into specific somatic cell types. *Cell Rep*. 2012;2:1178–1186. https://doi.org/10.1016/j.celrep.2012.09.020.
105. Robert VJ, Mercier MG, Bedet C, et al. The SET-2/SET1 histone H3K4 methyltransferase maintains pluripotency in the Caenorhabditis elegans germline. *Cell Rep*. 2014;9:443–450. https://doi.org/10.1016/j.celrep.2014.09.018.
106. Robert VJ, Knutson AK, Rechtsteiner A, et al. Caenorhabditis elegans SET1/COMPASS maintains germline identity by preventing transcriptional deregulation across generations. *Front Cell Dev Biol*. 2020;8. https://doi.org/10.3389/fcell.2020.561791.
107. Cavalli G, Heard E. Advances in epigenetics link genetics to the environment and disease. *Nature*. 2019;571:489–499. https://doi.org/10.1038/s41586-019-1411-0.
108. Moore RS, Kaletsky R, Murphy CT. Piwi/PRG-1 Argonaute and TGF-β mediate transgenerational learned pathogenic avoidance. *Cell*. 2019;177:1827–1841. https://doi.org/10.1016/j.cell.2019.05.024.
109. Klosin A, Casas E, Hidalgo-Carcedo C, Vavouri T, Lehner B. Transgenerational transmission of environmental information in C. elegans. *Science*. 2017;356:320–323. https://doi.org/10.1126/science.aah6412.

110. Perez MF, Lehner B. Intergenerational and transgenerational epigenetic inheritance in animals. *Nat Cell Biol.* 2019;21:143–151. https://doi.org/10.1038/s41556-018-0242-9.
111. Woodhouse RM, Ashe A. How do histone modifications contribute to transgenerational epigenetic inheritance in C. elegans? *Biochem Soc Trans.* 2020;48:1019–1034. https://doi.org/10.1042/BST20190944.
112. Fabrizio P, Garvis S, Palladino F. Histone methylation and memory of environmental stress. *Cells.* 2019;6(4):339. https://doi.org/10.3390/cells8040339.
113. Özdemir I, Steiner FA. Transmission of chromatin states across generations in C. elegans. *Semin Cell Dev Biol.* 2022;127:133–141. https://doi.org/10.1016/j.semcdb.2021.11.008.
114. Vastenhouw NL, Brunschwig K, Okihara KL, Muller F, Tijsterman M, Plasterk RHA. Gene expression: long-term gene silencing by RNAi. *Nature.* 2006;442:882. https://doi.org/10.1038/442882a.
115. Minkina O, Hunter CP. Intergenerational transmission of gene regulatory information in Caenorhabditis elegans. *Trends Genet.* 2018;34:54–64. https://doi.org/10.1016/j.tig.2017.09.012.
116. Buckley BA, Burkhart KB, Gu SG, et al. A nuclear Argonaute promotes multigenerational epigenetic inheritance and germline immortality. *Nature.* 2012;486:447. https://doi.org/10.1038/nature11352.
117. Mao H, Zhu C, Zong D, et al. The Nrde pathway mediates small-RNA-directed histone H3 lysine 27 trimethylation in Caenorhabditis elegans. *Curr Biol.* 2015;25:2398–2403. https://doi.org/10.1016/j.cub.2015.07.051.
118. Kalinava N, Ni JZ, Peterman K, Chen E, Gu SG. Decoupling the downstream effects of germline nuclear RNAi reveals that H3K9me3 is dispensable for heritable RNAi and the maintenance of endogenous siRNA-mediated transcriptional silencing in Caenorhabditis elegans. *Epigenet Chromatin.* 2017;10:1–14. https://doi.org/10.1186/s13072-017-0114-8.
119. Kalinava N, Ni JZ, Gajic Z, Kim M, Ushakov H, Gu SG. C. elegans heterochromatin factor SET-32 plays an essential role in transgenerational establishment of nuclear RNAi-mediated epigenetic silencing. *Cell Rep.* 2018;25:2273–2284.e3. https://doi.org/10.1016/j.celrep.2018.10.086.
120. Woodhouse RM, Buchmann G, Hoe M, et al. Chromatin modifiers SET-25 and SET-32 are required for establishment but not long-term maintenance of transgenerational epigenetic inheritance. *Cell Rep.* 2018;25:2259–2272. https://doi.org/10.1016/j.celrep.2018.10.085. e5.
121. Lev I, Seroussi U, Gingold H, Bril R, Anava S, Rechavi O. MET-2-Dependent H3K9 methylation suppresses transgenerational small RNA inheritance. *Curr Biol.* 2017;27:1138–1147. https://doi.org/10.1016/j.cub.2017.03.008.
122. Lev I, Gingold H, Rechavi O. H3K9me3 is required for inheritance of small RNAs that target a unique subset of newly evolved genes. *ELife.* 2019;8. https://doi.org/10.7554/eLife.40448.
123. McMurchy AN, Stempor P, Gaarenstroom T, et al. A team of heterochromatin factors collaborates with small RNA pathways to combat repetitive elements and germline stress. *ELife.* 2017;6:e21666. https://doi.org/10.7554/eLife.21666.
124. Ahringer J, Gasser SM. Repressive chromatin in Caenorhabditis elegans: establishment, composition, and function. *Genetics.* 2018;208:491.
125. Padeken J, Zeller P, Towbin B, et al. Synergistic lethality between BRCA1 and H3K9me2 loss reflects satellite derepression. *Genes Dev.* 2019;33:436–451. https://doi.org/10.1101/gad.322495.118.

CHAPTER

Epigenetic regulation of *cis*-regulatory elements and transcription factors during development

3

Shagufta Khan[1], Rakesh K. Mishra[1,2], Surabhi Srivastava[1,2]

[1]*CSIR - Centre for Cellular and Molecular Biology, Hyderabad, Telangana, India;* [2]*Tata Institute for Genetics and Society, Bangalore, Karnataka, India*

Epigenetic mechanisms govern developmental processes via genome regulation

Chromatin—the DNA and the amino acids of the histones around which it is wrapped—plays a crucial role in cellular function and provides a structure to package the genome in the nucleus. Only a small fraction of the nuclear genome is protein coding, while a large proportion—initially categorized as junk DNA—does not directly code for functional genes. For example, only about 2% of the human genome encodes the ~25,000 genes that govern all biological processes. Comparative genome analysis of several species suggests a discordance between organismal complexity and the number of coding genes; complexity instead corresponds with the extent of the noncoding component of the genome.[1] Over the past few years, noncoding sequences have been demonstrated to comprise the *cis*-regulatory elements or modules (CREs/CRMs)[2] that regulate the expression of coding genes in a tissue- and developmental stage-specific manner. Different CREs may be conserved in sequence[3] or functionally defined by the *trans*-acting factors, including transcription factors (TFs) encoded by about 5%–10% protein-coding genes, which bind to them and activate or suppress gene expression (Box 3.1).

Epigenetic factors work at the CREs and constitute the genome's machinery to bring about the precise outcomes encoded in the DNA sequence. They include nucleosome packaging, posttranslational modifications (PTMs) of histones and their variants, DNA methylation, and noncoding RNAs that control epigenetic and cellular memory.[30] Epigenetic mechanisms also serve to propagate changes that are not described in the genetic sequence but have been introduced during the life span of a cell or an individual.[31] The dynamically changing epigenetic landscape of the cell governs chromatin structure and conformation, which in turn drives gene expression, thus creating a complex highly interlinked system. Permissive chromatin states involve the setting of "activating" epigenetic marks that bring about transcription of associated genes. In contrast, repressive chromatin structure is

Perinatal and Developmental Epigenetics. https://doi.org/10.1016/B978-0-12-821785-6.00004-9

Box 3.1 *cis*-regulatory elements (CREs)

CREs are DNA sequences that contain multiple binding sites for interaction with diffusible *trans*-acting factors such as proteins, transcription factors (TFs), or noncoding RNAs. As the name suggests, CREs regulate genes on the same chromosome in a cell-specific manner to bring about distinct transcriptional outcomes. Typically spanning several hundred base pairs, they are categorized based on their relative location from their target gene(s) as promoters and promoter-proximal elements situated near the transcription start site (TSS) of a gene or distal regulatory regions that can influence gene expression from megabases away.

Promoters. Core promoters are DNA sequences spanning around 50 bp upstream and downstream of the TSS (at the 5′ end of the gene); they are adequate to initiate transcription of a gene upon binding of RNA polymerase and associated transcription factors, but with low basal activity. Core promoters either have a single TSS (focused or sharp promoters)—associated mainly with tissue-specific or terminally differentiated cell-specific genes—or multiple closely spaced TSSs (dispersed or broad promoters) that are associated with housekeeping genes and developmental transcription factors. Around 250 bp from the TSS lies the proximal promoter region or upstream activating sequence that harbors binding sites for sequence-specific TFs, which can activate/enhance transcription by aiding the recruitment of the transcriptional machinery. Together, the core promoter and proximal promoter can autonomously drive transcription in response to regulatory inputs. Promoter elements can be discerned from other types of CREs based on the signature motifs they contain (discussed in Section Identification and Characterization of cis-regulatory Elements and trans-acting Factors).

Enhancers. Basal-level transcriptional activity of the promoter is usually enhanced by distally located regulatory DNA sequences called enhancers. Enhancers contain multiple binding sites for TFs, which can recruit coactivators or corepressors through their effector domains, and significantly increase or diminish the transcription rate from target promoters in response to extracellular/intracellular cues. The outcome from an enhancer is a combinatorial effect of the activator and repressor concentration in each cell.[4] Enhancer was first identified as a modular element in simian vacuolating virus (SV40); a salient feature of transcriptional enhancers is that they can act independently of their orientation relative to the target promoter and from a considerable distance.[5,6] Now, it is well established that their distance from the target locus may span kilobases or even megabases. Astonishingly, an enhancer on one allele can also interact with the promoter of the other allele when autosomes pair during interphase in *Drosophila* in a specialized phenomenon called transvection.

Generally, a protein-coding gene is controlled by multiple enhancers in complex tissue- and stage-specific patterns that get misregulated if an enhancer is disrupted.[7] Given that there are fewer expressed genes than the number of enhancers in a given cell type and enhancers are non-uniformly arranged in the genome, some genes lie in genomic regions rich in enhancers. Such enhancer-rich regions are referred to as superenhancers. More often than not, superenhancers are associated with genes that determine the identity of a cell type, for example, genes that code for developmental TFs (such as OCT4 and SOX2) in murine embryonic stem cells.[8]

Insulators. Insulators, initially discovered in *Drosophila*, are DNA sequences that prevent nonspecific interactions between genomic regions and play a regulatory and architectural role in the overall spatial organization of the genome. They are functional when bound by specific DNA-binding proteins called insulator proteins and additional factors that interact with these proteins via their protein–protein interaction domains. The two experimentally characterized ways in which insulators prevent illegitimate interactions are by "enhancer blocking" and by acting as "barriers." Enhancer blocking insulators prevent enhancers from inappropriately interacting with noncognate promoters of neighboring genes, and barrier insulators prevent the encroachment of heterochromatin (repressed chromatin) into the euchromatin (active chromatin) domain.[9] Insulators have been very helpful to researchers in transgenic constructs and, recently, in genome-wide screens to identify novel CREs.[10]

Box 3.1 *cis*-regulatory elements (CREs)—cont'd

Although most mammalian insulators appear to be dependent on binding of the CCTC-binding factor (CTCF), a wide variety of insulator elements and their binding factors are known in *Drosophila*. Key examples include the *scs* and *scs'* sequences (flanking the heat shock *hsp70* locus), which are bound by Zeste-White 5 (ZW5)[11] and boundary element–associated factor 32 (BEAF32), respectively, to form chromatin domains,[12,13] the *gypsy* insulator discovered initially in the *gypsy* retrotransposon harboring binding sites for the suppressor of hairy-wing (Su(Hw)) protein.[14] The developmentally regulated cluster of homeotic genes in the *Hox* complex bears a complex collection of insulators in the bithorax complex of *Drosophila* that interact with each other and multiple CREs to bring about tightly regulated spatial and temporal regulation of the *Hox* genes in each segment of the developing embryo. These elements depend on precise and targeted binding of *trans*-factors including the *Drosophila* homolog of CTCF (dCTCF) and the *trithorax-like* gene-encoded GAF (GAGA-associated factor).[15,16]

Silencers. Negative regulation of transcription is an essential aspect of spatial and temporal genome modulation that can turn "off" genes in a cell- and stage-specific manner in the developing organism. A common mechanism involves the binding of short-range repressive factors at enhancer elements, thereby silencing the chromatin locally.[17] Local or short-range repression implies the control of only one enhancer by blocking or competing with its activator binding sites, without impeding the activity of other enhancers for the same promoter. Transcriptional "silencer" elements and alternative mechanisms of negatively regulating transcription via polycomb-mediated silencing are reviewed in Grossniklaus and Paro (2014).[18]

Silencer elements are *cis*-acting DNA sequences comprised of sequence-specific binding sites for repressive TFs or repressors that can suppress transcription from an otherwise active promoter.[19] Unlike short-range repression at enhancer elements, silencers can suppress enhancer and promoter activity from a long range. As a result of silencer–promoter interaction, the target promoter becomes impervious to all enhancers irrespective of their distance. Long-range repression by silencer elements can thus silence an entire chromosomal locus despite multiple CRMs regulating it.[20] The distinction of enhancers and silencers into two groups of CREs notwithstanding the development of new genome-based approaches has helped identify a handful of DNA elements that exhibit context-dependent bifunctionality—enhancers that can act as silencers and vice versa in different cell-types.[21]

Locus Control Regions. Locus control regions (LCRs) are CREs that greatly enhance the transcription of the target gene(s) at distal sites in a tissue-specific and copy number–dependent manner during development and lineage commitment. Typically composed of clustered sequence modules or response elements, they serve as binding sites for developmentally regulated factors, and their cumulative effect on the cognate locus brings about coordinated transcriptional control.[22] These modules are distinct from other types of CREs in that they often display characteristics of multiple elements. For instance, LCRs can enhance transcription like enhancers in parallel with protecting the locus from position effects of the opposing chromatin environment, a property of insulators.

LCRs contain multiple DNAse I hypersensitivity sites (DHS), bound by different factors to manifest enhancer or insulator activity in a cell type–specific manner. A large segment located upstream of the human beta-globin locus is the most studied example of LCR-mediated tissue-specific expression in erythroid cell lineages. Deletion of individual DHS of beta-globin LCR in transgenic mice results in diminished expression of globin genes, indicating that the full complement of LCR is essential for globin gene activation.[23]

Scaffold/Matrix Attachment Regions. In 1974, Berezney and Coffey found a fibrillar meshwork analogous to the cytoskeleton inside rat liver nuclei and termed it as the nuclear protein matrix. The NuMat has been identified in several other species including *Drosophila*, humans, and *Arabidopsis* and consists of nonnucleosomal proteins, RNA, and in tiny amounts, nuclease, and salt-resistant nuclear DNA called scaffold/matrix attachment regions (S/MARs). NuMat has been

Continued

Box 3.1 *cis*-regulatory elements (CREs)—cont'd

implicated in providing a structural framework and a functional platform on which several nuclear functions such as DNA replication, DNA damage repair, transcription, RNA processing, and splicing are executed. S/MARs are the chromatin regions associated with NuMat, one class remaining constant to provide the basic structure while the other is functionally dynamic in different cell lineages. They are envisioned as the point of contacts or anchors in the chromatin loops, inherent to higher-order genome organization inside the nucleus (Section Identification and Characterization of cis-regulatory Elements and trans-acting Factors).

In addition, genomes carry several other kinds of noncoding sequences that play a role in gene regulation. Seventy years ago, transposons were discovered as mobile elements that could replicate and move around the genome,[24] and early work in *Drosophila* proposed their role in controlling development.[25] Genome-wide studies have now revealed the extent to which transposable elements (TEs) confer CRE activity in host genomes.[26] Microsatellites or simple sequence repeats (SSRs) are short tandem repeats of 1–6 nt motifs present in almost all genomes.[27] Originally relegated to the junk DNA compartment, SSRs may play a role in cellular processes, chromosome packaging, and gene regulation and are also associated with gene regulatory disorders.[28] Polycomb/trithorax response elements (PREs/TREs) are targets of chromatin-modifying complexes or transcription factors and influence gene transcription by affecting chromatin structure.[29] Collectively, all these elements operate in harmony with one another to direct coordinated gene expression and contribute to organismal complexity via their interactions with *trans*-factors.

marked by a "suppressive" environment that prevents the binding of protranscription factors and the transcription machinery. The localized modulation of chromatin state further governs its accessibility—the opening and closing of the packaged genes and regulatory elements—to enable transcription and direct every process and function in the cell.

Development of multicellular organisms constitutes one of the most complex biological processes in nature. The union of a sperm cell (male gamete) and an egg cell (female gamete) gives rise to the zygote (fertilization), which inherits the genetic and epigenetic information from them. Fertilization triggers the initiation of a large number of growth and differentiation programs in the zygote, requiring precise and coordinated gene activation for enabling phenotypic and morphological complexity. The first step in the developmental cascade is a measured response to maternal gradients and the spatial patterns of morphogens defining neighboring cells in the developing embryo[32]; each cell's positional and functional identity is then refined as growth and development proceeds (spatial and temporal gene regulation). The precise timing and extent of usage of genes thus determines differential cell fate and gives rise to the multitude of cell types that arise from the single-cell zygote.

Gene expression can be modulated at various steps, from transcription to RNA processing, and translation to PTM of proteins. Regulation of most eukaryotic genes occurs primarily at the transcriptional level,[33] and achieving this fine-tuning requires the synergism of two complementary components, the CREs, and the *trans*-acting factors. This chapter explores the genomic and epigenomic features of these regulators, highlighting their role in cellular differentiation and organismal development. The genesis of next-generation sequencing (NGS)–based genome-wide approaches has dramatically improved our understanding of the intricacies of gene regulation across species as well as at the single-cell level, and we discuss some of the advances that allow the identification of novel CREs and regulatory factors in their genomic context. Despite the wealth of epigenetic information gathered in the past few decades, the regulatory pathways that direct developmental stages have not been delineated. Misregulation of early gene expression patterns can lead to severe outcomes, manifested in genetic disorders via mechanisms that are not yet understood. We summarize the conserved mechanisms known from stem and progenitor cell types and highlight some of the unsolved issues in the context of in vivo development and disease states.

cis-regulatory elements and *trans*-acting factors orchestrate cell type–specific gene transcription

The genome is a sequentially organized code of nucleotides shared by all cells in an organism, and CREs are a part of the fixed genomic sequence. This begs the question of how they may be differentially utilized in a cell state and developmental stage-specific manner. Gene regulation happens at multiple steps of transcription—the assembly of transcriptional machinery at CREs, followed by the activation, initiation, and elongation of transcription. At all these layers, *trans*-acting factors provide selectivity and specificity of the interactions between CREs. These factors are encoded by *trans*-regulatory elements encompassing diffusible molecules and *trans*-acting factors such as DNA-binding proteins and noncoding RNAs (ncRNAs) that affect the cell type–specific transcription of genes. They either directly bind to CREs via sequence-specific DNA-binding domains or influence the CRE indirectly by interacting with other DNA-bound factors in a complex. The outcome of these interactions depends on a diverse panoply of interlinked factors such as protein/RNA partners, PTMs, binding affinity to target site, expression level, and post-transcriptional regulation in the target tissue as well as the chromatin state and accessibility of the target CREs. In the following sections, we will focus mostly on the epigenetic mechanisms interlinking *cis*-elements and their *trans*-acting factors in the context of developmental regulation.

Trans-acting factors

Transcription factors

Transcription factors (TFs) are a major class of *trans*-acting proteins that decipher the genome by directly binding to specific DNA motifs, many of which are highly conserved in metazoans. Computational methods have predicted that the *Drosophila* and the human genome encode around 700 and 1600 TFs, respectively, belonging to characteristic DNA-binding domain families[34,35] that allow them to recognize specific CREs.[36] During early developmental stages when the chromatin is largely silent, master or pioneer TFs, such as maternal factor Zelda (ZLD), scan the nucleosomal DNA for their target motifs. The precise mechanisms are unclear, but this initiates the transition from closed to slightly accessible chromatin for lineage-specific TFs to bind.[4,37] Once primed, lineage-specific TFs then act during cell fate determination to activate cell-specific gene programs by recruiting the general transcriptional machinery to promoters and enhancers.[38] Transcription initiation requires the assembly of the RNA polymerase RNPII and general transcription factors (GTFs) to form the preinitiation complex or PIC at the core promoter elements (Box 3.1). The GTFs bind sequentially at the core promoter, starting with TFIID binding and followed by DNA duplex unwinding at the TSS to support RNA synthesis.[39] Loading of the general transcription machinery (RNPII and GTFs) at the core promoter results in the basal level of gene expression. However, binding of cell type–specific TFs to CREs such as enhancers or repressor elements changes the transcriptional outcome depending on their context-specific activator or repressor function.

A majority of TFs contain effector domains that can act as ligand-binding domains, protein–protein interaction domains, and domains for recruiting chromatin modifiers, thereby modulating their interaction with CREs. The genome exerts exquisite control over these interactions via controlling the timing and abundance of TF expression in a tissue-specific manner or at a particular developmental stage. Many TFs undergo PTMs that either activate or diminish their function by modulating their stability, DNA binding activity, or protein–protein interaction. For example, phosphorylation at Thr116 residue of SOX2 (HMG-box containing master TF) is necessary for its recruitment at the promoter of hedgehog acyltransferase (HHAT), critical for the development of a stem cell–like phenotype.[40] In contrast, SUMOylation at Lys247 residue hinders its binding to target CREs such as the Fgf4 enhancer, negatively regulating transcription.[41] Even after appropriate expression, TFs may be sequestered until relevant environmental or developmental cues are received to set off the correct cascade of binding events. The acetylation/deacetylation of SRY protein regulates its nuclear localization in a precise window during embryogenesis in mammals for proper male gonad development.[42] Lastly, some TFs can function as both activators or repressors depending on the context/microenvironment in which they bind. The determinants of such an activator–repressor switch include interacting cofactors, ligand binding in response to signaling cues, location, number and features of the binding site, and the relative concentration of the TFs and cofactors.[43] The interaction of TFs with CREs is multifaceted, and thus, their understanding requires the development of new context-sensitive approaches (discussed in Section Identification and Characterization of cis-regulatory Elements and trans-acting Factors).

Noncoding RNAs

Small- and long-noncoding RNAs (lncRNA, >200 bp) do not code for any proteins and constitute a large fraction of the transcribed genome. ncRNAs generally have a high degree of tissue-specific expression and are key members of the gene regulatory network.[44] Recent studies have identified a crucial role of ncRNAs in development and differentiation, including in maintenance and induction of stem cell pluripotency.[45,46]

In 2010, two reports[47,48] demonstrated that extragenic transcription sites enriched in RNPII with chromatin signatures unique to enhancer elements were extensively transcribed to produce a novel class of lncRNAs, now called as enhancer-derived ncRNAs (eRNAs). The levels of eRNA expression correspond with the target gene mRNA levels and often precede the target gene activation. Transcripts from enhancers are primarily 800 bp to <2 kb in length, unspliced, nonpolyadenylated, and unstable, unlike the transcripts from the core promoters of genes.[49] The functional role of eRNAs was initially unclear, and they were thought to be transcribed as a consequence of the proximity of enhancers with the active promoter. However, eRNA knockdown has been demonstrated to result in the loss of enhancer and promoter contacts,[50] highlighting their role in bringing CREs together for activation of genes. Transcription from extragenic and intragenic sites, including enhancer elements, is now well known to occur in multiple cell types and organisms in a stage-specific manner, explained via multiple models.[51] Consistent with chromosomal looping as a model for bringing enhancers into proximity with target promoters, long-range regulatory contacts are associated with the binding of looping factors near the interacting elements[52] (discussed further in Section Three-Dimensional Genome Organisation).

The grouping of regulatory enhancer elements is particularly relevant in early development where multiple genes controlling a specific lineage need to be coregulated. For example, the stage-specific coexpression of superenhancer RNAs with cardiac function–related genes, and the overlap in their TSS and the binding site of master TFs, promotes ESC differentiation to cardiomyocytes.[53] Adding further to the complex regulatory network, PTMs of ncRNAs also affect their regulatory function.

Making the connections: readers and writers program the epigenome during early development

As we have seen so far, *trans*-factors depend on localized chromatin accessibility at cognate CREs for successful binding in a developmental stage- or tissue-specific manner. However, specialized epigenetic mechanisms orchestrated by chromatin regulators are needed to define this "chromatin landscape" at CREs and the DNA sequence around them.[30] One such conserved process includes the dynamic modification of histones; chemical modifications of histone tails at specific residues are manipulated via the coordinated action of highly regulated enzymes recruited to the CREs. The lysine methyltransferase (KMTs/HMTs) class of histone-modifying enzymes are special "writers" that add one to three methyl groups on lysine residues. Methylation of histones H3 at positions K9 and K27 via SU(VAR) and the polycomb (PcG) proteins, respectively, is generally associated with repressive chromatin at intergenic regions as well as promoter elements. H3K4 and H3K36 methylation, on the other hand, correlates with active transcription at promoters and the gene body, respectively, and is brought about by the trithorax (trxG) proteins. Since epigenetic mechanisms operate throughout an organism's lifetime, most of the enzymes that set these marks are expressed across cell types and developmental stages. However, some writers are developmental stage-specific and critical to the germline and fetal programs. ESET[54] and Ezh2[55] knockout mice have early development defects with aberrant H3K9 and H3K27 methylation. While more ubiquitous KMTs such as *Set1*, *G9a*, *Pfm1*, and *Ezh2* have been detected in founder as well as somatic cells, *Blimp1* (B lymphocyte maturation-induced protein-1 or *Prdm1*) shows expression exclusively in the founder primordial germ cells (PGCs)[56] and is transcribed in early development in the mouse, where it represses the somatic program along with *Prdm14*. In zebrafish and *Xenopus*, *Blimp1* plays a role in lineage specification,[57,58] and it is also conserved in invertebrates. *Ezh2* also plays a key role in stem cell differentiation, discussed later.

Histone acetyltransferases (HATs) that acetylate the histone tails at specific residues are another important category of epigenetic writers. Acetylation of the lysine residues makes transcriptional binding sites more accessible, leading to increased gene expression, while histone deacetylases (HDACs) are associated with gene repression. Acetylation of the lysine residues of histone H3 (H3K9ac and H3K27ac) is important for generating a permissive transcriptional state at proximal gene promoters and distal enhancers, and depletion of HAT p300 or CBP leads to early embryonic lethality in mice.[59] p300 also binds with RNPII at nucleosome-depleted regions (NDR) of active enhancers to initiate bidirectional transcription, extending till the borders of the actively marked enhancer region.

Early reprogramming in the germ cell lineage also involves the erasure of repressive epigenetic marks to enable a pluripotent chromatin state (described in detail in Section trans-Regulatory Networks Govern Stem Cell Pluripotency and Differentiation). KDMs (lysine demethylases), such as *Utx* or *Kdm6a* that removes the H3K27 methylation marked by *Ezh2,* are crucial for expression of early developmental and pluripotency genes upon fertilization. Fertilization induces reprogramming via genome-wide DNA demethylation and replacement of the sperm protamines by histones, including variants like H3.3[60] followed by resetting of the histone modification marks. The oocyte genome already has high levels of both active and repressive histone modifications.[61] Acquisition of DNA methylation occurs via the *de novo* and maintenance DNA methyltransferases (DNMTs). Although the roles of epigenome modifiers are well established, the exact contribution of specific histone modifications and how they are set in different tissues to control phenotypic gene expression during lineage commitment is not well understood. However, many details of the so-called epigenetic code—the association of gene transcription status with chromatin modification marks at key regulatory elements—have been worked out and are described in subsequent sections.

Three-dimensional genome organization

The 3D organization of the eukaryotic genome inside the nucleus is closely linked to its function. The nuclear positioning of enhancer element and cognate promoter, for instance, is crucial in achieving the expression pattern of genes in a spatiotemporal fashion during development and for maintenance of these patterns over multiple cell divisions. It is well established that the distance of enhancer elements from their target locus may span kilobases, such as the enhancer of the gene *shavenbaby* that controls the development of larval trichomes in *Drosophila*,[62] or even megabases; for instance, at the *Shh* gene involved in limb bud development in mouse, the ZRS enhancer is approximately one megabase away from the TSS.[63] How do developmental enhancers, kilobases or megabases away, interact with their target promoter in the appropriate cells? Several models have been suggested for enhancer–promoter interaction—tracking, linking, looping, or some combination thereof, and mobilization to transcription factories/hubs.[64] The most widely accepted model is the looping model where factors that are bound to distant enhancer elements physically interact with protein complexes associated with the promoter resulting in extrusion of the DNA linking the two sites. Upon establishing contact with the enhancer, promoter activation results in the transcription of the associated gene. In contrast, many developmentally crucial enhancers from *Drosophila* to human are seen in proximity to their target promoter even before gene activation. Additionally, the interaction or physical proximity remains stable irrespective of the expression status in different cell type or developmental stages. A majority of these developmental genes are associated with paused RNPII at proximal promoter, indicating that they are in an "on your mark, get set" position poised for transcriptional activation as soon as appropriate developmental cues are received.[65]

Microscopic imaging studies including fluorescence in situ hybridization (FISH) and chromosome painting have revealed that each chromosome occupies a distinct position inside the nucleus, forming subnuclear territories called as chromosome territories (CT; for review, see Cremer and Cremer, 2010).[66] Chromosomes, in their respective territories, are further spatially segregated into transcriptionally active and inactive compartments (which corresponds to euchromatin and heterochromatin, respectively), signifying preferred long-range interactions amid chromatin with similar epigenetic features. At an even narrower scale, the genome is organized into topologically associating domains (TADs). TADs are genomic regions with preferred intradomain interactions compared with interdomain interactions. Genome-wide chromatin conformation capture studies have revealed that the arrangement of the genome into such domains is a conserved feature in multiple organisms, from TADs in humans, mouse, plant, and *Drosophila* to self-interacting domains called globules in *Schizosaccharomyces pombe* and TAD-like domains in bacteria *Caulobacter crescentus*.[67] Such a hierarchical organization of the genome into TADs, which is proposed to form through DNA looping, brings distal CREs into the vicinity of each other and mediates activation or repression of genes throughout the developmental program.

With thousands of enhancers and promoters in the nucleus and each TAD consisting of many transcription units and CREs, it is crucial to ensure that correct enhancer–promoter interaction is made to bring about cell type–specific gene expression. One such mechanism includes the positioning of nucleoprotein complexes called insulators that can safeguard genes from opposing environments. Studies have suggested that the TAD borders or boundaries are enriched in elements with insulator activity or binding sites for insulator proteins such as CTCF; strong TAD borders correspond with high occupancy of insulator proteins.[68,69] Although mammalian insulator sequences seem to rely extensively on the zinc finger protein CTCF (CCCTC-binding factor), numerous kinds of insulator DNA sequences and insulator-binding proteins have been studied in detail in *Drosophila*[70] (Box 3.1). Insulators demarcating contact domains between regulatory elements and CTCF motifs are often found in a convergent orientation, enabling pairing. The finding that at the majority of loop anchors, CTCF cooccurs with cohesin ring complex subunits (RAD21 and SMC3) favors the hypothesis that the genome is organized into topologically associated domains via DNA looping.[71] Thus, the role of insulators is believed to be crucial in the currently proposed model of dynamic genome interaction via TADs, and genetic or epigenetic alterations in insulator function can result in misregulation of genes, cancer, and developmental diseases,[70] discussed further in Section Epigenetic Programming in Perinatal and Early Development.

Recent studies have suggested "liquid–liquid phase separation (LLPS)" as one of the mechanisms by which chromatin compartmentalization plays a role in developmental genome regulation. Briefly, distinct phases separate from a mixture based on their physical and chemical properties termed LLPS when these phases resemble liquid-like states or form liquid droplets. Chromatin regions bearing similar features such as histone PTMs and other bound factors (described in Section Identification and Characterization of cis-regulatory Elements and trans-acting Factors) tend to self-interact, and emerging models envisage chromatin compartments/subcompartments as multiple liquid-like phase-separated droplets. For example, heterochromatin protein 1 (HP1)–mediated LLPS on H3K9me3 chromatin[72–75] and polycomb complex–mediated LLPS on H3K27me3 chromatin together constitute the inactive chromatin compartment.[76] Although less studied, intrinsically disordered regions of transcription factors and RNPII could also contribute to phase separation of active compartments of euchromatin.[77] The key drivers of phase separation are unclear, and the biochemical composition of these bodies is being explored. However, LLPS appears to be an intrinsic property of chromatin that can be disrupted upon acetylation of histone tails.[78,79] It is likely that hierarchical organization of the genome requires both loop extrusion and phase separation mechanisms linked to chromatin features such as histone modifications, associated factors, and their physicochemical properties.[80] Therefore, to maintain a tightly coordinated gene expression, an interplay between all these factors is imperative, particularly during development, which is a highly dynamic process and is marked by rapid changes in the transcriptional status of genes.

Identification and characterization of *cis*-regulatory elements and *trans*-acting factors

Emerging epigenomic models suggest that regulatory elements play a role not only in genome regulation that defines a species but also in genome evolution that gives rise to new form and functions.[81] CREs can be gained and lost over time, and their turnover and evolution are still being understood. As discussed in Box 3.1, eukaryotic genomes have a large proportion of transposable elements (TEs) that evolved with their own *cis*-sequences to utilize host factors for transcription and mobility. Intriguingly, though fewer than 1% of human TEs are active, their transcriptional activation increases in early development and has been associated with totipotent states, where chromatin is the most permissive and flexible.[82] There is also growing evidence that TEs are inducible in response to environmental cues and stressors, presenting an evolutionary source of CRE activity coopted for developmental regulation. Although most TEs are actively repressed in the genome, an emerging concept is that their mutations serve as material for novel *cis*-regulatory developmental mechanisms in conjunction with host *trans*-factors.[83]

Why is any specific module in a CRE important and how can we identify it? There are two main features that characterize CREs: (1) conservation in sequence or pattern of chromatin state and (2) binding sites for specific TFs. Regulatory sequences have traditionally been analyzed individually for these parameters, outside their native genomic context, which poses significant challenges in understanding their context-dependent role. Reporter gene assays or activity-based assays that measure enhancer, core promoter, or promoter activities via a reporter gene readout (Fig. 3.1) may be confounded by position effects from regulatory regions surrounding the integration site. Furthermore, the putative genomic element may lack its native complement of TF sites, boundary or PRE/TRE elements required for full activity in a particular developmental stage or tissue. Deleting or mutating candidate sequences in the endogenous locus to test for enhancer requirement may also affect gene expression in unexpected ways, especially if exonic regions are involved. Most methods prioritize the genomic proximity of the regulatory element to their closest genes to estimate effect on gene expression. However, many elements, such as enhancers and LCRs, can function across huge distances in a direction-independent manner, as described in the preceding sections. Newer approaches utilizing high-throughput assays have helped define new roles for CRE activity.[84] Only recently have whole-genome technologies advanced sufficiently to allow mapping of all noncoding elements and systematically analyze their evolution and contribution to gene regulation over a developmental or differentiation time course. The following sections outline how *cis*-regulatory function is encoded and conserved in the genome and how it can be dissected to identify and understand its role in developmental regulation.

Conserved sequences and chromatin features aid in the identification of regulatory elements

Sequence conservation and characteristic motifs

CREs contain short DNA sequences or motifs recognized and bound by *trans*-acting factors such as TFs and cofactors in response to activating or repressive signals. CREs such as promoters and enhancers often have multiple motifs of variable lengths; their order and arrangement differ between elements of similar function. The TATA box–binding protein (TBP) of the transcription factor IID (TFIID)

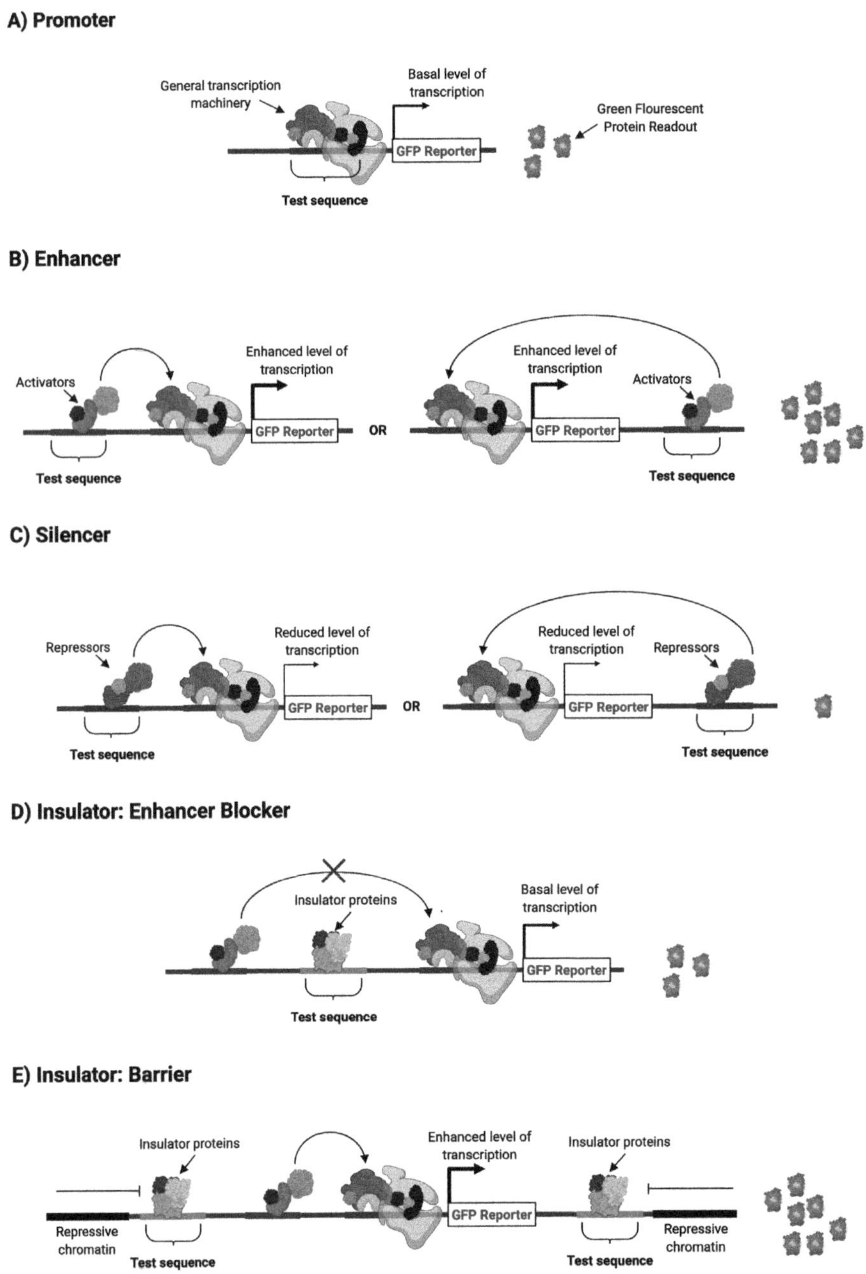

FIGURE 3.1 Reporter gene assays to identify and characterize regulatory elements.

(A) A candidate promoter sequence is cloned upstream of a reporter gene, such as GFP, lacking an endogenous promoter. The GFP readout in the assay indicates promoter complex recognizes and binds to the TATA box motif (~30 bp upstream of the focused core promoter) and helps in the assembly of the preinitiation complex (PIC) for transcription.[85] Promoters that lack a TATA box bear the initiator (Inr) motif and have been shown to be bound by transcription factor TFIID subunits 1 and 2 (TAF1 and TAF2) in *Drosophila*.[86] Inr motif overlaps with the TSS and cooccurs with the downstream promoter element (DPE) downstream of the TSS. Additionally, the motif ten element (MTE) in *Drosophila* and TFIIB recognition elements and downstream core elements (DCEs) in humans are also observed at a definite position close to the TSS (reviewed in Ref. 87). A majority of the promoters associated with housekeeping genes or developmentally essential genes involved in body patterning in vertebrates lack a signature motif. Such promoters are instead found within stretches of CpG dinucleotide repeats called CpG islands (CGIs).[49] Single nucleotide polymorphisms and genetic mutations in these core promoter motifs can affect the level of expression of genes by hampering the recruitment of GTFs and PIC assembly at the TSS. The sequence determinants of promoter elements often overlap with enhancer elements since they bind to similar TFs and are transcribed. Due to such similarities, sequence-based enhancer prediction tools can also end up predicting promoters and need to be interpreted and tested with caution.

Chromatin features

The arrangement of nucleosomes is irregular across the genome. While densely arranged in heterochromatin, nucleosomes are depleted at regulatory elements and transcribed gene bodies. Nucleosome-depleted regions (NDRs) are a prerequisite for the general transcription machinery and other chromatin-binding proteins to bind and regulate gene expression. For example, two well-positioned nucleosomes upstream and downstream of the TSS define NDR of many promoters, and a variety of factors maintain nucleosome exclusion within the NDR, including BRG1/BRM-associated factor (BAF) and proximally paused RNA polymerase.[88] Although NDRs are a sign of active CREs such as promoters and enhancers, recent studies in multiple cell types, including embryonic stem cells, have hinted that such loci may not be devoid of nucleosomes but occupied by biochemically less stable and easily displaceable nucleosomes containing the noncanonical histone variants H3.3 and H2A.Z.[89]

activity and strength. (B) A candidate enhancer sequence is cloned upstream or downstream of the reporter gene driven by a promoter. Increased GFP readout in the assay indicates the enhancer activity of the candidate sequence. (C) A candidate silencer sequence is cloned upstream or downstream of the reporter gene driven by a promoter. Decreased GFP readout in the assay indicates the silencer or repressor activity of the candidate sequence. (D) A candidate insulator sequence that blocks enhancer–promoter interaction when placed between them brings down the transcriptional output to basal levels, depending on the strength of its enhancer blocking activity. (E) Candidate insulator sequences that protect the reporter gene expression from the effect of opposing chromatin environment (or position effects) when flanking the reporter cassette highlight the extent of their barrier activity.

Early genome-wide investigations using high-throughput techniques such as chromatin immunoprecipitation and chromatin accessibility assays have unraveled different states of regulatory elements with distinguishable chromatin signatures, though both enhancers and promoters are characterized by the presence of bound RNPII. The active promoters of housekeeping genes and developmental TFs are found in CpG islands, with punctate marks of H3K4me3 around the TSS. Poised promoters are marked by H3K27me3 and H3K4me3, which probably primes them for activation in the correct cell lineage while silencing them in all other cells; the mechanisms of bivalency and its resolution in stem cells are discussed in Section Switching Chromatin States-Dynamic Epigenomic Landscape at Developmental CREs. There are, however, some general rules of an "epigenetic code" that can be extrapolated to most genes and cell types based on our current understanding. For instance, while H3K4me3 is the active promoter mark, the presence of only the polycomb-mediated H3K27me3 mark indicates an inactive or repressed promoter. Presence of monomethylation of histone H3 lysine 4 (H3K4me1) is a hallmark of enhancers, and the cooccurrence of H3K4me1, H3K4me2, H3K27ac, and, for a subset, the additional presence of H3K4me3 and H3K9ac defines active enhancers.[84] While both active promoters and enhancers are enriched for methylation at lysine 4, enhancers are generally associated with a greater degree of monomethylation.[90] In *Drosophila melanogaster,* the combination of H3K4me1, H3K27ac, and H3K79me3 has been reported to be a good predictor for developmental enhancers; superenhancers, the best examples for interlinked developmental gene regulation (Box 3.1), can be differentiated from typical enhancers by a high density of mediator complex binding sites and extended (>3 Kb) deposition of H3K27ac.[8] Despite these generalizations, it is important to note that chromatin is a highly complex entity and accessible or open chromatin need not always indicate active enhancers and/or promoters. It could also be occupied (and can thus be identified) by other regulatory proteins such as CP190 or CTCF (insulator proteins) or other repressive TFs.

Emergent approaches for the identification of regulatory elements

What are the indicators for understanding a *cis* element, its role, and structure? As discussed before, some of it can be predicted based on sequence features and captured using functional assays. However, the activity of a large majority of *cis*-elements depends on the genomic and cellular context, including the developmental stage, tissue type, chromatin state, and the factors (protein/RNA) that it can interact with. Very few CREs can be discovered using stand-alone approaches; indeed, constantly updated tools and techniques have now become essential for understanding genome regulation. High-throughput and computational technologies have come of age in the past couple of decades and have the potential to decode regulatory element complexities across species, cell types, and developmental stages.[91] As next-generation sequencing (NGS) costs decline rapidly, genome-wide association studies (GWAS) have become the method of choice for exploring the distribution of histone modifications that are predictive of gene transcription and the binding of master regulators to lineage differentiation genes.[92]

Many aspects of gene regulation can now be investigated at a genomic scale including sequence-based prediction of CREs, TF binding sites and their enrichment, chromatin compaction and long-range interactions, and the 3-D looping of regulatory elements. The development of RNA sequencing (RNA-seq) was the first step in quantitating and following expression-level changes across the entire genome.[93] Chromatin immunoprecipitation followed by high-throughput sequencing (ChIP-seq[94]) allowed locating the position and extent of binding of transcription factors on the DNA, thus revealing the dynamics of TF binding as well as potential CREs involved in gene regulation. Using these two technologies, it became possible to explore the genome-wide binding of TFs in multiple cell types including in embryonic stem cells[95] and develop a TF-binding code predictive of differential gene expression profiles.[96] Epigenetic marks such as histone modification are increasingly understood to govern gene expression by regulating the binding of TFs and maintaining target genes in an expressed or repressed transcriptional state. These modifications of the DNA or the histones around which it is wrapped have been assayed in the context of gene transcription levels to develop the epigenetic code that regulates chromatin state and directs functionality via modulation of gene expression.[31]

The dynamic relationship between chromatin accessibility around the promoter region and gene expression is well mapped by the new experimental methods arising out of whole-genome technologies described in Table 3.1. In conjunction with differential time course assays in stem cells/cell culture systems and in vivo models, where homogeneity and positional information are retained, they have helped follow chromatin changes during the differentiation of progenitor cells. Exploring the whole-genome 3D conformations of chromatin and resulting *cis*-regulatory interactions has clarified our view of regulation of lineage commitment and cell differentiation.[97,98] For instance, using a single-cell transcriptome approach on staged *Drosophila* embryos, a virtual map of spatiotemporal gene regulation in each of the 6000 cells of the fly embryo has been created,[99] offering an entirely new level of perspective into genome regulation during development. Single-cell epigenomics, single-molecule live imaging, and other superresolution methods are helping characterize individual cells during the process of differentiation and the molecular mechanisms governing master regulators that influence cell fate decisions.[100]

Computational methods for harnessing epigenomic data sets to predict regulatory elements

Genome-wide studies are characterized by the large-scale identification of putative CREs and their binding factors. However, they serve primarily as a repository of possible candidate regulatory elements and require the development of suitably scaled-up functional assays to validate these elements and understand their mechanism of action. Massively parallel reporter gene assays that leverage NGS technologies with high-throughput DNA synthesis have been developed recently for

Table 3.1 Techniques to identify and study regulatory elements.

Technique	Approach	References
Protein occupancy on chromatin		
DNA adenine methylase identification	DNA adenine methyltransferase (Dam) fused with the protein of interest methylates the adenines in GATC motif upon coming near the target DNA regions in vivo. The methylated sequences that represent the binding sites of the protein of interest are identified by enriching them using methyladenine-specific enzymes or antibodies followed by next-generation sequencing.	101
Chromatin immunoprecipitation sequencing	Antibodies against the protein of interest are used to target cross-linked DNA—protein complexes or chromatin fragments (100–400 bp) in vitro. DNA regions bound to the protein of interest are purified from the enriched DNA—protein complexes and sequenced.	102–104
ChIP-exo	The immunoprecipitated cross-linked DNA–protein complexes or ChIP DNA is treated with an exonuclease after first adaptor ligation, which digests the double-stranded DNA in a 5′-3′ direction until stopped by covalent interaction between protein and DNA. After purification, the DNA is PCR amplified, followed by ligation of the second adaptor. Libraries are prepared and sequenced. This method provides resolution up to single nucleotide.	105–106
Cleavage under targets and release using nuclease	Antibody-guided cleavage by the micrococcal nuclease (MNase) enzyme activity releases the DNA regions bound to the protein of interest. As a result, the protein–DNA complex diffuses out of the nuclei upon washing. The DNA fragments are purified from the complex and prepared for high-throughput sequencing. Since this method is performed in situ, limitations associated with cell lysis using harsh chemicals and chromatin fragmentation by sonication are bypassed. Furthermore, antibody-guided cleavage by MNase reduces the noise from nonspecific chromatin regions.	107

Table 3.1 Techniques to identify and study regulatory elements.—*cont'd*

Technique	Approach	References
RNA–chromatin interaction		
Chromatin isolation by RNA purification sequencing	Cross-linked and fragmented chromatin is hybridized with biotinylated oligo probes designed complementary to the RNA of interest. The hybridized RNA–DNA complexes are captured on streptavidin beads. DNA is purified and subjected to library preparation for next-generation sequencing.	108
Chromatin accessibility		
DNase I hypersensitive sites sequencing	DNase I selectively digests the accessible chromatin zones such as decondensed *cis*-regulatory or nucleosome-depleted regions in the intact permeabilized nuclei. Libraries are prepared using the purified DNA and sequenced. Here, the digested DNA represents DNase hypersensitive sites (DHSs) or open chromatin.	109–110
Formaldehyde-Isolation of regulatory elements	Chromatin is cross-linked in vivo using formaldehyde and sheared by sonication. Next, the phenol–chloroform extraction results in segregation of cross-linked protein–DNA complexes and non–cross-linked protein-free fragments in the organic and aqueous phase, respectively. Here, the non–cross-linked DNA represents the accessible chromatin. The basis for this method is that the nucleosome-bound DNA is efficiently cross-linked by formaldehyde compared with depleted regions.	111
Micrococcal nuclease sequencing	Chromatin (cross-linked or non–cross-linked) is digested with micrococcal nuclease enzyme, which binds and cleaves the protein-free regions of DNA on chromatin. As a result, DNA bound to histones or other proteins such as transcription factors remains undigested. The undigested DNA is then purified and sequenced using next-generation sequencing methods.	112–113
Assay for transposase-accessible chromatin using sequencing	Mutated hyperactive Tn5 transposase integrates the sequencing adapters into the accessible regions of chromatin of unfixed nuclei in vitro, a process called	114

Continued

Table 3.1 Techniques to identify and study regulatory elements.—*cont'd*

Technique	Approach	References
	tagmentation. This process generates transposed DNA fragments which are purified, PCR-amplified using primers against integrated adapters and sequenced.	
Spatial organization of genome		
Chromatin conformation capture	Intact nuclei are treated with formaldehyde to fix the interactions between genomic loci. The cross-linked genome is digested with frequently cutting restriction endonucleases and ligated under conditions where intramolecular ligation or ligation between proximal fragments is preferred over intermolecular. The cross-linking is reversed, and the ligation products thus obtained are quantified by PCR using known locus-specific primers. 3C allows the quantification of interactions between a single pair of genomic loci such as an enhancer–promoter.	115
Chromatin conformation capture-on-chip (4C)	4C or 3C-on-chip enables quantification of interactions of a given locus with all other genomic loci. The ligated products obtained after 3C procedure are digested with a second frequently cutting restriction enzyme and again ligated. The second ligation step results in circularized fragments, which are subsequently used for an inverse PCR utilizing primers of the locus of interest, resulting in amplification of the unknown ligated sequences.	116
Chromatin conformation capture carbon copy (5C)	The ligation products generated using 3C protocol are subjected to annealing with 5C primer pool (designed for predicted 3C junctions) and ligation. 5C forward and reverse primers get ligated if they anneal next to each other on the 3C ligation product. The 5C library thus generated is amplified using universal primers against the 5C primer tails and sequenced.	117
Hi-C	Cross-linked chromatin is digested with a restriction enzyme leaving 5′ overhangs. The overhangs are filled with biotinylated nucleotides resulting in ligated products with biotin at the junction. DNA is purified	118

Table 3.1 Techniques to identify and study regulatory elements.—*cont'd*

Technique	Approach	References
	and sheared. Biotin-labeled fragments are pulled down using streptavidin beads and sequenced. The final data represent genome-wide chromatin interactions.	
Chromatin immunoprecipitation loop (ChIP-loop)	This technique combines 3C with ChIP-seq to detect long-range chromatin interaction between two loci mediated by a protein. Briefly, the cross-linked chromatin digested with a restriction endonuclease is immunoprecipitated using an antibody against the protein of interest before ligation in the 3C procedure.	119
Chromatin interaction analysis by paired-end tag sequencing	This technique combines Hi-C with ChIP-seq to detect genome-wide chromatin interactions mediated by a protein. Briefly, the cross-linked chromatin digested with a restriction endonuclease is immunoprecipitated using an antibody against the protein of interest before ligation in the Hi-C procedure.	120
Mapping transcription start site		
Cap analysis gene expression	This technique involves cap trapping of the first-strand cDNA for high-throughput sequencing of the first 20 nucleotides from the 5′-ends of mRNAs.	121
Global run-on sequencing	Ongoing transcription is halted by keeping the cell nuclei on ice. Transcription is resumed in the presence of bromodeoxyuridine (BrdU) under conditions that hinder with the formation of new preinitiation complexes and allows elongation of RNA. RNA molecules that have incorporated BrdU are affinity-purified using an anti-BrdU antibody. The enriched nascent RNAs are then subjected to the library preparation and sequenced. This technique allows the identification and genome-wide mapping of nascent RNAs, including the less-abundant enhancer RNAs.	122
Precision nuclear run-on sequencing	PRO-seq utilizes biotin-labeled nucleotides for nuclear run-on reaction instead of BrdU. Addition of single biotin-labeled nucleotide halts further	123

Continued

Table 3.1 Techniques to identify and study regulatory elements.—*cont'd*

Technique	Approach	References
	elongation of nascent RNAs by RNA polymerase II. The RNA is extracted and pulled down using streptavidin beads postfragmentation. Purified RNA is further used for library preparation and sequencing. This technique results in base pair resolution.	
GRO-Cap	In GRO-cap, nascent RNAs prepared in a similar way as in GRO-seq are subjected to enzymatic treatments to enrich for RNAs with 7-methyl guanosine residue or the 5′ cap.	124
Self-transcribing active regulatory region sequencing	Sheared genomic DNA fragments are placed downstream of a minimal promoter in a reporter vector generating a reporter library. Cells are transfected with reporter library giving the candidate sequences an opportunity for self-transcription. RNA is extracted, reverse-transcribed, and used further for library preparation and sequencing. This technique is based on the fact that an active enhancer can induce transcription from minimal promoters.	125
DNA modifications		
Methylated DNA immunoprecipitation sequencing (MeDIP-Seq)	MeDIP-Seq utilizes antibody against methylated DNA (5-methylcytosine) to immunoprecipitate methylated genomic DNA fragments. The immunoprecipitated DNA is used as an input for library preparation and sequencing.	126
Whole genome bisulfite sequencing	First, genomic DNA is treated with sodium bisulfite, which converts unmethylated cytosine to uracil while the methylated cytosines remain unchanged. After the whole-genome sequencing of bisulfite converted DNA, the methylation status of cytosine is identified based on mismatches during the alignment of data to reference genome resulting from the conversion of unmethylated cytosine to uracil.	127–129
Reduced representation bisulfite sequencing (RRBS)	In RRBS, before bisulfite conversion, and sequencing, the genomic DNA is digested with methyl-insensitive enzymes followed by adapter ligation and size selection.	130

large-scale testing and functional validation of the thousands of potential candidate genomic elements.[131] Since many regulatory elements work in tandem, and via stochastic association with multiple epigenetic and transcription factors, such functional validations are often difficult to perform in isolation. The generation of large data sets from whole-genome approaches has thus necessitated the parallel development of specialized bioinformatics tools and algorithms to analyze them. The measurement of CRE activity in such approaches does not depend on a functional assay for transcriptional readout; rather, it remains indirect, based on correlating the transcription factor occupancy and chromatin structure modifications with the pattern of gene expression. High-end predictive algorithms rely on self-learning computational models to understand epigenomic regulation. Recent work using such approaches has described the concept of *cis*-regulatory element communities (coCREs) or genomic regions that show correlated patterns of activity and TF occupancy in embryonic stem cells.[132] Other methods employ TF binding and enhancer activity data to predict spatiotemporal *cis*-regulatory activity; these combinatorial data-driven approaches do not require previous knowledge of TF sequence affinity, function, or expression and can thus be applied across cell types or differentiation states.[92,133]

Statistical and machine learning computational approaches are based largely on prediction and building on rules for developmental programming that are not fully understood, despite intense work in the past few decades. Based on statistically significant clusters of predicted occurrences of motifs for TF binding and other known CRE-binding proteins, DNA motif prediction has been an early workhorse in understanding gene regulation. Motif recognition evolved early on to incorporate multiple parameters such as correlation with expression,[134] distance from target gene,[135] enrichment scores from ChIP,[136] conservation across species, and combinations of features.[137] Many databases offer collections of DNA motifs highlighting the binding preferences of TFs based on experimental evidence. However, TF binding is context specific and dependent on chromatin state of the *cis*-element, and very few motifs actually bind their matched TF in vivo in a particular cell or developmental stage. Recent work using a statistical approach shows that genomic loci characterized by high phylogenetic conservation, distinct epigenomic marks, and strong TF binding are the most predictive of gene regulatory elements.[132] It is unlikely that the complex networks governing cellular identity, tissue differentiation, and the development of an entire organism from a single cell can be truly unraveled in the near future; however, the coming together of multiple disciplines and domain expertise has triggered breakthroughs in our current understanding.

International consortia such as the ENCODE,[138] The Epigenomics Roadmap,[139] and the FANTOM5[140] provide genome-scale data and resources, enabling the development of new methods to predict *cis*-DNA interactions with their *trans*-regulatory factors. ENCODE Phase III encompasses cell and tissue repertoires of RNA transcription, chromatin structure and modification, DNA methylation, chromatin looping, and occupancy by transcription factors and RNA-binding proteins from the mouse and human genomes.[141] The new and distinct field of computational epigenomics that has emerged from these technological advances has transformed the study of epigenetic regulation and its manifestation as the chromatin state and is invaluable for understanding the transcriptional control of stem and progenitor cells. As a consequence, our understanding of developmental processes is being revolutionized at a pace faster than could have been anticipated even just a decade ago.

Epigenetic programming in perinatal and early development

The act of fertilization begins a chain of events subsequent to which a whole new organism is formed, a property termed as totipotency (from *totipotentia* in Latin meaning ability for all). As embryonic development proceeds in mammals, this property is lost and the cells in the inner cell mass (ICM) at the blastocyst stage become pluripotent, differentiating into all the somatic cells needed to make an entire fetus. With further progression in development, potency rapidly declines to form multipotent (give rise to several closely related cell types), unipotent (give rise to only one cell type), and terminally differentiated cells (acquire specialized functions but have lost the ability to proliferate). This decay in cellular plasticity is caused by a progressive induction of lineage-specific genes and suppression of stemness genes, which is in turn coordinated by the underlying epigenetic mechanisms and the chromatin state.[142]

The zygotic genome in the initial stages after fertilization is transcriptionally silent. Gradually, zygotic genome activation (ZGA) takes place, and maternally deposited factors are substituted by the products of zygotic transcription. For zygotic gene expression to occur, chromatin has to transition from closed to an open state. As described in the preceding sections, pioneer factors or master TFs bind to nucleosomal DNA and recruit additional factors for the displacement of nucleosomes, recruitment of chromatin remodelers, and modification of chromatin.[143] Epigenetic remodeling then enables a change in cell state, leading to body patterning and setting of the functional and positional identity of cells, and lineage commitment. Many of these mechanisms are conserved in all eukaryotes. Chromatin modifications have recently been investigated even in relatively simple multicellular organisms such as the sponge, where conserved and developmentally regulated landscapes could be identified.[144] Developmental regulation thus presents an ideal model to investigate epigenetic regulation and dynamic changes in the epigenetic landscape that lead to changes in gene transcription profiles.

Switching chromatin states—dynamic epigenomic landscape at developmental *cis*-regulatory elements

With increased understanding of the evolution of genomes in the past couple of decades, many *cis*-regulatory mechanisms are believed to have coevolved with developmental regulatory genes conserved across species.[145] But what are their cognate partners in this process and do they show differences in chromatin state?

Investigations of gametogenesis and embryogenesis in multiple vertebrate and invertebrate models have highlighted conserved similarities and species-specific differences.[146–148] Unlike terminally differentiated cells, where gene expression profiles are bound by the functional requirement of a particular cell/tissue type, stem cells can express a diverse set of genes for differentiation. Transcriptionally permissive chromatin states in stem cells are maintained by the presence of activating histone modifications at gene promoters and coding regions, such that they are primed for expression on receiving developmental signals and lineage cues. Upon lineage commitment, the chromatin state becomes increasingly restrictive, and distinct cell fate and function are subsequently maintained throughout development.

In mammalian development, intense epigenetic reprogramming occurs in the PGCs, the preimplantation embryo as well as during the differentiation of the three germ layers. Early development in the embryo entails the determination of the ICM cells via master regulatory transcription factors such as OCT4, NANOG, and SOX2 as well as commitment and maintenance of the outer trophectoderm cell fate via regulation of *CDX2, ELF5, GATA6*, and other modulatory TFs. Germ cell specification leading to formation of PGCs is one of the first steps of epigenetic marking but is not as well conserved across species, primarily because of the differences in the early vertebrate developmental program. In vitro models using pluripotent embryonic stem cells isolated from the ICM cells of the blastocyst are reminiscent of the epigenetic changes in the early embryo and have largely been used to understand developmental regulators. Mouse embryonic stem cells (mESCs) from the epiblast were established four decades ago to form somatic as well as the germline lineages.[149] Human embryonic stem cells (hESCs) derived from human blastocysts[150] were then found to be epigenetically distinct from mouse ES cells, with diverse transcriptome profiles and culture requirements, suggesting differences in their regulatory mechanisms.[151] Induced pluripotent stem cells (iPSCs) created from somatic cells, instead of from the developing embryo, have been critical to understanding the dynamic epigenetic mechanisms governing pluripotency in mice[152] and humans.[153]

The resetting of the established gene expression program in terminally differentiated cells to reapproach a highly plastic pluripotent-like state offers an insight into the strong regulatory control exerted by master TF genes (*OCT3/4*, *SOX2*, *KLF4*, and *c-MYC)* in iPSCs. Several other factors (SALL4, NANOG, ESRRB, and LIN28) have since been shown to induce somatic cell reprogramming.[154] While recapitulating pluripotency, ESCs and iPSCs display differences in their DNA methylation profile compared with ICM cells, which can be regained to some extent by the addition of leukemia inhibitory factor (LIF).[155,156] iPSCs retain traces of their parental cell DNA methylation patterns, unlike mESCs where these have been erased for reprogramming. DNA methylation at promoters has emerged as the strongest correlate of gene silencing and is critical for lineage-specific developmental expression, genomic imprinting, and X-chromosome inactivation.[157] Knockout mice lacking DNA methyltransferases DNMT1 or DNMT3a/3b are unable to complete fetal development resulting in gestational death.[158,159] DNMT1 is critical to the progenitor cell state, and its knockdown leads to defects in tissue homeostasis.[160] Murine ESCs, however, can self-renew and maintain pluripotency in culture even without DNMTs,[161] highlighting the need to be aware of the nature and degree of context-dependent control by epigenetic factors.

Approaches employing human pluripotent stem cells (hPSCs) to investigate developmental regulation recapitulate the in vivo perigastrulation stage.[162] SOX17 and BLIMP1 are necessary and sufficient for inducing PGC fate, and for initiating the germline-specific epigenetic program associated with epiblast development and gastrulation in human PGCs. *Blimp1* is a key lysine methyltransferase, expressed in lineage-restricted PGCs that helps set the epigenomic profile.[56] The activation of the germline-specific epigenetic program also involves a global reduction in DNMT-mediated DNA methylation, with concomitant increase in 5 hmC or 5-hydroxymethylcytosine.[163] Often called the sixth base in DNA, 5 hmC is created by the oxidation of 5-methylcytosine (5 mC) by the ten-eleven translocation methylcytosine dioxygenases or TET family enzymes. Various cells and tissue types express the TET enzymes; however, *TET1* and *TET2* are expressed in ESCs under the control of the OCT4/SOX2 network and both *TET1* and *TET2* contain binding sites for them.[164] 5 hm C production by TET is thus directly linked to the pluripotency network and is downregulated in OCT4/SOX2-depleted cells, further affecting the expression of lineage commitment TFs. *TET1* and DNMT3A binding has recently been demonstrated to be coregulated at target loci, where they also interact with polycomb-mediated H3K27me3 for repression of bivalently marked promoters.[165] About 60%–80% of CpGs in the genome undergo DNA methylation, but regions rich in developmental genes have been shown to exist in DNA methylation valleys—hypomethylated genomic regions that are associated with cancer when abnormally methylated.[166] Promoters and CpG islands (CGIs) also remain largely undermethylated as do enhancers that are associated with dynamic methylation changes.

The switching of epigenetic states to enable different gene regulatory outcomes is exemplified by the resolution of bivalent promoters during developmental progression. Trimethylation of both histone H3 lysine 4 and lysine 27 (H3K4me3/H3K27me3) at gene promoters—promoter bivalency—is a key regulator of development, proliferation, and differentiation in embryogenesis, marking genes that are not necessarily transcribed, but are "poised" to do so.[167] This unique chromatin state, that is simultaneously activating and repressive, marks key pluripotency and developmental genes in hESCs as well,[168] likely recapitulating the developmental chromatin state. Enhancers enriched for H3K27me3 in conjunction with H3K4me1 are associated with inactive genes, some of which subsequently get activated as development proceeds; these bear H3K27 acetylation along with the lysine 4 monomethylation. "Poised" enhancers thus allow premarking genes for rapid switching and activation during differentiation.[169,170] Methylation of H3K4 alone is a strong inverse marker for promoter DNA methylation at actively transcribing genes in mammalian cells; DNA methylation is disrupted by inhibiting DNMTs at sites of H3K4 methylation, while H3K27me3 marked regions exhibit *de novo* DNA methylation after differentiation to allow gene silencing.[171] Polycomb-mediated

H3K27me3 expansion marks a large number of developmental promoters, likely maintaining repression as differentiation proceeds. H3K27me3 domains also mark enhancers in a lineage-specific manner, forming a repressive chromatin state following lineage commitment.[172] By contrast, DNA methylation is widely present at enhancers of all stages but enriched at promoters of select lineage-restricted genes and some key master regulators in ESCs such as *Nanog* and *Oct4*.[166] OCT4 and NANOG repression is mediated via increased DNA methylation and H3K27me3 histone modification, while SOX2 acquires H3K27me3/H3K9me3 domains at the promoter. While overlap between methylated DNA and methylation of lysine 9 of the associated histones is less prevalent (in the context of repressed chromatin), H3K36me3 is generally associated with fully methylated DNA at transcribing exons. SOX2 is hyperacetylated at H3K27 in hESCs but enriched in H3K27me3 upon differentiation; H3K27 acetylation at active gene promoters is antagonistic to polycomb repression by preventing the histone methylation. Many promoters thus undergo a chromatin switch between methylation and acetylation at H3K27 for their activation. Such dual mechanisms ensure the fine-tuned response of pluripotency and cell fate decision genes, enabling early activation in early development, restricting dedifferentiation in lineage-committed cells and maintaining repression in terminally differentiated cells. The chromatin state switch is a unique feature of developmental and determination genes and requires the concerted action of the *cis*- and *trans*-regulatory machinery to bring about dynamic chromatin remodeling at these loci.

Trans-regulatory networks govern stem cell pluripotency and differentiation

Adult stem cells are critical for homeostasis, tissue repair, and regeneration in multicellular organisms. Decline in stem cell function brings about aging and disease and is associated with aberrations in stem cell regulation.[173] The core transcriptional network for hESCs was established early on[174] and subsequently compared with more differentiated stem and adult progenitor cell types to investigate changes during differentiation.[166] Epigenetic memory ensures that the chromatin state set in progenitor cells, with its accompanying histone modifications and lineage-specific transcriptional regulator profiles, is propagated and amplified during tissue differentiation. Lineage commitment is a stochastic process, and the adult stem cell niche offers a tractable model to understand lineage-specific contexts for genome regulation via the action of TFs.

Lineage-specific networks and their associated chromatin states have been explored in a variety of adult stem cell lineages using genome-wide and single-cell techniques described in Section Identification and Characterization of cis-regulatory Elements and trans-acting Factors. Chromatin accessibility correlated with RNA-seq analysis at the single-cell resolution revealed dynamic interactions between distal regulatory elements and their target genes in differentiation assays using hematopoietic stem cells (HSCs), including a key role for GATA motif in lineage priming via expression of master regulator GATA3.[175] Mesenchymal stem cells (MSCs), pluripotent mesodermal precursors derived from the bone marrow,[176] are a model for adult stem cell differentiation and regeneration, forming adipocytes, myoblasts, chondrocytes, osteoblasts, and even cardiac muscle and neurons given appropriate conditions.[177] Signaling cues via master regulators such as the Wnt developmental pathway, bone morphogenetic proteins (BMPs), and growth factors (TGFs and FGFs) have been shown to act on downstream tissue-specific transcription factors to influence the decision to commit to the formation of fat tissue, muscle, cartilage, or bone. Loss of H3K4me3 signals at proximal promoters of upregulated genes, concomitant with acquisition of acetyl modifications, drives the specification of each of these sublineages and plays an important role in skeletal development by activating master regulators.[178] In satellite cells, activation of muscle-specific transcription factors driven by DNA hypomethylation is a key attribute of commitment toward skeletal muscle lineage and myogenic regeneration.[179] Quiescent stem cells express polycomb repressive complex 2 (PRC2), causing H3K27me3-mediated silencing, but with promoters of stem cell regulators such as MyoD marked with the activating H3K4me3 mark.[180] MyoD is a master regulator of muscle differentiation and binds to target CREs called E-boxes located within many muscle-relevant genes including Myf5 and muscle regulatory factors or MRFs. In the quiescent state, these are hypoacetylated via HDACs, and MyoD is inhibited.[181] Upon induction of differentiation, MyoD works with HMTs to remove the repressive H3K27me3 and acquire H3K4me3.[180] A complex interplay of the precise E-box motif numbers, as well as their variant sequences and spatial arrangement in conjunction with regulatory histone modifications at the target CREs, governs the precise binding affinity and transcriptional activity of MyoD. HAT p300 further acetylates both MyoD and the downstream MRFs to create distinct "context-specific" chromatin states that can direct muscle differentiation and homeostasis.

The role of noncoding RNAs and miRNAs in regulating TF occupancy, and thereby the adult stem niche, is increasingly coming into focus and may have clinical applications for therapeutic regenerative approaches,[182–185] discussed elsewhere in this issue. Cellular phenotypes, including stem cell formation, maintenance, and differentiation, are controlled by combinatorial TF interactions. However, TF-binding events are governed by a host of dynamic epigenetic factors depending on context, genomic loci, and developmental stage. In ESCs, variation in gene expression patterns during differentiation is associated with master TFs (OCT4, NANOG, SOX2, SMAD1, STAT3, TCFCP2L1, and ESRRB) that can act as both activators and repressors depending on the target gene and the developmental context.[96] The loss of specific HOX factors via repressive chromatin domains at the developmentally regulated *Hox* clusters is associated with initiation of most differentiation pathways beyond the ESC state.[172] The *Hox* genes encode highly conserved homeodomain-containing TFs that determine anterior to posterior body axis in almost all bilaterians and are essential for initiating positional identity during development. Following lineage commitment, most *Hox* genes are associated with expanded H3K27me3 domains with H3K9me3/H3K27me3-marked promoters;

redistribution and expansion of repressive epigenomic blocks highlight increasing commitment from a pluripotent state. The repression of *Hox* genes underlies suppression of the somatic cell fate in germline founders as well and appears to be an important mechanism in lineage commitment. Transcription factor occupancy has now been identified at genome-wide scales across different stem cells and developmental stages; however, decoding this regulatory landscape to understand in vivo development still remains a daunting challenge.

Regulatory elements in early development inform adult health and disease states

In most organisms, the developmental program is initiated by maternal proteins in the oocyte, via epigenetic control of early developmental regulators. The *trans*-regulatory network in early development is extremely intricate, and its exact contributions to early cell fate decisions remain unclear. In mammals, including humans, the early developmental phase encompasses the period of gonadal maturation as well as fertilization and growth of the fetus and involves rapid transcriptional changes in response to differentiation signals. The periconceptional period (that includes a few months prior to conception, involving oocyte growth and maturation, and the fertilization event, as well as the first trimester of pregnancy) marks the resetting of the epigenetic landscape and the genome-wide initiation of growth and developmental cascades. This period is therefore crucial for understanding the long-term effects of internal and external environmental exposures on the developing fetus. Epigenetic memory is a key feature of PGCs, manifested as the erasure and resetting of DNA methylation during gametogenesis and fertilization, especially during X chromosome inactivation[186] and at imprinted loci.[187] Imprinted genes are strongly associated with early development, regulating placental inputs to the fetus. Their transcription is governed in a parent-of-origin-specific manner, allowing reciprocal monoallelic expression from the parental alleles. DNA methylation at cytosine bases within differentially methylated regions (DMRs) and imprinting control regions (ICRs) at imprinted loci typically controls the binding of methylation sensitive factors, such as the insulator protein CTCF, which prevent transcriptional activation. Imprinted genes control fetal growth and early development and are thus prime candidates to investigate epigenetic regulation in development.

Epigenetic factors have been understood relatively recently to be a global and dynamic mechanism to respond to early developmental flux. Such responses have the potential to become a part of the inheritance of the individual in adulthood and can even be passed on to their future progeny via transgenerational epigenetic mechanisms. Indeed, epigenetic factors govern perinatal development to the extent of enabling future responsiveness to environmental factors, the genome's "epigenetic inheritance."[31] The precise mechanisms by which this regulation is fine-tuned are not yet understood but are believed to initiate through placental signaling. Placental–fetal exchange is a very complex phenomenon; besides providing nutrition and appropriate signaling cues for differentiation, growth, and development, it also allows exposure to external conditions via placental sampling. This leads to resetting of DNA and histone methylation patterns at gene regulatory elements, thereby influencing the *trans*-factor binding machinery and downstream gene expression cascades, in the developing fetus (reviewed in Ref. [188]). Lifelong effects are brought about, in part, via a more permanent resetting of epigenetic and thus gene expression cascades. In recent years, the discovery of the role of the microbiome, our "second genome," has added another layer of complexity.[189] The influence of the maternal microbiome via its metabolites and the development and interaction of the fetal and neonatal microbiome with its genome is the latest niche in the study of epigenetic regulation.[190,191] Recent studies have focused on the genome-scale identification of regulatory elements in placental tissues that can govern all these interactions.[192] The final consequences of these epigenetic changes can be measured as phenotypic responses to early environmental stimuli, termed "predictive adaptive responses" (PARs), that provide fitness at a later stage of development and are crucial to understanding the role of early developmental influences on adult human health.[193]

The hypothesis for fetal origins of adult disease has been demonstrated to hold true for a large number of noncommunicable diseases that manifest in adult life. With a large body of work outlining the influence of the early developmental environment (intrauterine postconception to early childhood postnatal cues), the phenomenon has been concretized into the concept of Developmental Origins of Health and Disease (DOHaD) and has brought the study of epigenetic regulation in development to the forefront.[194] DOHaD suggests that nonideal developmental conditions faced by a fetus (poor nutrition, toxic exposures, infections, and hormonal disturbances) can prime it to adapt by changing its gene expression profile to support optimal survival, using the mechanisms of early developmental plasticity. This can have far-reaching implications for the future health and well-being of the individual after birth and in adulthood, manifested in their increased likelihood for developing diseases. Misregulation of epigenetic mechanisms in early development, including during gametogenesis and early gestation, can have exponentially severe consequences in later life. Studies of the "Hunger Winter" caused by the Dutch famine in 1944–45, where pregnant and lactating mothers faced severe nutritional deficiencies, have documented the adverse effects seen in this cohort. The babies conceived in this period showed hypomethylation at the *IGF2* locus—an imprinted growth factor gene encoding the insulin-like growth factor II hormone—in adulthood.[195] IGF2 plays a critical role in development, and the *IGF2/H19* locus is known to be epigenetically misregulated in a large number of disorders.[196] Further work suggested altered methylation patterns at multiple CREs—differentially methylated regions or DMRs—that control the imprinted gene expression patterns of the *IGF2* and *H19* genes at this locus. *IGF2* is paternally expressed, while *H19* has maternal expression; only the unmethylated maternal insulator binds CTCF to block enhancer–promoter interaction at *IGF2* and keep the *H19* promoter hypomethylated at the same time. The *IGF2* DMRs were found to be methylated in individuals with periconceptional famine exposure compared with unexposed same-sex siblings, possibly reflecting decreased *IGF2* transcription to adapt for reduced intrauterine growth.[197] There was, however, no change in the global DNA methylation,[198] suggesting a tightly coordinated and highly specialized epigenetic response at a few key genes in response to a prenatal insult.

Interestingly, the precise timing of the inappropriate cues—during periconception or late in gestation—may play a role in the extent and persistence of DNA methylation changes at different loci.[197] Gender-dependent associations can add further complexity due to the interactions with sex hormones, making many of the downstream effects gender-specific. Prenatal alcohol exposure is a well-studied example of toxic insult during early development that can alter the epigenome of the developing fetus, exacerbated by gender, as explored in many human and animal models.[199] Mothers consuming alcohol during pregnancy risk having growth-retarded babies with birth defects as well as later-life ill health, the so-called fetal alcohol spectrum disorders (FASDs[200]). Alcohol exposure has long been known to inhibit fetal DNA methylation, including at the *IGF2/H19* locus,[201] but several studies over the past decade have revealed widespread epigenetic gestational reprogramming in adults due to early alcohol exposure,[202,203] both via maternal influences and through the paternal germline.[204] The epigenetic aberrations are not restricted to DNA methylation but also involve altered histone modification patterns.[205] The mechanisms of aberrant epigenetic patterns in disease and disordered states are numerous and will be detailed in another chapter in this issue.

Conclusion

The interdependence of genome regulation and developmental decisions sets up epigenetic cascades that last through the lifetime of an organism and, as is becoming increasingly apparent from recent reports, even beyond. This is because developmental events, as a result of cellular signaling, give rise to specific programs of gene expression that form feedback loops and networks with the epigenetic machinery, for example, the regulation of imprinted regions in germ cells via DNA methylation patterns. This sets subsequent gene expression patterns during an individual's life span such that the influence of the cellular or environmental cues becomes epigenetically fixed. Developmentally regulated epigenetic patterns in the germline can have an even longer-lasting impact, as they can extend to the next generation and possibly beyond that. Alterations in developmental pathways are also correlated with epigenetic deregulation leading to cancer,[206] showing many parallels with stem cell biology. Cancer progenitor/stem cells are believed to arise from unbalanced epigenetic mechanisms predisposing cells to immortality under the influence of their environment.

Technological advances in NGS have enabled the precise mapping of epigenetic modifications, illustrating the epigenome as the outcome of developmental processes. Obtaining a homogenous population of cells to define epigenomic features unambiguously remains a limitation of current approaches. However, the advent of single cell genomics and more sensitive techniques requiring a very small number of cells is making it increasingly possible to uncover the epigenetic profile in different cell types during development and aging. Further refinement of these techniques along with increasingly more efficient analytics holds great promise for understanding epigenome biology. The epigenome is now understood to be a reflection as well as a record of the prenatal environment.[207] Epigenetic mechanisms play a large and crucial role in development, allowing for effects of "nurture" or the environment to work in tandem with those inherited through the DNA sequence and its mutations or "nature," and cannot be ignored if we are to understand human disease and its origins. [208]

The expression of eukaryotic genes is a tightly regulated process. It is crucial for genes to be expressed in the correct cell type to an appropriate level and at the correct time during cell differentiation and development in response to internal and external signals. All regulatory elements and many of their interacting factors are present in all the cells in the body—it then becomes crucial to ask how different components interact with each other to control gene expression, not only at a single gene locus, but globally across the genome governing biological and transcriptional programs in a cell type–specific manner. The emerging view is that epigenetic mechanisms play a key role in this process. Understanding how gene expression is regulated—from early development and all through our lifespan—is essential to fully decode the function of our genome and to search for therapeutic remedies for genetic diseases. Beyond the CREs, however, lies a complex overlay of PTMs of the histones that frame the chromatin structure, the DNA-binding proteins that modulate it, and the covalent modifications of ncRNAs. Phenomenal progress in delineating these factors in recent years promises major insights into the mechanisms and understanding of the biology of development, including aging and disease.

References

1. Taft RJ, Pheasant M, Mattick JS. The relationship between non-protein-coding DNA and eukaryotic complexity. *Bioessays*. 2007;29(3):288–299. https://doi.org/10.1002/bies.20544.
2. Davidson EH. *The Regulatory Genome: Gene Regulatory Networks in Development and Evolution*. 2006.
3. Nelson AC, Wardle FC. Conserved non-coding elements and cis regulation: actions speak louder than words. *Development*. 2013;140(7):1385–1395. https://doi.org/10.1242/dev.084459. Cambridge.
4. Small S, Arnosti DN. Transcriptional enhancers in *Drosophila*. *Genetics*. 2020;216(1):1–26. https://doi.org/10.1534/genetics.120.301370.
5. Banerji J, Rusconi S, Schaffner W. Expression of a β-globin gene is enhanced by remote SV40 DNA sequences. *Cell*. 1981;27(2):299–308. https://doi.org/10.1016/0092-8674(81)90413-X.
6. Moreau P, Hen R, Wasylyk B, Everett R, Gaub MP, Chambon P. The SV40 72 base repair repeat has a striking effect on gene expression both in SV40 and other chimeric

recombinants. *Nucleic Acids Res.* 1981;9(22):6047–6068. https://doi.org/10.1093/nar/9.22.6047.

7. Smith E, Shilatifard A. Enhancer biology and enhanceropathies. *Nat Struct Mol Biol.* 2014;21(3):210–219. https://doi.org/10.1038/nsmb.2784.
8. Denes H, Brian JA, Tong Ihn L, et al. Super-enhancers in the control of cell identity and disease. *Cell.* 2013:934–947. https://doi.org/10.1016/j.cell.2013.09.053.
9. West AG, Gaszner M, Felsenfeld G. Insulators: many functions, many mechanisms. *Gene Dev.* 2002;16(3):271–288. https://doi.org/10.1101/gad.954702.
10. Bessa J, Luengo M, Rivero-Gil S, et al. A mobile insulator system to detect and disrupt cis-regulatory landscapes in vertebrates. *Genome Res.* 2014;24(3):487–495. https://doi.org/10.1101/gr.165654.113.
11. Gaszner M, Vazquez J, Schedl P. The Zw5 protein, a component of the scs chromatin domain boundary, is able to block enhancer-promoter interaction. *Gene Dev.* 1999; 13(16):2098–2107. https://doi.org/10.1101/gad.13.16.2098.
12. Kellum R, Schedl P. A group of scs elements function as domain boundaries in an enhancer- blocking assay. *Mol Cell Biol.* 1992;12(5):2424–2431. https://doi.org/10.1128/mcb.12.5.2424.
13. Zhao K, Hart CM, Laemmli UK. Visualization of chromosomal domains with boundary element-associated factor BEAF-32. *Cell.* 1995;81(6):879–889. https://doi.org/10.1016/0092-8674(95)90008-X.
14. Geyer PK, Corces VG. DNA position-specific repression of transcription by a *Drosophila* zinc finger protein. *Gene Dev.* 1992;6(10):1865–1873. https://doi.org/10.1101/gad.6.10.1865.
15. Gyurkovics H, Gausz J, Kummer J, Karch F. A new homeotic mutation in the *Drosophila* bithorax complex remopves a boundary separating two domains of regulation. *EMBO J.* 1990;9(8):2579–2585. https://doi.org/10.1002/j.1460-2075.1990.tb07439.x.
16. Karch F, Galloni M, Sipos L, Gausz J, Gyurkovics H, Sched P. Mcp and Fab-7: molecular analysis of putative boundaries of cis-regulatory domains in the bithorax complex of *Drosophila melanogaster. Nucleic Acids Res.* 1994;22(15):3138–3146. https://doi.org/10.1093/nar/22.15.3138.
17. Gray S, Levine M. Transcriptional repression in development. *Curr Opin Cell Biol.* 1996;8(3):358–364. https://doi.org/10.1016/S0955-0674(96)80010-X.
18. Grossniklaus U, Paro R. Transcriptional silencing by polycomb-group proteins. *Cold Spring Harbor Perspect Biol.* 2014;6(11):1–26. https://doi.org/10.1101/cshperspect.a019331.
19. Ogbourne S, Antalis TM. Transcriptional control and the role of silencers in transcriptional regulation in eukaryotes. *Biochem J.* 1998;331(1):1–14. https://doi.org/10.1042/bj3310001.
20. Courey AJ, Jia S. Transcriptional repression: the long and the short of it. *Gene Dev.* 2001;15(21):2786–2796.
21. Gisselbrecht SS, Palagi A, Kurland JV, et al. Transcriptional silencers in *Drosophila* serve a dual role as transcriptional enhancers in alternate cellular contexts. *Mol Cell.* 2020;77(2). https://doi.org/10.1016/j.molcel.2019.10.004, 324-337.e8.
22. Li Q, Peterson KR, Fang X, Stamatoyannopoulos G. Locus control regions. *Blood.* 2002;100(9):3077–3086. https://doi.org/10.1182/blood-2002-04-1104.
23. Hardison R, Slightom JL, Gumucio DL, Goodman M, Stojanovic N, Miller W. Locus control regions of mammalian β-globin gene clusters: combining phylogenetic analyses

and experimental results to gain functional insights. *Gene*. 1997;205(1–2):73–94. https://doi.org/10.1016/S0378-1119(97)00474-5.
24. McClintock B. The origin and behavior of mutable loci in maize. *Proc Natl Acad Sci USA*. 1950;36(6):344–355. https://doi.org/10.1073/pnas.36.6.344.
25. Hoover KK, Chien AJ, Corces VG. Effects of transposable elements on the expression of the forked gene of *Drosophila melanogaster*. *Genetics*. 1993;135(2):507–526.
26. Chuong E, Elde N, Feschotte C. Regulatory activities of transposable elements: from conflicts to benefits. *Nat Rev Genet*. 2017;18:71–86.
27. Srivastava S, Avvaru AK, Sowpati DT, Mishra RK. Patterns of microsatellite distribution across eukaryotic genomes. *BMC Genom*. 2019;20(1). https://doi.org/10.1186/s12864-019-5516-5.
28. Iglesias AR, Kindlund E, Tammi M, Wadelius C. Some microsatellites may act as novel polymorphic cis-regulatory elements through transcription factor binding. *Gene*. 2004; 341(1–2):149–165. https://doi.org/10.1016/j.gene.2004.06.035.
29. Ringrose L, Paro R. Polycom/Trithorax response elements and epigenetic memory of cell identity. *Development*. 2007;134(2):223–232. https://doi.org/10.1242/dev.02723.
30. Kouzarides T. Chromatin modifications and their function. *Cell*. 2007;128(4):693–705. https://doi.org/10.1016/j.cell.2007.02.005.
31. Jaenisch R, Bird A. Epigenetic regulation of gene expression: how the genome integrates intrinsic and environmental signals. *Nat Genet*. 2003;33(3S):245–254. https://doi.org/10.1038/ng1089.
32. Wolpert L. Positional information and the spatial pattern of cellular differentiation. *J Theor Biol*. 1969;25(1):1–47. https://doi.org/10.1016/S0022-5193(69)80016-0.
33. Levine M, Tjian R. Transcription regulation and animal diversity. *Nature*. 2003; 424(6945):147–151. https://doi.org/10.1038/nature01763.
34. Hammonds AS, Bristow CA, Fisher WW, et al. Spatial expression of transcription factors in *Drosophila* embryonic organ development. *Genome Biol*. 2013;14(12). https://doi.org/10.1186/gb-2013-14-12-r140.
35. Lambert SA, Jolma A, Campitelli LF, et al. The human transcription factors. *Cell*. 2018; 172(4):650–665. https://doi.org/10.1016/j.cell.2018.01.029.
36. Weirauch MT, Hughes TR. A catalogue of eukaryotic transcription factor types, their evolutionary origin, and species distribution. *Sub Cell Biochem*. 2014;52:25–73. https://doi.org/10.1007/978-90-481-9069-0_3.
37. Zaret KS, Mango SE. Pioneer transcription factors, chromatin dynamics, and cell fate control. *Curr Opin Genet Dev*. 2016;37:76–81. https://doi.org/10.1016/j.gde.2015.12.003.
38. Mayran A, Sochodolsky K, Khetchoumian K, et al. Pioneer and nonpioneer factor cooperation drives lineage specific chromatin opening. *Nat Commun*. 2019;10(1). https://doi.org/10.1038/s41467-019-11791-9.
39. Thomas MC, Chiang CM. The general transcription machinery and general cofactors. *Crit Rev Biochem Mol Biol*. 2006;41(3):105–178. https://doi.org/10.1080/10409230600648736.
40. Justilien V, Walsh A, Thompson E, Murray N, Fields AP. The PRKCI and SOX2 oncogenes are coamplified and cooperate to activate Hedgehog signaling in lung squamous cell carcinoma. *Cancer Cell*. 2014;25(2). https://doi.org/10.1016/j.ccr.2014.01.008. PMID: 24525231.

41. Tsuruzoe S, Ishihara K, Uchimura Y, et al. Inhibition of DNA binding of Sox2 by the SUMO conjugation. *Biochem Biophys Res Commun*. 2006;351(4):920–926. https://doi.org/10.1016/j.bbrc.2006.10.130.
42. Thevenet L, Méjean C, Moniot B, et al. Regulation of human SRY subcellular distribution by its acetylation/deacetylation. *EMBO J*. 2004;23(16):3336–3345. https://doi.org/10.1038/sj.emboj.7600352.
43. Ma J. Crossing the line between activation and repression. *Trends Genet*. 2005;21(1): 54–59. https://doi.org/10.1016/j.tig.2004.11.004.
44. Rinn JL, Chang HY. Genome regulation by long noncoding RNAs. *Annu Rev Biochem*. 2012;81:145–166. https://doi.org/10.1146/annurev-biochem-051410-092902.
45. Guttman M, Donaghey J, Carey BW, et al. LincRNAs act in the circuitry controlling pluripotency and differentiation. *Nature*. 2011;477(7364):295–300. https://doi.org/10.1038/nature10398.
46. Fico A, Fiorenzano A, Pascale E, Patriarca EJ, Minchiotti G. Long non-coding RNA in stem cell pluripotency and lineage commitment: functions and evolutionary conservation. *Cell Mol Life Sci*. 2019;76(8):1459–1471. https://doi.org/10.1007/s00018-018-3000-z.
47. Kim TK, Hemberg M, Gray JM, et al. Widespread transcription at neuronal activity-regulated enhancers. *Nature*. 2010;465(7295):182–187. https://doi.org/10.1038/nature09033.
48. de Santa F, Barozzi I, Mietton F, et al. A large fraction of extragenic RNA Pol II transcription sites overlap enhancers. *PLoS Biol*. 2010;8(5). https://doi.org/10.1371/journal.pbio.1000384.
49. Haberle V, Stark A. Eukaryotic core promoters and the functional basis of transcription initiation. *Nat Rev Mol Cell Biol*. 2018;19(10):621–637. https://doi.org/10.1038/s41580-018-0028-8.
50. Pnueli L, Rudnizky S, Yosefzon Y, Melamed P. RNA transcribed from a distal enhancer is required for activating the chromatin at the promoter of the gonadotropin α-subunit gene. *Proc Natl Acad Sci USA*. 2015;112(14):4369–4374. https://doi.org/10.1073/pnas.1414841112.
51. Li W, Notani D, Ma Q, et al. Functional roles of enhancer RNAs for oestrogen-dependent transcriptional activation. *Nature*. 2013;498(7455):516–520. https://doi.org/10.1038/nature12210.
52. Whalen S, Truty RM, Pollard KS. Enhancer-promoter interactions are encoded by complex genomic signatures on looping chromatin. *Nat Genet*. 2016;48(5):488–496. https://doi.org/10.1038/ng.3539.
53. Chang HC, Huang HC, Juan HF, Hsu CL. Investigating the role of super-enhancer RNAs underlying embryonic stem cell differentiation. *BMC Genom*. 2019;20. https://doi.org/10.1186/s12864-019-6293-x.
54. Dodge JE, Kang YK, Beppu H, Lei H, Li E. Histone H3-K9 methyltransferase ESET is essential for early development. *Mol Cell Biol*. 2004;24(6):2478–2486. https://doi.org/10.1128/MCB.24.6.2478-2486.2004.
55. O'Carroll D, Erhardt S, Pagani M, Barton SC, Surani MA, Jenuwein T. The Polycomb-group gene Ezh2 is required for early mouse development. *Mol Cell Biol*. 2001;21(13): 4330–4336. https://doi.org/10.1128/MCB.21.13.4330-4336.2001.
56. Ohinata Y, Payer B, O'Carroll D, et al. Blimp1 is a critical determinant of the germ cell lineage in mice. *Nature*. 2005;436(7048):207–213. https://doi.org/10.1038/nature03813.

57. De Souza FSJ, Gawantka V, Gómez AP, Delius H, Ang SL, Niehrs C. The zinc finger gene Xblimp1 controls anterior endomesodermal cell fate in Spemann's organizer. *EMBO J*. 1999;18(21):6062–6072. https://doi.org/10.1093/emboj/18.21.6062.
58. Roy S, Ng T. Blimp-1 specifies neural crest and sensory neuron progenitors in the zebrafish embryo. *Curr Biol*. 2004;14(19):1772–1777. https://doi.org/10.1016/j.cub.2004.09.046.
59. Yao TP, Oh SP, Fuchs M, et al. Gene dosage-dependent embryonic development and proliferation defects in mice lacking the transcriptional integrator p300. *Cell*. 1998; 93(3):361–372. https://doi.org/10.1016/S0092-8674(00)81165-4.
60. Henikoff S, Smith MM. Histone variants and epigenetics. *Cold Spring Harbor Perspect Biol*. 2015;7(1). https://doi.org/10.1101/cshperspect.a019364.
61. Morgan HD, Santos F, Green K, Dean W, Reik W. Epigenetic reprogramming in mammals. *Hum Mol Genet*. 2005;14(1):R47–R58. https://doi.org/10.1093/hmg/ddi114.
62. Frankel N, Davis GK, Vargas D, Wang S, Payre F, Stern DL. Phenotypic robustness conferred by apparently redundant transcriptional enhancers. *Nature.*. 2010; 466(7305):490–493. https://doi.org/10.1038/nature09158.
63. LL A, W I, DP S, K F, D J, HR E. Development of five digits is controlled by a bipartite long-range cis-regulator. *Development*. 2014:1715–1725. https://doi.org/10.1242/dev.095430.
64. Furlong EEM, Levine M. Developmental enhancers and chromosome topology. *Science*. 2018;361(6409):1341–1345. https://doi.org/10.1126/science.aau0320.
65. Ghavi-Helm Y, Klein FA, Pakozdi T, et al. Enhancer loops appear stable during development and are associated with paused polymerase. *Nature*. 2014;512(1):96–100. https://doi.org/10.1038/nature13417.
66. Cremer T, Cremer M. Chromosome territories. *Cold Spring Harbor Perspect Biol*. 2010;2.
67. Szabo Q, Bantignies F, Cavalli G. Principles of genome folding into topologically associating domains. *Sci Adv*. 2019;5(4). https://doi.org/10.1126/sciadv.aaw1668.
68. Van Bortle K, Nichols MH, Li L, et al. Insulator function and topological domain border strength scale with architectural protein occupancy. *Genome Biol*. 2014;15(6):R82. https://doi.org/10.1186/gb-2014-15-5-r82.
69. Dixon JR, Selvaraj S, Yue F, et al. Topological domains in mammalian genomes identified by analysis of chromatin interactions. *Nature*. 2012;485(7398):376–380. https://doi.org/10.1038/nature11082.
70. Yang J, Corces VG. Chromatin insulators: a role in nuclear organization and gene expression. In: *Advances in Cancer Research*. 110. United States: Academic Press Inc.; 2011:43–76. https://doi.org/10.1016/B978-0-12-386469-7.00003-7.
71. Rao SSP, Huntley MH, Durand NC, et al. A 3D map of the human genome at kilobase resolution reveals principles of chromatin looping. *Cell*. 2014;159(7):1665–1680. https://doi.org/10.1016/j.cell.2014.11.021.
72. Larson AG, Elnatan D, Keenen MM, et al. Liquid droplet formation by HP1α suggests a role for phase separation in heterochromatin. *Nature*. 2017;547(7662):236–240. https://doi.org/10.1038/nature22822.
73. Sanulli S, Trnka MJ, Dharmarajan V, et al. HP1 reshapes nucleosome core to promote phase separation of heterochromatin. *Nature*. 2019;575(7782):390–394. https://doi.org/10.1038/s41586-019-1669-2.

74. Strom AR, Emelyanov AV, Mir M, Fyodorov DV, Darzacq X, Karpen GH. Phase separation drives heterochromatin domain formation. *Nature*. 2017;547(7662):241–245. https://doi.org/10.1038/nature22989.
75. Wang L, Gao Y, Zheng X, et al. Histone modifications regulate chromatin compartmentalization by contributing to a phase separation mechanism. *Mol Cell*. 2019;76(4). https://doi.org/10.1016/j.molcel.2019.08.019, 646-659.e6.
76. Plys AJ, Davis CP, Kim J, et al. Phase separation of polycomb-repressive complex 1 is governed by a charged disordered region of CBX2. *Gene Dev*. 2019;33(13–14): 799–813. https://doi.org/10.1101/gad.326488.119.
77. Guo Y, Manteiga J, Henninger J, et al. Pol II phosphorylation regulates a switch between transcriptional and splicing condensates. *Nature*. 2019;572(7770). https://doi.org/10.1038/s41586-019-1464-0. Epub.
78. Narlikar GJ. Phase-separation in chromatin organization. *J Biosci*. 2020;45(1). https://doi.org/10.1007/s12038-019-9978-z.
79. Gibson BA, Doolittle LK, Schneider MWG, et al. Organization of chromatin by intrinsic and regulated phase separation. *Cell*. 2019;179(2). https://doi.org/10.1016/j.cell.2019.08.037, 470-484.e21.
80. Hildebrand EM, Dekker J. Mechanisms and functions of chromosome compartmentalization. *Trends Biochem Sci*. 2020;45(5):385–396. https://doi.org/10.1016/j.tibs.2020.01.002.
81. Noah D, Yi-Fei H, Adam S, Sergei KP. Phylogenetic modeling of regulatory element turnover based on epigenomic data. *Mol Biol Evol*. 2020:2137–2152. https://doi.org/10.1093/molbev/msaa073.
82. Kunarso G, Chia NY, Jeyakani J, et al. Transposable elements have rewired the core regulatory network of human embryonic stem cells. *Nat Genet*. 2010;42(7):631–634. https://doi.org/10.1038/ng.600.
83. Miguel B, Edward C. Crossroads between transposons and gene regulation. *Phil Trans R Soc*. 2020.
84. Dao LTM, Galindo-Albarrán AO, Castro-Mondragon JA, et al. Genome-wide characterization of mammalian promoters with distal enhancer functions. *Nat Genet*. 2017;49(7): 1073–1081. https://doi.org/10.1038/ng.3884.
85. Hahn S, Buratowski S, Sharp PA, Guarente L. Yeast TATA-binding protein TFIID binds to TATA elements with both consensus and nonconsensus DNA sequences. *Proc Natl Acad Sci USA*. 1989;86(15):5718–5722. https://doi.org/10.1073/pnas.86.15.5718.
86. Chalkley GE, Verrijzer CP. DNA binding site selection by RNA polymerase II TAFs: a TAF(II)250-TAF(II)150 complex recognizes the initiator. *EMBO J*. 1999;18(17): 4835–4845. https://doi.org/10.1093/emboj/18.17.4835.
87. Smale ST, Kadonaga JT. The RNA polymerase II core promoter. *Annu Rev Biochem*. 2003;72:449–479. https://doi.org/10.1146/annurev.biochem.72.121801.161520.
88. Gilchrist DA, Dos Santos G, Fargo DC, et al. Pausing of RNA polymerase II disrupts DNA-specified nucleosome organization to enable precise gene regulation. *Cell*. 2010;143(4):540–551. https://doi.org/10.1016/j.cell.2010.10.004.
89. Jin C, Zang C, Wei G, et al. H3.3/H2A.Z double variant-containing nucleosomes mark "nucleosome-free regions" of active promoters and other regulatory regions. *Nat Genet*. 2009;41(8):941–945. https://doi.org/10.1038/ng.409.
90. Calo WJ. Transcriptional enhancers: from properties to genome-wide predictions. *Nat Rev Genet*. 2013;49(5). https://doi.org/10.1016/j.molcel.2013.01.038. PMID: 23473601; PMCID: PMC3857148. Shlyueva D.

91. Li Y, Chen Cy, Kaye AM, Wasserman WW. The identification of cis-regulatory elements: a review from a machine learning perspective. *Biosystems*. 2015;138:6–17. https://doi.org/10.1016/j.biosystems.2015.10.002.
92. Zinzen RP, Girardot C, Gagneur J, Braun M, Furlong EEM. Combinatorial binding predicts spatio-temporal cis-regulatory activity. *Nature*. 2009;462(7269):65–70. https://doi.org/10.1038/nature08531.
93. Mortazavi A, Williams BA, McCue K, Schaeffer L, Wold B. Mapping and quantifying mammalian transcriptomes by RNA-Seq. *Nat Methods*. 2008;5(7):621–628. https://doi.org/10.1038/nmeth.1226.
94. Johnson DS, Mortazavi A, Myers RM, Wold B. Genome-wide mapping of in vivo protein-DNA interactions. *Science*. 2007;316(5830):1497–1502. https://doi.org/10.1126/science.1141319.
95. Cloonan N, Forrest ARR, Kolle G, et al. Stem cell transcriptome profiling via massive-scale mRNA sequencing. *Nat Methods*. 2008;5(7):613–619. https://doi.org/10.1038/nmeth.1223.
96. Ouyang Z, Zhou Q, Wong WH. ChIP-Seq of transcription factors predicts absolute and differential gene expression in embryonic stem cells. *Proc Natl Acad Sci USA*. 2009; 106(51):21521–21526. https://doi.org/10.1073/pnas.0904863106.
97. Freire-Pritchett P, Schoenfelder S, Várnai C, et al. Global reorganisation of cis-regulatory units upon lineage commitment of human embryonic stem cells. *Elife*. 2017;6. https://doi.org/10.7554/eLife.21926.
98. Dekker J, Marti-Renom MA, Mirny LA. Exploring the three-dimensional organization of genomes: interpreting chromatin interaction data. *Nat Rev Genet*. 2013;14(6): 390–403. https://doi.org/10.1038/nrg3454.
99. Karaiskos N, Wahle P, Alles J, et al. The *Drosophila* embryo at single-cell transcriptome resolution. *Science*. 2017;358(6360):194–199. https://doi.org/10.1126/science.aan3235.
100. Kelsey G, Stegle O, Reik W. Single-cell epigenomics: recording the past and predicting the future. *Science*. 2017;358(6359):69–75. https://doi.org/10.1126/science.aan6826.
101. van Steensel B., Henikoff S. Identification of in vivo DNA targets of chromatin proteins using tethered dam methyltransferase. Nat Biotechnol. 2000;18(4):424–428. https://doi.org/10.1038/74487. PMID: 10748524.
102. Mikkelsen T.S., Ku M., Jaffe D.B., Issac B., Lieberman E., Giannoukos G., Alvarez P., Brockman W., Kim T.K., Koche R.P., Lee W., Mendenhall E., O'Donovan A., Presser A., Russ C., Xie X., Meissner A., Wernig M., Jaenisch R., Nusbaum C., Lander E.S., Bernstein B.E. Genome-wide maps of chromatin state in pluripotent and lineage-committed cells. Nature. 2007;448(7153):553–560. https://doi.org/10.1038/nature06008. Epub 2007 Jul 1. PMID: 17603471.
103. Johnson D.S., Mortazavi A., Myers R.M., Wold B. Genome-wide mapping of in vivo protein-DNA interactions. Science. 2007;316(5830):1497–1502. https://doi.org/10.1126/science.1141319. Epub 2007 May 31. PMID: 17540862.
104. Barski A, Cuddapah S, Cui K, Roh TY, Schones DE, Wang Z, Wei G, Chepelev I, Zhao K. High-resolution profiling of histone methylations in the human genome. Cell. 2007; 129(4):823–837. https://doi.org/10.1016/j.cell.2007.05.009. PMID: 17512414.
105. Rhee H.S., Pugh B.F. Comprehensive genome-wide protein-DNA interactions detected at single-nucleotide resolution. Cell. 2011;147(6):1408–1419. https://doi.org/10.1016/j.cell.2011.11.013. PMID: 22153082; PMCID: PMC3243364.

106. Rhee H.S., Pugh B.F. Genome-wide structure and organization of eukaryotic pre-initiation complexes. Nature. 2012;483(7389):295–301. https://doi.org/10.1038/nature10799. Erratum in: Nature. 2012 Jul 5;487(7405):128. PMID: 22258509; PMCID: PMC3306527.
107. Skene PJ, Henikoff S. An efficient targeted nuclease strategy for high-resolution mapping of DNA binding sites. Elife. 2017;6:e21856. https://doi.org/10.7554/eLife.21856. PMID: 28079019; PMCID: PMC5310842
108. Chu C, Qu K, Zhong FL, Artandi SE, Chang HY. Genomic maps of long noncoding RNA occupancy reveal principles of RNA-chromatin interactions. Mol cell. 2011; 44(4):667–678. https://doi.org/10.1016/j.molcel.2011.08.027. Epub 2011 Sep 29. PMID: 21963238; PMCID: PMC3249421.
109. Crawford G.E., Davis S., Scacheri P.C., Renaud G., Halawi M.J., Erdos M.R., Green R., Meltzer P.S., Wolfsberg T.G,. Collins F.S. DNase-chip: a high-resolution method to identify DNase I hypersensitive sites using tiled microarrays. Nat methods. 2006; 3(7):503–9. https://doi.org/10.1038/nmeth888. PMID: 16791207; PMCID: PMC2698431
110. Boyle A.P., Davis S., Shulha H.P., Meltzer P., Margulies E.H., Weng Z., Furey T.S., Crawford G.E. High-resolution mapping and characterization of open chromatin across the genome. Cell. 2008;132(2):311–322. https://doi.org/10.1016/j.cell.2007.12.014. PMID: 18243105; PMCID: PMC2669738.
111. Giresi P.G., Kim J., McDaniell R.M., Iyer V.R., Lieb J.D. FAIREs (Formaldehyde-Assisted isolation of regulatory elements) isolates active regulatory elements from human chromatin. Genome Res. 2007;17(6):877–885. https://doi.org/10.1101/gr.5533506. Epub 2006 Dec 19. PMID: 17179217; PMCID: PMC1891346
112. Schones D.E., Cui K., Cuddapah S., Roh T.Y., Barski A., Wang Z., Wei G., Zhao K. Dynamic regulation of nucleosome positioning in the human genome. Cell. 2008; 132(5):887–898. https://doi.org/10.1016/j.cell.2008.02.022. PMID: 18329373.
113. Cui K, Zhao K. Genome-wide approaches to determining nucleosome occupancy in metazoans using MNase-Seq. Methods Mol Biol. 2012;833:413–419. https://doi.org/10.1007/978-1-61779-477-3_24. PMID: 22183607; PMCID: PMC3541821.
114. Buenrostro J.D., Giresi P.G., Zaba L.C., Chang H.Y., Greenleaf W.J. Transposition of native chromatin for fast and sensitive epigenomic profiling of open chromatin, DNA-binding proteins and nucleosome position. Nat methods 2013;10(12): 1213–1218. https://doi.org/10.1038/nmeth.2688. Epub 2013 Oct 6. PMID: 24097267; PMCID: PMC3959825.
115. Dekker J, Rippe K, Dekker M, Kleckner N. Capturing chromosome conformation. Science. 200215;295(5558):1306–1311. https://doi.org/10.1126/science.1067799. PMID: 11847345.
116. Simonis M., Klous P., Splinter E., Moshkin Y., Willemsen R., de Wit E., van Steensel B., de Laat W. Nuclear organization of active and inactive chromatin domains uncovered by chromosome conformation capture-on-chip (4C). Nat genet. 2006;38(11): 1348–1354. https://doi.org/10.1038/ng1896. Epub 2006 Oct 8. PMID: 17033623
117. Dostie J., Richmond T.A., Arnaout R.A., Selzer R.R., Lee W.L., Honan T.A., Rubio E.D., Krumm A., Lamb J., Nusbaum C., Green R.D., Dekker J. Chromosome conformation capture Carbon copy (5C): a Massively parallel solution for mapping interactions between genomic elements. Genome Res. 2006;16(10):1299–309. https://doi.org/10.1101/gr.5571506. Epub 2006 Sep 5. PMID: 16954542; PMCID: PMC1581439.
118. Lieberman-Aiden E, van Berkum NL, Williams L, Imakaev M, Ragoczy T, Telling A, Amit I, Lajoie BR, Sabo PJ, Dorschner MO, Sandstrom R, Bernstein B, Bender MA, Groudine M, Gnirke A, Stamatoyannopoulos J, Mirny LA, Lander ES, Dekker J.

Comprehensive mapping of long-range interactions reveals folding principles of the human genome. Science. 2009;326(5950):289–293. https://doi.org/10.1126/science.1181369. PMID: 19815776; PMCID: PMC2858594

119. Horike S, Cai S, Miyano M, Cheng JF, Kohwi-Shigematsu T. Loss of silent-chromatin looping and impaired imprinting of DLX5 in Rett syndrome. Nat genet. 2005;37(1):31–40. https://doi.org/10.1038/ng1491. Epub 2004 Dec 19. PMID: 15608638
120. Fullwood M.J., Liu M.H., Pan Y.F., Liu J, Xu H, Mohamed Y.B., Orlov Y.L., Velkov S, Ho A, Mei P.H., Chew E.G., Huang P.Y,. Welboren W.J., han Y, Ooi H.S., Ariyaratne PN, Vega VB, Luo Y, Tan P.Y., Choy P.Y., Wansa K.D., Zhao B., Lim K.S., Leow S.C., Yow J.S., Joseph R, Li H, Desai KV, Thomsen J.S., Lee Y.K., Karuturi R.K., Herve T, Bourque G, Stunnenberg H.G., Ruan X., Cacheux-Rataboul V., Sung W.K., Liu E.T., Wei C.L., Cheung E., Ruan Y. An oestrogen-receptor-alpha-bound human chromatin interactome. Nature. 2009;462(7269):58–64. https://doi.org/10.1038/nature08497. PMID: 19890323; PMCID: PMC2774924
121. Shiraki T., Kondo S., Katayama S., Waki K., Kasukawa T., Kawaji H., Kodzius R., Watahiki A., Nakamura M., Arakawa T., Fukuda S., Sasaki D., Podhajska A., Harbers M., Kawai J., Carninci P., Hayashizaki Y. Cap analysis gene expression for high-throughput analysis of transcriptional starting point and identification of promoter usage. Proc Natl Acad Sci U S A. 2003;100(26):15,776–781. https://doi.org/10.1073/pnas.2136655100. Epub 2003 Dec 8. PMID: 14663149; PMCID: PMC307644.
122. Core L.J., Waterfall J.J., Lis J.T. Nascent RNA sequencing reveals widespread pausing and divergent initiation at human promoters. Science. 2008;322(5909):1845–1848. https://doi.org/10.1126/science.1162228. Epub 2008 Dec 4. PMID: 19056941; PMCID: PMC2833333.; Gardini A. Global Run-on sequencing (GRO-seq). Methods Mol Biol. 2017;1468:111–1120. https://doi.org/10.1007/978-1-4939-4035-6_9. PMID: 27662873; PMCID: PMC5880289.
123. Kwak H., Fuda N.J., Core L.J., Lis J.T. Precise maps of RNA polymerase reveal how promoters direct initiation and pausing. Science. 2013;339(6122):950–953. https://doi.org/10.1126/science.1229386. PMID: 23430654; PMCID: PMC3974810
124. Kruesi W.S., Core L.J., Waters C.T., Lis J.T., Meyer B.J. Condensin controls recruitment of RNA polymerase II to achieve nematode X-chromosome dosage compensation. Elife. 2013;2:e00808. https://doi.org/10.7554/eLife.00808. PMID: 23795297; PMCID: PMC3687364
125. Arnold C.D., Gerlach D., Stelzer C., Boryń Ł.M., Rath M., Stark A. Genome-wide quantitative enhancer activity maps identified by STARR-seq. Science. 2013; 339(6123):1074–1077. https://doi.org/10.1126/science.1232542. Epub 2013 Jan 17. PMID: 23328393
126. Weber M., Davies J.J., Wittig D., Oakeley E.J., Haase M., Lam W.L., Schübeler D. Chromosome-wide and promoter-specific analyses identify sites of differential DNA methylation in normal and transformed human cells. Nat genet. 2005;37(8):853–862. https://doi.org/10.1038/ng1598. Epub 2005 Jul 10. PMID: 16007088
127. Frommer M., McDonald L.E., Millar D.S., Collis C.M., Watt F., Grigg G.W., Molloy P.L., Paul C.L. A genomic sequencing protocol that yields a positive display of 5-methylcytosine residues in individual DNA strands. Proc Natl Acad Sci U S A. 1992;89(5):1827–1831. https://doi.org/10.1073/pnas.89.5.1827. PMID: 1,542,678; PMCID: PMC48546
128. Cokus S.J., Feng S., Zhang X., Chen Z., Merriman B., Haudenschild C.D., Pradhan S., Nelson S.F., Pellegrini M., Jacobsen S.E. Shotgun bisulphite sequencing of the

Arabidopsis genome reveals DNA methylation patterning. Nature. 2008;452(7184): 215–219. https://doi.org/10.1038/nature06745. Epub 2008 Feb 17. PMID: 18278030; PMCID: PMC2377394

129. Laurent L., Wong E., Li G., Huynh T., Tsirigos A., Ong C.T., Low H.M., Kin Sung K.W., Rigoutsos I., Loring J, Wei C.L. Dynamic changes in the human methylome during differentiation. Genome Res. 2010;20(3):320–331. https://doi.org/10.1101/gr.101907.109. Epub 2010 Feb 4. PMID: 20133333; PMCID: PMC2840979
130. Meissner A., Gnirke A., Bell G.W., Ramsahoye B., Lander E.S., Jaenisch R. Reduced representation bisulfite sequencing for comparative high-resolution DNA methylation analysis. Nucleic Acids Res. 2005;33(18):5868–5877. https://doi.org/10.1093/nar/gki901. PMID: 16224102; PMCID: PMC1258174
131. White MA. Understanding how cis-regulatory function is encoded in DNA sequence using massively parallel reporter assays and designed sequences. *Genomics*. 2015; 106(3):165–170. https://doi.org/10.1016/j.ygeno.2015.06.003.
132. Vijayabaskar MS, Goode DK, Obier N, et al. Identification of gene specific cis-regulatory elements during differentiation of mouse embryonic stem cells: an integrative approach using high-throughput datasets. *PLoS Comput Biol*. 2019;15(11). https://doi.org/10.1371/journal.pcbi.1007337.
133. Wilson NK, Foster SD, Wang X, et al. Combinatorial transcriptional control in blood stem/progenitor cells: genome-wide analysis of ten major transcriptional regulators. *Cell Stem Cell*. 2010;7(4):532–544. https://doi.org/10.1016/j.stem.2010.07.016.
134. Bussemaker HJ, Li H, Siggia ED. Regulatory element detection using correlation with expression. *Nat Genet*. 2001;27(2):167–171. https://doi.org/10.1038/84792.
135. Keleş S, Van Der Laan M, Eisen MB. Identification of regulatory elements using a feature selection method. *Bioinformatics*. 2002;18(9):1167–1175. https://doi.org/10.1093/bioinformatics/18.9.1167.
136. Gao FG, Foat BCF, Bussemaker HJ. Defining transcriptional networks through integrative modeling of mRNA expression and transcription factor binding data. *BMC Bioinf*. 2004;5. https://doi.org/10.1186/1471-2105-5-31.
137. Conlon EM, Liu XS, Lieb JD, Liu JS. Integrating regulatory motif discovery and genome-wide expression analysis. *Proc Natl Acad Sci USA*. 2003;100(6): 3339–3344. https://doi.org/10.1073/pnas.0630591100.
138. ENCODE Project Consortium. *Science*. 2004;306(5696):636–640.
139. Romanoski CE, Glass CK, Stunnenberg HG, Wilson L, Almouzni G. Epigenomics: Roadmap for regulation. *Nature*. 2015;518(7539):314–316. https://doi.org/10.1038/518314a.
140. Lizio M, Harshbarger J, Shimoji H, et al. Gateways to the FANTOM5 promoter level mammalian expression atlas. *Genome Biol*. 2015;16(1). https://doi.org/10.1186/s13059-014-0560-6.
141. MJ E, PM J, PH E, et al. Expanded encyclopaedias of DNA elements in the human and mouse genomes. *Nature*. 2020:699–710. https://doi.org/10.1038/s41586-020-2493-4.
142. Reik W, Azim Surani M. Germline and pluripotent stem cells. *Cold Spring Harbor Perspect Biol*. 2015;7(11). https://doi.org/10.1101/cshperspect.a019422.
143. Schulz KN, Harrison MM. Mechanisms regulating zygotic genome activation. *Nat Rev Genet*. 2019;20(4):221–234. https://doi.org/10.1038/s41576-018-0087-x.
144. Gaiti F, Jindrich K, Fernandez-Valverde SL, Roper KE, Degnan BM, Tanurdžić M. Landscape of histone modifications in a sponge reveals the origin of animal cis-regulatory complexity. *Elife*. 2017;6. https://doi.org/10.7554/eLife.22194.

145. Vavouri T, Walter K, Gilks WR, Lehner B, Elgar G. Parallel evolution of conserved non-coding elements that target a common set of developmental regulatory genes from worms to humans. *Genome Biol.* 2007;8(2). https://doi.org/10.1186/gb-2007-8-2-r15.
146. Beaster-Jones L. Cis-regulation and conserved non-coding elements in amphioxus. *Brief Funct Genom.* 2012;11(2):118–130. https://doi.org/10.1093/bfgp/els006.
147. Driever W, Thoma G, Nüsslein-Volhard C. Determination of spatial domains of zygotic gene expression in the *Drosophila* embryo by the affinity of binding sites for the bicoid morphogen. *Nature.* 1989;340(6232):363–367. https://doi.org/10.1038/340363a0.
148. Ke Y, Xu Y, Chen X, et al. 3D chromatin structures of mature gametes and structural reprogramming during mammalian embryogenesis. *Cell.* 2017;170(2). https://doi.org/10.1016/j.cell.2017.06.029, 367-381.e20.
149. Evans MJ, Kaufman MH. Establishment in culture of pluripotential cells from mouse embryos. *Nature.* 1981;292(5819):154–156. https://doi.org/10.1038/292154a0.
150. Thomson JA. Embryonic stem cell lines derived from human blastocysts. *Science.* 1998;282(5391):1145–1147. https://doi.org/10.1126/science.282.5391.1145.
151. Weinberger L, Ayyash M, Novershtern N, Hanna JH. Dynamic stem cell states: Naive to primed pluripotency in rodents and humans. *Nat Rev Mol Cell Biol.* 2016;17(3): 155–169. https://doi.org/10.1038/nrm.2015.28.
152. Takahashi K, Yamanaka S. Induction of pluripotent stem cells from mouse embryonic and adult fibroblast cultures by defined factors. *Cell.* 2006;126(4):663–676. https://doi.org/10.1016/j.cell.2006.07.024.
153. Takahashi K, Tanabe K, Ohnuki M, et al. Induction of pluripotent stem cells from adult human fibroblasts by defined factors. *Cell.* 2007;131(5):861–872. https://doi.org/10.1016/j.cell.2007.11.019.
154. Buganim Y, Markoulaki S, Van Wietmarschen N, et al. The developmental potential of iPSCs is greatly influenced by reprogramming factor selection. *Cell Stem Cell.* 2014; 15(3):295–309. https://doi.org/10.1016/j.stem.2014.07.003.
155. Hackett JA, Azim Surani M. Regulatory principles of pluripotency: from the ground state up. *Cell Stem Cell.* 2014;15(4):416–430. https://doi.org/10.1016/j.stem.2014.09.015.
156. Yagi M, Yamanaka S, Yamada Y. Epigenetic foundations of pluripotent stem cells that recapitulate in vivo pluripotency. *Lab Invest.* 2017;97(10):1133–1141. https://doi.org/10.1038/labinvest.2017.87.
157. Bird A. DNA methylation patterns and epigenetic memory. *Gene Dev.* 2002;16(1): 6–21. https://doi.org/10.1101/gad.947102.
158. Li E, Bestor TH, Jaenisch R. Targeted mutation of the DNA methyltransferase gene results in embryonic lethality. *Cell.* 1992;69(6):915–926. https://doi.org/10.1016/0092-8674(92)90611-F.
159. Okano M, Bell DW, Haber DA, Li E. DNA methyltransferases Dnmt3a and Dnmt3b are essential for de novo methylation and mammalian development. *Cell.* 1999;99(3): 247–257. https://doi.org/10.1016/S0092-8674(00)81656-6.
160. Sen GL, Reuter JA, Webster DE, Zhu L, Khavari PA. DNMT1 maintains progenitor function in self-renewing somatic tissue. *Nature.* 2010;463(7280):563–567. https://doi.org/10.1038/nature08683.
161. Tsumura A, Hayakawa T, Kumaki Y, et al. Maintenance of self-renewal ability of mouse embryonic stem cells in the absence of DNA methyltransferases Dnmt1, Dnmt3a and Dnmt3b. *Gene Cell.* 2006;11(7):805–814. https://doi.org/10.1111/j.1365-2443.2006.00984.x.

162. Murry CE, Keller G. Differentiation of embryonic stem cells to clinically relevant populations: lessons from embryonic development. *Cell.* 2008;132(4):661–680. https://doi.org/10.1016/j.cell.2008.02.008.
163. Tang WWC, Dietmann S, Irie N, et al. A unique gene regulatory network resets the human germline epigenome for development. *Cell.* 2015;161(6):1453–1467. https://doi.org/10.1016/j.cell.2015.04.053.
164. Koh KP, Yabuuchi A, Rao S, et al. Tet1 and Tet2 regulate 5-hydroxymethylcytosine production and cell lineage specification in mouse embryonic stem cells. *Cell Stem Cell.* 2011;8(2):200–213. https://doi.org/10.1016/j.stem.2011.01.008.
165. Gu T, Lin X, Cullen SM, et al. DNMT3A and TET1 cooperate to regulate promoter epigenetic landscapes in mouse embryonic stem cells. *Genome Biol.* 2018;19(1). https://doi.org/10.1186/s13059-018-1464-7.
166. Xie W, Schultz MD, Lister R, et al. Epigenomic analysis of multilineage differentiation of human embryonic stem cells. *Cell.* 2013;153(5):1134–1148. https://doi.org/10.1016/j.cell.2013.04.022.
167. Vastenhouw NL, Schier AF. Bivalent histone modifications in early embryogenesis. *Curr Opin Cell Biol.* 2012;24(3):374–386. https://doi.org/10.1016/j.ceb.2012.03.009.
168. Hawkins RD, Hon GC, Yang C, et al. Dynamic chromatin states in human ES cells reveal potential regulatory sequences and genes involved in pluripotency. *Cell Res.* 2011;21(10):1393–1409. https://doi.org/10.1038/cr.2011.146.
169. Alvaro R-I, Ruchi B, Tomek S, BS A, FR A, Joanna W. A unique chromatin signature uncovers early developmental enhancers in humans. *Nature.* 2011:279–283. https://doi.org/10.1038/nature09692.
170. CM P, CA W, WG G, et al. Histone H3K27ac separates active from poised enhancers and predicts developmental state. *Proc Natl Acad Sci USA.* 2010:21931–21936. https://doi.org/10.1073/pnas.1016071107.
171. Zhu J, Adli M, Zou JY, et al. Genome-wide chromatin state transitions associated with developmental and environmental cues. *Cell.* 2013;152(3):642–654. https://doi.org/10.1016/j.cell.2012.12.033.
172. Hawkins RD, Hon GC, Lee LK, et al. Distinct epigenomic landscapes of pluripotent and lineage-committed human cells. *Cell Stem Cell.* 2010;6(5):479–491. https://doi.org/10.1016/j.stem.2010.03.018.
173. Ermolaeva M, Neri F, Ori A, Rudolph KL. Cellular and epigenetic drivers of stem cell ageing. *Nat Rev Mol Cell Biol.* 2018;19(9):594–610. https://doi.org/10.1038/s41580-018-0020-3.
174. Boyer LA, Tong IL, Cole MF, et al. Core transcriptional regulatory circuitry in human embryonic stem cells. *Cell.* 2005;122(6):947–956. https://doi.org/10.1016/j.cell.2005.08.020.
175. Buenrostro JD, Corces MR, Lareau CA, et al. Integrated single-cell analysis maps the continuous regulatory landscape of human hematopoietic differentiation. *Cell.* 2018; 173(6). https://doi.org/10.1016/j.cell.2018.03.074, 1535-1548.e16.
176. Vodyanik MA, Yu J, Zhang X, et al. A mesoderm-derived precursor for mesenchymal stem and endothelial cells. *Cell Stem Cell.* 2010;7(6):718–729. https://doi.org/10.1016/j.stem.2010.11.011.
177. Murphy MB, Moncivais K, Caplan AI. Mesenchymal stem cells: environmentally responsive therapeutics for regenerative medicine. *Exp Mol Med.* 2013;45(11). https://doi.org/10.1038/emm.2013.94.

178. Bradley EW, Carpio LR, Van Wijnen AJ, McGee-Lawrence ME, Westendorf JJ. Histone deacetylases in bone development and skeletal disorders. *Physiol Rev.* 2015; 95(4):1359–1381. https://doi.org/10.1152/physrev.00004.2015.
179. Carrió E, Díez-Villanueva A, Lois S, et al. Deconstruction of DNA methylation patterns during myogenesis reveals specific epigenetic events in the establishment of the skeletal muscle lineage. *Stem Cell.* 2015;33(6):2025–2036. https://doi.org/10.1002/stem.1998.
180. Cui H, Bansal V, Grunert M, et al. Muscle-relevant genes marked by stable H3K4me2/3 profiles and enriched MyoD binding during myogenic differentiation. *PLoS One.* 2017; 12(6). https://doi.org/10.1371/journal.pone.0179464.
181. Puri PL, Iezzi S, Stiegler P, et al. Class I histone deacetylases sequentially interact with MyoD and pRb during skeletal myogenesis. *Mol Cell.* 2001;8(4):885–897. https://doi.org/10.1016/S1097-2765(01)00373-2.
182. Mortada I, Mortada R. Epigenetic changes in mesenchymal stem cells differentiation. *Eur J Med Genet.* 2018;61(2):114–118. https://doi.org/10.1016/j.ejmg.2017.10.015.
183. Li H, Cui D, Wu S, et al. Epigenetic regulation of gene expression in epithelial stem cells fate. *Curr Stem Cell Res Ther.* 2018;13(1):46–51. https://doi.org/10.2174/1574888X12666171115123241.
184. Shen Q, Jin H, Wang X. Epidermal stem cells and their epigenetic regulation. *Int J Mol Sci.* 2013;14(9):17861–17880. https://doi.org/10.3390/ijms140917861.
185. Breuls N, Giacomazzi G, Sampaolesi M. Epigenetic modifications in myogenic stem cells: from novel insights to therapeutic perspectives. *Cells.* 2019;8.
186. Pasque V, Plath K. X chromosome reactivation in reprogramming and in development. *Curr Opin Cell Biol.* 2015;37:75–83. https://doi.org/10.1016/j.ceb.2015.10.006.
187. Hackett JA, Sengupta R, Zylicz JJ, et al. Germline DNA demethylation dynamics and imprint erasure through 5-hydroxymethylcytosine. *Science.* 2013;339(6118):448–452. https://doi.org/10.1126/science.1229277.
188. Sailasree SP, Srivastava S, Mishra RK. The placental gateway of maternal transgenerational epigenetic inheritance. *J Genet.* 2017;96(3):465–482. https://doi.org/10.1007/s12041-017-0788-5.
189. Grice EA, Segre JA. The human microbiome: our second genome. *Annu Rev Genom Hum Genet.* 2012;13:151–170. https://doi.org/10.1146/annurev-genom-090711-163814.
190. Celluzzi A, Masotti A. How our other genome controls our epi-genome. *Trends Microbiol.* 2016;24(10):777–787. https://doi.org/10.1016/j.tim.2016.05.005.
191. Indrio F, Martini S, Francavilla R, et al. Epigenetic matters: the link between early nutrition, microbiome, and long-term health development. *Front Pediatr.* 2017;5. https://doi.org/10.3389/fped.2017.00178.
192. Abdulghani M, Jain A, Tuteja G. Genome-wide identification of enhancer elements in the placenta. *Placenta.* 2019;79:72–77. https://doi.org/10.1016/j.placenta.2018.09.003.
193. Gluckman PD, Hanson MA, Spencer HG. Predictive adaptive responses and human evolution. *Trends Ecol Evol.* 2005;20(10):527–533. https://doi.org/10.1016/j.tree.2005.08.001.
194. Martin-Gronert MS, Ozanne SE. Mechanisms underlying the developmental origins of disease. *Rev Endocr Metab Disord.* 2012;13(2):85–92. https://doi.org/10.1007/s11154-012-9210-z.
195. Heijmans BT, Tobi EW, Stein AD, et al. Persistent epigenetic differences associated with prenatal exposure to famine in humans. *Proc Natl Acad Sci USA.* 2008;105(44): 17046–17049. https://doi.org/10.1073/pnas.0806560105.

196. Chao W, D'Amore PA. IGF2: epigenetic regulation and role in development and disease. *Cytokine Growth Factor Rev.* 2008;19(2):111–120. https://doi.org/10.1016/j.cytogfr.2008.01.005.
197. Tobi EW, Lumey LH, Talens RP, et al. DNA methylation differences after exposure to prenatal famine are common and timing- and sex-specific. *Hum Mol Genet.* 2009; 18(21):4046–4053. https://doi.org/10.1093/hmg/ddp353.
198. Lumey LH, Terry MB, Delgado-Cruzata L, et al. Adult global DNA methylation in relation to pre-natal nutrition. *Int J Epidemiol.* 2012;41(1):116–123. https://doi.org/10.1093/ije/dyr137.
199. Driscoll CD, Streissguth AP, Riley EP. Prenatal alcohol exposure: comparability of effects in humans and animal models. *Neurotoxicol Teratol.* 1990;12(3):231–237. https://doi.org/10.1016/0892-0362(90)90094-S.
200. Ramsay M. Genetic and epigenetic insights into fetal alcohol spectrum disorders. *Genome Med.* 2010;2(4). https://doi.org/10.1186/gm148.
201. Haycock PC, Ramsay M. Exposure of mouse embryos to ethanol during preimplantation development: effect on DNA methylation in the H19 imprinting control region. *Biol Reprod.* 2009;81(4):618–627. https://doi.org/10.1095/biolreprod.108.074682.
202. Liu Y, Balaraman Y, Wang G, Nephew KP, Zhou FC. Alcohol exposure alters DNA methylation profiles in mouse embryos at early neurulation. *Epigenetics.* 2009;4(7): 500–511. https://doi.org/10.4161/epi.4.7.9925.
203. Kaminen-Ahola N, Ahola A, Maga M, et al. Maternal ethanol consumption alters the epigenotype and the phenotype of offspring in a mouse model. *PLoS Genet.* 2010; 6(1). https://doi.org/10.1371/journal.pgen.1000811.
204. Ouko LA, Shantikumar K, Knezovich J, Haycock P, Schnugh DJ, Ramsay M. Effect of alcohol consumption on CpG methylation in the differentially methylated regions of H19 and IG-DMR in male gametes - implications for fetal alcohol spectrum disorders. *Alcohol Clin Exp Res.* 2009;33(9):1615–1627. https://doi.org/10.1111/j.1530-0277.2009.00993.x.
205. Chater-Diehl EJ, Laufer BI, Singh SM. Changes to histone modifications following prenatal alcohol exposure: an emerging picture. *Alcohol.* 2017;60:41–52. https://doi.org/10.1016/j.alcohol.2017.01.005.
206. Feinberg AP, Koldobskiy MA, Göndör A. Epigenetic modulators, modifiers and mediators in cancer aetiology and progression. *Nat Rev Genet.* 2016;17(5):284–299. https://doi.org/10.1038/nrg.2016.13.
207. Heijmans BT, Tobi EW, Lumey LH, Slagboom PE. The epigenome: archive of the prenatal environment. *Epigenetics.* 2009;4(8):526–531. https://doi.org/10.4161/epi.4.8.10265.
208. Petronis A. Epigenetics as a unifying principle in the aetiology of complex traits and diseases. *Nature.* 2010;465(7299):721–727. https://doi.org/10.1038/nature09230.

CHAPTER

4 Genomic imprinting and developmental physiology: intrauterine growth and postnatal period

Vanessa Yu[1,§], Zoe Stamoulis[1,§], Keran Chen[1,§], Jiachun Jiang[1,§], Zhengying He[1], Guy A. Rutter[1,2,3], Steven J. Millership[1]

[1]*Section of Cell Biology and Functional Genomics, Department of Metabolism, Digestion and Reproduction, Faculty of Medicine, Imperial College London, London, United Kingdom;* [2]*CHUM Research Center and University of Montréal, Montréal, QC, Canada;* [3]*Lee Kong Chian School of Medicine, Nanyang Technological University, Singapore, Singapore*

Genomic imprinting: discovery and early work

First described in the 1940s, the term epigenetics describes the underlying mechanisms of gene expression regulation above the level of the DNA sequence.[1–3] Though its definition has been revised over the decades, epigenetic mechanisms are recognized as phenomena controlling the reading, and processing of, genetic information and include DNA methylation, histone modifications (e.g. acetylation, phosphorylation and methylation), chromatin remodeling, and non-coding RNA-mediated gene regulation.[4–7] Differential epigenomic modifications between cell/tissue-specific systems and throughout development therefore provide dynamic spatiotemporal control of gene expression throughout the life span of the organism. Additionally, epigenetic modifications can be maintained throughout, or even across, multiple generations such as in the case of permanently silenced genomic regions, as seen with transposable elements.[4,8–11]

A major epigenetic modification in the mammalian genome is DNA methylation. This primarily occurs on the 5′-cytosine of cytosine–guanine dinucleotide-rich genomic regions to create 5-methylcytosine (5 mC) at so-called CpG "islands." This process is performed by DNA methyltransferases (DNMT's), which typically methylate cytosine residues of CpG's in a symmetrical manner.[12,13] Non-CpG methylation is also possible, though it is a rare occurrence in specialized mammalian cells.[14–17] The removal of 5 mC methylation is achieved either passively through methylation dilution in the absence of DNMT's during DNA replication or actively via oxidation of methyl groups to aid their enzymatic removal via TET (ten–eleven

§ These authors contributed equally.

Perinatal and Developmental Epigenetics. https://doi.org/10.1016/B978-0-12-821785-6.00011-6

translocation) proteins. CpG methylation at key genomic regulatory regions, e.g., promoters, alongside histone modifications, mediates silencing of either the paternal or maternal allele of a particular subset of genes within the genome of therian mammals (eutherians, i.e., placental mammals, marsupials, and monotremes) known as imprinted genes, via a process known as genomic imprinting.[18,19] This epigenetically driven process therefore alters gene expression dosage from a typically biallelic to monoallelic, and parent-of-origin, expression pattern. Many imprinted genes exist in genomic clusters alongside other imprinted, as well as non-imprinted genes, which can be regulated as a cluster via an imprinting control region (ICR) that acts as a differentially methylation region (DMR), whereby one parental allele undergoes significant methylation for allelic repression.[20–23] Furthermore, monoallelic expression of imprinted genes is maintained throughout embryonic development and adult life and also transgenerationally via resetting and reestablishing of epigenomic marks at imprinted *loci* in germ cells.[24,25]

Pronuclear transplantation into fertilized mouse oocytes was used to create uniparental disomy (UPD) embryos with either diploid gynogenetic or androgenetic genomes (i.e., mouse embryos with either two maternal copies or two paternal copies of the genome, respectively.[26–30]) Zygotes with UPD did not survive past early gestation, indicating that both the maternal and paternal genomes were individually required for embryonic development and, therefore, were functionally distinct. Although this phenomenon of genomic imprinting was initially reported as exclusive to therian organisms, studies have identified similar gene expression dosage changes in flowering plants and chickens, although with significantly fewer imprinted genes reported than in therians.[31–34] Currently, around 150 and 100 imprinted genes have been identified in mice and humans, respectively, making it a highly conserved phenomenon.[1,35]

Imprinted gene functions converge on neuronal and metabolic processes that govern early development: the cost of altered imprinted gene dosage

Although genomic imprinting is conserved across therian mammals, the functions of imprinted gene products vary significantly, ranging from cell cycle regulators to determinants of intracellular protein turnover.[36–38] Despite these functional differences, imprinted genes share common expression traits. For example, their expression often peaks during the prenatal and/or postnatal period of development, with generally much lower expression levels during adulthood. Moreover, the expression pattern of imprinted genes is typically found in neuroendocrine and metabolic tissues such as the brain, adipose tissue, and pancreatic beta cells (as well as in the placenta during pregnancy)—systems that govern resource allocation and are highly metabolically active.[39–42] Collectively, the temporospatial expression pattern of imprinted genes explain why they are critical for placental growth and maintenance, and consequently for fetal, neonatal, and postnatal growth and development.

As imprinted genes display monoallelic expression, this renders them more vulnerable to the effect of genetic mutations in comparison with biallelically expressed genes, which benefit from a reserve copy if one allele incurs a pathogenic mutation.[43] Chromosomal modifications (including duplications) and specific mutations at imprinted *loci* can therefore have severe consequences to early development, with many human imprinting disorders displaying classical developmental delay phenotypes, as observed in Silver–Russell, Prader–Willi, Angelman, and Beckwith–Wiedemann syndromes as well as a transient form of neonatal diabetes (TNDM).[43–47] Patients with these disorders often display metabolic abnormalities in adulthood, suggesting that classical "developmental" genes also have important pre/postnatal roles in adult metabolism—and are discussed in greater detail later in the chapter.

The presence of imprinted genes in mammals, their predominant expression during early development, and the severe consequences of loss of function or restoration of biallelic expression (termed "loss of imprinting") have spawned several theories aiming to explain the likely evolutionary drive for genomic imprinting. The parental conflict model suggests a disparate survival interest between the maternal and paternal genomes, where the maternal genome (and thus maternally expressed genes) prioritizes conserving the mother's resources.[48] This benefits all pups across all litters, enabling the mother (who is related to all her offspring) to provide her offspring with enough resources. In contrast, paternal genomes (and paternally expressed genes) are "opportunistic" in character, aiming to maximize resource extraction. This is explained by the fact that a father may not be related to all litters, meaning that this promotion of maternal resource extraction aims to benefit the father's offspring. This hypothesis is supported by the general observation that maternally expressed genes (MEG's) are implicated in resource conservation, while paternally expressed genes (PEG's) are implicated in resource extraction, although it is important to note that there are several exceptions to this trend[48–51] as previously discussed in Millership et al.[52] Conversely, a separate "coadaptation" model outlines that imprinted genes benefit both the mother and offspring to aid offspring survival. Here, imprinted genes simultaneously stimulate the offspring's own survival, while also impacting the mother which, in turn, fuels maternal care of the offspring.[53] However, similar to the parental conflict model, many imprinted genes also do not abide by their expected behavior as outlined by the coadaptation model. This has resulted in a growing number of imprinted genes being observed as exceptions to these rules of parental conflict or coadaptation, casting doubt on whether their monoallelic expression is only important for these phenomena, or whether they exist because of other, unexplored, evolutionary pressures.[54–56]

Despite conflicting views over the true evolutionary drive and biological requirement for genomic imprinting, the importance of imprinted genes during and beyond early development has become increasingly appreciated since their discovery. This has been driven, in part, by a growing understanding of the relationship between imprinted genes implicated in developmental processes and in adult metabolism. It should be noted, however, that the pre-, neo-, and postnatal roles of imprinted genes may differ significantly between organisms, particularly when considering differing parent–offspring relationships in eutherians compared with marsupials, as reviewed extensively in Stringer et al.[19] An example is the marked difference between the maternal investments required for offspring survival between rodents and marsupials. Here, the placental attachment period in marsupials is significantly shorter, meaning that marsupial mothers have greater postnatal lactation care demands than rodent mothers (who may have greater demands during the placental period). This is reflected by genomic imprinting in the mammary gland, whereby the same genes can show vastly different expression patterns in marsupials versus rodents. Here, we will discuss the current understanding of imprinted genes in the context of the mammalian life span, focusing on their pre-, neo-, and postnatal roles in placental and embryonic development, and how this affects adult metabolism.

Genomic imprinting-mediated control of fetal growth and placental resources

A central role for imprinted genes in appropriate embryonic development was first recognized by the developmental defects of uniparental conceptuses, with androgenetic transplants resulting in poor embryonic development, while gynogenetic conceptus lacked extraembryonic tissues. These findings suggested major effects of genomic imprinting in normal embryonic development and growth,[27–30,57] with subsequent studies revealing indirect roles of imprinted genes in fetal growth through their contributions to placental development (reviewed in Cleaton et al. and Peters[40,58]). Approaches to assess alterations to dosage of imprinted genes typically involve mutant mice mimicking loss of function (via gene knockout, point mutations, or ICR deletion) or restoration of biallelic expression (using BAC-mediated overexpression of the imprinted locus). Early work defining the contribution of how imprinted genes govern a broad fetal growth effect included characterizing the paternally expressed *Igf2* and maternally expressed *Igf2r* genes.[59–61] *Igf2* encodes a potent growth hormone that promotes fetal growth, where disruption to exon 4 within the *Igf2* gene was shown to cause fetal growth restriction in mice, with neonates born approximately 60% compared to wild-type littermates.[60,62] Conversely, overexpression of *Igf2* is associated with fetal growth enhancement.[63] Maternally expressed *Igf2r* negatively regulates *Igf2*, where it targets IGF2 protein for lysosomal degradation.[64] Mice carrying targeted inactivation of *Igf2r* therefore exhibit prenatal overgrowth with perinatal lethality,[65–68] while this growth phenotype is rescued in *Igf2* and *Igf2r* double knockout mutants, confirming that the fetal overgrowth is caused by increased circulating IGF2 levels.[69]

Another imprinted gene that plays an important role in fetal growth is the maternally expressed *Grb10* gene. Mice null for *Grb10* on the maternal allele exhibit embryonic overgrowth, with mice born 30% larger than their wild-type littermates.[70,71]

Ablation of the *Grb10* ICR causes biallelic-based *Grb10* overexpression and severe growth deficiency upon paternal transmission, indicative of a growth inhibitory role of *Grb10*.[72] Mutant mice have also demonstrated the contributions of paternally expressed genes *Plagl1/ZAC1* and *Dlk1* in fetal growth, where deletion of *Plagl1/ZAC1* leads to intrauterine growth retardation and partial neonatal mortality[73] and BAC-mediated overexpression of the TNDM locus resulting in reduced fetal endocrine/islet development.[74] Deletion or overexpression of *Dlk1* led to pre- and postnatal growth restriction or growth-enhanced embryos, respectively.[75,76] In addition to the apparent functions in balancing fetal growth, the participation of imprinted genes in the development of metabolically important organs has also been documented, including the liver, adipose tissue, kidney, and adrenal gland.[77–81] Abnormal brain development is also observed in mouse chimeras following the introduction of gynogenetic or androgenetic cells in early embryogenesis including non-uniform spatial contribution of uniparental cells across specific brain regions, indicative of key roles of genomic imprinting in neurodevelopment (PMID: 7479883).[82,83]

Aside from the direct effects in fetal growth, imprinted genes also regulate early development through their contributions to placental development (reviewed in Bressan et al.[79]). A significant number of imprinted genes are reported to be highly expressed in the human placenta[71,84] with the importance of their dosage in the placenta demonstrated by loss of function or loss of imprinting of several individual imprinted genes causing severe placental abnormalities and subsequent fetal growth defects or even lethality. At the interface between the maternal and developing fetus, the placenta functions to support fetal growth in utero. It is responsible for hormone production in pregnancy and nutrient and waste exchange between the mother and fetus. In general, the growth of the placenta is consistent with the prediction of the parental conflict theory, with growth restriction or enhancement of the placenta observed when expression of paternally or maternally expressed genes is disrupted, respectively.[85] The altered placental size can result in changes in placental efficiency, which, in turn, influence fetal growth.[85,86] This is demonstrated by the fact that deletion of the placental-specific *Igf2* P0 gene causes major placental and fetal growth perturbation[87] with fetal growth restriction via placental-specific knockout of *Igf2* being worsened upon global deletion of *Igf2*.[87,88] Similarly, targeted deletion of either the paternally expressed genes *Peg3* or *Rtl1* leads to reduced placental growth with low birth weight,[42,89–91] with other imprinted genes also being shown to modulate fetal growth via an effect in both placental and embryonic tissues, i.e., *Grb10* and *Dlk1*.[39,70,75,76,92]

Additionally, several imprinted genes are central to the normal development, rather than growth, of placental tissues. For instance, the maternally expressed *Cdkn1c* and *Phlda2* are crucial for correct formation of the spongiotrophoblast,[93–96] a major placental endocrine lineage that supports maternal–fetal interaction.[85] Mice deficient in *Ascl2* or *Peg10* lack spongiotrophoblast,[97,98] leading to early embryonic lethality as a consequence of the failure to establish a viable placenta.[85,97] The paternally expressed imprinted gene *Mest* is also essential for proper trophoblast angiogenesis in the placenta and in its absence, placental growth restriction ensues.[71] Furthermore, some imprinted genes have been implicated in modulating placental functions by regulating nutrient supply. For example, the capacity to transport monoamines via the extraneuronal monoamine transporter (uptake-2) system is impaired in *Slc22a3* knockout mice, which encodes a transplacental solute transporter.[40,99] Similarly, mice null for the imprinted gene *Gab1* show defective formation of the placental syncytiotrophoblast[100,101] disrupting the maternal–fetal nutrient exchange and resulting in prenatal growth restriction.[102]

Genomic imprinting and human imprinting disorders

In addition to the use of mutant mouse models, the importance of imprinted genes in intrauterine growth is also highlighted by the clinical syndromes associated with imprinting alterations in humans, including Silver–Russell (SRS) and Beckwith–Wiedemann syndromes (BWS). SRS is a growth disorder characterized by placental and intrauterine undergrowth, a small triangular face and body asymmetry (for reviews, see Wakeling et al. and Girardot et al.[103,104]). SRS most frequently arises as a result of hypomethylation of the *H19/IGF2* ICR in the imprinting cluster 11p15.5. Loss of imprinting at this locus leads to elevated *H19* expression and downregulation of *IGF2* expression and therefore, is often associated with impaired fetal growth.[105] Less common underlying mechanisms of SRS include maternal uniparental disomy of chromosome 7 and duplication of the 7p11.2–p13 region.[106] Genes responsible for the SRS phenotype on chromosome 7 remain to be determined, but studies[103,107] have suggested *GRB10* and *MEST* as strong imprinting candidates contributing to SRS.

Contrary to SRS, BWS is a fetal overgrowth imprinting disorder. Clinical features include macroglossia, facial dysmorphisms and, in many cases, hemihypertrophy (reviewed in Girardot et al., and Soejima and Higashimoto[104,108]). Genetic evidence has linked BWS to two imprinting clusters on chromosome 11p15, specifically the *H19/IGF2* and the *CDKN1C/KCNQ1OT1* imprinted clusters.[109] Loss of methylation at the (maternal) *CDKN1C/KCNQ1OT1* ICR (designated ICR2) accounts for up to 60% of all BWS cases, and this locus is also associated with type 2 diabetes vulnerability via GWAS.[110-112] This loss of methylation is accompanied with biallelic reexpression of the normally suppressed maternal copy of *Kcnq1ot1*, which in turn promotes repression of the cis-regulated imprinted genes within the domain.[113–115] *Cdkn1c* encodes a cyclin-dependent kinase inhibitor that negatively regulates cell proliferation (PMID: 7729683).[116,117] Disruption of its expression can lead to altered cell proliferation and overgrowth, and the alterations to ICR2 described before are phenocopied in mice with knockout of maternal *Cdkn1c*, recreating the overgrowth phenotype observed in BWS patients.[41] Hypermethylation of the *H19/IGF2* imprinted cluster is also observed in a subset of BWS cases, where methylation of the normally unmethylated maternal *H19* DMR leads to upregulation of *IGF2* and a reduction in *H19* expression, with subsequent fetal overgrowth.[79,112] Altered imprinted gene dosage is also observed in neurodevelopmental disorders, namely the Prader–Willi syndrome (PWS) and Angelman syndrome (AS), with imprinting defects at the Prader–Willi Angelman Critical Region (15q11-q13) found in 1% and 3% patients, respectively.[79,107,118,119] Maternal UPD, paternal deletion, or DMR alterations at the imprinted 14q32 locus (containing *Dlk1* and *Gtl2/MEG3*) cause low birth weight, with growth and developmental complications in early life (and later life metabolic disease) as part of a condition called Temple syndrome.[120] Overall, the aforementioned work demonstrates the severe consequences of disrupted imprinting gene dosage in humans and mouse models in terms of fetal growth and development.

Imprinted genes regulate multiple maternal–fetal interactions and are essential for early survival

The neonatal period is a critical part of perinatal period, usually referring to the first 28 days after the birth of an infant.[121] During that period, a direct connection between the parents and the child is created, including the establishment of feeding patterns, which have profound effects on infant survival, growth, and development.[122] Several paternally expressed imprinted genes play a key role in feeding behavior, metabolic processes, and survival during the neonatal period. Mice with targeted deletions at the imprinted *Gnas* locus have poor suckling and thus inadequate nutrient acquisition during the neonatal nesting period, which results in high postnatal lethality.[123] Specifically, deletion of *Gnas* exon 2 disrupts multiple protein-coding transcripts including *Gnasxl*, encoding the paternally expressed *XLαs*,[124] resulting in a significant decrease in survival of mutant mice, with surviving pups showing metabolic deficits later in life, including abnormalities in glucose and energy homeostasis.[123] Similarly, paternal disruption of *Magel2* results in temporal changes in pup suckling behavior, which leads to abnormal feeding patterns and reduced neonatal survival.[125] Neonatal mice with *Magel2* deletion are also growth retarded[126] and have decreased overall appetite, which continues after weaning in mice.[127] In humans, *MAGEL2* is inactivated in PWS,[128] and consistent with the neonatal feeding difficulties observed in PWS patients, *Magel2*-null mice have a decreased proportion of leptin-responsive hypothalamic proopiomelanocortin (POMC) neurons, likely leading to the observed leptin resistance and reduced appetite.[129] Since the leptin/melanocortin pathway is heavily involved in the regulation of food intake and satiety, *Magel2* imprinting is crucial for adequate neonate nutrient acquisition.[130] Poor feeding in *Magel2* mutant mice has also been demonstrated to arise due to a reduction in hypothalamic oxytocin levels, supported by findings that a similar feeding deficit is also observed in wild-type mice treated with an oxytocin antagonist.[125]

A number of imprinted genes have been shown to be involved in regulating the behavioral activities of newborns and are linked to the parental conflict hypothesis.

For example, several imprinted genes have been shown to be involved in the behavior of mammalian offspring in terms of their huddling for warmth for the benefit of common heat production, a behavioral activity that improves pup survival. The maternally expressed imprinted genes *Grb10* and *Cdkn1c* and the paternally expressed genes *Dlk1* and *Igf2* support this hypothesis, which play a role in promoting thermogenesis[131,132] and suppressing heat production,[133,134] respectively. In contrast, imprinted genes do not always conform to this hypothesis: for example, the paternally expressed imprinted gene *Peg3* promotes heat production in pups during the huddling process.[135] The coadaptation theory suggests that there is a genetic symbiosis between offspring growth and maternal care and that the evolution of a particular locus has facilitated the adaptive coherence of maternity and offspring, in favor of offspring growth and development.[53] Interestingly, *Peg3* has been shown to be important for maternal care as well as nutrition supply in the mother, on top of a simultaneously (positive) role in suckling behavior of the offspring.[136] *Peg3* increases maternal food intake in response to pregnancy-stimulated placental hormone release, providing the additional energy required for postpartum milk letdown, with *Peg3* mutant mothers therefore being unable to provide adequate nutrients, resulting in increased offspring mortality.[136] *Peg3* is also important in other maternal behaviors in mice such as nest building, crouching over pups, and pup retrieval,[89] with similar mechanisms of imprinted gene function (in terms of maternal behavior) observed for the paternally expressed gene, *Peg1/Mest*.[137] An interesting case describing the physiological impact of genomic imprinting on offspring nutrient acquisition is the imprinted gene *Grb10*. *Grb10* is widely maternally expressed in tissues of mesodermal and endodermal origin but paternally expressed in parts of the CNS.[138] Echoing the parental conflict theory, *Grb10* is maternally expressed in the placenta and its expression limits fetal growth,[39] with maternal *Grb10* loss of function leading to fetal overgrowth in mice.[70] In the neonatal period, disruption of maternal *Grb10* expression affects offspring nutrient acquisition via effects on both the mother and the offspring, influencing early growth and body composition.[139] Specifically, deficiency of maternally expressed *Grb10* in neonatal offspring desuppresses their demands for nutrients; however, in the mother, maternal *Grb10* disruption alters lactating mammary gland function and prevents adequate milk letdown, which exacerbates the offspring feeding defect.[139]

Thermoregulation is essential for offspring survival in the neonatal nesting period. Pups can maintain their body temperature from two major factors: the external huddling behavior of other individuals in the nest, and their internal thermogenesis. According to the parental conflict theory, MEG's are expected to contribute to social thermogenesis while PEG's are predicted to restrict contribution to communal heating.[140] Because nonshivering thermogenesis is dependent on brown adipose tissue (BAT) formation and function, the effect of imprinted genes on the production, recruitment, and activation of BAT reflects the specific organism's physiological contribution to huddling for warmth.[141] Under this assumption, multiple imprinted genes comply with the parent-of-origin expression predicted by the parental conflict theory including *Gnas*, *Grb10*, *Cdkn1c*, and *Dlk1*.[76,131,132,140] However, outliers such as *Peg3* exist whereby its deletion leads to absent maternal huddling behavior[136] as well as reduced BAT-mediated thermogenesis and reduced metabolic rate in offspring, leading to neonatal fatality following cold exposure.[135] Dosage alterations of the paternally expressed *Dlk1-Dio3* imprinted gene cluster (via disruption of the ICR) are also associated with postnatal thermogenesis and growth defects, impacting the BAT-mediated thermogenesis required for the transition from nesting neonates to independent adult life.[133] Within the *Dlk1-Dio3* cluster, *Dlk1* modulates adipocyte differentiation from preadipocytes,[142] and mutation of *Dlk1* significantly restricts neonatal growth,[75] whereas overexpression results in abnormal organ development and increased neonatal mortality.[76] Similarly, ablation of *Dio3* in newborns exhibits a reduction in pancreatic beta cell mass,[143] alongside neonatal thyrotoxicosis and a persistent defect in thyroid hormone function signaling. The latter pathway is critical for brown adipogenesis and maintaining adequate thermogenesis,[144] with disruption at the *Dlk1/Dio3* locus leading to hypothermia.[145] Consequently, the functions of *Dlk1* and *Dio3* converge on managing BAT maturation to direct postnatal thermoregulation,[133] with appropriate dosage control via genomic imprinting dictating the developmental timing required for postnatal thriving and/or survival.

Postnatal control of growth and development is mediated by imprinted genes via key metabolic systems and is crucial for preserving adult metabolic health

The generation of white adipose tissue (WAT) from preadipocytes signifies the lipolytic-to-lipogenic transition in metabolism, as the offspring move away from fat-rich milk to a carbohydrate-rich diet during early mammalian development as part of the weaning process.[146] This fine-tuning of white adipocyte number (hyperplasia) and size (hypertrophy) in the postnatal period in mice involves multiple imprinted genes including *Dlk1*, *Ndn*, *Peg1*, and *Cdkn1c*.[76,132,147,148] Under adipogenic stimulation, *Ndn* functions to restrict excessive white adipose generation, where deficiency of *Ndn* causes increased preadipocyte proliferation and adult obesity.[147] Increased dosage of *Peg1* expression has been shown to lead to enlarged adipocytes and expanded adipose tissue mass and is also strongly associated with adult obesity.[148,149] Loss of imprinting of *Cdkn1c* causes neonatal hypoglycemia and extreme lack of body fat via increased browning of WAT, which appears to be protective against diet- and age-induced obesity in adulthood.[132] Interestingly, one of the two forms of insulin found in mice, *Ins2*, has biallelic expression postnatally under a pancreatic enhancer but is imprinted in a tissue-specific manner in utero.[40,150] Deletion of both of the insulin genes (or ablating the insulin receptor) does not cause major disruptions to glucose homeostasis during the gestational period in mice[151]; however, these mutants rapidly develop neonatal diabetes and have a lower survival rate than their wild type counterparts.[151] Several imprinted genes affect organismal metabolism through modulation of pancreatic beta cell proliferation and function, including the paternally expressed genes *Nnat*, *Plagl1*, and

Rasgrf. Paternally expressed *Nnat* is essential to maintain adequate insulin content available for glucose-stimulated insulin secretion (GSIS), via maturation of early insulin precursors.[152] *Plagl1* is a paternally expressed gene encoding a zinc finger transcription factor that causes a transient form of neonatal diabetes in humans (TNDM) if overexpressed[74,153] also via a role in mediating GSIS.[154] The degree of methylation at the *Plagl1* locus correlates with glucose homeostasis and growth in human populations, with loss of imprinting of *Plagl1* in mice demonstrating that the *Plagl1*-mediated inhibitory effect on glucose-stimulated insulin secretion is through reduction of *Rasgrf1* expression.[155] Paternally expressed *Rasgrf1* itself functionally influences beta cell function via a CTCF-dependent enhancer blocking mechanism,[156] with significant defects in beta cell neogenesis and proliferation observed in *Rasgrf1*-deficient mice, leading to reduced beta cell mass, hypoinsulinemia, and impaired glucose tolerance in adulthood.[157,158]

The hypothalamus integrates sensory inputs from peripheral tissues to coordinate metabolic and physiological outcomes, particularly in the context of feeding behavior and energy expenditure. Multiple imprinted genes are highly expressed in the hypothalamus including *Magel2*, *Ndn*, *Grb10*, *Peg3*, *Nnat*, and those at the *Gnas* locus.[89,127,135,139,159,160] Altered imprinted gene expression disrupts hypothalamic function and perturbs metabolic homeostasis, particularly during early development. One of the *Gnas* gene products, *Gsα*, is expressed under a brain-specific promoter and governs melanocortin-mediated energy expenditure.[161] Disruption of *Gsα* expression in mice subsequently manifests as impaired melanocortin action, a pathway known to modulate glucose tolerance, insulin resistance, and obesity.[161,162] Similarly, deletion of *Nnat* leads to stunted postnatal growth as well as reducing metabolic rate, physical activity, leptin sensitivity, and energy expenditure.[163] Mice with *Nnat* ablation also display a subsequent period of "catch-up" growth after weaning, following this postnatal growth defect,[163] a characteristic phenotype that is observed in several mouse models with targeted disruption at imprinted *loci* including *Snord116*, *Rasgrf1*, *Dlk1*, and *Magel2* (reviewed in Peters[58]). Environmental factors during early mammalian development such as nutrient availability have long been proposed to be associated with increased risk for metabolic diseases in later life. Epigenetic modifiers, such as CpG methylation, have been proposed to be responsible for the correlation between early-life adversities (such as over- or undernutrition) and later-life metabolic complications,[164] and this appears to act, at least partly, via expression of imprinted genes. DNA methylation is dependent on the one-carbon metabolism in mammals, and this system requires methyl donors from dietary sources; hence the diet can have significant effects on methylation.[165] Mice fed a diet lacking in folic acid had altered DNA methylation at the *Igf2* locus, which was maintained after reintroduction of dietary folic acid, indicative of long-term adjustments to imprinted gene expression via the diet.[166] Folic acid supplementation has also been shown to rescue the loss of imprinting observed at the *Cdkn1c* locus following feeding with a low protein diet.[167] Due to obvious ethical concerns, human studies on the effect of nutritional exposure are not performed; however, famines such as the Chinese Great Famine (1959–61) have demonstrated the effects of early life malnutrition on later life metabolic complications, such as an increased risk of dyslipidemia in women with early-life exposure[168] potentially via DNA methylation at imprinted *loci*[169] and are discussed in greater detail in Millership et al., and Villanueva-Hayes and Millership.[52,170]

Final comments

This chapter explores the roles of imprinted genes in the context of growth and development during the in utero, neonatal, and postnatal periods and in adult life. Altered expression of imprinted genes have major consequences on mammalian fetal growth and development and affect social behavior and metabolism in both the offspring and the mother. Previous studies have consistently shown that early-life adversity, e.g., nutritional changes during the pre- and postnatal period is correlated to metabolic syndrome (and in particular, lipid and glucose metabolism) leading to later-life diseases such as obesity and diabetes, as discussed above. As early-life nutrient acquisition is linked with changes in the methylation status at imprinted *loci*, imprinted genes have been proposed to be a mechanism linking early-life environment exposures to later-life metabolic disease. Therefore, it is important that these mechanisms be fully interrogated to understand the genetic and metabolic pathways involved, to aid the development of novel treatments for the clinical management of imprinting diseases as well as earlier intervention in the process of developing adult metabolic syndrome. Therefore, with further investigation into how imprinted genes serve as the bridge between early development and adult metabolic disease, this will deepen our understanding of the role of the epigenome in the wider context of environment−organism interactions.

Acknowledgments

Work in the S.J.M. Lab was supported by a Wellcome Trust ISSF Fellowship [PS3619_WREC] and a Society for Endocrinology Early Career Grant (P93206). G.A.R. was supported by a Wellcome Trust Investigator (WT212625/Z/18/Z) Award, an MRC Programme grant (MR/R022259/1), a John R. Evans Leaders Award from Innovation Canada, an MRC Project Grant (MR/R010676/1), and has received funding from the European Union's Horizon 2020 research and innovation program via the Innovative Medicines Initiative 2 Joint Undertaking under grant agreement No 115881 (RHAPSODY) to G.A.R. The authors declare no conflict of interest.

References

1. Ferguson-Smith AC. Genomic imprinting: the emergence of an epigenetic paradigm. *Nat Rev Genet*. 2011;12(8):565−575. https://doi.org/10.1038/nrg3032.
2. Tronick E, Hunter RG. Waddington, dynamic systems, and epigenetics. *Front Behav Neurosci*. 2016;10:107. https://doi.org/10.3389/fnbeh.2016.00107.
3. Waddington CH. The epigenotype. *Endeavour*. 1942;1:18−20.

4. Adalsteinsson BT, Ferguson-Smith AC. Epigenetic control of the genome-lessons from genomic imprinting. *Genes*. 2014;5(3):635–655. https://doi.org/10.3390/genes5030635.
5. Bannister AJ, Kouzarides T. Regulation of chromatin by histone modifications. *Cell Res*. 2011;21(3):381–395. https://doi.org/10.1038/cr.2011.22.
6. Chen Q, Yan W, Duan E. Epigenetic inheritance of acquired traits through sperm RNAs and sperm RNA modifications. *Nat Rev Genet*. 2016;17(12):733–743. https://doi.org/10.1038/nrg.2016.106.
7. Jenuwein T, Allis CD. Translating the histone code. *Science (New York, NY.)*. 2001; 293(5532):1074–1080, 293/5532/1074 [pii].
8. Gan Y, Tao H, Zou G, Yan C, Guan J. Dynamic epigenetic mode analysis using spatial temporal clustering. *BMC Bioinf*. 2016;17(Suppl 17):537. https://doi.org/10.1186/s12859-016-1331-z.
9. He Y, Hariharan M, Gorkin DU, et al. Spatiotemporal DNA methylome dynamics of the developing mouse fetus. *Nature*. 2020;583(7818):752–759. https://doi.org/10.1038/s41586-020-2119-x.
10. Lacal I, Ventura R. Epigenetic inheritance: concepts, mechanisms and perspectives. *Front Mol Neurosci*. 2018;11:292. https://doi.org/10.3389/fnmol.2018.00292.
11. Shaban HA, Seeber A. Monitoring the spatio-temporal organization and dynamics of the genome. *Nucleic Acids Res*. 2020;48(7):3423–3434. https://doi.org/10.1093/nar/gkaa135.
12. Okano M, Xie S, Li E. Cloning and characterization of a family of novel mammalian DNA (cytosine-5) methyltransferases. *Nat Genet*. 1998;19(3):219–220. https://doi.org/10.1038/890.
13. Okano M, Bell DW, Haber DA, Li E. DNA methyltransferases Dnmt3a and Dnmt3b are essential for de novo methylation and mammalian development. *Cell*. 1999;99(3): 247–257. S0092-8674(00)81656-6 [pii].
14. Gowher H, Jeltsch A. Enzymatic properties of recombinant Dnmt3a DNA methyltransferase from mouse: the enzyme modifies DNA in a non-processive manner and also methylates non-CpG [correction of non-CpA] sites. *J Mol Biol*. 2001;309(5): 1201–1208. S0022283601947102 [pii].
15. Lister R, Pelizzola M, Dowen RH, et al. Human DNA methylomes at base resolution show widespread epigenomic differences. *Nature*. 2009;462(7271):315–322. https://doi.org/10.1038/nature08514.
16. Ramsahoye BH, Biniszkiewicz D, Lyko F, Clark V, Bird AP, Jaenisch R. Non-CpG methylation is prevalent in embryonic stem cells and may be mediated by DNA methyltransferase 3a. *Proc Natl Acad Sci USA*. 2000;97(10):5237–5242, 97/10/5237 [pii].
17. Ziller MJ, Müller F, Liao J, et al. Genomic distribution and inter-sample variation of non-CpG methylation across human cell types. *PLoS Genet*. 2011;7(12):e1002389. https://doi.org/10.1371/journal.pgen.1002389.
18. Monk D, Mackay DJG, Eggermann T, Maher ER, Riccio A. Genomic imprinting disorders: lessons on how genome, epigenome and environment interact. *Nat Rev Genet*. 2019;20(4):235–248. https://doi.org/10.1038/s41576-018-0092-0.
19. Stringer JM, Pask AJ, Shaw G, Renfree MB. Post-natal imprinting: evidence from marsupials. *Heredity*. 2014;113(2):145–155. https://doi.org/10.1038/hdy.2014.10.
20. Onyango P, Miller W, Lehoczky J, et al. Sequence and comparative analysis of the mouse 1-megabase region orthologous to the human 11p15 imprinted domain. *Genome Res*. 2000;10(11):1697–1710. https://doi.org/10.1101/gr.161800.

21. Villar AJ, Pedersen RA. Parental imprinting of the Mas protooncogene in mouse. *Nat Genet*. 1994;8(4):373–379. https://doi.org/10.1038/ng1294-373.
22. Zemel S, Bartolomei MS, Tilghman SM. Physical linkage of two mammalian imprinted genes, H19 and insulin-like growth factor 2. *Nat Genet*. 1992;2(1):61–65. https://doi.org/10.1038/ng0992-61.
23. Zwart R, Sleutels F, Wutz A, Schinkel AH, Barlow DP. Bidirectional action of the Igf2r imprint control element on upstream and downstream imprinted genes. *Genes Dev*. 2001;15(18):2361–2366. https://doi.org/10.1101/gad.206201.
24. Hajkova P, Ancelin K, Waldmann T, et al. Chromatin dynamics during epigenetic reprogramming in the mouse germ line. *Nature*. 2008;452(7189):877–881. https://doi.org/10.1038/nature06714.
25. Surani MA, Hayashi K, Hajkova P. Genetic and epigenetic regulators of pluripotency. *Cell*. 2007;128(4):747–762. S0092-8674(07)00191-2 [pii].
26. Barton SC, Surani MA, Norris ML. Role of paternal and maternal genomes in mouse development. *Nature*. 1984;311(5984):374–376. https://doi.org/10.1038/311374a0.
27. McGrath J, Solter D. Nuclear transplantation in the mouse embryo by microsurgery and cell fusion. *Science (New York, NY)*. 1983;220(4603):1300–1302. https://doi.org/10.1126/science.6857250.
28. McGrath J, Solter D. Completion of mouse embryogenesis requires both the maternal and paternal genomes. *Cell*. 1984;37(1):179–183, 0092-8674(84)90313-1 [pii].
29. Surani MA, Barton SC. Development of gynogenetic eggs in the mouse: implications for parthenogenetic embryos. *Science (New York, NY)*. 1983;222(4627):1034–1036. https://doi.org/10.1126/science.6648518.
30. Surani MA, Barton SC, Norris ML. Development of reconstituted mouse eggs suggests imprinting of the genome during gametogenesis. *Nature*. 1984;308(5959):548–550. https://doi.org/10.1038/308548a0.
31. Frésard L, Leroux S, Servin B, et al. Transcriptome-wide investigation of genomic imprinting in chicken. *Nucleic Acids Res*. 2014;42(6):3768–3782. https://doi.org/10.1093/nar/gkt1390.
32. Kermicle JL. Dependence of the R-mottled aleurone phenotype in maize on mode of sexual transmission. *Genetics*. 1970;66(1):69–85. https://doi.org/10.1093/genetics/66.1.69.
33. Koski LB, Sasaki E, Roberts RD, Gibson J, Etches RJ. Monoalleleic transcription of the insulin-like growth factor-II gene (Igf2) in chick embryos. *Mol Reprod Dev*. 2000;56(3):345–352. https://doi.org/10.1002/1098-2795(200007)56:33.0.CO;2-1.
34. Pires ND, Grossniklaus U. Different yet similar: evolution of imprinting in flowering plants and mammals. *F1000prime reports*. 2014;6:63. https://doi.org/10.12703/P6-63.
35. Chen Z, Hagen DE, Wang J, et al. Global assessment of imprinted gene expression in the bovine conceptus by next generation sequencing. *Epigenetics*. 2016;11(7):501–516. https://doi.org/10.1080/15592294.2016.1184805.
36. Avrahami D, Li C, Yu M, et al. Targeting the cell cycle inhibitor p57Kip2 promotes adult human β cell replication. *J Clin Investig*. 2014;124(2):670–674, 69519 [pii].
37. Kamiya M, Judson H, Okazaki Y, et al. The cell cycle control gene ZAC/PLAGL1 is imprinted–a strong candidate gene for transient neonatal diabetes. *Hum Mol Genet*. 2000;9(3):453–460, 045 [pii].
38. Lopez SJ, Segal DJ, LaSalle JM. UBE3A: an E3 ubiquitin ligase with genome-wide impact in neurodevelopmental disease. *Front Mol Neurosci*. 2019;11:476. https://doi.org/10.3389/fnmol.2018.00476.

39. Charalambous M, Cowley M, Geoghegan F, et al. Maternally-inherited Grb10 reduces placental size and efficiency. *Dev Biol*. 2010;337(1):1–8. https://doi.org/10.1016/j.ydbio.2009.10.011.
40. Cleaton MA, Edwards CA, Ferguson-Smith AC. Phenotypic outcomes of imprinted gene models in mice: elucidation of pre- and postnatal functions of imprinted genes. *Annu Rev Genom Hum Genet*. 2014;15:93–126. https://doi.org/10.1146/annurev-genom-091212-153441.
41. Tunster SJ, Van de Pette M, John RM. Fetal overgrowth in the Cdkn1c mouse model of Beckwith-Wiedemann syndrome. *Dis Models Mech*. 2011;4(6):814–821. https://doi.org/10.1242/dmm.007328.
42. Tunster SJ, Jensen AB, John RM. Imprinted genes in mouse placental development and the regulation of fetal energy stores. *Reproduction (Cambridge, England)*. 2013;145(5):117. https://doi.org/10.1530/REP-12-0511.
43. Soellner L, Begemann M, Mackay DJ, et al. Recent advances in imprinting disorders. *Clin Genet*. 2017;91(1):3–13. https://doi.org/10.1111/cge.12827.
44. Boonen SE, Mackay DJ, Hahnemann JM, et al. Transient neonatal diabetes, ZFP57, and hypomethylation of multiple imprinted loci: a detailed follow-up. *Diabetes Care*. 2013;36(3):505–512. https://doi.org/10.2337/dc12-0700.
45. Eggermann T, Perez de Nanclares G, Maher ER, et al. Imprinting disorders: a group of congenital disorders with overlapping patterns of molecular changes affecting imprinted loci. *Clin Epigenet*. 2015;7:123–128. https://doi.org/10.1186/s13148-015-0143-8. eCollection 2015.
46. Netchine I, Rossignol S, Dufourg MN, et al. 11p15 imprinting center region 1 loss of methylation is a common and specific cause of typical Russell-Silver syndrome: clinical scoring system and epigenetic-phenotypic correlations. *J Clin Endocrinol Metabol*. 2007;92(8):3148–3154. jc.2007-0354 [pii].
47. Yazdi PG, Su H, Ghimbovschi S, et al. Differential gene expression reveals mitochondrial dysfunction in an imprinting center deletion mouse model of Prader-Willi syndrome. *Clin Transl Sci*. 2013;6(5):347–355. https://doi.org/10.1111/cts.12083.
48. Moore T, Haig D. Genomic imprinting in mammalian development: a parental tug-of-war. *Trends Genet: TIG (Trends Genet)*. 1991;7(2):45–49, 0168-9525(91)90230-N [pii].
49. Dent CL, Humby T, Lewis K, et al. Impulsive choice in mice lacking paternal expression of Grb10 suggests intragenomic conflict in behavior. *Genetics*. 2018;209(1):233–239. https://doi.org/10.1534/genetics.118.300898.
50. Haig D. *Annu Rev Genet*. 2004;38:553–585. https://doi.org/10.1146/annurev.genet.37.110801.142741.
51. Wilkins JF, Haig D. What good is genomic imprinting: the function of parent-specific gene expression. *Nat Rev Genet*. 2003;4(5):359–368. nrg1062 [pii].
52. Millership SJ, Van de Pette M, Withers DJ. Genomic imprinting and its effects on postnatal growth and adult metabolism. *Cell Mol Life Sci: CMLS*. 2019;76(20):4009–4021. https://doi.org/10.1007/s00018-019-03197-z.
53. Wolf JB, Hager R. A maternal-offspring coadaptation theory for the evolution of genomic imprinting. *PLoS Biol*. 2006;4(12):e380.
54. Ashbrook DG, Hager R. Empirical testing of hypotheses about the evolution of genomic imprinting in mammals. *Front Neuroanat*. 2013;7:6. https://doi.org/10.3389/fnana.2013.00006.

55. Spencer HG, Clark AG. Non-conflict theories for the evolution of genomic imprinting. *Heredity*. 2014;113(2):112–118. https://doi.org/10.1038/hdy.2013.129.
56. Wilkins JF. Genomic imprinting of Grb10: coadaptation or conflict? *PLoS Biol*. 2014; 12(2):e1001800. https://doi.org/10.1371/journal.pbio.1001800.
57. McGrath J, Solter D. Nuclear transplantation in mouse embryos. *J Exp Zool*. 1983; 228(2):355–362. https://doi.org/10.1002/jez.1402280218.
58. Peters J. The role of genomic imprinting in biology and disease: an expanding view. *Nat Rev Genet*. 2014;15(8):517–530. https://doi.org/10.1038/nrg3766.
59. Barlow DP, Stöger R, Herrmann BG, Saito K, Schweifer N. The mouse insulin-like growth factor type-2 receptor is imprinted and closely linked to the Tme locus. *Nature*. 1991;349(6304):84–87. https://doi.org/10.1038/349084a0.
60. DeChiara TM, Robertson EJ, Efstratiadis A. Parental imprinting of the mouse insulin-like growth factor II gene. *Cell*. 1991;64(4):849–859, 0092-8674(91)90513-X [pii].
61. Ferguson-Smith AC, Cattanach BM, Barton SC, Beechey CV, Surani MA. Embryological and molecular investigations of parental imprinting on mouse chromosome 7. *Nature*. 1991;351(6328):667–670. https://doi.org/10.1038/351667a0.
62. DeChiara TM, Efstratiadis A, Robertson EJ. A growth-deficiency phenotype in heterozygous mice carrying an insulin-like growth factor II gene disrupted by targeting. *Nature*. 1990;345(6270):78–80. https://doi.org/10.1038/345078a0.
63. Leighton PA, Ingram RS, Eggenschwiler J, Efstratiadis A, Tilghman SM. Disruption of imprinting caused by deletion of the H19 gene region in mice. *Nature*. 1995;375(6526): 34–39. https://doi.org/10.1038/375034a0.
64. Smith FM, Garfield AS, Ward A. Regulation of growth and metabolism by imprinted genes. *Cytogenet Genome Res*. 2006;113(1–4):279–291, 90843 [pii].
65. Lau MM, Stewart CE, Liu Z, Bhatt H, Rotwein P, Stewart CL. Loss of the imprinted IGF2/cation-independent mannose 6-phosphate receptor results in fetal overgrowth and perinatal lethality. *Genes Dev*. 1994;8(24):2953–2963. https://doi.org/10.1101/gad.8.24.2953.
66. Ludwig T, Eggenschwiler J, Fisher P, D'Ercole AJ, Davenport ML, Efstratiadis A. Mouse mutants lacking the type 2 IGF receptor (IGF2R) are rescued from perinatal lethality in Igf2 and Igf1r null backgrounds. *Dev Biol*. 1996;177(2):517–535. S0012-1606(96)90182-7 [pii].
67. Wang ZQ, Fung MR, Barlow DP, Wagner EF. Regulation of embryonic growth and lysosomal targeting by the imprinted Igf2/Mpr gene. *Nature*. 1994;372(6505): 464–467. https://doi.org/10.1038/372464a0.
68. Wylie AA, Pulford DJ, McVie-Wylie AJ, et al. Tissue-specific inactivation of murine M6P/IGF2R. *Am J Pathol*. 2003;162(1):321–328. S0002-9440(10)63823-0 [pii].
69. Filson AJ, Louvi A, Efstratiadis A, Robertson EJ. Rescue of the T-associated maternal effect in mice carrying null mutations in Igf-2 and Igf2r, two reciprocally imprinted genes. *Development (Cambridge, England)*. 1993;118(3):731–736. https://doi.org/10.1242/dev.118.3.731.
70. Charalambous M, Smith FM, Bennett WR, Crew TE, Mackenzie F, Ward A. Disruption of the imprinted Grb10 gene leads to disproportionate overgrowth by an Igf2-independent mechanism. *Proc Natl Acad Sci USA*. 2003;100(14):8292–8297, 1532175100 [pii].
71. Frost JM, Moore GE. The importance of imprinting in the human placenta. *PLoS Genet*. 2010;6(7):e1001015. https://doi.org/10.1371/journal.pgen.1001015.

72. Shiura H, Nakamura K, Hikichi T, et al. Paternal deletion of Meg1/Grb10 DMR causes maternalization of the Meg1/Grb10 cluster in mouse proximal chromosome 11 leading to severe pre- and postnatal growth retardation. *Hum Mol Genet.* 2009;18(8): 1424–1438. https://doi.org/10.1093/hmg/ddp049.
73. Varrault A, Gueydan C, Delalbre A, et al. Zac1 regulates an imprinted gene network critically involved in the control of embryonic growth. *Dev Cell.* 2006;11(5): 711–722. S1534-5807(06)00396-0 [pii].
74. Ma D, Shield JP, Dean W, et al. Impaired glucose homeostasis in transgenic mice expressing the human transient neonatal diabetes mellitus locus. *TNDM. J Clin Invest.* 2004;114(3):339–348, 19876 [pii].
75. Moon YS, Smas CM, Lee K, et al. Mice lacking paternally expressed Pref-1/Dlk1 display growth retardation and accelerated adiposity. *Mol Cell Biol.* 2002;22(15): 5585–5592, 1760 [pii].
76. da Rocha ST, Charalambous M, Lin SP, et al. Gene dosage effects of the imprinted delta-like homologue 1 (dlk1/pref1) in development: implications for the evolution of imprinting. *PLoS Genet.* 2009;5(2):e1000392. https://doi.org/10.1371/journal.pgen.1000392.
77. Argyraki M, Damdimopoulou P, Chatzimeletiou K, et al. In-utero stress and mode of conception: impact on regulation of imprinted genes, fetal development and future health. *Hum Reprod Update.* 2019;25(6):777–801. https://doi.org/10.1093/humupd/dmz025.
78. Lambertini L, Marsit CJ, Sharma P, et al. Imprinted gene expression in fetal growth and development. *Placenta.* 2012;33(6):480–486. https://doi.org/10.1016/j.placenta.2012.03.001.
79. Bressan FF, De Bem TH, Perecin F, et al. Unearthing the roles of imprinted genes in the placenta. *Placenta.* 2009;30(10):823–834. https://doi.org/10.1016/j.placenta.2009.07.007.
80. Zhang P, Liégeois NJ, Wong C, et al. Altered cell differentiation and proliferation in mice lacking p57KIP2 indicates a role in Beckwith-Wiedemann syndrome. *Nature.* 1997;387(6629):151–158. https://doi.org/10.1038/387151a0.
81. Bartolomei MS, Ferguson-Smith AC. Mammalian genomic imprinting. *Cold Spring Harbor Perspect Biol.* 2011;3(7):a002592. https://doi.org/10.1101/cshperspect.a002592.
82. Barton SC, Ferguson-Smith AC, Fundele R, Surani MA. Influence of paternally imprinted genes on development. *Development (Cambridge, England).* 1991;113(2): 679–687. https://doi.org/10.1242/dev.113.2.679.
83. Keverne EB, Fundele R, Narasimha M, Barton SC, Surani MA. Genomic imprinting and the differential roles of parental genomes in brain development. *Dev Brain Res.* 1996;92(1):91–100, 016538069500209X [pii].
84. Monk D. Genomic imprinting in the human placenta. *Am J Obstet Gynecol.* 2015;213(4 Suppl):152. S0002-9378(15)00632-8 [pii].
85. Fowden AL, Coan PM, Angiolini E, Burton GJ, Constancia M. Imprinted genes and the epigenetic regulation of placental phenotype. *Prog Biophys Mol Biol.* 2011;106(1): 281–288. https://doi.org/10.1016/j.pbiomolbio.2010.11.005.
86. Fowden AL, Sferruzzi-Perri AN, Coan PM, Constancia M, Burton GJ. Placental efficiency and adaptation: endocrine regulation. *J Physiol.* 2009;587(Pt 14):3459–3472. https://doi.org/10.1113/jphysiol.2009.173013.

87. Constância M, Hemberger M, Hughes J, et al. Placental-specific IGF-II is a major modulator of placental and fetal growth. *Nature*. 2002;417(6892):945–948. nature00819 [pii].
88. Baker J, Liu JP, Robertson EJ, Efstratiadis A. Role of insulin-like growth factors in embryonic and postnatal growth. *Cell*. 1993;75(1):73–82, 0092-8674(93)90680-O [pii].
89. Li L, Keverne EB, Aparicio SA, Ishino F, Barton SC, Surani MA. Regulation of maternal behavior and offspring growth by paternally expressed Peg3. *Science (New York, NY)*. 1999;284(5412):330–333. https://doi.org/10.1126/science.284.5412.330.
90. Hiby SE, Lough M, Keverne EB, Surani MA, Loke YW, King A. Paternal monoallelic expression of PEG3 in the human placenta. *Hum Mol Genet*. 2001;10(10):1093–1100. https://doi.org/10.1093/hmg/10.10.1093.
91. Sekita Y, Wagatsuma H, Nakamura K, et al. Role of retrotransposon-derived imprinted gene, Rtl1, in the feto-maternal interface of mouse placenta. *Nat Genet*. 2008;40(2): 243–248. https://doi.org/10.1038/ng.2007.51.
92. Cleaton MA, Dent CL, Howard M, et al. Fetus-derived DLK1 is required for maternal metabolic adaptations to pregnancy and is associated with fetal growth restriction. *Nat Genet*. 2016;48(12):1473–1480. https://doi.org/10.1038/ng.3699.
93. Frank D, Fortino W, Clark L, et al. Placental overgrowth in mice lacking the imprinted gene Ipl. *Proc Natl Acad Sci USA*. 2002;99(11):7490–7495, 122039999 [pii].
94. Takahashi K, Kobayashi T, Kanayama N. p57(Kip2) regulates the proper development of labyrinthine and spongiotrophoblasts. *Mol Hum Reprod*. 2000;6(11):1019–1025. https://doi.org/10.1093/molehr/6.11.1019.
95. Tunster SJ, Tycko B, John RM. The imprinted Phlda2 gene regulates extraembryonic energy stores. *Mol Cell Biol*. 2010;30(1):295–306. https://doi.org/10.1128/MCB.00662-09.
96. Tunster SJ, Van De Pette M, John RM. Isolating the role of elevated Phlda2 in asymmetric late fetal growth restriction in mice. *Dis Models Mech*. 2014;7(10): 1185–1191. https://doi.org/10.1242/dmm.017079.
97. Guillemot F, Nagy A, Auerbach A, Rossant J, Joyner AL. Essential role of Mash-2 in extraembryonic development. *Nature*. 1994;371(6495):333–336. https://doi.org/10.1038/371333a0.
98. Ono R, Nakamura K, Inoue K, et al. Deletion of Peg10, an imprinted gene acquired from a retrotransposon, causes early embryonic lethality. *Nat Genet*. 2006;38(1): 101–106. ng1699 [pii].
99. Zwart R, Verhaagh S, Buitelaar M, Popp-Snijders C, Barlow DP. Impaired activity of the extraneuronal monoamine transporter system known as uptake-2 in Orct3/Slc22a3-deficient mice. *Mol Cell Biol*. 2001;21(13):4188–4196, 0262 [pii].
100. Itoh M, Yoshida Y, Nishida K, Narimatsu M, Hibi M, Hirano T. Role of Gab1 in heart, placenta, and skin development and growth factor- and cytokine-induced extracellular signal-regulated kinase mitogen-activated protein kinase activation. *Mol Cell Biol*. 2000;20(10):3695–3704, 1759 [pii].
101. Hanna CW. Placental imprinting: emerging mechanisms and functions. *PLoS Genet*. 2020;16(4):e1008709. https://doi.org/10.1371/journal.pgen.1008709.
102. Woods L, Perez-Garcia V, Hemberger M. Regulation of placental development and its impact on fetal growth-new insights from mouse models. *Front Endocrinol*. 2018;9: 570. https://doi.org/10.3389/fendo.2018.00570.

103. Wakeling EL, Brioude F, Lokulo-Sodipe O, et al. Diagnosis and management of Silver-Russell syndrome: first international consensus statement. *Nat Rev Endocrinol.* 2017; 13(2):105–124. https://doi.org/10.1038/nrendo.2016.138.
104. Girardot M, Feil R, Llères D. Epigenetic deregulation of genomic imprinting in humans: causal mechanisms and clinical implications. *Epigenomics.* 2013;5(6): 715–728. https://doi.org/10.2217/epi.13.66.
105. Bliek J, Terhal P, van den Bogaard MJ, et al. Hypomethylation of the H19 gene causes not only Silver-Russell syndrome (SRS) but also isolated asymmetry or an SRS-like phenotype. *Am J Hum Genet.* 2006;78(4):604–614. S0002-9297(07)63699-4 [pii].
106. Abu-Amero S, Monk D, Frost J, Preece M, Stanier P, Moore GE. The genetic aetiology of Silver-Russell syndrome. *J Med Genet.* 2008;45(4):193–199. jmg.2007.053017 [pii].
107. Eggermann T, Netchine I, Temple IK, et al. Congenital imprinting disorders: EUCID.net—a network to decipher their aetiology and to improve the diagnostic and clinical care. *Clin Epigenet.* 2015;7(1). https://doi.org/10.1186/s13148-015-0050-z, 23-z. eCollection 2015.
108. Soejima H, Higashimoto K. Epigenetic and genetic alterations of the imprinting disorder Beckwith-Wiedemann syndrome and related disorders. *J Hum Genet.* 2013;58(7): 402–409. https://doi.org/10.1038/jhg.2013.51.
109. Chang S, Bartolomei MS. Modeling human epigenetic disorders in mice: Beckwith-Wiedemann syndrome and Silver-Russell syndrome. dmm044123 *Dis Models Mech.* 2020;13(5). https://doi.org/10.1242/dmm.044123.
110. Travers ME, Mackay DJ, Dekker Nitert M, et al. Insights into the molecular mechanism for type 2 diabetes susceptibility at the KCNQ1 locus from temporal changes in imprinting status in human islets. *Diabetes.* 2013;62(3):987–992. https://doi.org/ 10.2337/db12-0819.
111. Hiramoto M, Udagawa H, Ishibashi N, et al. A type 2 diabetes-associated SNP in KCNQ1 (rs163184) modulates the binding activity of the locus for Sp3 and Lsd1/ Kdm1a, potentially affecting CDKN1C expression. *Int J Mol Med.* 2018;41(2): 717–728. https://doi.org/10.3892/ijmm.2017.3273.
112. Eggermann T, Algar E, Lapunzina P, et al. Clinical utility gene card for: Beckwith-Wiedemann syndrome. *Eur J Hum Genet: EJHG (Eur J Hum Genet).* 2014;22(3). https://doi.org/10.1038/ejhg.2013.132. Epub 2013 Jul 3.
113. Pandey RR, Mondal T, Mohammad F, et al. Kcnq1ot1 antisense noncoding RNA mediates lineage-specific transcriptional silencing through chromatin-level regulation. *Mol Cell.* 2008;32(2):232–246. https://doi.org/10.1016/j.molcel.2008.08.022.
114. Mohammad F, Mondal T, Guseva N, Pandey GK, Kanduri C. Kcnq1ot1 noncoding RNA mediates transcriptional gene silencing by interacting with Dnmt1. *Development (Cambridge, England).* 2010;137(15):2493–2499. https://doi.org/10.1242/dev.048181.
115. Mancini-Dinardo D, Steele SJ, Levorse JM, Ingram RS, Tilghman SM. Elongation of the Kcnq1ot1 transcript is required for genomic imprinting of neighboring genes. *Genes Dev.* 2006;20(10):1268–1282, 20/10/1268 [pii].
116. Hatada I, Mukai T. Genomic imprinting of p57KIP2, a cyclin-dependent kinase inhibitor, in mouse. *Nat Genet.* 1995;11(2):204–206. https://doi.org/10.1038/ng1095-204.
117. Matsuoka S, Thompson JS, Edwards MC, et al. Imprinting of the gene encoding a human cyclin-dependent kinase inhibitor, p57KIP2, on chromosome 11p15. *Proc Natl Acad Sci USA.* 1996;93(7):3026–3030. https://doi.org/10.1073/pnas.93.7.3026.

118. Horsthemke B, Ludwig M. Assisted reproduction: the epigenetic perspective. *Hum Reprod Update*. 2005;11(5):473–482. mi022 [pii].
119. Amor DJ, Halliday J. A review of known imprinting syndromes and their association with assisted reproduction technologies. *Hum Reprod (Oxf)*. 2008;23(12): 2826–2834. https://doi.org/10.1093/humrep/den310.
120. Ioannides Y, Lokulo-Sodipe K, Mackay DJ, Davies JH, Temple IK. Temple syndrome: improving the recognition of an underdiagnosed chromosome 14 imprinting disorder: an analysis of 51 published cases. *J Med Genet*. 2014;51(8):495–501. https://doi.org/10.1136/jmedgenet-2014-102396.
121. Pomerance HH. Nelson textbook of pediatrics. *Arch Pediatr Adolesc Med*. 1997;151(3): 324, 518318 [pii].
122. Gabbe SG, Niebyl JR, Simpson JL, et al. *Obstetrics: Normal and Problem Pregnancies e-Book*. Elsevier Health Sciences; 2016.
123. Plagge A, Gordon E, Dean W, et al. The imprinted signaling protein XL alpha s is required for postnatal adaptation to feeding. *Nat Genet*. 2004;36(8):818–826. ng1397 [pii].
124. Xie T, Plagge A, Gavrilova O, et al. The alternative stimulatory G protein alpha-subunit XLalphas is a critical regulator of energy and glucose metabolism and sympathetic nerve activity in adult mice. *J Biol Chem*. 2006;281(28):18989–18999. S0021-9258(19)46458-2 [pii].
125. Schaller F, Watrin F, Sturny R, Massacrier A, Szepetowski P, Muscatelli F. A single postnatal injection of oxytocin rescues the lethal feeding behaviour in mouse newborns deficient for the imprinted Magel2 gene. *Hum Mol Genet*. 2010;19(24):4895–4905. https://doi.org/10.1093/hmg/ddq424.
126. Bischof JM, Stewart CL, Wevrick R. Inactivation of the mouse Magel2 gene results in growth abnormalities similar to Prader-Willi syndrome. *Hum Mol Genet*. 2007;16(22): 2713–2719. m225 [pii].
127. Kozlov SV, Bogenpohl JW, Howell MP, et al. The imprinted gene Magel2 regulates normal circadian output. *Nat Genet*. 2007;39(10):1266–1272. ng2114 [pii].
128. Lee S, Kozlov S, Hernandez L, et al. Expression and imprinting of MAGEL2 suggest a role in Prader-Willi syndrome and the homologous murine imprinting phenotype. *Hum Mol Genet*. 2000;9(12):1813–1819. https://doi.org/10.1093/hmg/9.12.1813.
129. Pravdivyi I, Ballanyi K, Colmers WF, Wevrick R. Progressive postnatal decline in leptin sensitivity of arcuate hypothalamic neurons in the Magel2-null mouse model of Prader-Willi syndrome. *Hum Mol Genet*. 2015;24(15):4276–4283. https://doi.org/10.1093/hmg/ddv159.
130. Mercer RE, Michaelson SD, Chee MJ, Atallah TA, Wevrick R, Colmers WF. Magel2 is required for leptin-mediated depolarization of POMC neurons in the hypothalamic arcuate nucleus in mice. *PLoS Genet*. 2013;9(1):e1003207. https://doi.org/10.1371/journal.pgen.1003207.
131. Liu M, Bai J, He S, et al. Grb10 promotes lipolysis and thermogenesis by phosphorylation-dependent feedback inhibition of mTORC1. *Cell Metabol*. 2014; 19(6):967–980. S1550-4131(14)00123-5 [pii].
132. Van De Pette M, Tunster SJ, McNamara GI, et al. Cdkn1c boosts the development of Brown adipose tissue in a murine model of Silver Russell syndrome. *PLoS Genet*. 2016;12(3):e1005916. https://doi.org/10.1371/journal.pgen.1005916.

133. Charalambous M, Ferron SR, da Rocha ST, et al. Imprinted gene dosage is critical for the transition to independent life. *Cell Metabol.* 2012;15(2):209–221. https://doi.org/10.1016/j.cmet.2012.01.006.
134. Borensztein M, Viengchareun S, Montarras D, et al. Double Myod and Igf2 inactivation promotes brown adipose tissue development by increasing Prdm16 expression. *Faseb J: Off Publ Feder Am Soc Exp Biol.* 2012;26(11):4584–4591. https://doi.org/10.1096/fj.12-208496.
135. Curley JP, Pinnock SB, Dickson SL, et al. Increased body fat in mice with a targeted mutation of the paternally expressed imprinted gene Peg3. *Faseb J: Off Publ Feder Am Soc Exp Biol.* 2005;19(10):1302–1304, 04-3216fje [pii].
136. Curley JP, Barton S, Surani A, Keverne EB. Coadaptation in mother and infant regulated by a paternally expressed imprinted gene. *Proc Biol Sci.* 2004;271(1545):1303–1309, 6A65N9DT2AP4W9UJ [pii].
137. Lefebvre L, Viville S, Barton SC, Ishino F, Keverne EB, Surani MA. Abnormal maternal behaviour and growth retardation associated with loss of the imprinted gene Mest. *Nat Genet.* 1998;20(2):163–169. https://doi.org/10.1038/2464.
138. Garfield AS, Cowley M, Smith FM, et al. Distinct physiological and behavioural functions for parental alleles of imprinted Grb10. *Nature.* 2011;469(7331):534–538. https://doi.org/10.1038/nature09651.
139. Cowley M, Garfield AS, Madon-Simon M, et al. Developmental programming mediated by complementary roles of imprinted Grb10 in mother and pup. *PLoS Biol.* 2014;12(2):e1001799. https://doi.org/10.1371/journal.pbio.1001799.
140. Haig D. Huddling: brown fat, genomic imprinting and the warm inner glow. *Curr Biol: CB.* 2008;18(4):172. https://doi.org/10.1016/j.cub.2007.12.040.
141. Cannon B, Nedergaard J. Brown adipose tissue: function and physiological significance. *Physiol Rev.* 2004;84(1):277–359, 84/1/277 [pii].
142. Hudak CS, Sul HS. Pref-1, a gatekeeper of adipogenesis. *Front Endocrinol.* 2013;4:79. https://doi.org/10.3389/fendo.2013.00079.
143. Medina MC, Molina J, Gadea Y, et al. The thyroid hormone-inactivating type III deiodinase is expressed in mouse and human beta-cells and its targeted inactivation impairs insulin secretion. *Endocrinology.* 2011;152(10):3717–3727. https://doi.org/10.1210/en.2011-1210.
144. Bianco AC, McAninch EA. The role of thyroid hormone and brown adipose tissue in energy homoeostasis. *Lancet Diabetes Endocrinol.* 2013;1(3):250–258. S2213-8587(13)70069-X [pii].
145. Hernandez A, Garcia B, Obregon MJ. Gene expression from the imprinted Dio3 locus is associated with cell proliferation of cultured brown adipocytes. *Endocrinology.* 2007;148(8):3968–3976. en.2007-0029 [pii].
146. Herrera E, Amusquivar E. Lipid metabolism in the fetus and the newborn. *Diabetes/Metabol Res Rev.* 2000;16(3):202–210. https://doi.org/10.1002/1520-7560(200005/06)16:33.0.CO;2-#.
147. Fujiwara K, Hasegawa K, Ohkumo T, Miyoshi H, Tseng YH, Yoshikawa K. Necdin controls proliferation of white adipocyte progenitor cells. *PLoS One.* 2012;7(1):e30948. https://doi.org/10.1371/journal.pone.0030948.
148. Nikonova L, Koza RA, Mendoza T, Chao PM, Curley JP, Kozak LP. Mesoderm-specific transcript is associated with fat mass expansion in response to a positive energy balance. *Faseb J: Off Publ Feder Am Soc Exp Biol.* 2008;22(11):3925–3937. https://doi.org/10.1096/fj.08-108266.

149. Takahashi M, Kamei Y, Ezaki O. Mest/Peg1 imprinted gene enlarges adipocytes and is a marker of adipocyte size. *Am J Physiol Endocrinol Metab.* 2005;288(1):117, 00244.2004 [pii].
150. Deltour L, Montagutelli X, Guenet JL, Jami J, Páldi A. Tissue- and developmental stage-specific imprinting of the mouse proinsulin gene. *Ins2 Dev Biol.* 1995;168(2): 686−688. S0012-1606(85)71114-1 [pii].
151. Duvillié B, Cordonnier N, Deltour L, et al. Phenotypic alterations in insulin-deficient mutant mice. *Proc Natl Acad Sci USA.* 1997;94(10):5137−5140, 0682 [pii].
152. Millership SJ, Da Silva Xavier G, Choudhury AI, et al. Neuronatin regulates pancreatic beta cell insulin content and secretion. *J Clin Invest.* 2018;128(8):3369−3381. https://doi.org/10.1172/JCI120115.
153. Iglesias-Platas I, Court F, Camprubi C, et al. Imprinting at the PLAGL1 domain is contained within a 70-kb CTCF/cohesin-mediated non-allelic chromatin loop. *Nucleic Acids Res.* 2013;41(4):2171−2179. https://doi.org/10.1093/nar/gks1355.
154. Du X, Ounissi-Benkalha H, Loder MK, Rutter GA, Polychronakos C. Overexpression of ZAC impairs glucose-stimulated insulin translation and secretion in clonal pancreatic beta-cells. *Diabetes/Metabol Res Rev.* 2012;28(8):645−653. https://doi.org/10.1002/dmrr.2325.
155. Hoffmann A, Spengler D. Transient neonatal diabetes mellitus gene Zac1 impairs insulin secretion in mice through Rasgrf1. *Mol Cell Biol.* 2012;32(13):2549−2560. https://doi.org/10.1128/MCB.06637-11.
156. Yoon B, Herman H, Hu B, et al. Rasgrf1 imprinting is regulated by a CTCF-dependent methylation-sensitive enhancer blocker. *Mol Cell Biol.* 2005;25(24):11184−11190, 25/24/11184 [pii].
157. Manyes L, Arribas M, Gomez C, Calzada N, Fernandez-Medarde A, Santos E. Transcriptional profiling reveals functional links between RasGrf1 and Pttg1 in pancreatic beta cells. *BMC Genom.* 2014;15. https://doi.org/10.1186/1471-2164-15-1019, 1019-1019.
158. Font de Mora J, Esteban LM, Burks DJ, et al. Ras-GRF1 signaling is required for normal beta-cell development and glucose homeostasis. *EMBO J.* 2003;22(12): 3039−3049. cdg280 [pii].
159. Kuwajima T, Nishimura I, Yoshikawa K. Necdin promotes GABAergic neuron differentiation in cooperation with Dlx homeodomain proteins. *J Neurosci: Off J Soc Neurosci.* 2006;26(20):5383−5392, 26/20/5383 [pii].
160. Chen M, Gavrilova O, Liu J, et al. Alternative Gnas gene products have opposite effects on glucose and lipid metabolism. *Proc Natl Acad Sci USA.* 2005;102(20):7386−7391, 0408268102 [pii].
161. Chen M, Shrestha YB, Podyma B, et al. Gsα deficiency in the dorsomedial hypothalamus underlies obesity associated with Gsα mutations. *J Clin Investig.* 2017;127(2): 500−510, 88622 [pii].
162. Huszar D, Lynch CA, Fairchild-Huntress V, et al. Targeted disruption of the melanocortin-4 receptor results in obesity in mice. *Cell.* 1997;88(1):131−141. S0092-8674(00)81865-6 [pii].
163. Millership SJ, Tunster SJ, Van de Pette M, et al. Neuronatin deletion causes postnatal growth restriction and adult obesity in 129S2/Sv mice. *Mol Metabol.* 2018;18:97−106. S2212-8778(18)30804-4 [pii].
164. McMillen IC, Robinson JS. Developmental origins of the metabolic syndrome: prediction, plasticity, and programming. *Physiol Rev.* 2005;85(2):571−633, 85/2/571 [pii].

165. Anderson OS, Sant KE, Dolinoy DC. Nutrition and epigenetics: an interplay of dietary methyl donors, one-carbon metabolism and DNA methylation. *J Nutr Biochem.* 2012; 23(8):853–859. https://doi.org/10.1016/j.jnutbio.2012.03.003.
166. Waterland RA, Garza C. Early postnatal nutrition determines adult pancreatic glucose-responsive insulin secretion and islet gene expression in rats. *J Nutr.* 2002;132(3): 357–364. https://doi.org/10.1093/jn/132.3.357.
167. Van de Pette M, Abbas A, Feytout A, et al. Visualizing changes in Cdkn1c expression links early-life adversity to imprint mis-regulation in adults. *Cell Rep.* 2017;18(5): 1090–1099. S2211-1247(17)30028-1 [pii].
168. Wang Z, Li C, Yang Z, Ma J, Zou Z. Fetal and infant exposure to severe Chinese famine increases the risk of adult dyslipidemia: results from the China health and retirement longitudinal study. *BMC Publ Health.* 2017;17(1):488. https://doi.org/10.1186/s12889-017-4421-6.
169. Shen L, Li C, Wang Z, et al. Early-life exposure to severe famine is associated with higher methylation level in the IGF2 gene and higher total cholesterol in late adulthood: the Genomic Research of the Chinese Famine (GRECF) study. *Clin Epigenet.* 2019; 11(1):88. https://doi.org/10.1186/s13148-019-0676-3.
170. Villanueva-Hayes C, Millership SJ. Imprinted genes impact upon beta cell function in the current (and potentially next) generation. *Front Endocrinol.* 2021;12:660532. https://doi.org/10.3389/fendo.2021.660532.

CHAPTER

Role of RNA epigenetics in development

5

Akin Cayir

Vocational Health College, Canakkale Onsekiz Mart University, Canakkale, Turkey

Introduction

Modifications in DNA and proteins with different chemical groups play critical roles in important biological processes.[1] DNA 5-methyl cytosine (m5C) is a type of DNA modification and affects many biological processes by regulating gene expression.[2] These modifications are dynamic and can be altered by external and internal factors such as environmental exposures. Addition or removal of the chemical groups is an essential phenomenon during cellular processes. Altering levels of modifications plays an important role in the development of diseases due to their functions in cellular processes.[3] The modifications that occur without any change in the DNA sequence are expressed with the term epigenetics, which was defined by Conrad Waddington in 1942.[4]

Recently, the term epitranscriptomics has been suggested to refer to modifications in RNAs as analogues of term epigenetics.[5] Although the first modifications in RNA were discovered in the 1950s and 1960s, today, more than >150 modifications have been identified.[6] Such modifications formed by different chemical groups can be observed in all bases. Studies conducted in the past 10 years have reported that different bases are modified in transfer RNA (tRNA) and ribosomal RNA (rRNA). Thanks to advanced technology such as m6A-specific methylated RNA immunoprecipitation (MeRIP) with next-generation sequencing (RNA-Seq), several types of RNA modifications have also been identified in mRNA and in other types of RNAs. Various chemical groups, including aminoacyl group, aminoalkyl, glycosyl group, heavy atoms, hydrocarbon, methyl group, ring modifications, and sulfur, modify the RNA bases. According to the MODOMICS database, there are 35 different modifications of adenosine, 20 different modifications of cytidine, 26 modifications of guanosine, 59 modifications of uridine, and 11 modifications of guanosine reported in different organisms.[7] Several modifications of mRNAs in eukaryotic organisms have been reported, including 6-methyladenosine (m6A), N6,2′-*O*-dimethyladenosine, 5-methylcytidine (m5C), 5-hydroxylmethylcytidine (hm5C), inosine, pseudouridine (Ψ), and N1-methyladenosine (m1A).[8–11]

Here, this chapter focuses on recent findings within the context of the current knowledge focusing on the association between RNA modifications,

Perinatal and Developmental Epigenetics. https://doi.org/10.1016/B978-0-12-821785-6.00003-7

RNA-modifying genes in terms of regulation of development processes. Among RNA modifications, RNA m6A, RNA m5C, and RNA A-to-I editing were evidenced for their role in regulatory processes of development.

RNA modifications

RNA N6-methyladenosine modification (m6A)

The m6A modification was first detected in the poly-A tails of messenger RNAs in the 1970s.[12] In particular, it was reported that m6A was found in the form of Pu (G>A) M6AC (U>A>C) (Pu stands for a purine base), and each mRNA had 3 m6A residues.[13] Later studies have shown that m6A was common in mRNAs and noncoding long-chain RNAs of highly organized eukaryotes. Substantial findings indicate that RNA m6A is the most common modification found in eukaryotic RNA molecules. The large-scale transcriptome studies performed by two different research groups in 2012 provided transcriptome-wide pattern of m6A.[9] In the deep-sequencing studies, the presence of 12000 m6A regions in transcripts and 300 noncoding genome regions, which are the products of about 7000 genes in general, was detected. Until now, m6A RNA modification has been detected and profiled in organisms including humans, mice, yeasts, and plants.[14] The findings obtained have shown that, in general, m6A methylation in mRNA is preserved, common, and dynamic. Previous and recent studies have revealed three properties of m6A: (1) the presence of sequences within a certain motif; (2) not evenly distributed at the transcriptome level, their distribution is not uniform and they are generally located in the stop codon, 3′UTR, and long internal exons; and (3) m6A-modified genes of human and mouse embryo were shown to be well preserved between stem cells and somatic cells.[8,9]

Current evidence suggests that RNA modifications are regulated by different types of enzymes. Three types of enzymes are suggested for the modifications. The writer is called methyltransferase enzyme, which catalyzes the addition reaction of chemical group to the RNA bases. The erasers are called demethylase enzymes, which catalyze the removing of chemical groups from the modified bases. Finally, readers can recognize and utilize the modified bases. RNA m6A is the first RNA modification that was identified for its three types of enzymes indicating the reversible and its dynamic properity. Writer enzymes of m6A are METTL3 (methyltransferase like 3), METTL14 (methyltransferase like 14), METTL16 (methyltransferase like 16), RBM15 (RNA-binding motif protein 15), WTAP (Wilm's tumor 1-associating protein), and KIAA1429 (vir like m6A methyltransferase associated). Several proteins have been identified as reader of m6A such as HNRNPC (heterogenous nuclear ribonucleoprotein C), HNRNPA2B1 (heterogenous nuclear ribonucleoprotein A2/B1), YTHDF2 (YTH N6-methyladenosine RNA-binding protein 2), YTHDF1 (YTH N6-methyladenosine RNA-binding protein 1), and eIF3 (eukaryotic initiation factor 3). Finally, two proteins have been suggested to remove the methyl group from the adenosine, including FTO/ALKBH9 (alpha-ketoglutarate-dependent dioxygenase) and ALKBH5 (alkb homolog 5, RNA demethylase) for m6A.[15]

The METTL3, a 70-kDa protein, was identified by Bokar in 1997 and isolated from the 200-kDa methyltransferase complex.[16] In phylogenetic analysis, it has been reported that METTL14 is homologous to METTL3, as it was approximately 43% similar.[17] Suppression of METTL14 resulted in a significant reduction in the level of m6A in HeLa and human HEK293 cells. Although METTL14 resulted in more methylation than METTL3, the co-existence of both proteins has shown a synergistic effect in terms of altering the m6A modification level. Currently, the METTL14−METLL3 heterodimer structure appears to be an important component of the mammalian m6A methyltransferase complex. WTAP has been identified as the third component of the m6A methyltransferase complex,[18] which binds to Wilm's tumor 1 protein and plays a regulatory role in the cell cycle and embryionic development. Studies have shown that WTAP easily binds to the METTL14−METTL3 heterodimer complex and does not play a role in m6A accumulation alone, but acts together with the heterodimer structure. M6A eraser in mammals, FTO, was first discovered as the enzyme that demethylates m6A oxidatively in mRNA. FTO belongs to the AlkB subfamily consisting of 8 different members. In a study conducted by Jia in 2011, it was revealed that FTO has the potential to remove m6A marks with demethylase activity in nuclear RNA.[19] Subsequent studies have shown that FTO oxidizes m6A, forming two new structures previously unknown. These are N6-hydroxymethyladenosine (hm6A) and N6-formyladenosine (f6A) structures. With this feature, it is equivalent to the conversion of DNA to m5C (methylcytosine), hm5C (hydroxymethylcytosine), and f5C (5 formylcytosine) via TET (10−11 translocation) enzyme.[20] Although later changes occurring in DNA are quite stable, intermediate products formed in RNA and formed by the oxidation of FTO have a short (approximately 3 h) life under physiological conditions. Like FTO, ALKBH5 has recently been discovered as m6A RNA demethylase enzyme, and it has been shown that m6A can be removed from RNAs.[21] Like FTO, ALKBH5 protein binds to single-stranded RNA. The difference from FTO is that ALKBH5 directly converts m6A to adenosine. ALKBH5 mainly co-exists with nucleus speckles and is involved in the transfer of mRNA to the cytoplasm and, thus, RNA metabolism.[22] ALKBH5 expression increased in testes with impaired male fertility in knockout mice.[22] In addition, there are several proteins that recognize and intearct with m6A during cellular processes. Studies conducted in 2012 identified several m6A readers. YTHDF1-3 family proteins have been found as m6A readers in the cytoplasm.[23] Another reader protein, YTHDC1, was discovered by Xu in 2014. A recent study determined that a nuclear ribonuleoprotein (hnRNPs) was indirectly a nuclear m6A reader. m6A modification was associated with RNA splicing[22,24,25] and RNA stability.[26,27] In addition, RNA m6A has been reported to be involved in mRNA degradation (half-life).[15] RNA m6A modifiying genes are involved in many biological processes. For example, WTAP is involved in RNA editing.[22,24,25] YTHDF1 and eIF3 were associated with the translation process,[26,28] and YTHDF2 plays an important role in mRNA degradation.[15]

RNA 5-methylcytosine (m5C) modification

RNA methylation occurs at a variety of atoms, nucleotides, sequences, and tertiary structures. Methylation of different RNA species includes tRNA, rRNA, mRNA, tmRNA, snRNA, snoRNA, miRNA, and viral RNA. Among the various modifications, a privileged position is held by m5C. This residue has been identified in several cellular RNAs and complex enzymatic machinery because its synthesis was found in representative organisms from all kingdoms of life. Currently, numerous methyltransferase enzymes are known that catalyze the addition of methyl group to the fifth position of cytosine. Enzymes are distributed in different subcellular localizations including nucleus, nucleolus, cytoplasm, and mitochondria. Although the main targets are tRNAs, other types of RNAs are also target of enzymes that catalyze the methyl group addition at various.

The presence of m5C has been shown in 240 regions belonging to tRNAs and 10275 mRNA and noncoding RNA regions.[10] It was determined that m5C was a common mark especially in the TC-stem region, and cytosine at the 48th and 49th positions are methylated in tRNAs. The m5C involved in the tRNA is typically located in the variable region and antiloop region of the tRNA. Cytosine at two positions (4413/4 and 3761) of human 28S rRNA have been reported to be methylated. However, so far, no methylation has been observed in 18S rRNA.[29] DNA methyltransferase 2 (DNMT2) is the first enzyme catalyzing addition of methyl group to $tRNA^{Asp}$ at the 38th position.[30] NSUN2 is another m5C methyltransferase enzyme that was discovered later. NSUN2 has the potential to add methyl group to long noncoding RNAs. Evidence indicate that m5C modification has multiple functions in cellular response and tRNA stability,[31] response to cellular stress,[32,33] and was reported to be involved in cell growth.[34] On the other hand, it was stated that tRNA modifications increase ribosome-binding affinity, reduce false readings, and thus play a role in translation.[35] It was reported that the RNA m5C modification stabilized the secondary structure of the tRNA, affects amino-acetylation, and took part in codon recognition.[36] rRNA m5C modifications has a role in translational fidelity and tRNA recognition.[36] Overall, such findings provide that cytosine methylation in various RNAs has critical roles in gene regulation.

Although there are several methyltransferase enzymes that have been identified, as far as we know, no demethylase or reader enzymes have been identified for RNA m5C modifications. Enzymes, belonging to various families available in nucleolus, nucleus, mitochondria, cytoplasm, and golgi, act as methyltransferase enzyme of various types of RNAs.[37] In human, DNMT2 has been identified to catalyze methyl group to the fifth position of cytosine in $tRNA^{Asp,\ Gly,\ and\ Val}$.[38] In addition, NSUN1, NSUN2, NSUN3, NSUN4, NSUN5, NSUN6, and NSUN7 have been identified in a number of RNAs in *Homo sapiens*.[29] NSUN1 and NSUN5 catalyze the addition of methyl group to the 28S rRNA at positions of 4413 and 3761, respectively.[39,40]

NSUN2 localizes in nucleus and nucleolus to catalyze the addition of methyl group to the tRNAs.[41,42] NSUN3 and NSUN4 localize in mitochondria and have catalytic function to methylate mt-tRNA[43] and mt-12S rRNA.[44] NSUN requires for the methylation of tRNA at the position of 72.[45]

RNA A-to-I editing

RNA editing occurs in the specific bases of RNAs and is a posttranscriptional modification that resulted in RNA sequencing alteration. After RNA editing, new sequence is different from its encoded DNA sequences.[46] Two forms of RNA editing including adenosine to inosine (A-to-I) and cytidine (C) and uridine (U) conversion are common. It was discovered in *Xenopus laevis* embryos[47,48] and identified in human tissues in 1987 as cytidine (C) to uridine (U) conversion apolipoprotein-B48 gene's transcript.[49] Currently, several enzymes, which catalyze the conversation of A-to-I editing event in mammals, have been identified. The members of adenosine deaminase that act on the RNA (ADAR enzymes) family have been identified for catalyzing the editing adenosine deamination to inosine. After conversion, inosine is recognized as guanosine,[50] which is base-paired with cytidine in a double-stranded RNA.[51]

It has been known that RNA A-to-I editing event is a common event in various types of RNA. Furthermore, RNA A-to-I editing has been observed in more than 30 tissues.[52] Human mRNAs contain RNA A-to-I editing at more than 14500 sites.[53] However, the most common RNA A-to-I editing events were determined in Alu elements edited in 1.6 million sites.[54] RNA A-to-I editing was also detected in the precursors of microRNAs. RNA A-to-I editing was observed in the primary transcripts of several microRNAs in human brain, which was associated with miRNA processing.[55] It was reported that the presence of RNA A-to-I editing in pre-miRNA affected the interaction with the dicer complex.[56] Specifically, RNA A-to-I editing in mir-376 alters the gene expression by affecting the binding of miRNA to the targets.[56,57] RNA A-to-I editing—modifying enzymes have been associated with the function of the miRNA. It was reported that ADAR enzymes affect miRNA processing and silencing processes mediated by miRNA.[58,59] RNA editing occurs following the reaction catalyzed by adenosine deaminase RNA-specific (ADAR) genes, including ADAR1, ADARB1, and ADARB2. Although the expression of three enzymes varies depending on tissues, ADARB2 is particularly expressed in brain.[60–62] ADAR1 enzyme converts A-to-I in double-helical RNA substrates in humans.[60,63] Except for RNA editing—modifying genes, adenosine deaminase domain—containing proteins (ADAD1 and ADAD2) were identified as double-strand RNA-binding proteins.[64]

Role of RNA modifications in development

Substantial evidences indicate that RNA modifications play critical roles in RNA metabolisim and functions. Studies with different organisms have shown that

RNA modifications are involved in processes such as development and gametogenesis. m6A is the most identified RNA modification on mRNAs, and it is a reversible modification in mammalian cells. Therefore, topology of RNA m6A modification in mRNAs can be altered depending on different stages of cells, diseases, exposure etc. The global level of m6A has been associated with various cellular processes such as mRNA splicing, translation, expression, and degradation of mRNA.[65] Results obtained from transcriptome mapping studies revealed that approximately 10000 m6A residues were determined in 5578 transcripts, including 5461 mRNAs and 117 lncRNAs in mouse embryonic stem cells (ESCs). Similarly, the presence of m6A was observed in human ESCs, which include 16943 m6A marks representing 7530 coding and 341 noncoding RNAs. Depending on differentiation, such as endoderm differentiation, the number of m6A peaks altered.[66] Inactivation of METTL3, which is one of the writer enzymes of m6A, induced the disappearing of m6A in several genes and prolonged Nanog expression, which involved ESCs proliferation, renewal, and pluripotency during differentiation. Also, it affected ESCs exit from self-renewal to differentiation, all of which indicate the requirement of m6A for the differentiation of stem cell.[66] Mettl3 knockout in mice was associated with the early embryonic lethality.[27] The knockout of Mettl3 or Mettl14 (another writer enzymes of m6A) in mouse spermatogenic cells resulted in the depletion of spermatogonia stem cells, which are required for the proliferation/differentiation of spermatogonial stem cells. mRNA m6A methylation affects the translation at different stages of spermatogenesis and thus, has a crucial role in germline development.[67] METTL3 was reported as an essential m6A protein for male fertility and spermatogenesis.[68] In another study, the RNA m6A was reported to play a critical role in terms of coordinating different stages of spermatogenesis by regulating gene expression.[67]

RNA m6A has been reported as a necessary modification for stem cells to differentiate into a specific lineage. It was shown that inactivation of Mettl3 enzyme affected human embryonic stem cells exit from self-renewal into differentiation of several lineages.[66] Knockdown of two methyltransferase enzymes, Mettl3 and Mettl14, resulted in loss of self-renewal capability in mouse embryonic stem cells without a self-renewal capability and a decreased level of RNA m6A that correlated with the mRNA stability and gene expression of developmental regulators.[26] During cell fate transition, the transcript modified by m6A facilitates their turnover and thus has a role on the transition of cellular states.[69] In an in-vivo study, the absence of Mettl3 protein affected the T cells differentiation. In addition, T cell remained in naïve states for up to 12 weeks in Mettl3-deficient T cells. It was observed that the mRNAs of SOCS family genes were modified by RNA m6A. These modified mRNAs showed a slower mRNA decay and also increased the expression of protein in naïve T cells lacking Mettl3.[70] The role of epitranscriptomic marks in mammalian cortical neurogenesis was also shown in embryonic mouse brain. Knockout of Mettl3 was associated with cell cycle progression time and resulted in delaying the cortical neurogenesis. In the model, loss of Mettl3 was associated with mRNA decays that had functions related to neurogenesis and neuronal differentiation.[11]

The knockout of Ythdf2 in mice, a reader protein of m6A, resulted in lethality, and neural development did not occur properly in embryos. Reducing the proliferation and differentiation potential of several cell types in embryonic neocortex was significantly affected by the depletion of Ythdf2. More importantly, it was suggested that Ythdf2 regulated the neural development through RNA m6A-dependent way.[70,71] Similar observations were observed in zebrafish embryos exhibiting lower decay of maternal mRNA modified by m6A due to the depletion of Ythdf2. Such embryos showed several irregularities such as the maternal-to-zygotic transition initiation and delaying larval life development.[72] Mettl16, an another RNA m6A writer, was associated with a decreased level of m6A in mouse embryos cells. Thus, m6A writer enzymes have been suggested as a critical protein for mouse embryonic development.[73,74] Zinc finger protein 217 (ZFP217) has a potential to activate several genes. It plays a role with METTL3 to form m6A in the transcripts. The depletion of ZFP217 was associated with an increase in m6A and ESC self-renewal and somatic cell reprogramming. Furthermore, the author reported accumulation of m6A in pluripotency regulators such as the Nanog, Sox2, Klf4, and Myc.[75]

Genes involved in RNA m6A regulation are associated with developmental process and they could negatively affect some processes in spermatogenesis and/or oogenesis. For example, ALKBH5 is the mammalian demethylase of m6A, and its deficiency in male mice resulted in the decrease of m6A in mRNA, which was associated with impaired infertility due to the failure in meiotic metaphase stage spermatocytes.[22] Therefore, it can be concluded that the demethylation of m6A has the potential to affect the development process in mammalian cells. In addition, Alkbh5 in mice was associated with infertility due to the defect of division of spermatocytes.[22] It was reported that YTHDC2 had a role in cell fate transition, including from mitosis to meiosis. Besides, expression of several meiotic markers was affected by the knockout of YTHDC2 in male germ cells.[76] YTHDC2 was also shown to be an essential protein for male and female infertility by controlling the gene expression through meiosis[74] and regulated the transition from different stages, including proliferation to differentiation. The conserved RNA helicase YTHDC2 regulates the transition from proliferation to differentiation in the germline.[76,77] YTHDF2 was also associated with infertility in female[78] and spermatogenesis.[79]

RNA modifications are also present in noncoding RNAs such as tRNAs, rRNAs, long-noncoding RNAs (lncRNAs), circular RNA (circRNAs), and microRNAs (miRNAs). NSUN2 is an RNA m5C methyltransferase enzyme that was associated with cellular stress and neurodevelopment disorders[41], which was a critical protein for stem cell fate by affecting the stem cell self-renewal and differentiation in skin.[80] DNMT2 catalyzed the addition of methyl group to the C38 in several tRNAs associated with hematopoiesis. It was observed that the number of hematopoietic stem and progenitor cell decreased following the knocking out of Dnmt2 in mice.[81] In a similar study, the knockdown of Dnmt2 in zebra fish embryos was associated with anomalies in several organs including brain, liver, and retina. It was observed that Dnmt2 protein in cytoplasm was needed for proper organ differentiation and thus suggested as critical components for zebrafish development.[82] In a knockdown mouse model with Dnmt2 and Nsun2, mouse vitality was not affected due to the the lack of both proteins. However, underdevelopment and altered cellular differentiation in double mutant were observed when both enzyme models characterized with the loss of tRNA m5C lacked. Overall, the tRNA m5C modification was suggested as the component of mouse development.[42] Interestingly, PUS7 is an enzyme of pseudouridylation in tRNA, and the inactivation of PUS7 altered the translation and the protein synthesis impaired cell layer in embryonic stem cells.[83] Dyskerin (Dkc1) is a protein catalyzing pseudouridylation in rRNA[84] and is mutated in X-linked dyskeratosis congenital. Mutation in Dkc1 also implicates embryonic lethality in mice.[85] The DKC1 complex, which is the responsible protein for the pseudouridylation of rRNA, regulates pluripotency genes, such as OCT4, SOX2, and NANOG, requiring for self-renewal in embryonic stem cells. Furthermore, lack of DKC1 was associated with a decreased efficiency to generate pluripotent stem (iPS) cells.[86]

RNA editing is one of the RNA modifications that have a crucial roles in biological processes. Several enzymes have been identified for catalyzing the RNA A-to-I editing event. Such enzymes have been linked with development. ADAR1 belongs to the ADAR (adenosine deaminase acting on RNA) gene family, and the ablation of ADAR1 includes embryos lethality with defects in the hematopoietic system.[87] Similarly, mutant mice (ADAR2−/−) have been associated with postnatal lethality.[88] The depletion of ADAR1 in mice was associated with lethality in mice embryos with activated double-strand RNA (dsRNA). RNA A-to-I editing in dsRNA is critical for the repression of the activation of the cytosolic dsRNA response.[89] RNA editing A-to-I editing genes were associated with fertility and gametogenesis. ADARB1, RNA A-to-I editing modification enzyme, was the responsible enzyme for editing event in testicular events and was associated with fertility in male.[90] RNA A-to-I editing is regulated by adenosine deaminase (AD) domain such as ADAD1 and ADAD2. It was reported that the two proteins were essential for male sterility and that ADAD2 was also required for the germ cell progression.[91] In mouse embryos, ADAR1 was a critical protein for the proper development of red blood cells.[92] ADAR1 was also reported as a crucial protein for normal erythropoiesis.[93] In humans, three RNA A-to-I editing patterns were observed during cortical development. Expression of ADAR was associated with different patterns of RNA A-to-I editing. The authors reported that increasing pattern of RNA A-to-I editing event was implicated with neuronal maturation.[94] In pigs, ADAR1 was determined as the most expressed enzymes of the ADAR family during postnatal pineal gland development. It was observed that RNA editing level changed depending on developmental stages.[95] Adar1 and Adar2, catalyzing RNA A-to-I editing, were associated with embryonic and postnatal lethality in mice. The depletion of Adar1 resulted in normal development; however, after E10.5, cellular apoptosis was observed in multiple tissues and reported to have late-stage embryonic lethality.[96,97] The depletion of Adar1 in cardiomyocytes in mice resulted in death at birth, indicating the critical role of Adar1 in early embryonic development. The authors summarized that Adar1 was required for the survival and proliferation of embryonic cardiomyocytes.[97]

Conclusion and future perspective

Recent advances have highlighted the association between RNA modifications and developmental processes Taken together, studies discussed above, indicate that there is a network composed of RNA modifications and RNA modification–modifying genes in different stages of cells to regulate the developmental processes. Our current understanding is very limited since we have limited data about most of the RNA modifications and their role in developmental processes. Although there are more than 150 RNA modifications which have been identified, due to the limited data, we focused the role of a few RNA modifications on developmental processes. Much of our current understanding is based on RNA modification–modifying genes such as writers, readers, and erasers. Very few studies investigated the profiling of RNA modifications in specific RNAs such as coding and noncoding RNAs. The main conclusion is that modification in mRNA and noncoding RNAs has unique roles, which affects the gene expression and, thus, affects the processes from gametogenesis to subsequent developmental stages.

The studies discussed here contribute to our understanding the role of epitranscriptomics field in developmental processes, which is still in its infancy. We need more studies to determine how the whole network of RNA modifications and/or RNA modification–modifying genes affects the developmental stages. Future studies could unravel how RNA modifications profiling in different RNAs can alter depending on developmental stages. For many modifications, very little information is available regarding how the stability and dynamics of RNA modifications are controlled through different cell types in developmental stages. We also need to determine how readers, writers, and erasers of reversible RNA modifications contribute to the cells differentiations and which enzymes have functional roles in which stages of development. Furthermore, it will be essential to understand what the dynamics of the epitranscriptome are over the differentiation of cells. In this context, the ratio of modified and unmodified any bases in RNAs types might be critical for the development. Therefore, determination of the ratio of modified and unmodified RNA modification, for example, reversible RNA m6A modification, could explain the dynamic properties of the RNA modification. RNA modifications and RNA modification–modifying enzymes are present in different locations in cells. It would be interesting to study how RNA modifications in various subcellular localizations including cytoplasm, nucleus, nucleolus, and mitochondria could contribute to the differentiation of different cell types across development. Understanding the position of the modification(s) and changing the position or the quantity of the modification depending on the cell types in development will also contribute to reveal the role of RNA modifications. Finally, the studies providing the association between RNA modifications and development indicated that various RNA modifications have a distinct role in various stages. Further chracterization of this combinatorial code of RNA modifications may shed light on the role of RNA modifications in fine tuning gene expression during various developmental stages.

References

1. Liyanage VRB, Jarmasz JS, Murugeshan N, Bigio MRD, Rastegar M, Davie JR. DNA modifications: function and applications in normal and disease states. *Biology*. 2014;3(4):670–723. https://doi.org/10.3390/biology3040670.
2. Phillips T. The role of methylation in gene expression. *Nat Educ*. 2008;1(1):116.
3. Moosavi A, Ardekani AM. Role of epigenetics in biology and human diseases. *Iran Biomed J*. 2016;20(5):246–258. https://doi.org/10.22045/ibj.2016.01.
4. Waddington CH. The epigenotype. *Endeavour*. 1942;1:18–20.
5. Saletore Y, Meyer K, Korlach J, Vilfan ID, Jaffrey S, Mason CE. The birth of the Epitranscriptome: deciphering the function of RNA modifications. *Genome Biol*. 2012;13(10):175. https://doi.org/10.1186/gb-2012-13-10-175.
6. Boccaletto P, Stefaniak F, Ray A, et al. MODOMICS: a database of RNA modification pathways. 2021 update. *Nucleic Acids Res*. 2022;50(D1):D231–D235. https://doi.org/10.1093/nar/gkab1083.
7. Boccaletto P, MacHnicka MA, Purta E, et al. MODOMICS: a database of RNA modification pathways. 2017 update. *Nucleic Acids Res*. 2018;46(1):D303–D307. https://doi.org/10.1093/nar/gkx1030.
8. Dominissini D, Moshitch-Moshkovitz S, Schwartz S, et al. Topology of the human and mouse m6A RNA methylomes revealed by m6A-seq. *Nature*. 2012;485(7397):201–206. https://doi.org/10.1038/nature11112.
9. Meyer KD, Saletore Y, Zumbo P, Elemento O, Mason CE, Jaffrey SR. Comprehensive analysis of mRNA methylation reveals enrichment in 3′ UTRs and near stop codons. *Cell*. 2012;149(7):1635–1646. https://doi.org/10.1016/j.cell.2012.05.003.
10. Squires JE, Patel HR, Nousch M, et al. Widespread occurrence of 5-methylcytosine in human coding and non-coding RNA. *Nucleic Acids Res*. 2012;40(11):5023–5033. https://doi.org/10.1093/nar/gks144.
11. Yoon KJ, Ringeling FR, Vissers C, et al. Temporal control of mammalian cortical neurogenesis by m6A methylation. *Cell*. 2017;171(4):877–889.e17. https://doi.org/10.1016/j.cell.2017.09.003.
12. Dubin DT, Taylor RH. The methylation state of poly A-containing messenger RNA from cultured hamster cells. *Nucleic Acids Res*. 1975;2:1653–1668. https://doi.org/10.1093/nar/2.10.1653.
13. Rottman F, Shatkin AJ, Perry RP. Sequences containing methylated nucleotides at the 5′ termini of messenger RNAs: possible implications for processing. *Cell*. 1974;3(3):197–199. https://doi.org/10.1016/0092-8674(74)90131-7.
14. Schwartz S, Agarwala SD, Mumbach MR, et al. High-resolution mapping reveals a conserved, widespread, dynamic mRNA methylation program in yeast meiosis. *Cell*. 2013;155(6):1409–1421. https://doi.org/10.1016/j.cell.2013.10.047.
15. Cao G, Li HB, Yin Z, Flavell RA. Recent advances in dynamic m6A RNA modification. *Open Biol*. 2016;6(4). https://doi.org/10.1098/rsob.16.0003.

16. Bokar JA, Shambaugh ME, Polayes D, Matera AG, Rottman FM. Purification and cDNA cloning of the AdoMet-binding subunit of the human mRNA (N6-adenosine)-methyltransferase. *RNA*. 1997;3(11):1233–1247.
17. Bujnicki JM, Feder M, Radlinska M, Blumenthal RM. Structure prediction and phylogenetic analysis of a functionally diverse family of proteins homologous to the MT-A70 subunit of the human mRNA:m6A methyltransferase. *J Mol Evol*. 2002;55(4): 431–444. https://doi.org/10.1007/s00239-002-2339-8.
18. Yue Y, Liu J, He C. RNA N6-methyladenosine methylation in post-transcriptional gene expression regulation. *Genes Dev*. 2015;29(13):1343–1355. https://doi.org/10.1101/gad.262766.115.
19. Jia G, Fu Y, Zhao X, et al. N6-methyladenosine in nuclear RNA is a major substrate of the obesity-associated FTO. *Nat Chem Biol*. 2011;7(12):885–887. https://doi.org/10.1038/nchembio.687.
20. Tahiliani M, Koh KP, Shen Y, et al. Conversion of 5-methylcytosine to 5-hydroxymethylcytosine in mammalian DNA by MLL partner TET1. *Science*. 2009;324(5929): 930–935. https://doi.org/10.1126/science.1170116.
21. Niu Y, Zhao X, Wu YS, Li MM, Wang XJ, Yang YG. N6-methyl-adenosine (m6A) in RNA: an old modification with a novel epigenetic function. *Dev Reprod Biol*. 2013; 11(1):8–17. https://doi.org/10.1016/j.gpb.2012.12.002.
22. Zheng G, Dahl JA, Niu Y, et al. ALKBH5 is a mammalian RNA demethylase that impacts RNA metabolism and mouse fertility. *Mol Cell*. 2013;49(1):18–29. https://doi.org/10.1016/j.molcel.2012.10.015.
23. Wan Y, Qu K, Zhang QC, et al. Landscape and variation of RNA secondary structure across the human transcriptome. *Nature*. 2014;505(7485):706–709. https://doi.org/10.1038/nature12946.
24. Little NA, Hastie ND, Davies RC. Identification of WTAP, a novel Wilms' tumour 1-associating protein. *Hum Mol Genet*. 2000;9(15):2231–2239. https://doi.org/10.1093/oxfordjournals.hmg.a018914.
25. Ping XL, Sun BF, Wang L, et al. Mammalian WTAP is a regulatory subunit of the RNA N6-methyladenosine methyltransferase. *Cell Res*. 2014;24(2):177–189. https://doi.org/10.1038/cr.2014.3.
26. Wang Y, Li Y, Toth JI, Petroski MD, Zhang Z, Zhao JC. N6-methyladenosine modification destabilizes developmental regulators in embryonic stem cells. *Nat Cell Biol*. 2014; 16(2):191–198. https://doi.org/10.1038/ncb2902.
27. Geula S, Moshitch-Moshkovitz S, Dominissini D, et al. m6A mRNA methylation facilitates resolution of naïve pluripotency toward differentiation. *Science*. 2015;347(6225): 1002–1006. https://doi.org/10.1126/science.1261417.
28. Meyer KD, Patil DP, Zhou J, et al. 5′ UTR m6A promotes cap-independent translation. *Cell*. 2015;163(4):999–1010. https://doi.org/10.1016/j.cell.2015.10.012.
29. Motorin Y, Lyko F, Helm M. 5-methylcytosine in RNA: detection, enzymatic formation and biological functions. *Nucleic Acids Res*. 2009;38(5):1415–1430. https://doi.org/10.1093/nar/gkp1117.
30. Trixl L, Lusser A. The dynamic RNA modification 5-methylcytosine and its emerging role as an epitranscriptomic mark. *Wiley Interdiscip Rev RNA*. 2019;10(1):e1510.
31. Alexandrov A, Chernyakov I, Gu W, et al. Rapid tRNA decay can result from lack of nonessential modifications. *Mol Cell*. 2006;21(1):87–96. https://doi.org/10.1016/j.molcel.2005.10.036.

32. Begley U, Dyavaiah M, Patil A, et al. Trm9-catalyzed tRNA modifications link translation to the DNA damage response. *Mol Cell*. 2007;28(5):860–870. https://doi.org/10.1016/j.molcel.2007.09.021.
33. Netzer N, Goodenbour JM, David A, et al. Innate immune and chemically triggered oxidative stress modifies translational fidelity. *Nature*. 2009;462(7272):522–526. https://doi.org/10.1038/nature08576.
34. Emilsson V, Näslund AK, Kurlad CG. Thiolation of transfer RNA in *Escherichia coli* varies with growth rate. *Nucleic Acids Res*. 1992;20(17):4499–4505. https://doi.org/10.1093/nar/20.17.4499.
35. Chan CTY, Pang YLJ, Deng W, et al. Reprogramming of tRNA modifications controls the oxidative stress response by codon-biased translation of proteins. *Nat Commun*. 2012;3. https://doi.org/10.1038/ncomms1938.
36. Liebers R, Rassoulzadegan M, Lyko F. Epigenetic regulation by heritable RNA. *PLoS Genet*. 2014;10(4). https://doi.org/10.1371/journal.pgen.1004296.
37. Katherine B, Claudia H, Markus B. Eukaryotic 5-methylcytosine (m5C) RNA methyltransferases: mechanisms, cellular functions, and links to disease. *Genes*. 2019:102. https://doi.org/10.3390/genes10020102.
38. Goll MG, Kirpekar F, Maggert KA, et al. Methylation of tRNAAsp by the DNA methyltransferase homolog Dnmt2. *Science*. 2006;311(5759):395–398. https://doi.org/10.1126/science.1120976.
39. Bourgeois G, Ney M, Gaspar I, et al. Eukaryotic rRNA modification by yeast 5-methylcytosine-methyltransferases and human proliferation-associated antigen p120. *PLoS One*. 2015;10(7). https://doi.org/10.1371/journal.pone.0133321.
40. Schosserer M, Minois N, Angerer TB, et al. Methylation of ribosomal RNA by NSUN5 is a conserved mechanism modulating organismal lifespan. *Nat Commun*. 2015;6. https://doi.org/10.1038/ncomms7158.
41. Blanco S, Dietmann S, Flores JV, et al. Aberrant methylation of tRNAs links cellular stress to neuro-developmental disorders. *EMBO J*. 2014;33(18):2020–2039. https://doi.org/10.15252/embj.201489282.
42. Tuorto F, Liebers R, Musch T, et al. RNA cytosine methylation by Dnmt2 and NSun2 promotes tRNA stability and protein synthesis. *Nat Struct Mol Biol*. 2012;19(9):900–905. https://doi.org/10.1038/nsmb.2357.
43. Van Haute L, Dietmann S, Kremer L, et al. Deficient methylation and formylation of mt-tRNAMet wobble cytosine in a patient carrying mutations in NSUN3. *Nat Commun*. 2016;7. https://doi.org/10.1038/ncomms12039.
44. Metodiev MD, Spåhr H, Loguercio Polosa P, et al. NSUN4 is a dual function mitochondrial protein required for both methylation of 12S rRNA and coordination of mitoribosomal assembly. *PLoS Genet*. 2014;10(2). https://doi.org/10.1371/journal.pgen.1004110.
45. Liu RJ, Long T, Li J, Li H, Wang ED. Structural basis for substrate binding and catalytic mechanism of a human RNA:m5C methyltransferase NSun6. *Nucleic Acids Res*. 2017;45(11):6684–6697. https://doi.org/10.1093/nar/gkx473.
46. Uchida S, Jones SP. RNA editing unexplored opportunities in the cardiovascular system. *Circ Res*. 2018;122(3):399–401. https://doi.org/10.1161/CIRCRESAHA.117.312512.
47. Bass BL, Weintraub H. A developmentally regulated activity that unwinds RNA duplexes. *Cell*. 1987;48(4):607–613. https://doi.org/10.1016/0092-8674(87)90239-X.

48. Rebagliati MR, Melton DA. Antisense RNA injections in fertilized frog eggs reveal an RNA duplex unwinding activity. *Cell.* 1987;48(4):599–605. https://doi.org/10.1016/0092-8674(87)90238-8.
49. Powell LM, Wallis SC, Pease RJ, Edwards YH, Knott TJ, Scott J. A novel form of tissue-specific RNA processing produces apolipoprotein-B48 in intestine. *Cell.* 1987;50(6):831–840. https://doi.org/10.1016/0092-8674(87)90510-1.
50. Bass BL. RNA editing by adenosine deaminases that act on RNA. *Annu Rev Biochem.* 2002;71:817–846. https://doi.org/10.1146/annurev.biochem.71.110601.135501.
51. Nishikura K. A-to-I editing of coding and non-coding RNAs by ADARs. *Nat Rev Mol Cell Biol.* 2016;17(2):83–96. https://doi.org/10.1038/nrm.2015.4.
52. Kim DDY, Kim TTY, Walsh T, et al. Widespread RNA editing of embedded Alu elements in the human transcriptome. *Genome Res.* 2004;14(9):1719–1725. https://doi.org/10.1101/gr.2855504.
53. Alekos A, Alexander R, Stefan M. Widespread A-to-I RNA editing of Alu-containing mRNAs in the human transcriptome. *PLoS Biol.* 2004:e391. https://doi.org/10.1371/journal.pbio.0020391.
54. Bazak L, Haviv A, Barak M, et al. A-to-I RNA editing occurs at over a hundred million genomic sites, located in a majority of human genes. *Genome Res.* 2014;24(3):365–376. https://doi.org/10.1101/gr.164749.113.
55. Kawahara Y, Megraw M, Kreider E, et al. Frequency and fate of microRNA editing in human brain. *Nucleic Acids Res.* 2008;36(16):5270–5280. https://doi.org/10.1093/nar/gkn479.
56. Kawahara Y, Zinshteyn B, Chendrimada TP, Shiekhattar R, Nishikura K. RNA editing of the microRNA-151 precursor blocks cleavage by the Dicer—TRBP complex. *EMBO Rep.* 2007;8(8):763–769. https://doi.org/10.1038/sj.embor.7401011.
57. Kawahara Y, Zinshteyn B, Sethupathy P, Iizasa H, Hatzigeorgiou AG, Nishikura K. Redirection of silencing targets by adenosine-to-inosine editing of miRNAs. *Science.* 2007;315(5815):1137–1140. https://doi.org/10.1126/science.1138050.
58. Öhman M. A-to-I editing challenger or ally to the microRNA process. *Biochimie.* 2007;89(10):1171–1176. https://doi.org/10.1016/j.biochi.2007.06.002.
59. Mannion N, Arieti F, Gallo A, Keegan LP, O'Connell MA. New insights into the biological role of mammalian ADARs; the RNA editing proteins. *Biomolecules.* 2015;5(4):2338–2362. https://doi.org/10.3390/biom5042338.
60. O'Connell MA, Krause S, Higuchi M, et al. Cloning of cDNAs encoding mammalian double-stranded RNA-specific adenosine deaminase. *Mol Cell Biol.* 1995;15(3):1389–1397. https://doi.org/10.1128/mcb.15.3.1389.
61. Melcher T, Maas S, Herb A, Sprengel R, Higuchi M, Seeburg PH. RED2, a brain-specific member of the RNA-specific adenosine deaminase family. *J Biol Chem.* 1996;271(50):31795–31798. https://doi.org/10.1074/jbc.271.50.31795.
62. Chen CX, Cho DSC, Wang Q, Lai F, Carter KC, Nishikura K. A third member of the RNA-specific adenosine deaminase gene family, ADAR3, contains both single- and double-stranded RNA binding domains. *RNA.* 2000;6(5):755–767. https://doi.org/10.1017/S1355838200000170.
63. Hough RF, Bass BL. Purification of the *Xenopus laevis* double-stranded RNA adenosine deaminase. *J Biol Chem.* 1994;269(13):9933–9939.
64. Schumacher JM, Lee K, Edelhoff S, Braun RE. Distribution of Tenr, an RNA-binding protein, in a lattice-like network within the spermatid nucleus in the mouse. *Biol Reprod.* 1995;52(6):1274–1283. https://doi.org/10.1095/biolreprod52.6.1274.

65. Meyer KD, Jaffrey SR. The dynamic epitranscriptome: N6-methyladenosine and gene expression control. *Nat Rev Mol Cell Biol*. 2014;15(5):313–326. https://doi.org/10.1038/nrm3785.
66. Batista PJ, Molinie B, Wang J, et al. M6A RNA modification controls cell fate transition in mammalian embryonic stem cells. *Cell Stem Cell*. 2014;15(6):707–719. https://doi.org/10.1016/j.stem.2014.09.019.
67. Lin Z, Hsu PJ, Xing X, et al. Mettl3-/Mettl14-mediated mRNA N 6-methyladenosine modulates murine spermatogenesis. *Cell Res*. 2017;27(10):1216–1230. https://doi.org/10.1038/cr.2017.117.
68. Xu K, Yang Y, Feng GH, et al. Mettl3-mediated m 6 A regulates spermatogonial differentiation and meiosis initiation. *Cell Res*. 2017;27(9):1100–1114. https://doi.org/10.1038/cr.2017.100.
69. Frye M, Harada BT, Behm M, He C. RNA modifications modulate gene expression during development. *Science*. 2018;361(6409):1346–1349. https://doi.org/10.1126/science.aau1646.
70. Li HB, Tong J, Zhu S, et al. M 6 A mRNA methylation controls T cell homeostasis by targeting the IL-7/STAT5/SOCS pathways. *Nature*. 2017;548(7667):338–342. https://doi.org/10.1038/nature23450.
71. Li M, Zhao X, Wang W, et al. Ythdf2-mediated m 6 A mRNA clearance modulates neural development in mice. *Genome Biol*. 2018;19(1):16.
72. Zhao BS, Wang X, Beadell AV, et al. M6 A-dependent maternal mRNA clearance facilitates zebrafish maternal-to-zygotic transition. *Nature*. 2017;542(7642):475–478. https://doi.org/10.1038/nature21355.
73. Mendel M, Chen, Homolka P, et al. Methylation of structured RNA by the m6A writer METTL16 is essential for mouse embryonic development. *Mol Cell*. 2018;71(6): 986–1000, 1011.
74. Wojtas MN, Pandey RR, Mendel M, Homolka D, Sachidanandam R, Pillai RS. Regulation of m6A transcripts by the 3′→5′ RNA helicase YTHDC2 is essential for a successful meiotic program in the mammalian germline. *Mol Cell*. 2017;68(2). https://doi.org/10.1016/j.molcel.2017.09.021, 374-387.e12.
75. Aguilo F, Zhang F, Sancho A, et al. Coordination of m6A mRNA methylation and gene transcription by ZFP217 regulates pluripotency and reprogramming. *Cell Stem Cell*. 2015;17(6):689–704. https://doi.org/10.1016/j.stem.2015.09.005.
76. Bailey AS, Batista PJ, Gold RS, et al. The conserved RNA helicase YTHDC2 regulates the transition from proliferation to differentiation in the germline. *eLife*. 2017;6. https://doi.org/10.7554/eLife.26116.
77. Jain D, Puno MR, Meydan C, et al. Ketu mutant mice uncover an essential meiotic function for the ancient RNA helicase YTHDC2. *eLife*. 2018;7. https://doi.org/10.7554/eLife.30919.
78. Ivanova I, Much C, Giacomo MD, et al. The RNA m6A reader YTHDF2 is essential for the post-transcriptional regulation of the maternal transcriptome and oocyte competence. *Mol Cell*. 2017;67(6):1059–1067, 1054.
79. Hsu PJ, Zhu Y, Ma H, et al. Ythdc2 is an N6-methyladenosine binding protein that regulates mammalian spermatogenesis. *Cell Res*. 2017;27(9):1115–1127. https://doi.org/10.1038/cr.2017.99.
80. Blanco S, Kurowski A, Nichols J, Watt FM, Benitah SA, Frye M. The RNA–methyltransferase Misu (NSun2) poises epidermal stem cells to differentiate. *PLoS Genet*. 2011;7(12):e1002403.

81. Tuorto F, Herbst F, Alerasool N, et al. The tRNA methyltransferase Dnmt2 is required for accurate polypeptide synthesis during haematopoiesis. *EMBO J.* 2015;34(18): 2350–2362. https://doi.org/10.15252/embj.201591382.
82. Rai K, Chidester S, Zavala CV, et al. Dnmt2 functions in the cytoplasm to promote liver, brain, and retina development in zebrafish. *Genes Dev.* 2007;21(3):261–266. https://doi.org/10.1101/gad.1472907.
83. Guzzi N, Cieśla M, Ngoc PCT, et al. Pseudouridylation of tRNA-derived fragments steers translational control in stem cells. *Cell.* 2018;173(5):1204–1216, 1226.
84. Lafontaine DLJ, Bousquet-Antonelli C, Henry Y, Caizergues-Ferrer M, Tollervey D. The box H + ACA snoRNAs carry Cbf5p, the putative rRNA pseudouridine synthase. *Genes Dev.* 1998;12(4):527–537. https://doi.org/10.1101/gad.12.4.527.
85. He J, Navarrete S, Jasinski M, et al. Targeted disruption of Dkc1, the gene mutated in X-linked dyskeratosis congenita, causes embryonic lethality in mice. *Oncogene.* 2002; 21(50):7740–7744. https://doi.org/10.1038/sj.onc.1205969.
86. Fong YW, Ho JJ, Inouye C, Tjian R. The dyskerin ribonucleoprotein complex as an OCT4/SOX2 coactivator in embryonic stem cells. *Elife.* 2014;3:e03573.
87. Wang Q, Khillan J, Gadue P, Nishikura K. Requirement of the RNA editing deaminase ADAR1 gene for embryonic erythropoiesis. *Science.* 2000;290(5497):1765–1768. https://doi.org/10.1126/science.290.5497.1765.
88. Higuchi M, Maas S, Single FN, et al. Point mutation in an AMPA receptor gene rescues lethality in mice deficient in the RNA-editing enzyme ADAR2. *Nature.* 2000;406(6791): 78–81. https://doi.org/10.1038/35017558.
89. Liddicoat BJ, Piskol R, Chalk AM, et al. RNA editing by ADAR1 prevents MDA5 sensing of endogenous dsRNA as nonself. *Science.* 2015;349(6252):1115–1120. https://doi.org/10.1126/science.aac7049.
90. Snyder EM, Licht K, Braun RE. Testicular adenosine to inosine RNA editing in the mouse is mediated by ADARB1. *Biol Reprod.* 2017;96(1):244–253. https://doi.org/10.1095/biolreprod.116.145151.
91. Snyder E, Chukrallah L, Seltzer K, Goodwin L, Braun RE. ADAD1 and ADAD2, testis-specific adenosine deaminase domain-containing proteins, are required for male fertility. *Sci Rep.* 2020;10(1). https://doi.org/10.1038/s41598-020-67834-5.
92. Keegan LP, Gallo A, O'Connell MA. Survival is impossible without an editor. *Science.* 2000;290(5497):1707–1709. https://doi.org/10.1126/science.290.5497.1707.
93. Liddicoat BJ, Hartner JC, Piskol R, et al. Adenosine-to-inosine RNA editing by ADAR1 is essential for normal murine erythropoiesis. *Exp Hematol.* 2016;44(10):947–963. https://doi.org/10.1016/j.exphem.2016.06.250.
94. Hwang T, Park CK, Leung AKL, et al. Dynamic regulation of RNA editing in human brain development and disease. *Nat Neurosci.* 2016;19(8):1093–1099. https://doi.org/10.1038/nn.4337.
95. Rong Z, Wenye Y, Chundi X, et al. Developmental stage-specific A-to-I editing pattern in the postnatal pineal gland of pigs (*Sus scrofa*). *J Anim Sci Biotechnol.* 2020. https://doi.org/10.1186/s40104-020-00495-6.
96. Wang Q, Miyakoda M, Yang W, et al. Stress-induced apoptosis associated with null mutation of ADAR1 RNA editing deaminase gene. *J Biol Chem.* 2004;279(6):4952–4961. https://doi.org/10.1074/jbc.M310162200.
97. Moore JB, Sadri G, Fischer AG, et al. The A-to-I RNA editing enzyme Adar1 is essential for normal embryonic cardiac growth and development. *Circ Res.* 2020;127(4): 550–552. https://doi.org/10.1161/CIRCRESAHA.120.316932.

SECTION 2

Epigenetic programming of placenta regulates short-term and long-term health outcomes

CHAPTER

Epigenetic regulation of placental function

6

Kumari Nishi, Nafisa H. Balasinor*

National Institute for Research in Reproductive and Child Health, Department of Neuroendocrinology, Mumbai, Maharashtra, India

**Corresponding Author.*

Introduction

Placenta is a unique transient exchange organ, considered as the largest organ of fetal origin and the first to develop. It is central to the in utero healthy development of the fetus.[1] A diversity in the functions of different placental cell types is required for implantation, uterine remodeling of vasculature, hormone synthesis, maternal immunomodulation, and nutrient transport to the developing fetus.[2]

The signals arising from placenta not only govern fetal growth but also help in altering maternal physiology during pregnancy. Taken together, all these processes require a tight regulation of gene expression under the influence of the maternal uterine environment of pregnancy. These "critical windows" of gene–environment interface determine the formation and development of placenta and fetal health[3]

Recent advances in the field of epigenetics and pregnancy have unraveled a plethora of information on the establishment of epigenetic marks and how these marks affect trophoblast differentiation and functions, respectively. Moreover, it has a potential role in addressing the role of various environment and genetic factors affecting pregnancy outcomes.[4] Epigenetic regulation of gene expression involves modifications in chromatin, which are reversible in nature. These modifications include DNA methylation, posttranslational modifications of histones, and posttranslational modifications of ncRNAs.[5]

This chapter highlights the placental epigenetic changes occurring during zygotic phases, characteristics of placental epigenome, and epigenetic regulation of placental functions. The epigenetic changes associated with pregnancy-related disorders are also discussed.

Epigenetic regulation during placentation

Epigenetics and human placenta development

The placenta develops from trophectoderm, a single-celled polarized outer layer of the preimplantation embryo, and the first epithelium formed at about 5 days after

Perinatal and Developmental Epigenetics. https://doi.org/10.1016/B978-0-12-821785-6.00006-2

fertilization. The differentiation of blastocyst cells results in the formation of two lineages, i.e., inner cell mass (embryo proper which later becomes developing fetus) and trophectoderm from which the extraembryonic tissue develops. This entire process of differentiation involves gene expression changes facilitated by distinct epigenetic events.[4,6]

The trophectoderm interacts with the uterine luminal epithelium leading to implantation (6–7 days postconception), and the development of placenta starts. Once the implantation has taken place, the differentiation of early mononucleated cytotrophoblasts and multinucleated primitive syncytial trophoblasts takes place by the end of day 8 of conception. It is the primitive syncytium where the invasive placental cells are present which later invade the maternal decidua.[7]

The expansion of primitive syncytium starts at around day 10 postconception followed by the development and morphogenesis of placental villi as the cytotrophoblast buds break through the syncytial mass. The villi are now known as the primitive villi. The primitive villi extend to the vessels and glands in the uterus. The extraembryonic mesodermal cells migrate to the primary villi and transforms primitive villi to secondary villi.[7,8] Furthermore, the villous cytotrophoblasts proliferate and fuse together asymmetrically, leading to the expansion and branching of the epithelium. This is the key event for the formation of syncytiotrophoblast layer responsible for the endocrine functions of the placenta. At around day 17, the secondary villi differentiate into the tertiary villi containing placental blood vessels. By the third week of gestation, the tertiary villi get surrounded by the double trophoblast layers, i.e., syncytiotrophoblast and cytotrophoblast layer. Thus, by the end of 4 weeks, the basic structure of placenta gets completed.[7]

At term, the human placenta is a discoidal organ of diameter about 15–20 cm, weighing about 500 g. The chorionic plate and basal plate are two surfaces of placenta. The chorionic plate is on the fetal side of placenta where the fetus is attached through umbilical cord and the basal plate is the surface facing maternal side. The blood-filled cavity between chorionic plate and basal plate is the intervillous space harboring chorionic villi. Chorionic villi consisting of fibroblasts, mesenchymal cells, endothelial cells, Hofbauer cells, and fetal–placental blood vessels form the structural and functional unit of placenta. The villous cytotrophoblasts form the inner cellular layer of villi. These cytotrophoblasts differentiate and form the outer layer of villous syncytiotrophoblasts or proximal cell column trophoblasts (pCCTs). The pCCTs differentiate into distal cell column trophoblasts which further detach from the dCCTs and invade the maternal tissues.[1,9]

The mechanism of placental invasion is similar to the growth of the cancer cells. The key enzymes regulating placental invasion and cancer cell growth are endopeptidases degrading extracellular matrix (ECM) known as matrix metalloproteinases (MMPs).[10] The endogenous tissue inhibitors of metalloproteinases regulate the activities of MMPs by regulating transcription, translation, and activation.[11,12] The placental invasion includes migration and invasion of stroma and arteries by trophoblasts, epithelial to mesenchyme transition, heterotypic interactions, etc. These *MMP*s specially *MMP2* and *MMP9* are involved in the degradation of the components of maternal basement membrane, ECM-like gelatins, laminin, and collagens (type V and XI). These *MMPs* have also shown to mediate trophoblast invasion in vitro.[13,14]

To maintain a normal pregnancy, a proper balance between *MMP2*, *MMP9*, and their inhibitors *TIMP2* and *TIMP1* is necessary. This balance is regulated epigenetically. The invasiveness and metastasis of trophoblasts are associated with *MMP*s activity, while *TIMP*s activity inhibits metastasis. With an advancing gestation, the activities of *MMP2* and *MMP9* reduce and are associated with reduced trophoblast invasiveness. The CpG site promoter region DNA methylation at promoter regions and histone methylation studies have revealed hypomethylation at promoter regions of *MMP2* and *MMP9* in placental villi of first trimester. In comparison with the first-trimester villi, the *H3K9me3* and *H3K27me3* levels were increased at *MMP2* and *MMP9* promoter region in third-trimester villi. The levels of *H3K9/27me3* were decreased at *TIMP1* and *TIMP2* promoter regions in third-trimester placental villi as compared with first-trimester placental villi. This underlines the role of histone modifications in regulating the activities of *MMP*s during normal gestation. Any disturbance in this regulating pathway may result in placental disorders.[15]

Another interesting factor that is epigenetically regulated during pregnancy is *maspin*, a tumor suppressor gene. *Maspin (SerpinB5)*, a member of noninhibitory Serpin family, is differentially expressed during human placenta development. It is involved in the regulation of various processes such as cell motility, invasion, apoptosis, and angiogenesis.[16,17]

Studies have indicated that the expression of *maspin* in human placenta is epigenetically regulated. *Maspin* is expressed predominantly in cytotrophoblast and syncytiotrophoblast layers of placenta, and its expression increases with advancing gestation. This increased expression is correlated with multiple histone modifications with the advancement of gestation. Studies have also shown that the switching of methylation to acetylation of H3 at K9 increases methylation of H3 and K4, and reduced methylation of H3 at K27 is responsible for the increased maspin levels throughout pregnancy. DNA methylation of maspin promoter is not involved in the process.[16,17]

In human placenta, the trophoblasts differentiate into invasive extravillous trophoblast (EVTs) during early placentation. After the formation of mature placental villi, the cytotrophoblasts in the anchoring villi tip differentiate to form EVTs. In the villi, the proximal cell column trophoblast rows develop, which serve as the progenitor cell pool of differentiated EVTs. This further leads to the formation of distal cell column trophoblasts (dCCTs). At this stage, the cessation of mitosis takes place, but exit from the cell cycle does not take place. The dCCTs enter endoreduplicative cycles and undergo polyploidization and senescence once they differentiate into EVTs. The interstitial cytotrophoblasts and endovascular cytotrophoblasts formed after 15–16 day of conception colonize maternal spiral artery.[18,19]

The interstitial cytotrophoblast cells interact with decidual stromal cells, macrophages, and uterine natural killer cells so as to facilitate immunological acceptance of the developing fetal allograft. Once the interstitial cells reach underlying myometrium, they lose their invasiveness and differentiate into multinucleated cells.[18,19] The migration of EVTs into the maternal spiral artery and spiral artery remodeling is another step in human placentation.[9,20]

The homeobox gene family members are also expressed in placenta. These genes have their role in embryogenesis, organogenesis and regulation of trophoblast migration, proliferation, apoptosis, and epithelial—mesenchymal transition.[21,22] For the most part, the *hox* genes do not show any methylation changes during gestation. Most of the homeobox genes have low methylation at first trimester and stay in a hypomethylated state during gestation. The genes *HOXB9* and *MSX1* show significant hypomethylation, while *DLX5, TLX1, HOXA10, TGIF*, and *HOXB6* show an increased methylation throughout gestation. The methylation levels of genes *TLX1, HOXA10,* and *DLX5* were increased with a subsequent decreased expression with advancing gestation.[23]

Among these homeobox genes, the largest differentially methylated region (DMR) is found in the DLX5 (distal-less homeobox5) with 5CpG sites showing higher expression in second and third trimesters. HOXA10 and TLX1 RNA expression was also reduced in second and third trimesters with increased DNA methylation indicating regulation of these genes during gestation. In placental villi, these genes are primarily localized in the syncytiotrophoblasts. Loss-of-function studies for these genes show a loss of proliferation and altered differentiation of trophoblasts.[4,23]

Ras association domain family 1A (*RASSF1A*), a tumor-suppressing gene, has been found to be hypermethylated in human placental tissue in all three trimesters. *RASSF1A* plays an important role in placental development possibly in cytotrophoblats development by modulating ID2 (inhibitor of DNA binding 1).[24–26]

The trophoblastic shells formed during first phase of invasion form intravascular plugs blocking the maternal artery. Low oxygen level leads to the disaggregation of these plugs during 9—10 weeks, which leads to the more free flow of maternal blood into the intervillous spaces resulting in increased oxygen pressure in intervillous space, which signals for EVT growth and differentiation. A low oxygen concentration prompts the expression of the hypoxia-inducible factor (*HIF1α*), which activates transcription of various genes involved in angiogenesis and metastasis.[27] Oxygen levels impact placental development partly by modifying histone methylation codes. HIF-1 recruits and regulates HDACs[28,29] and induces expression of H3K9 demethylases by binding to specific sites on promoter. HIF-1 also activates *JMJD1A* and *JMJD2A* to remove dimethyl marks on *H3K9me2*, *JMJD2B* to remove trimethyl marks (*H3K9me3*), and *JMJD2C* to converts *H3K9me3* to *me2*.[30–32] Animal studies have confirmed the influences of oxygen concentrations on epigenetic regulation of TSCs differentiation[29,33]. A second wave of invasion starts at 11—13 weeks. During this phase, the formation of hemochorial placenta takes place.[34] The interstitial cytotrophoblasts invade decidua and fuse to form multinuclear cells. Moreover, the vasodilators such as nitric oxide (NO) and carbon monoxide (CO) act on the spiral arteries. The endovascular cytotrophoblasts invasion for the spiral artery remodeling results in the loss of smooth muscle layer and vasomotor control. To mimic the adherent endothelial cell, the cell−cell adhesion molecules in cytotrophoblasts switch their expression. The adhesion receptors such as integrin α6β4 and E-cadherin characteristic of endothelial cells get downregulated, while analogous receptors such as integrins α1β1and αVβ3 and VE-cadherin get upregulated.[35] In addition to invasion, cytotrophoblast cells exchange the endothelial lining of the arteries, which results in the formation of pseudo-endothelium and arterial remodeling from the high-resistant to a low resistant one and with increased intervillous blood flow to promote fetal growth.[34,36]

During first trimester, placenta shows a distinct methylation profile of different trophoblast populations, i.e., trophoblast stem cell side population, CTB, and EVTs. In trophoblast stem cells, the genes involved in potency and cell cycle regulation are differentially methylated, while in EVTs and CTBs, the genes involved in EMT, trophoblast differentiation, and invasion are differentially methylated. Moreover, the 5-methylcytosine (5-mC) levels also differ in cytotrophoblasts and syncytiotrophoblats.[37]

Other genes regulated by DNA methylation and playing role in placental development are CREB5,[38,39] MEG8,[40] TIMP3,[41] Oct4,[42] adenomatous polyposis coli,[43] c-myc[38] etc. Noncoding RNAs regulate gene expression at different levels of protein synthesis without being translated into proteins. Though different types of ncRNAs, i.e., miRs, lncRNA, siRNAs, etc. take part in epigenetic regulation of gene expression, most studied in placental development are miRs and lncRNAs.

MicroRNAs (MiRs) are small RNA molecules (single stranded) that have been studied for their role in trophoblast differentiation and proliferation. They have also been investigated for their role in vasculogenesis and angiogenesis.[4]

Epigenetics and mouse placental development

A cascade of complex genetic and epigenetic interactions leads to the development of functional trophoblasts. Defects in the development, differentiation, and function of trophoblasts may be due to loss of epigenetic factors that can be lethal to the embryo.[44] Studies on ICM-derived cells, trophoblasts from embryonic day E3.5 and E6.5, trophectoderm stem cells, and embryonic stem cells have shown establishment of extraembryonic-specific DNA methylation immediately after the separation of ICM and TS cell lineages.[45] In mice, the mural trophectoderm cells, which are not in contact with the inner cell mass, do not proliferate after their formation, but the endoreduplication of their DNA continues, which results in the formation of polyploid trophoblast giant cells. These trophoblast giant cells are important for implantation, maternal decidual remodeling for fetal allograft, and nutrient transfer.[46] Extraembryonic ectoderm and ectoplacental cone are formed by the proliferation of polar trophectoderm cells overlying inner cell mass. Furthermore, at embryonic day 8.5, chorioallantoic attachment takes place in which the chorion derived from extraembryonic ectoderm and allantois derived from extraembryonic mesoderm make contact. To enable well-organized gas and nutrient exchange, the trophoblasts cells of the chorion undergo extensive branching to form a dense villi structure known as labyrinth and the fetoplacental vascular network is derived from the allantois. The spongiotrophoblast cell layer originating from the ectoplacental cone provides structural support to the labyrinth. The secondary trophoblast giant cells are also derived from ectoplacental cone.[47,48]

An epigenetic asymmetry exists in the embryonic and extraembryonic lineages due to a differential global DNA methylation establishment. The trophectoderm is relatively hypomethylated as compared with ICM, and these methylation levels remain the same later in embryo and trophoblasts.[49,50]

In mice and humans, an extensive DNA methylation occurs at the time of fertilization, while an extensive undermethylation in mouse ectoplacental cone and extraembryonic ectoderm has been reported as compared with the embryonic tissues.[51] More recently, genome-wide promoter analysis of ESC and TSC confirmed the asymmetry in DNA methylation levels but suggested that the majority of the differences are likely to be within intergenic regions, repeat sequences, and centromeric heterochromatin, whereas promoter-associated regions are methylated at similar levels between embryonic and extraembryonic cells.[52]

The increased level of DNA methylation is required to prevent the conversion of embryonic stem cells toward a trophoblast cell type suggesting its role in lineage maintenance, and the ESCs with a reduced DNA methylation levels were found to readily differentiate into trophoblast giant cells or TSC.[53,54] In embryonic stem cells of embryonic day E9.5 from *Dnmt1*$^{-/-}$ embryos, the genetic inactivation of *Dnmt1* leads to the ectopic expression of trophoblast genes such as *Cdx2* and *Eomes*. In *Dnmt1*$^{-/-}$ ES cells, the *Elf5* promoter is hypomethylated resulting in higher expression of Elf5, which further upregulates the expression of trophoblast marker genes.[53] Moreover, studies have shown that either the removal of *Dnmt1* or its inactivation by 5-azacytidine (DNA methyltransferase inhibitor) resulting in low levels of DNA methylation in ESC favors their differentiation into TGC and adds to the trophectoderm population in aggregation chimeras.[53,54]

Earlier studies have shown that hypermethylation of *Elf5* an E74-like factor 5 (ETS domain transcription factor) promoter in embryonic tissues and hypomethylation in trophoblast tissues is important in the cellular adoption of trophoblast fate. *Elf5* encodes an Ets family transcription factor vital for trophoblast development.[55] *Elf5* forms a regulatory loop with other key transcription factors in trophoblast. DNA methylation is important to regulate the transcription of the genes needed for placental formation, growth, trophoblast invasion, and placental function.[53]

Embryonic lethality due to genetic inactivation of *Dnmt1* and *Dnmt3a/Dnmt3b* and faulty chorioallantoic development due to the mutation in these genes has been reported,[56,57] while the ectopic expression of *Pou5f1* in the placentae, which were *Dnmt1*-deficient suggests, the need of epigenetic repression of a subset of genes in trophoblast cells.[58] Moreover a higher chorionic expression of *Dnmt3l* favors labyrinth formation in placenta, and *Dnmt3l* mutation results in poor development of spongiotrophoblast layer as well as excess TGC formation.[59] Studies on

Dnmt1/Dnmt3a/Dnmt3b mutant TSC showing properties of that of wild-type TSC with undetectable levels of DNA methylation have shown that self-renewal and long-term proliferation of undifferentiated TSC do not need DNA methylation.[54] A number of imprinted genes are also involved in the placental growth. They encode for a number of growth factors that drive placental growth.[60]

Transcriptional factors and trophoblast lineage differentiation

The transcription factors play an important role in deciding the cell fate during placenta development. Most of the information about the role of these transcription factors originated from mouse gene knockout studies, and the mTSC studies are instrumental to understand the trophoblast proliferation and differentiation.

The transcription factors such as OCT4 (POU5F1) and CDX2 are responsible for initiation of cell fate decision and establishment of embryonic and trophoblast lineages at the blastocysts stage. In mouse, the blastomere expresses OCT4 until the formation of early blastocyst, while the CDX2 gets restricted to the outer cells of the morula stage embryo at E2.5.[61] The establishment of a trophoblast transcription program depends on the restriction of CDX2 expression initiated by HIPPO signaling pathway. The HIPPO signaling plays a role in the integration of cell positions and cell–cell contact.[62]

In the outer polarized cells of morula, the nuclear translocation of a YAP1 (a transcriptional coactivator and HIPPO component) along with another transcription factor TEAD4 activates cdx2 expression. This explains the formation of TE from outer polar cells, which are CDX2+ and inner ICM from CDX2−apolar cells and hence biased lineage fate allocation.[63]

The periimplantation stage earmarks the establishment of irreversible cell fate assignment at around mouse E4.5, a phase where rapid establishment of DNA methylation patterns takes place.[63] In a situation similar to the exit from TSC niche, the vCTB cells first express *cdx2* in placenta of early first trimester, which then gets downregulated.[64,65] The role of TEAD4 is also supposed to be conserved between mice and humans.[66,67] In the course of early postimplantation development in mouse, these transcription factors, i.e., TEAD4 and CDX2 along with other transcription factors such as EOMES, ELF5, ETS2, ESRRB, GATA3, SOX2, and TEAP2C act together to form the functional trophoblast compartment.[68,69] Though these factors are not specific to trophoblast lineage and also present in the embryo, they serve different functions in both the compartments owing to different protein environments in both tissue types. Moreover, the binding of these transcription factors is influenced by DNA methylation, e.g., CDX2 binds to the methylated sequence elements, while others such as ELF5, TFAP2C, and ETS2 bind to the unmethylated elements.[70,71] Chromatin environment and DNA methylation influence the capacity of different transcription factors to bind to DNA as is evidenced by in vitro studies,[72,73] and DNA methylation affects the binding affinity of about 60% of transcription factors in human.[74] Major transcription factors involved in cell proliferation prefer binding to the regions, which are unmethylated and important for trophoblasts.[71,75,76]

On the other hand, factors such as OCT4 and CDX2, which are enriched for gene ontology terms with respect to reproduction, prefer to bind the methylated regions. Most of the times, the presence of a single CpG in the binding motif plays a subtle role in determining transcription factor–binding preferences. This suggests that protein–DNA interactions are potentially influenced by the state of methylation. In addition to this, the CpG density determining the composition of underlying sequence alone can enhance and is sufficient for transcription factor binding and promoter activity. The binding of a great number of transcription factors in an epigenome-dependent manner makes them more complex in terms of functionality, as the same factors can perform different functions in different types of cells.[76–78]

DNA methylation is an important mechanism to regulate the endogenous retroviral long terminal repeat (LTR) elements. In placenta, the suppression of LTR is low, and there is abundance of binding sites for transcription factors on LTRs in extraembryonic tissues.[79,80] This makes them to have unique functions in placental cell types. In mice and human placental trophoblasts, these LTRs are hypomethylated, which helps LTRs to acquire enhancer of promoter functions.[81,82] This further leads to the acquisition of binding sites on the LTRs for transcription factors such as CDX2, EOMES, and ELF5 in mouse trophoblasts and accountable for the regulatory activity of LTRs.[83] Moreover, the LTRs can also act as methylation-sensitive alternative promoters for their neighboring genes in both mice and human trophoblasts. Interestingly, during mammalian evolution, the mice and human placental LTR expression has led to the unique acquisition of retroviral gene functions such as "syncytialization" mediated by the "syncytins," the retrovirus-derived envelope genes.[83–85] This suggests that trophoblast LTRs with unique methylation landscape can regulate mammalian placental development.

Epigenetics and regulation of trophoblast functions

There are two phases of epigenome reprogramming, i.e., (1) during generation of gametes before fertilization and (2) at the time of early embryo development, i.e., postfertilization. The erasal of DNA methylation during these two events takes place. At the time of primordial germ cell differentiation, genome-wide loss of methylation occurs to allow the expression of genes required for meiosis and resetting of imprinted genes in a parent of origin manner with the erasal of acquired epimutations.[86] A subset of LTRs, which resists demethylation, is potent enough to carry forward the epigenetic information across generations. As the use of single-cell epigenetic techniques have paved way toward epigenetic profiling, the information about the CpG islands (CGIs) retaining some DNA methylation marks and not associated with imprinted genes has come up suggesting the importance of the maintenance of DNA methylation in trophoblast development.[87] One such example is *Scml2* gene locus at X chromosome in the mouse encoding polycomb group of proteins. Its transcriptional activity is regulated by a DNA methylation mark inherited by oocyte. This mark is maintained only in the trophoblast and not in the embryonic lineage. Loss of oocyte methylation results in the aberrant expression of Scml2, which is associated with defects in placenta.[88]

Epigenetic mechanisms regulating placental development and function

Placenta is a vital organ for the developing fetus, and any dysregulation in the expression of genes involved in placental metabolic, endocrine functions, nutrients, and waste product transport activities in response to in utero exposures can result in abnormal fetal programming.

DNA methylation

DNA methylation is the best studied epigenetic modification, which comprises the adding of a methyl group to a cytosine at cytosine–guanine dinucleotide, i.e., CpG site by enzymes known as DNA methyltransferases. A hypomethylated state (2.5%–3% of global DNA methylation) is the characteristic of placenta, which differentiates it from other somatic tissues (4%–5% of global DNA methylation) in terms of DNA methylation levels.[4] There is a lack of complete understanding of functional relevance of placental hypomethylation, but the identification of active cryptic transposon-derived promoters suggests its relevance in the regulation of "genome plasticity." Identification of several active *SINE* (short interspersed nuclear elements) retrotransposon-derived alternative promoters expressing placenta-specific transcripts, potassium voltage-gated channel subfamily H member 5 (KCNH5), and placenta-specific *IL-2* receptor beta expression[81,89,90] also indicates the functional relevance of placental hypomethylation.

The initial concept of specific hypomethylation of retrotransposons in placenta has been challenged by the recent findings of Schroeder and colleagues in which they have shown that large stretches of (up to 100 kb) of partially methylated domains (PMDs) in gene-poor regions are the main sites of DNA Hypomethylation. These hypomethylated regions are interspersed with higher DNA methylation regions. The PMDs cover nearly about 37% of placental genome and are rich in gene specific to placental functions. The LINE-1 elements in PMDs had similar methylation levels (40%) to the nearby regions, while the LINE-1 elements outside PMDs had up to 80% methylation.[91] The understanding of the epigenetic regulation of PMDs and their role of in placental functions warrants further investigation.

DNA methylation in trophoblast

The dynamics of DNA methylation in embryonic and extraembryonic tissues is regulated by the expression of DNA methyltransferases, i.e., DNMT3A, DNMT3B, and cofactor DNMT3L.[92,93] The trophoblast compartment is globally hypomethylated, i.e., 40%–50% of CpG methylation in comparison with epiblast with ~80% of methylated CpGs. The levels of methylation in human placenta increase with gestation indicating gradual loss of cellular plasticity. Moreover, prevalence of intermediately methylated domains spanning transcriptionally repressed polycomb targeted regions is another characteristic of trophoblast epigenome.[71]

In a similar way, the levels of 5-hydroxymethylcytosine (5hmC) are also lower in the trophoblast as compared with epiblast. The 5hmC is produced by the oxidation of 5 mC by TET enzymes and, in mouse TSCs, is associated with stem cell state. The expression of Tet1 and Tet2 gets downregulated, as mouse TSCs differentiate. Moreover, the in vivo ablation of Tet1 affects metabolic genes in the trophoblast. TET1 is localized at putative enhancer regions specifically in mouse TSCs and may have a noncatalytic function regulating expression of genes.[94]

DNA methylation and regulation of placental nutrient transfer

Nutrient transport across placenta and metabolic programming are highly coordinated processes, and any perturbation can be detrimental to fetal growth. Recent studies have shown that large for gestational age (LGA) new-born babies were associated with a significantly higher global DNA methylation in placenta as compared with SGA and AGA babies.[95]

Filiberto et al.[96] have also shown that differential promoter methylation of placental glucocorticoid receptor is associated with fetal weight. They showed that increased DNA methylation of the exon1F of glucocorticoid receptor (GR) is associated with LGA status. The glucocorticoid receptor mediates glucocorticoid signaling and is epigenetically regulated by DNA methylation.[96] An altered glucocorticoid signaling in placenta may lead to a variety of metabolic diseases.

Moreover, an increased WNT2 promoter methylation has been found to be significantly associated with low birth weight.[97] The WNT2 gene is a member of secreted glycoproteins family binding to frizzled (FZD) family of receptors, and the binding of WNT2 to its receptor activates signaling pathway involved in cell proliferation, migration, cell death, etc.[98] The higher expression of WNT2 in all stages of developing placenta indicates its role in fetal growth.[97] Furthermore, the levels of vitamin D during pregnancy are also regulated epigenetically as indicated by low methylation levels in vitamin D receptor and renal1 alpha-hydroxylase CYP27B1 genes and partial methylation of 24-hydroxylase CYP24A1. In nonpregnant state, the vitamin D levels are regulated by feedback regulation of CYP27B1 (positive feedback) and CYP24A1 (catabolic enzyme for vitamin D). This feedback mechanism is attenuated during pregnancy to meet nutritional demands of vitamin D.[99]

Genomic imprinting

Presence of imprinted genes in the genome is one major difference between placental and nonplacental species of mammals.[100] An accurate establishing and maintenance of genome imprinting is vital to the normal placental growth, development, and functions.[101] Imprinted genes are characterized by their parent of origin-specific monoallelic expression. Till date, more than 100 imprinted genes have been identified in mice and human, and some of them are essential for the placentation and fetal growth. The placenta and brain show expression of the highest number of imprinted genes.[102] The monoallelic expression of imprinted genes is regulated by the differentially methylated regions inherited from germline, i.e., gDMRs, and about 24 imprinted gDMRs have been identified. Many of these gDMRs act as imprinting control regions (ICRs).[103] These ICRs regulate the monoallelic expression of many gene clusters through promoter methylation. The regulation of imprinting of gene clusters is achieved by (1) antisense gene-mediated disruption of transcription,[104] (2) targeting enhancers/insulator elements in a distinct manner,[105] (3) different usage of polyadenylation signals,[106] and (4) targeting epigenetic modifiers in an allele-specific manner.[107] The DMRs that are not inherited from germline but gain differential methylation during embryonic development are known as secondary DMRs, majority of which are located distal to gDMRs and also regulated by them.[108] The differential methylation at sDMRs is essential for imprinted gene regulation.[109]

The evolution of imprinted genes in mammals is considered to be a response to the parental genome conflict in the transplacental transfer of nutrients. It has been suggested that the genes that are paternally expressed intensely favor the use of maternal resources to support feta development and the genes, which are maternally expressed try to preserve the maternal resources.[100,102] Gene imprinting is crucial for placental development as knockout studies of some paternally/maternally imprinted genes are associated with abnormal placental development and related pregnancy disorders.[110]The dependence of embryo development on some of the maternal genes and of extraembryonic tissue on active paternal genes is indicative of parental genomic regulation of resource allocation.[111]

Imprinted genes perform a major function of nutrient transport regulation in the placenta as evidenced by the studies on insulin-like growth receptor 2 (*Igf2*) gene. *Igf2* is paternally expressed, and maternally silenced encodes IGF2, a potent fetoplacental growth factor.[112]

Two imprinted genes, i.e., IGF2 and H19 regulating fetoplacental growth are expressed opposite to each other. These genes are located at 11p15.5 adjacent to each other and have same transcription regulatory epigenetic mechanisms. The methylation of DMR upstream of H19 (containing sequences that bind zinc finger protein CTCF) on paternal chromosome inhibits CTCF binding.[113] As a result of which, the IGF2 promoter accesses the enhancers present downstream to H19 and expresses IGF2, while on maternal chromosome, the nonmethylation of H19 results in the binding of CTCF, which allows H19 promoter access to enhancers and subsequent transcription of H19. High expression of IGF2 in mouse and human placenta is involved in the nutrient transfer to the fetus and regulation of placental size.[113,114]

Recent studies have indicated an increase in the number of gDMRs in placenta than what was thought earlier. These gDMRs are involved in the maternal resource allocation as said before. In a study by Hanna et al. 144 gDMRs were identified in human placenta maintaining the maternal DNA methylation patterns.[115] A similar observation was made by Hamada et al. in which approximately 1800 mDMRs were identified by using combined whole-genome bisulfite sequencing and RNA-seq. These DMRs were linked to potential paternal and maternal imprinted genes.[116]

In placenta, two imprinted gene clusters, i.e., *KCNQ1/Kcnq1* on chromosome 7 and *IGF2R/Igf2r* on chromosome 17 contain a subset of genes with imprinted expression, and each of these imprinted domains has a maternal gDMR acting as an ICR.[117,118]

For more than two decades, DNA methylation was considered as the only known germline imprinting mark[108,119,120] until the gDMR independent imprinted genes were identified.[121–123] This DNA methylation-independent imprinting mechanism is known as noncanonical imprinting. Inoue et al. revealed the presence of high oocyte-specific trimethylation at lysine 27 in histone H3, i.e., H3K27me3 levels in noncanonically imprinted genes, and a loss of imprinting was associated with a loss of H3K27me3.[123] The H3K27me3-dependent imprinting is thought to be established during oogenesis, while the DNA methylation–dependent imprinting is majorly maintained in both embryonic and extraembryonic lineages.[115,121–123] The *Sfmbt2, Gab1*, and *Slc38a* are placenta-specific genes primarily imprinted by maternal allele–specific H3K27me3. The SCNT, i.e., somatic cell nuclear transfer embryos generated from oocytes and single cells are reported to have abnormal placental development.[124] The H3K27me3 mark is completely lost in the extraembryonic lineages of SCNT mouse embryos,[125] and a biallelic expression of these three (*Sfmbt2, Gab1*, and *Slc38a*) placenta-specific genes was observed due to loss of imprinting in absence of H3K27me3.[122] Moreover, loss of H3K27me3 imprinting was found to result in unusual placental enlargement as well as low birth rates following SCNT in mouse.[126]

One of the main functions of placenta is to provide nutrition to developing fetus, which is under the control of hormones produced by placenta. Genomic imprinting plays a vital role in the regulation of placental cell lineages involved in synthesis and secretion of placental hormone.[127]

Genomic imprinting and regulation of placental endocrine function

Placental hormones play a pivotal role during gestation to fulfill the energy demands of the fetus nutrient demands and facilitate maternal fetal exchanges, adjustment of maternal physiology to fetal allograft, etc. The epigenetic regulation on placental hormone production is mostly achieved by genome imprinting. Most of the epigenetic-related work in relation to placental endocrine function has been done on mice.[127] Some of the imprinted genes involved in the regulation of placental hormones in mice are discussed in Table 6.1.[127–138]

The alteration in the expression of these genes results in the changes in the placental cell lineages involved in hormone synthesis and thus affects the placental functions regulated by these hormones.

Some nonclassically imprinted genes are also involved in the regulation of endocrine lineages such as *Esx1* (extraembryonic spermatogenesis, homeobox 1) expressed in chorionic ectoderm, trophoblasts of labyrinth, and visceral yolk sac

Table 6.1 The table enlists the imprinted genes involved in the regulation of placental hormones in mice.

Imprinted gene	Parent of origin	Effects on the loss of expression	Reference
Ascl2 (Achaete-Scute family BHLH transcription factor 2) also known as Mash2	Maternally expressed	**1.** Complete absence of spongiotrophoblast lineage **2.** Reduced labyrinth trophoblast **3.** Expansion of parietal TGCs layer **4.** Placenta failure at E10.5 irrespective of chorioallantoic fusion **5.** Transgenic mice with 50% expression produce viable embryos at term with no spongiotrophoblast layer, no glycogen cells, and expansion of parietal TGCs layer	[127,128,134,137]
Phlda2/Ipl (pleckstrin homology-like domain family A, member 2)	Maternally expressed	**1.** Increased junctional zone and glycogen staining	[127,130–133,138]
Peg3 (paternally expressed gene 3)	Paternally expressed	**1.** Decreased embryonic and placental weight **2.** Alteration in the expression of five prolactin-like genes (Prl8a8, Prl3c1, Prl7b1, Prl7a2, and Prl8a6) **3.** Altered expression of spongiotrophoblast/glycogen and giant cell marker Tpbpa (trophoblast-specific protein β) at E12.5 **4.** Altered endocrine function	[127,129,139]
Peg10 (paternally expressed gene 10)	Paternally expressed	**1.** Embryo lethality in $Peg10^{+/-}$ embryo at E10.5 **2.** Lack of spongiotrophoblasts in placenta **3.** No expansion of giant cell population	[127,135]

Continued

Table 6.1 The table enlists the imprinted genes involved in the regulation of placental hormones in mice.—*cont'd*

Imprinted gene	Parent of origin	Effects on the loss of expression	Reference
Igf2	Paternally expressed	**1.** Reduced labyrinth and junctional zone **2.** Loss of glycogen cells and TGCs	[127,140]
Cdkn1c (cyclin-dependent kinase inhibitor 1c)	Maternally expressed	**1.** Placentomegaly **2.** Reduced spongiotrophoblast cell lineage **3.** Reduced sinusoidal TGCs **4.** Reduced expression of prolactin genes	[127,138]

endoderm. An increased expression of *Esx1* is related to a reduced spongiotrophoblast differentiation,[141] and *Esx1* deficiency is related to retardation in embryonic growth.[142] Another gene involved in placental endocrine function is *Cited1* (Cbp/P300 interacting transactivator with Glu/Asp-rich carboxy-terminal domain 1), expressed in placenta in all trophoblast-derived cells as evident from animal studies. *Cited1*$^{-/+}$ placentae were associated with increased expression of *Prl2c2* and increased spongiotrophoblast as well as TGC lineages.[143]

Nrk (Nil-related kinase) expressed in placental junctional zone encodes a serine/threonine kinase. *Nrk*$^{-/+}$ placentae also had an increased spongiotrophoblast and glycogen cell lineages.[144] Similarly, *Plac1*$^{-/+}$ (placenta-specific protein 1) placentae have expansion in junctional zone owing to spongiotrophoblasts encroaching labyrinth and increased TGCs. *Plac1* deficiency is related to growth retardation in embryo.[145]

Histone modifications and placental functions

Histone modifications, i.e., acetylation, methylation, and phosphorylation, are important epigenetic marks and play a crucial role in the regulation of gene expression. A very few studies are available elucidating the role of histone modifications in placental development and functions. Most of the studies are in in vitro or mouse models. Methylation and acetylation are the two most characterized histone modifications.[146,147] Different enzymes that catalyze histone modifications are histone methyltransferases, acetyltransferases, kinases, and ubiquitylases. While the marks from the histone tail are removed by demethylase, deacetylases, phosphatases, and deubiquitylases. Histone "H3" is the most modified histone.[147] A higher H3 arginine methylation in blastocysts is required in blastomeres to form pluripotent cells in ICM as compared with those that form trophoblast lineage.[148] The differentiation of extraembryonic tissues largely depends on the repressive histone modifications as a mechanism to control gene expression in the trophoblast cells by using these modifications as a complementary to DNA methylation.[149] The role of DNA methylation in regulating the placental cell fate is well known. In addition to DNA methylation, histone modifications also influence cell fate decisions as evident from different animal studies.[150,151] SETDB1, a methyltransferase when ablated in the mouse embryonic stem cells, results in the cellular differentiation toward trophoblast cell types with upregulated trophoblast genes, e.g., cdx2. SETDB1 is a methyltransferase mediating H3K9 trimethylation.[152,153] Similar changes have been observed in the mESCs deficient for KDM1A, a lysine demethylase that carries out demethylation of H3K9me2/3 and is related to the derepression of placenta-specific LTR elements.[154] Furthermore, a failed chorioallantoic fusion leading to embryonic lethality due to ablation of EHMT2 or G9A is another example of failed H3K9 demethylation. EHMT2 mediates H3K9 demethylation.[149]

Moreover, in murine trophoblasts, the H3K27me3 derived from oocyte results in the silencing of maternal alleles of some genes in a parent of origin manner and thus carries out noncanonical type of imprinting[123,126] as discussed before.

Signals regulating nutrient availability and transport are important to program the epigenetic machinery in placenta. One such mechanism is OGT (O-linked *N*-acetyl glucosamine transferase; O-GlcNAc transferase) signaling pathway. The enzyme OGT regulates nutritional signals and chromatin modification in nutrient responsive manner as it catalyzes O-GlcNAc addition to serine and threonine residues of various proteins. The OGT-mediated O-GlcNAcylation is an important mechanism of histone modification as it directly modifies core histone proteins. The placental OGT senses changes in maternal energy and alters placental signaling so as to aid in fetal programming.[155]

Noncoding RNA

Noncoding RNAs (ncRNAs) regulate gene expression at different levels of protein synthesis without being translated into proteins. Though different types of ncRNAs, i.e., microRNAs, lncRNAs, small interfering RNAs, etc. take part in epigenetic regulation of gene expression, most studied ncRNAs in placental development are miRs and lncRNAs.

miRs

These small RNA molecules (single stranded) have been studied for their roles in trophoblast differentiation and proliferation. Studies have also been done to understand their role in the invasion and migration processes, vasculogenesis, and angiogenesis.

The three miRNA clusters largely or absolutely expressed during pregnancy have their origin in three evolutionarily dissimilar chromosomal regions, i.e., C19MC,

C14MC, and C2MC (or Sfmbt2 miRNA cluster). The expression of C19MC and C2MC is restricted to paternally inherited chromosome and of C14MC to its maternal allele.[156]

Chromosome 19 microRNA cluster (C19MC) at human Chr19q13 is primate-specific cluster with 46 miRNA genes in a ~100-kb-long region. Physiologically, C19MC is primarily identified in placenta, undifferentiated ESCs, and germ cells.[157]

C19MC in placental functions

The C19MC members have been reported to be involved in trophoblast physiology. The miR-515 family members are downregulated during the differentiation of cytotrophoblasts to syncytiotrophoblasts in vitro, while an overexpression of miR-515-5p inhibits differentiation. Moreover, an attenuated cell migration in HTR-8/SVneo cell line which does not express C19MC was observed on forced expression of C19MC. Furthermore, in extravillous trophoblasts, mi-RNAs from C19MC have also been found to modulate migration and invasion.[158–161] The involvement of C19MC in placental metabolism mainly in adaptation to hypoxia and maintenance of undifferentiated state of ESCs is also evolving.[162–164]

Interestingly, high C19MC expression in term primary human trophoblasts offers resistance to various DNA and RNA viruses such as human cytomegalovirus (CMV), vesicular stomatitis virus (VSV), herpes simplex virus1, vaccinia virus, poliovirus, and others.[165–167] The C19MC miRNAs have been proposed to be packaged into exosomes. These exosomes are transported to the recipient cells (maternal or fetal) through the bloodstream. After incorporation of these exosomes in the recipient cells, the antiviral response is elicited by upregulating the autophagy pathways.[165]

C14MC in placental function

The imprinted Dlk1-Dio3 domain is specific to the eutherians and is crucial for the process of placentation.[168,169] This domain is of ~1 Mb size and located at human 14q32 (distal mouse chromosome 12). There are three paternally expressed genes in this region, i.e., delta-like homolog 1 (Dlk1), retrotransposon-like 1 (Rtl1), and type III iodothyronine deiodinase (Dio3). Other than this, a set of maternally expressed noncoding RNA transcripts including Meg3, as well as numerous small RNAs belonging to the SNORD and miRNA families, are also expressed. In humans, the cluster of miRNAs is collectively called as C14MC (chromosome 14 microRNA cluster). Of these, most of the miRNAs are categorized into two clusters, i.e., the miR-127/miR-136 and the miR-379-miR-410 clusters located upstream and downstream to the SNORD gene cluster. The studies on paternally expressed Rtl1 gene have shown its association with embryonic and placental development as it regulated the growth of fetal capillaries in labyrinth. On the contrary, the miR-127/miR136 cluster, and mainly miR-127, has an antagonizing effect on placental growth as it contributes in the Rtl1 mRNA silencing.[170,171]The understanding of the functions of other miRNAs at the Dlk1-Dio3 in placenta warrants more investigations due to paucity in the information. Labialle and colleagues observed ~30% partially penetrant neonatal lethality owing to the maternal deletion of entire miR-379/miR-410 cluster upon maternal inheritance possibly due to defective activation of neonatal hepatic gene resulting in compromised neonatal metabolism.[172] In vitro studies with HTR8/SVneo showed that gain-of-function of miR-376c, a member of the miR-379/miR-410 cluster, is associated with increased proliferation, migration, and invasion.[173,174]

C2MC in placental functions

Chromosome 2 microRNA cluster (C2MC) is rodent specific, and its expression is limited to the paternal allele.[175] It is present at the intron 10 of the Sfmbt2 (Sex combs on the midleg with four MBT domains).[176] The C2MC miRNAs are supposed to regulate proliferation, migration, and invasion in the placenta as indicated by microarray analyses in C2MC-deficient placentas at E8.5 and E11.5.[177]

Other miRNAs involved in placental development and functions are enlisted in Table 6.2.[158,159,173,174,177–204]

Long noncoding RNAs

lncRNAs also do not encode for a protein. lncRNAs are more than 200 nucleotides long involved in epigenetic modifications and imprinting in addition to other processes such as posttranslational modifications, X-chromosome inactivation, RNA splicing, and editing. These lncRNAs have been found to be involved in trophoblast proliferation, migration, and invasion. Table 6.3[190,207–214] shows the lncRNAs and their role in placental development. Some long-intergenic noncoding RNAs are also expressed in placenta, which take part in placental development.

Epigenetics and placenta-related pathologies

Many investigators have tried to examine the epigenetic cause of various placenta-related pathologies such as preeclampsia, gestational diabetes mellitus, and intrauterine growth restriction. Some of them are discussed in the following.

Epigenetics and preeclampsia

About ~2%–5% of pregnancies are affected by preeclampsia (PE), a disease characterized by hypertension and proteinuria resulting in maternal and fetal morbidity and mortality. Preeclampsia, a multifactorial disorder, can be detrimental to both maternal and fetal health.[215] Experimental evidence suggest the role of epigenetics in the etiology of PE though the precise mechanism is still not clear and under investigation. Many studies have shown that DNA methylation is involved in the regulation of genes involved in fetal development and growth. The expression of these genes is influenced by the environmental exposures. These epigenetic changes are

Table 6.2 Role of miRNAs in placental development.

Process/phase of placental development	Study model and miRs regulating the process	References
Trophoblast differentiation	miR-378a-5p (human placenta and BeWo cells), miR-17-92 cluster (human placenta), miR-106a-363 cluster (human placenta), chromosome 19 miR cluster C19MC (human placenta)	[177–181]
Trophoblast growth and proliferation	miR-675 (human placenta), miR-376c (human placenta), miR-378a-5p (human placenta), miR-141 (choriocarcinoma cell line), miR-155 (HTR8/SVneo cells), miR-144 (human trophoblast cells), miR-377 (human placenta), let-7a (human placenta), miR-195 (human placenta), miR-20a (human placenta), miR-101 (HTR8/SVneo cells), miR-128a (HTR8/SVneo cells), miR-29b (human placenta), miR-30a-3p (human placenta), miR-21 (HTR8/SVneo cells/JEG-3 cells), miR-290 cluster (mice placenta)	[158,173,177–179, 182–188,190,191,194,205]
Trophoblast migration and invasion	miR-376c (human placenta), miR-378-5p (human placenta and BeWo cells), miR-155 (HTR8/SVneo cells), miR-195 (human placenta), miR-20a (human placenta, BeWo cells), miR-29b (human placenta), miR-30a-3p (human placenta), C19MC cluster (human placenta), miR-210 (human placenta), miR-204 (BeWo and JEG-3 trophoblast like cells), miR-520g (human placenta), miR-34a (human choriocarcinoma cells), miR-135b (HTR8/SVneo cells),miR-21 (HTR8/SVneo cells/JEG-3 choriocarcinoma cells), miR-18a (human placenta, JEG-3, (choriocarcinoma cells), miR-346 (JAR, HTR8/SVneo cells), miR-582-3p (JAR/HTR8/SVneo cells)	[159,173,177–181, 184,187–192,194–199, 201,204,206]
Placental angiogenesis	miR-20a (human placenta, BeWo cells), miR-29b (human placenta), miR-290 luster (mice placenta), miR-20b (human placenta), miR136 (human placenta), miR-126 (human placenta), miR16 (human placenta)	[183,185,189,192,193,200,202]

Table 6.3 lncRNAs involved in trophoblast proliferation, migration, and invasion.

lncRNA	Placental developmental process	Associated pathology
H19	Proliferation	Abnormal placental development[190,209]
MALAT1	Invasion, migration, apoptosis	Preeclampsia, placenta previa/increta/percreta[208]
SPRY4-IT1	Proliferation, migration, apoptosis	Preeclampsia[214]
RPAIN	Proliferation, invasion	Preeclampsia[210]
MEG 3	Apoptosis, migration	Preeclampsia[213]
TUG1	Proliferation, apoptosis	Preeclampsia[211]
PVT1	Proliferation, apoptosis	Preeclampsia[211]
FLT1P1, LOC284100, CEACAMP8	Placental lipid metabolism?	Preeclampsia[207]
lincRNA		
MIR503HG	Invasion, migration	Not known[208]
LINC00629	Invasion, migration	Not known[208]

characterized by changes in oxygen levels.[4,160,216] Low oxygen induces the expression of hypoxia-inducible factor (HIF1). The binding of HIF1 to hypoxia responsive element activates various genes involved in angiogenesis and metastasis. This abnormal oxygen signaling further increases reactive oxygen species (ROS) and, thus, oxidative stress that speeds up the process of trophoblast aging.[217] Many genome-wide studies have shown the existence of different methylation patterns and hence different etiology in early- and late-onset PE.[204,218]

In a study Yeung and coworkers, 303 DMRs in PE were recognized. As per the KEGG pathway, the hypermethylated regions were near to the genes involved in ATP transport and Reactome pathways. The homeobox, Wnt2 signaling, fertilization and implantation genes, and cell adhesion genes were also altered.[216] Leavey and coworkers also found a similar differential methylation and gene alteration in PE placentae.[219]

A differential expression of noncoding RNAs has also been observed in PE. Many studies have been carried out to identify potential biomarkers useful for diagnostic purposes to predict onset of PE. The most studied ncRNAs in PE are lncRNA and miRNA, while some work has been done on circular RNA to study development of PE.[190,209–211,213,220] Some studies have focused on the histone modification in PE. In a study by Chakraborty and colleagues, hypoxia-derived placental HIF–KDM3A–MMP12 pathway involved in the trophoblast invasion and spiral artery remodeling was found to be altered in a model of PE.[221] A mitigated histone demethylase JMJD6 (jumonji domain–containing protein 6) activity was also observed in PE placenta under hypoxic conditions.[222] A reduced expression of HDAC9 at mRNA and protein levels was found in syncytiotrophoblasts of PE placentae.[223] The imprinted genes are also differentially expressed in PE placenta. A disturbance in the regulation of imprinted genes such as *GATA3* and *DLX5* has been proved in PE placenta. *GATA3* is involved in the inhibition of trophoblast invasion and could be a role player in the development of PE, while *DLX5* is involved in the organ development processes in embryo.[224] van Dijk and coworkers have identified STOX1 (Storkhead box 1) as a preeclampsia susceptibility gene, and methylation levels of differentially methylated regions in intron 1 of this gene coincide with STOX1 expression. A high STOX1 expression was found to have a negative effect on trophoblast invasive potential.[225]

Epigenetics and gestational diabetes

There is very less information available on the placental epigenetic changes associated with gestational diabetes mellitus. A recent study revealed upregulation of 35 and downregulation of 10 imprinted genes in microarray analysis of placental RNA of mouse model of uterine hyperglycemia. The imprinted genes were altered in the placentae of two filial generations. *Dlk1* that promotes insulin/IGF1 signaling pathway activation was downregulated and *Gtl2*, a TGF-β and notch signaling pathway regulator, was upregulated.[226] Reichetzeder and colleagues have shown the association of placental global DNA hypermethylation with GDM in human.[227] A link between maternal hyperglycemia (GDM) and alterations in placental DNA methylation at *LEP* (leptin) and *ADIPOQ* (adiponectin) genes has been investigated by Bouchard and colleagues.[228,229] A low methylation of LEP on the fetal side and high on maternal side was correlated with higher glucose values after 2 hours of administration. A low methylation at ADIPOQ in second and third trimesters was correlated with a high insulin resistance. Both these genes are involved in glucose metabolism, and their altered methylation levels may later cause metabolic disturbances in both mother and child.[228,229] In another study, a low methylation level at *MEST* gene (maternally imprinted) was observed in GDM placentae.[230] A low methylation at placental lipoprotein lipase (*LPL*) gene, 1.6-fold higher expression, and glucose levels in GDM women has also been reported.[231] An association between varied DNA methylation levels in placenta at genes involved in regulation of brown and beige tissues and glucose homeostasis was demonstrated.[232]

Epigenetics and intrauterine growth restriction

Another pregnancy-related disorder where placental epigenetic dysregulation has been investigated is intrauterine growth restriction (IUGR). The IUGR placentae are characterized by a comparatively smaller placentae and abnormal vascularization. The causes of IUGR can be both maternal and fetal. Placental insufficiency is the cause of IUGR in about 10%–15% of pregnancies[233,234] In a recent genome-wide methylation study, 22 potential sites with differential DNA methylation in IUGR placentae were identified.[235] Similar to Russell–Silver syndrome, the loss of imprinting owing to abnormal methylation at H19-DMR and IGF2-DMRs can lead to restricted placental and fetal growth resulting in IUGR.[236–238] In preterm placenta and cord blood, DNA methylation patterns specific to IUGR have been observed. Epigenetic silencing of *Pdx-1* has been unraveled in IUGR placentae.[239] Moreover, in IUGR, placentae imprinting defects have also been reported such as hypomethylation of H19 (maternally expressed), and mRNA overexpression is associated with IUGR.[5] Similar to PE placentae, the hypermethylation of Ras association domain family 1 isoform A and hypomethylation of an inflammation-induced protease inhibitor serpin peptidase inhibitor A3 has been reported in IUGR.[240] Reduced hepatic and serum levels of IGF-1 related to altered histone code have been observed in vivo.[241] Differential methylation in placentae of Wnt1, Dlk1, cadherin pathway genes, genes related to beta oxidation of fatty acids, and genes regulating transcription, i.e., DECR1, ZNF300, DNAJA4, CCL28, LEPR, HSPA1A/L, GSTO1, and GNE has been reported in IUGR cases.[242]

Conclusion

Recent advances in the field of epigenetics have widened the window of regulatory mechanisms of placental functions be it nutrient transfer, metabolic programming, or hormone synthesis. Epigenetic mechanisms, i.e., DNA methylation, genomic imprinting, histone modifications, and ncRNAs regulate key processes involved in placental development. Furthermore, any disturbances in these mechanisms can be detrimental to placental as well as fetal health. The association of disturbed epigenetic changes and various placenta-related disorders has set another platform for diagnosis and therapeutics. The role of noncanonical imprinting conferred through maternally inherited repressive histone modification H3K27me3 in extraembryonic lineages needs to be explored further. Future studies on the noncanonically imprinted genes regulating different signaling pathways will give a better picture of how they influence fetal growth and maternal adaptations to pregnancy.

References

1. Burton GJ, Fowden AL. The placenta: a multifaceted, transient organ. *Phil Trans Biol Sci*. 2015;(1663):370. https://doi.org/10.1098/rstb.2014.0066.
2. Turco MY, Moffett A. Development of the human placenta. *Development (Camb)*. 2019; 146(22). https://doi.org/10.1242/dev.163428.
3. Ilekis JV, Tsilou E, Fisher S, et al. Placental origins of adverse pregnancy outcomes: potential molecular targets: an Executive Workshop summary of the Eunice Kennedy

Shriver National Institute of Child health and human development. *Am J Obstet Gynecol.* 2016;215(1):S1−S46. https://doi.org/10.1016/j.ajog.2016.03.001.
4. Apicella C, Ruano CSM, Méhats C, Miralles F, Vaiman D. The role of epigenetics in placental development and the etiology of preeclampsia. *Int J Mol Sci.* 2019;20(11). https://doi.org/10.3390/ijms20112837.
5. Januar V, Desoye G, Novakovic B, Cvitic S, Saffery R. Epigenetic regulation of human placental function and pregnancy outcome: considerations for causal inference. *Am J Obstet Gynecol.* 2015;213(4):S182−S196. https://doi.org/10.1016/j.ajog.2015.07.011.
6. Herzog M. A contribution to our knowledge of the earliest known stages of placentation and embryonic development in man. *Am J Anat.* 1909;9(1):361−400. https://doi.org/10.1002/aja.1000090112.
7. James JL, Carter AM, Chamley LW. Human placentation from nidation to 5 weeks of gestation. Part I: what do we know about formative placental development following implantation? *Placenta.* 2012;33(5):327−334. https://doi.org/10.1016/j.placenta.2012.01.020.
8. Boyd JD, Hamilton WJ. *The Human Placenta.* 212. 1970. https://doi.org/10.1002/tera.1420080118.
9. Pollheimer J, Vondra S, Baltayeva J, Beristain AG, Knöfler M. Regulation of placental extravillous trophoblasts by the maternal uterine environment. *Front Immunol.* 2018;9. https://doi.org/10.3389/fimmu.2018.02597.
10. Soundararajan R, Rao AJ. Trophoblast "pseudo-tumorigenesis": significance and contributory factors. *Reprod Biol Endocrinol.* 2004;2. https://doi.org/10.1186/1477-7827-2-15.
11. Visse R, Nagase H. Matrix metalloproteinases and tissue inhibitors of metalloproteinases: structure, function, and biochemistry. *Circ Res.* 2003;92(8):827−839. https://doi.org/10.1161/01.RES.0000070112.80711.3D.
12. Nagaset H, Woessner JF. Matrix metalloproteinases. *J Biol Chem.* 1999;274(31):21491−21494. https://doi.org/10.1074/jbc.274.31.21491.
13. Staun-Ram E, Shalev E. Human trophoblast function during the implantation process. *Reprod Biol Endocrinol.* 2005;3. https://doi.org/10.1186/1477-7827-3-56.
14. Nagase H, Visse R, Murphy G. Structure and function of matrix metalloproteinases and TIMPs. *Cardiovasc Res.* 2006;69(3):562−573. https://doi.org/10.1016/j.cardiores.2005.12.002.
15. Rahat B, Sharma R, Bagga R, Hamid A, Kaur J. Imbalance between matrix metalloproteinases and their tissue inhibitors in preeclampsia and gestational trophoblastic diseases. *Reproduction.* 2016;152(1):11−22. https://doi.org/10.1530/REP-16-0060.
16. Dokras A, Gardner LMG, Kirschmann DA, Seftor EA, Hendrix MJC. The tumour suppressor gene maspin is differentially regulated in cytotrophoblasts during human placental development. *Placenta.* 2002;23(4):274−280. https://doi.org/10.1053/plac.2001.0784.
17. Dokras A, Coffin J, Field L, et al. Epigenetic regulation of maspin expression in the human placenta. *Mol Hum Reprod.* 2006;12(10):611−617. https://doi.org/10.1093/molehr/gal074.
18. Pijnenborg R, Dixon G, Robertson WB, Brosens I. Trophoblastic invasion of human decidua from 8 to 18 weeks of pregnancy. *Placenta.* 1980;1(1):3−19. https://doi.org/10.1016/S0143-4004(80)80012-9.

19. Pijnenborg R, Vercruysse L, Hanssens M. The uterine spiral arteries in human pregnancy: Facts and Controversies. *Placenta.* 2006;27(9–10):939–958. https://doi.org/10.1016/j.placenta.2005.12.006.
20. Moffett A, Chazara O, Colucci F. Maternal allo-recognition of the fetus. *Fertil Steril.* 2017;107(6):1269–1272. https://doi.org/10.1016/j.fertnstert.2017.05.001.
21. Acampora D, Pannese M, D'esposito M, Simeone A, Boncinelli E. Human homoeobox-containing genes in development. *Hum Reprod.* 1987;2(5):407–414. https://doi.org/10.1093/oxfordjournals.humrep.a136559.
22. Oudejans CBM, Pannese M, Simeone A, Meijer CJLM, Boncinelli E. The three most downstream genes of the Hox-3 cluster are expressed in human extraembryonic tissues including trophoblast of androgenetic origin. *Development.* 1990;108(3):471–477.
23. Novakovic B, Fournier T, Harris LK, et al. Increased methylation and decreased expression of homeobox genes TLX1, HOXA10 and DLX5 in human placenta are associated with trophoblast differentiation. *Sci Rep.* 2017;7(1). https://doi.org/10.1038/s41598-017-04776-5.
24. Bellido ML, Radpour R, Lapaire O, et al. MALDI-TOF mass array analysis of RASSF1A and SERPINB5 methylation patterns in human placenta and plasma. *Biol Reprod.* 2010;82(4):745–750. https://doi.org/10.1095/biolreprod.109.082271.
25. Salvianti F, Inversetti A, Smid M, et al. Prospective evaluation of RASSF1A cell-free DNA as a biomarker of pre-eclampsia. *Placenta.* 2015;36(9):996–1001. https://doi.org/10.1016/j.placenta.2015.07.003.
26. Tsui DWY, Chan KCA, Chim SSC, et al. Quantitative aberrations of hypermethylated RASSF1A gene sequences in maternal plasma in pre-eclampsia. *Prenat Diagn.* 2007; 27(13):1212–1218. https://doi.org/10.1002/pd.1897.
27. Horiuchi A, Hayashi T, Kikuchi N, et al. Hypoxia upregulates ovarian cancer invasiveness via the binding of HIF-1α to a hypoxia-induced, methylation-free hypoxia response element of S100A4 gene. *Int J Cancer.* 2012;131(8):1755–1767. https://doi.org/10.1002/ijc.27448.
28. Charron CE, Chou PC, Coutts DJC, et al. Hypoxia-inducible factor 1α induces corticosteroid-insensitive inflammation via reduction of histone deacetylase-2 transcription. *J Biol Chem.* 2009;284(52):36047–36054. https://doi.org/10.1074/jbc.M109.025387.
29. Maltepe E, Krampitz GW, Okazaki KM, et al. Hypoxia-inducible factor-dependent histone deacetylase activity determines stem cell fate in the placenta. *Development.* 2005; 132(15):3393–3403. https://doi.org/10.1242/dev.01923.
30. Pollard PJ, Loenarz C, Mole DR, et al. Regulation of Jumonji-domain-containing histone demethylases by hypoxia-inducible factor (HIF)-1α. *Biochem J.* 2008;416(3): 387–394. https://doi.org/10.1042/BJ20081238.
31. Wellmann S, Bettkober M, Zelmer A, et al. Hypoxia upregulates the histone demethylase JMJD1A via HIF-1. *Biochem Biophys Res Commun.* 2008;372(4):892–897. https://doi.org/10.1016/j.bbrc.2008.05.150.
32. Xia M, Yao L, Zhang Q, et al. Long noncoding RNA HOTAIR promotes metastasis of renal cell carcinoma by up-regulating histone H3K27 demethylase JMJD3. *Oncotarget.* 2017;8(12):19795–19802. https://doi.org/10.18632/oncotarget.15047.
33. Maltepe E, Bakardjiev AI, Fisher SJ. The placenta: transcriptional, epigenetic, and physiological integration during development. *J Clin Invest.* 2010;120(4):1016–1025. https://doi.org/10.1172/JCI41211.

34. Lyall F, Bulmer JN, Duffie E, Cousins F, Theriault A, Robson SC. Human trophoblast invasion and spiral artery transformation: the role of PECAM-1 in normal pregnancy, preeclampsia, and fetal growth restriction. *Am J Pathol.* 2001;158(5):1713–1721. https://doi.org/10.1016/S0002-9440(10)64127-2.
35. Zhou Y, Fisher SJ, Janatpour M, et al. Human cytotrophoblasts adopt a vascular phenotype as they differentiate: a strategy for successful endovascular invasion? *J Clin Invest.* 1997;99(9):2139–2151. https://doi.org/10.1172/JCI119387.
36. Lunghi L, Ferretti ME, Medici S, Biondi C, Vesce F. Control of human trophoblast function. *Reprod Biol Endocrinol.* 2007;5. https://doi.org/10.1186/1477-7827-5-6.
37. Gamage TKJB, Schierding W, Hurley D, et al. The role of DNA methylation in human trophoblast differentiation. *Epigenetics.* 2018;13(12):1154–1173. https://doi.org/10.1080/15592294.2018.1549462.
38. Rahat B, Hamid A, Najar RA, Bagga R, Kaur J. Epigenetic mechanisms regulate placental c-myc and hTERT in normal and pathological pregnancies; c-myc as a novel fetal DNA epigenetic marker for pre-eclampsia. *Mol Hum Reprod.* 2014;20(10):1026–1040. https://doi.org/10.1093/molehr/gau053.
39. Yu M, Du G, Xu Q, et al. Integrated analysis of DNA methylome and transcriptome identified CREB5 as a novel risk gene contributing to recurrent pregnancy loss. *EBioMedicine.* 2018;35:334–344. https://doi.org/10.1016/j.ebiom.2018.07.042.
40. Sheng F, Sun N, Ji Y, et al. Aberrant expression of imprinted lncRNA MEG8 causes trophoblast dysfunction and abortion. *J Cell Biochem.* 2019;120(10):17378–17390. https://doi.org/10.1002/jcb.29002.
41. Feng H, Cheung ANY, Xue WC, et al. Down-regulation and promoter methylation of tissue inhibitor of metalloproteinase 3 in choriocarcinoma. *Gynecol Oncol.* 2004;94(2):375–382. https://doi.org/10.1016/j.ygyno.2004.04.019.
42. Zhang HJ, Siu MKY, Wong ESY, et al. Oct4 is epigenetically regulated by methylation in normal placenta and gestational trophoblastic disease. *Placenta.* 2008;29(6):549–554. https://doi.org/10.1016/j.placenta.2008.03.003.
43. Wong NC, Novakovic B, Weinrich B, et al. Methylation of the adenomatous polyposis coli (APC) gene in human placenta and hypermethylation in choriocarcinoma cells. *Cancer Lett.* 2008;268(1):56–62. https://doi.org/10.1016/j.canlet.2008.03.033.
44. Knöfler M, Haider S, Saleh L, Pollheimer J, Gamage TKJB, James J. Human placenta and trophoblast development: key molecular mechanisms and model systems. *Cell Mol Life Sci.* 2019;76(18):3479–3496. https://doi.org/10.1007/s00018-019-03104-6.
45. Nakanishi MO, Hayakawa K, Nakabayashi K, Hata K, Shiota K, Tanaka S. Trophoblast-specific dna methylation occurs after the segregation of the trophectoderm and inner cell mass in the mouse periimplantation embryo. *Epigenetics.* 2012;7(2):173–182. https://doi.org/10.4161/epi.7.2.18962.
46. Cross JC, Werb Z, Fisher SJ. Implantation and the placenta: key pieces of the development puzzle. *Science.* 1994;266(5190):1508–1518. https://doi.org/10.1126/science.7985020.
47. Rossant J, Cross JC. Placental development: lessons from mouse mutants. *Nat Rev Genet.* 2001;2(7):538–548. https://doi.org/10.1038/35080570.
48. Watson ED, Cross JC. Development of structures and transport functions in the mouse placenta. *Physiology.* 2005;20(3):180–193. https://doi.org/10.1152/physiol.00001.2005.

49. Hemberger M, Dean W, Reik W. Epigenetic dynamics of stem cells and cell lineage commitment: digging Waddington's canal. *Nat Rev Mol Cell Biol.* 2009;10(8): 526–537. https://doi.org/10.1038/nrm2727.
50. Santos F, Hendrich B, Reik W, Dean W. Dynamic reprogramming of DNA methylation in the early mouse embryo. *Dev Biol.* 2002;241(1):172–182. https://doi.org/10.1006/dbio.2001.0501.
51. Rugg-Gunn PJ. Epigenetic features of the mouse trophoblast. *Reprod Biomed Online.* 2012;25(1):21–30. https://doi.org/10.1016/j.rbmo.2012.01.012.
52. Farthing CR, Ficz G, Ng RK, et al. Global mapping of DNA methylation in mouse promoters reveals epigenetic reprogramming of pluripotency genes. *PLoS Genet.* 2008; 4(6). https://doi.org/10.1371/journal.pgen.1000116.
53. Ng RK, Dean W, Dawson C, et al. Epigenetic restriction of embryonic cell lineage fate by methylation of Elf5. *Nat Cell Biol.* 2008;10(11):1280–1290. https://doi.org/10.1038/ncb1786.
54. Sakaue M, Ohta H, Kumaki Y, et al. DNA methylation is dispensable for the growth and survival of the extraembryonic lineages. *Curr Biol.* 2010;20(16):1452–1457. https://doi.org/10.1016/j.cub.2010.06.050.
55. Donnison M, Beaton A, Davey HW, Broadhurst R, L'Huillier P, Pfeffer PL. Loss of the extraembryonic ectoderm in Elf5 mutants leads to defects in embryonic patterning. *Development.* 2005;132(10):2299–2308. https://doi.org/10.1242/dev.01819.
56. Li E, Bestor TH, Jaenisch R. Targeted mutation of the DNA methyltransferase gene results in embryonic lethality. *Cell.* 1992;69(6):915–926. https://doi.org/10.1016/0092-8674(92)90611-F.
57. Masaki O, W BD, A HD, En L. DNA methyltransferases Dnmt3a and Dnmt3b are essential for de Novo methylation and mammalian development. *Cell.* 1999: 247–257. https://doi.org/10.1016/s0092-8674(00)81656-6.
58. Hattori N, Nishino K, Ko YG, et al. Epigenetic control of mouse Oct-4 gene expression in embryonic stem cells and trophoblast stem cells. *J Biol Chem.* 2004;279(17): 17063–17069. https://doi.org/10.1074/jbc.M309002200.
59. Arima T, Hata K, Tanaka S, et al. Loss of the maternal imprint in Dnmt3Lmat-/- mice leads to a differentiation defect in the extraembryonic tissue. *Dev Biol.* 2006;297(2): 361–373. https://doi.org/10.1016/j.ydbio.2006.05.003.
60. Constância M, Hemberger M, Hughes J, et al. Placental-specific IGF-II is a major modulator of placental and fetal growth. *Nature.* 2002;417(6892):945–948. https://doi.org/10.1038/nature00819.
61. Niwa H, Toyooka Y, Shimosato D, et al. Interaction between Oct3/4 and Cdx2 determines trophectoderm differentiation. *Cell.* 2005;123(5):917–929. https://doi.org/10.1016/j.cell.2005.08.040.
62. Lorthongpanich C, Issaragrisil S. Emerging role of the Hippo signaling pathway in position sensing and lineage specification in mammalian preimplantation embryos. *Biol Reprod.* 2015;92(6). https://doi.org/10.1095/biolreprod.114.127803.
63. Nishioka N, Inoue Ki, Adachi K, et al. The Hippo signaling pathway components Lats and Yap pattern Tead4 activity to distinguish mouse trophectoderm from inner cell mass. *Dev Cell.* 2009;16(3):398–410. https://doi.org/10.1016/j.devcel.2009.02.003.
64. Horii M, Li Y, Wakeland AK, et al. Human pluripotent stem cells as a model of trophoblast differentiation in both normal development and disease. *Proc Natl Acad Sci USA.* 2016;113(27):E3882–E3891. https://doi.org/10.1073/pnas.1604747113.

65. Soncin F, Khater M, To C, et al. Comparative analysis of mouse and human placentae across gestation reveals species-specific regulators of placental development. *Development*. 2018;145(2). https://doi.org/10.1242/dev.156273. Cambridge.
66. Home P, Saha B, Ray S, et al. Altered subcellular localization of transcription factor TEAD4 regulates first mammalian cell lineage commitment. *Proc Natl Acad Sci USA*. 2012;109(19):7362–7367. https://doi.org/10.1073/pnas.1201595109.
67. Yagi R, Kohn MJ, Karavanova I, et al. Transcription factor TEAD4 specifies the trophectoderm lineage at the beginning of mammalian development. *Development*. 2007;134(21):3827–3836. https://doi.org/10.1242/dev.010223.
68. Kidder BL, Palmer S. Examination of transcriptional networks reveals an important role for TCFAP2C, SMARCA4, and EOMES in trophoblast stem cell maintenance. *Genome Res*. 2010;20(4):458–472. https://doi.org/10.1101/gr.101469.109.
69. Latos PA, Hemberger M. Review: the transcriptional and signalling networks of mouse trophoblast stem cells. *Placenta*. 2014;35:S81–S85. https://doi.org/10.1016/j.placenta.2013.10.013.
70. Hemberger M, Udayashankar R, Tesar P, Moore H, Burton GJ. ELF5-enforced transcriptional networks define an epigenetically regulated trophoblast stem cell compartment in the human placenta. *Hum Mol Genet*. 2010;19(12):2456–2467. https://doi.org/10.1093/hmg/ddq128.
71. Hemberger M, Hanna CW, Dean W. Mechanisms of early placental development in mouse and humans. *Nat Rev Genet*. 2020;21(1):27–43. https://doi.org/10.1038/s41576-019-0169-4.
72. Hellman LM, Fried MG. Electrophoretic mobility shift assay (EMSA) for detecting protein-nucleic acid interactions. *Nat Protoc*. 2007;2(8):1849–1861. https://doi.org/10.1038/nprot.2007.249.
73. Jolma A, Yan J, Whitington T, et al. DNA-binding specificities of human transcription factors. *Cell*. 2013;152(1–2):327–339. https://doi.org/10.1016/j.cell.2012.12.009.
74. Bell O, Tiwari VK, Thomä NH, Schübeler D. Determinants and dynamics of genome accessibility. *Nat Rev Genet*. 2011;12(8):554–564. https://doi.org/10.1038/nrg3017.
75. Hu S, Wan J, Su Y, et al. DNA methylation presents distinct binding sites for human transcription factors. *Elife*. 2013;2013(2). https://doi.org/10.7554/eLife.00726.
76. Yin Y, Morgunova E, Jolma A, et al. Impact of cytosine methylation on DNA binding specificities of human transcription factors. *Science*. 2017;356(6337). https://doi.org/10.1126/science.aaj2239.
77. Hartl D, Krebs AR, Grand RS, et al. CG dinucleotides enhance promoter activity independent of DNA methylation. *Genome Res*. 2019;29(4):554–563. https://doi.org/10.1101/gr.241653.118.
78. Knott JG, Paul S. Transcriptional regulators of the trophoblast lineage in mammals with hemochorial placentation. *Reproduction*. 2014;148(6):R121–R136. https://doi.org/10.1530/REP-14-0072.
79. Bourque G, Leong B, Vega VB, et al. Evolution of the mammalian transcription factor binding repertoire via transposable elements. *Genome Res*. 2008;18(11):1752–1762. https://doi.org/10.1101/gr.080663.108.
80. Pavlicev M, Hiratsuka K, Swaggart KA, Dunn C, Muglia L. Detecting endogenous retrovirus-driven tissue-specific gene transcription. *Genome Biol Evol*. 2015;7(4):1082–1097. https://doi.org/10.1093/gbe/evv049.

81. Cohen CJ, Rebollo R, Babovic S, Dai EL, Robinson WP, Mager DL. Placenta-specific expression of the Interleukin-2 (IL-2) receptor β subunit from an endogenous retroviral promoter. *J Biol Chem*. 2011;286(41):35543–35552. https://doi.org/10.1074/jbc.M111.227637.
82. Macfarlan TS, Gifford WD, Agarwal S, et al. Endogenous retroviruses and neighboring genes are coordinately repressed by LSD1/KDM1A. *Gene Dev*. 2011;25(6):594–607. https://doi.org/10.1101/gad.2008511.
83. Chuong EB, Rumi MAK, Soares MJ, Baker JC. Endogenous retroviruses function as species-specific enhancer elements in the placenta. *Nat Genet*. 2013;45(3):325–329. https://doi.org/10.1038/ng.2553.
84. Sha M, Xinhua L, Xiang-ping L, et al. Syncytin is a captive retroviral envelope protein involved in human placental morphogenesis. *Nature*. 2000:785–789. https://doi.org/10.1038/35001608.
85. Dupressoir A, Vernochet C, Harper F, et al. A pair of co-opted retroviral envelope syncytin genes is required for formation of the two-layered murine placental syncytiotrophoblast. *Proc Natl Acad Sci USA*. 2011;108(46):E1164–E1173. https://doi.org/10.1073/pnas.1112304108.
86. Messerschmidt DM, Knowles BB, Solter D. DNA methylation dynamics during epigenetic reprogramming in the germline and preimplantation embryos. *Gene Dev*. 2014; 28(8):812–828. https://doi.org/10.1101/gad.234294.113.
87. Brind'Amour J, Kobayashi H, Richard Albert J, et al. LTR retrotransposons transcribed in oocytes drive species-specific and heritable changes in DNA methylation. *Nat Commun*. 2018;9(1). https://doi.org/10.1038/s41467-018-05841-x.
88. Branco MR, King M, Perez-Garcia V, et al. Maternal DNA methylation regulates early trophoblast development. *Dev Cell*. 2016;36(2):152–163. https://doi.org/10.1016/j.devcel.2015.12.027.
89. MacAulay EC, Weeks RJ, Andrews S, Morison IM. Hypomethylation of functional retrotransposon-derived genes in the human placenta. *Mamm Genome*. 2011; 22(11–12):722–735. https://doi.org/10.1007/s00335-011-9355-1.
90. Reiss D, Zhang Y, Mager DL. Widely variable endogenous retroviral methylation levels in human placenta. *Nucleic Acids Res*. 2007;35(14):4743–4754. https://doi.org/10.1093/nar/gkm455.
91. Schroeder DI, Blair JD, Lott P, et al. The human placenta methylome. *Proc Natl Acad Sci USA*. 2013;110(15):6037–6042. https://doi.org/10.1073/pnas.1215145110.
92. Laird PW. The power and the promise of DNA methylation markers. *Nat Rev Cancer*. 2003;3(4):253–266. https://doi.org/10.1038/nrc1045.
93. Herman JG, Baylin SB. Gene silencing in cancer in association with promoter hypermethylation. *N Engl J Med*. 2003;349(21):2042–2054. https://doi.org/10.1056/NEJMra023075.
94. Khoueiry R, Sohni A, Thienpont B, et al. Lineage-specific functions of TET1 in the postimplantation mouse embryo. *Nat Genet*. 2017;49(7):1061–1072. https://doi.org/10.1038/ng.3868.
95. Dwi Putra SE, Reichetzeder C, Hasan AA, et al. Being born large for gestational age is associated with increased global placental DNA methylation. *Sci Rep*. 2020;10(1). https://doi.org/10.1038/s41598-020-57725-0.
96. Filiberto AC, Maccani MA, Koestler D, et al. Birthweight is associated with DNA promoter methylation of the glucocorticoid receptor in human placenta. *Epigenetics*. 2011; 6(5):566–572. https://doi.org/10.4161/epi.6.5.15236.

97. Ferreira JC, Choufani S, Grafodatskaya D, et al. WNT2 promoter methylation in human placenta is associated with low birthweight percentile in the neonate. *Epigenetics*. 2011; 6(4):440–449. https://doi.org/10.4161/epi.6.4.14554.
98. Zeng CM, Chen Z, Fu L. Frizzled receptors as potential therapeutic targets in human cancers. *Int J Mol Sci*. 2018;19(5). https://doi.org/10.3390/ijms19051543.
99. Novakovic B, Sibson M, Ng HK, et al. Placenta-specific methylation of the vitamin D 24-hydroxylase gene. Implications for feedback autoregulation of active vitamin D levels at the fetomaternal interface. *J Biol Chem*. 2009;284(22):14838–14848. https://doi.org/10.1074/jbc.M809542200.
100. Barlow DP, Bartolomei MS. Genomic imprinting in mammals. *Cold Spring Harbor Perspect Biol*. 2014;6(2). https://doi.org/10.1101/cshperspect.a018382.
101. Hemberger M. Epigenetic landscape required for placental development. *Cell Mol Life Sci*. 2007;64(18):2422–2436. https://doi.org/10.1007/s00018-007-7113-z.
102. Babak T, Deveale B, Tsang EK, et al. Genetic conflict reflected in tissue-specific maps of genomic imprinting in human and mouse. *Nat Genet*. 2015;47(5):544–549. https://doi.org/10.1038/ng.3274.
103. White CR, MacDonald WA, Mann MRW. Conservation of DNA methylation programming between mouse and human gametes and preimplantation embryos. *Biol Reprod*. 2016;95(3). https://doi.org/10.1095/biolreprod.116.140319.
104. Latos PA, Pauler FM, Koerner MV, et al. Airn transcriptional overlap, but not its lncRNA products, induces imprinted Igf2r silencing. *Science*. 2012;338(6113): 1469–1472. https://doi.org/10.1126/science.1228110.
105. Hark AT, Schoenherr CJ, Katz DJ, Ingram RS, Levorse JM, Tilghman SM. CTCF mediates methylation-sensitive enhancer-blocking activity at the H19/Igf2 locus. *Nature*. 2000;405(6785):486–489. https://doi.org/10.1038/35013106.
106. Wood AJ, Schulz R, Woodfine K, et al. Regulation of alternative polyadenylation by genomic imprinting. *Gene Dev*. 2008;22(9):1141–1146. https://doi.org/10.1101/gad.473408.
107. Terranova R, Yokobayashi S, Stadler MB, et al. Polycomb group proteins Ezh2 and Rnf2 direct genomic contraction and imprinted repression in early mouse embryos. *Dev Cell*. 2008;15(5):668–679. https://doi.org/10.1016/j.devcel.2008.08.015.
108. Ferguson-Smith AC. Genomic imprinting: the emergence of an epigenetic paradigm. *Nat Rev Genet*. 2011;12(8):565–575. https://doi.org/10.1038/nrg3032.
109. Fan T, Hagan JP, Kozlov SV, Stewart CL, Muegge K. Lsh controls silencing of the imprinted Cdkn1c gene. *Development*. 2005;132(4):635–644. https://doi.org/10.1242/dev.01612.
110. Frost JM, Moore GE. The importance of imprinting in the human placenta. *PLoS Genet*. 2010;6(7):1–9. https://doi.org/10.1371/journal.pgen.1001015.
111. Moore T, Haig D. Genomic imprinting in mammalian development: a parental tug-of-war. *Trends Genet*. 1991;7(2):45–49. https://doi.org/10.1016/0168-9525(91)90230-N.
112. Sibley CP, Coan PM, Ferguson-Smith AC, et al. Placental-specific insulin-like growth factor 2 (Igf2) regulates the diffusional exchange characteristics of the mouse placenta. *Proc Natl Acad Sci USA*. 2004;101(21):8204–8208. https://doi.org/10.1073/pnas.0402508101.
113. Kurukuti S, Tiwari VK, Tavoosidana G, et al. CTCF binding at the H19 imprinting control region mediates maternally inherited higher-order chromatin conformation to restrict enhancer access to Igf2. *Proc Natl Acad Sci USA*. 2006;103(28): 10684–10689. https://doi.org/10.1073/pnas.0600326103.

114. Koukoura O, Sifakis S, Spandidos DA. DNA methylation in the human placenta and fetal growth (review). *Mol Med Rep*. 2012;5(4):883–889. https://doi.org/10.3892/mmr.2012.763.
115. Hanna CW, Pérez-Palacios R, Gahurova L, et al. Endogenous retroviral insertions drive non-canonical imprinting in extra-embryonic tissues. *Genome Biol*. 2019;20(1). https://doi.org/10.1186/s13059-019-1833-x.
116. Hamada H, Okae H, Toh H, et al. Allele-specific methylome and transcriptome analysis reveals Widespread imprinting in the human placenta. *Am J Hum Genet*. 2016;99(5):1045–1058. https://doi.org/10.1016/j.ajhg.2016.08.021.
117. Sleutels F, Barlow DP. The origins of genomic imprinting in mammals. In: Dunlap JC, Wu CT, eds. *Advances in Genetics,* Academic Press. 2002;46:119–163. https://doi.org/10.1016/s0065-2660(02)46006-3.
118. Fitzpatrick GV, Soloway PD, Higgins MJ. Regional loss of imprinting and growth deficiency in mice with a targeted deletion of KvDMR1. *Nat Genet*. 2002;32(3):426–431. https://doi.org/10.1038/ng988.
119. Bartolomei MS, Webber AL, Brunkow ME, Tilghman SM. Epigenetic mechanisms underlying the imprinting of the mouse H19 gene. *Gene Dev*. 1993;7(9):1663–1673. https://doi.org/10.1101/gad.7.9.1663.
120. Stöger R, Kubička P, Liu CG, et al. Maternal-specific methylation of the imprinted mouse Igf2r locus identifies the expressed locus as carrying the imprinting signal. *Cell*. 1993;73(1):61–71. https://doi.org/10.1016/0092-8674(93)90160-R.
121. Okae H, Hiura H, Nishida Y, et al. Re-investigation and RNA sequencing-based identification of genes with placenta-specific imprinted expression. *Hum Mol Genet*. 2012;21(3):548–558. https://doi.org/10.1093/hmg/ddr488.
122. Okae H, Matoba S, Nagashima T, et al. RNA sequencing-based identification of aberrant imprinting in cloned mice. *Hum Mol Genet*. 2014;23(4):992–1001. https://doi.org/10.1093/hmg/ddt495.
123. Inoue A, Jiang L, Lu F, Suzuki T, Zhang Y. Maternal H3K27me3 controls DNA methylation-independent imprinting. *Nature*. 2017;547(7664):419–424. https://doi.org/10.1038/nature23262.
124. Yang X, Smith SL, Tian XC, Lewin HA, Renard JP, Wakayama T. Nuclear reprogramming of cloned embryos and its implications for therapeutic cloning. *Nat Genet*. 2007;39(3):295–302. https://doi.org/10.1038/ng1973.
125. Matoba S, Wang H, Jiang L, et al. Loss of H3K27me3 imprinting in somatic cell nuclear transfer embryos Disrupts post-implantation development. *Cell Stem Cell*. 2018;23(3). https://doi.org/10.1016/j.stem.2018.06.008, 343-354.e5.
126. Inoue K, Ogonuki N, Kamimura S, et al. Loss of H3K27me3 imprinting in the Sfmbt2 miRNA cluster causes enlargement of cloned mouse placentas. *Nat Commun*. 2020;11(1). https://doi.org/10.1038/s41467-020-16044-8.
127. John RM. Epigenetic regulation of placental endocrine lineages and complications of pregnancy. *Biochem Soc Trans*. 2013;41(3):701–709. https://doi.org/10.1042/BST20130002.
128. Guillemot F, Nagy A, Auerbach A, Rossant J, Joyner AL. Essential role of Mash-2 in extraembryonic development. *Nature*. 1994;371(6495):333–336. https://doi.org/10.1038/371333a0.
129. Curley JP, Barton S, Surani A, Keverne EB. Coadaptation in mother and infant regulated by a paternally expressed imprinted gene. *Proc Biol Sci*. 2004;271(1545):1303–1309. https://doi.org/10.1098/rspb.2004.2725.

130. Dunwoodie SL, Beddington RSP. The expression of the imprinted gene Ipl is restricted to extra-embryonic tissues and embryonic lateral mesoderm during early mouse development. *Int J Dev Biol*. 2002;46(4):459–466.
131. Frank D, Fortino W, Clark L, et al. Placental overgrowth in mice lacking the imprinted gene Ipl. *Proc Natl Acad Sci USA*. 2002;99(11):7490–7495. https://doi.org/10.1073/pnas.122039999.
132. Frank D, Mendelsohn CL, Ciccone E, Svensson K, Ohlsson R, Tycko B. A novel pleckstrin homology-related gene family defined by Ipl/Tssc3, TDAG51, and Tih1: tissue-specific expression, chromosomal location, and parental imprinting. *Mamm Genome*. 1999;10(12):1150–1159. https://doi.org/10.1007/s003359901182.
133. Kawase T, Ohki R, Shibata T, et al. PH domain-only protein PHLDA3 is a p53-regulated repressor of Akt. *Cell*. 2009;136(3):535–550. https://doi.org/10.1016/j.cell.2008.12.002.
134. Oh-McGinnis R, Bogutz AB, Lefebvre L. Partial loss of Ascl2 function affects all three layers of the mature placenta and causes intrauterine growth restriction. *Dev Biol*. 2011;351(2):277–286. https://doi.org/10.1016/j.ydbio.2011.01.008.
135. Ono R, Nakamura K, Inoue K, et al. Deletion of Peg10, an imprinted gene acquired from a retrotransposon, causes early embryonic lethality. *Nat Genet*. 2006;38(1):101–106. https://doi.org/10.1038/ng1699.
136. Qian N, Frank D, O'Keefe D, et al. The IPL gene on chromosome 11p15.5 is imprinted in humans and mice and is similar to TDAG51, implicated in Fas expression and apoptosis. *Hum Mol Genet*. 1997;6(12):2021–2029. https://doi.org/10.1093/hmg/6.12.2021.
137. Tanaka M, Gertsenstein M, Rossant J, Nagy A. Mash2 acts cell autonomously in mouse spongiotrophoblast development. *Dev Biol*. 1997;190(1):55–65. https://doi.org/10.1006/dbio.1997.8685.
138. Tunster SJ, Van De Pette M, John RM. Fetal overgrowth in the Cdkn1c mouse model of Beckwith-Wiedemann syndrome. *DMM Dis Model Mech*. 2011;4(6):814–821. https://doi.org/10.1242/dmm.007328.
139. Broad DB, Keverne BK. Placental protection of the fetal brain during short-term food deprivation. *Proc Natl Acad Sci USA*. 2011;108(37):15237–15241. https://doi.org/10.1073/pnas.1106022108.
140. Esquiliano DR, Guo W, Liang L, et al. Placental glycogen stores are increased in mice with H19 null mutations but not in those with insulin or IGF type 1 receptor mutations. *Placenta*. 2009;30:693–699. https://doi.org/10.1016/j.placenta.2009.05.004.
141. Pan Z, Sikandar S, Witherspoon M, et al. Impaired placental trophoblast lineage differentiation in Alkbh1 -/- mice. *Dev Dynam*. 2008;237(2):316–327. https://doi.org/10.1002/dvdy.21418.
142. Li Y, Behringer RR. Esx1 is an X-chromosome-imprinted regulator of placental development and fetal growth. *Nat Genet*. 1998;20(3):309–311. https://doi.org/10.1038/3129.
143. Rodriguez TA, Sparrow DB, Scott AN, et al. Cited1 is required in trophoblasts for placental development and for embryo growth and survival. *Mol Cell Biol*. 2004;24(1):228–244. https://doi.org/10.1128/MCB.24.1.228-244.2004.
144. Denda K, Nakao-Wakabayashi K, Okamoto N, et al. Nrk, an X-linked protein kinase in the germinal center kinase family, is required for placental development and fetoplacental induction of labor. *J Biol Chem*. 2011;286(33):28802–28810. https://doi.org/10.1074/jbc.M111.258160.

145. Jackman SM, Kong X, Fant ME. Plac1 (placenta-specific 1) is essential for normal placental and embryonic development. *Mol Reprod Dev.* 2012;79(8):564–572. https://doi.org/10.1002/mrd.22062.
146. Audia JE, Campbell RM. Histone modifications and cancer. *Cold Spring Harbor Perspect Biol.* 2016;8(4). https://doi.org/10.1101/cshperspect.a019521.
147. Bannister AJ, Kouzarides T. Regulation of chromatin by histone modifications. *Cell Res.* 2011;21(3):381–395. https://doi.org/10.1038/cr.2011.22.
148. Torres-Padilla ME, Parfitt DE, Kouzarides T, Zernicka-Goetz M. Histone arginine methylation regulates pluripotency in the early mouse embryo. *Nature.* 2007; 445(7124):214–218. https://doi.org/10.1038/nature05458.
149. Tachibana M, Sugimoto K, Nozaki M, et al. G9a histone methyltransferase plays a dominant role in euchromatic histone H3 lysine 9 methylation and is essential for early embryogenesis. *Gene Dev.* 2002;16(14):1779–1791. https://doi.org/10.1101/gad.989402.
150. O'Carroll D, Erhardt S, Pagani M, Barton SC, Surani MA, Jenuwein T. The Polycomb-group gene Ezh2 is required for early mouse development. *Mol Cell Biol.* 2001;21(13): 4330–4336. https://doi.org/10.1128/MCB.21.13.4330-4336.2001.
151. Pasini D, Bracken AP, Jensen MR, Denchi EL, Helin K. Suz12 is essential for mouse development and for EZH2 histone methyltransferase activity. *EMBO J.* 2004;23(20): 4061–4071. https://doi.org/10.1038/sj.emboj.7600402.
152. Dodge JE, Kang YK, Beppu H, Lei H, Li E. Histone H3-K9 methyltransferase ESET is essential for early development. *Mol Cell Biol.* 2004;24(6):2478–2486. https://doi.org/10.1128/MCB.24.6.2478-2486.2004.
153. Yuan P, Han J, Guo J, et al. Eset partners with Oct4 to restrict extraembryonic trophoblast lineage potential in embryonic stem cells. *Gene Dev.* 2009;23(21):2507–2520. https://doi.org/10.1101/gad.1831909.
154. Macfarlan TS, Gifford WD, Driscoll S, et al. Embryonic stem cell potency fluctuates with endogenous retrovirus activity. *Nature.* 2012;487(7405):57–63. https://doi.org/10.1038/nature11244.
155. Nugent BM, Bale TL. The omniscient placenta: metabolic and epigenetic regulation of fetal programming. *Front Neuroendocrinol.* 2015;39:28–37. https://doi.org/10.1016/j.yfrne.2015.09.001.
156. Malnou EC, Umlauf D, Mouysset M, Cavaillé J. Imprinted microRNA gene clusters in the evolution, development, and functions of mammalian placenta. *Front Genet.* 2019; 10. https://doi.org/10.3389/fgene.2018.00706.
157. Bortolin-Cavaille ML, Dance M, Weber M, Cavaille J. C19MC microRNAs are processed from introns of large Pol-II, non-protein-coding transcripts. *Nucleic Acids Res.* 2009;37(10):3464–3473. https://doi.org/10.1093/nar/gkp205.
158. Ding J, Huang F, Wu G, et al. MiR-519d-3p suppresses invasion and migration of trophoblast cells via targeting MMP-2. *PLoS One.* 2015;10(3). https://doi.org/10.1371/journal.pone.0120321.
159. Jiang L, Long A, Tan L, et al. Elevated microRNA-520g in pre-eclampsia inhibits migration and invasion of trophoblasts. *Placenta.* 2017;51:70–75. https://doi.org/10.1016/j.placenta.2017.02.001.
160. Anton L, Olarerin-George AO, Hogenesch JB, Elovitz MA. Placental expression of miR-517a/b and miR-517c contributes to trophoblast dysfunction and preeclampsia. *PLoS One.* 2015;10(3). https://doi.org/10.1371/journal.pone.0122707.

161. Takahashi H, Ohkuchi A, Kuwata T, et al. Endogenous and exogenous miR-520c-3p modulates CD44-mediated extravillous trophoblast invasion. *Placenta*. 2017;50: 25–31. https://doi.org/10.1016/j.placenta.2016.12.016.
162. Donker RB, Mouillet JF, Chu T, et al. The expression profile of C19MC microRNAs in primary human trophoblast cells and exosomes. *Mol Hum Reprod*. 2012;18(8): 417–424. https://doi.org/10.1093/molehr/gas013.
163. Nguyen PNN, Choo KB, Huang CJ, Sugii S, Cheong SK, Kamarul T. MIR-524-5p of the primate-specific C19MC miRNA cluster targets TP53IPN1-and EMT-Associated genes to regulate cellular reprogramming. *Stem Cell Res Ther*. 2017;8(1). https://doi.org/10.1186/s13287-017-0666-3.
164. Nguyen PNN, Huang CJ, Sugii S, Cheong SK, Choo KB. Selective activation of miRNAs of the primate-specific chromosome 19 miRNA cluster (C19MC) in cancer and stem cells and possible contribution to regulation of apoptosis. *J Biomed Sci*. 2017; 24(1). https://doi.org/10.1186/s12929-017-0326-z.
165. Delorme-Axford E, Donker RB, Mouillet JF, et al. Human placental trophoblasts confer viral resistance to recipient cells. *Proc Natl Acad Sci USA*. 2013;110(29): 12048–12053. https://doi.org/10.1073/pnas.1304718110.
166. Bayer A, Delorme-Axford E, Sleigher C, et al. Human trophoblasts confer resistance to viruses implicated in perinatal infection. *Am J Obstet Gynecol*. 2015;212(1). https://doi.org/10.1016/j.ajog.2014.07.060, 71-71.e8.
167. Bayer A, Lennemann NJ, Ouyang Y, et al. Chromosome 19 microRNAs exert antiviral activity independent from type III interferon signaling. *Placenta*. 2018;61:33–38. https://doi.org/10.1016/j.placenta.2017.11.004.
168. Rocha STd, Edwards CA, Ito M, Ogata T, Ferguson-Smith AC. Genomic imprinting at the mammalian Dlk1-Dio3 domain. *Trends Genet*. 2008;24(6):306–316. https://doi.org/10.1016/j.tig.2008.03.011.
169. Edwards CA, Mungall AJ, Matthews L, et al. The evolution of the DLK1-DIO3 imprinted domain in mammals. *PLoS Biol*. 2008;6(6):1292–1305. https://doi.org/10.1371/journal.pbio.0060135.
170. Sekita Y, Wagatsuma H, Nakamura K, et al. Role of retrotransposon-derived imprinted gene, Rtl1, in the feto-maternal interface of mouse placenta. *Nat Genet*. 2008;40(2): 243–248. https://doi.org/10.1038/ng.2007.51.
171. Ito M, Sferruzzi-Perri AN, Edwards CA, et al. A trans-homologue interaction between reciprocally imprinted miR-127 and Rtl1 regulates placenta development. *Development (Camb)*. 2015;142(14):2425–2430. https://doi.org/10.1242/dev.121996.
172. Labialle S, Marty V, Bortolin-Cavaillé ML, et al. The miR-379/miR-410 cluster at the imprinted Dlk1-Dio3 domain controls neonatal metabolic adaptation. *EMBO J*. 2014; 33(19):2216–2230. https://doi.org/10.15252/embj.201387038.
173. Fu G, Brkić J, Hayder H, Peng C. MicroRNAs in human placental development and pregnancy complications. *Int J Mol Sci*. 2013;14(3):5519–5544. https://doi.org/10.3390/ijms14035519.
174. Fu G, Ye G, Nadeem L, et al. MicroRNA-376c impairs transforming growth factor-β and nodal signaling to promote trophoblast cell proliferation and invasion. *Hypertension*. 2013;61(4):864–872. https://doi.org/10.1161/HYPERTENSIONAHA.111.203489.
175. Inoue K, Hirose M, Inoue H, et al. The rodent-specific MicroRNA cluster within the Sfmbt2 gene is imprinted and essential for placental development. *Cell Rep*. 2017; 19(5):949–956. https://doi.org/10.1016/j.celrep.2017.04.018.

176. Lehnert S, Kapitonov V, Thilakarathne PJ, Schuit FC. Modeling the asymmetric evolution of a mouse and rat-specific microRNA gene cluster intron 10 of the Sfmbt2 gene. *BMC Genom.* 2011;12. https://doi.org/10.1186/1471-2164-12-257.
177. Luo L, Ye G, Nadeem L, et al. Microrna-378a-5p promotes trophoblast cell survival, migration and invasion by targeting Nodal. *J Cell Sci.* 2012;125(13):3124–3132. https://doi.org/10.1242/jcs.096412.
178. Nadeem U, Ye G, Salem M, Peng C. MicroRNA-378a-5p targets cyclin G2 to inhibit fusion and differentiation in BeWo cells. *Biol Reprod.* 2014;91(3). https://doi.org/10.1095/biolreprod.114.119065.
179. Kumar P, Luo Y, Tudela C, Alexander JM, Mendelson CR. The c-myc-regulated microRNA-17~92 (mir-17~92) and miR-106a~363 clusters target hCYP19A1 and hGCM1 to inhibit human trophoblast differentiation. *Mol Cell Biol.* 2013;33(9): 1782–1796. https://doi.org/10.1128/MCB.01228-12.
180. Zhang M, Muralimanoharan S, Wortman AC, Mendelson CR. Primate-specific miR-515 family members inhibit key genes in human trophoblast differentiation and are upregulated in preeclampsia. *Proc Natl Acad Sci USA.* 2016;113(45):E7069–E7076. https://doi.org/10.1073/pnas.1607849113.
181. Xie L, Mouillet JF, Chu T, et al. C19MC microRNAs regulate the migration of human trophoblasts. *Endocrinology.* 2014;155(12):4975–4985. https://doi.org/10.1210/en.2014-1501.
182. Keniry A, Oxley D, Monnier P, et al. The H19 lincRNA is a developmental reservoir of miR-675 that suppresses growth and Igf1r. *Nat Cell Biol.* 2012;14(7):659–665. https://doi.org/10.1038/ncb2521.
183. Paikari A, Belair CD, Saw D, Blelloch R. The eutheria-specific miR-290 cluster modulates placental growth and maternal-fetal transport. *Development (Camb).* 2017; 144(20):3731–3743. https://doi.org/10.1242/dev.151654.
184. Chaiwangyen W, Ospina-Prieto S, Photini SM, Schleussner E, Markert UR, Morales-Prieto DM. Dissimilar microRNA-21 functions and targets in trophoblastic cell lines of different origin. *Int J Biochem Cell Biol.* 2015;68:187–196. https://doi.org/10.1016/j.biocel.2015.08.018.
185. Li P, Guo W, Du L, et al. MicroRNA-29b contributes to pre-eclampsia through its effects on apoptosis, invasion and angiogenesis of trophoblast cells. *Clin. Sci.* 2013; 124(1):27–40. https://doi.org/10.1042/CS20120121.
186. Morales-Prieto DM, Schleussner E, Markert UR. Reduction in miR-141 is induced by leukemia inhibitory factor and inhibits proliferation in choriocarcinoma cell line JEG-3. *Am J Reprod Immunol.* 2011;66(1):57–62. https://doi.org/10.1111/j.1600-0897.2011.01037.x.
187. Dai Y, Qiu Z, Diao Z, et al. MicroRNA-155 inhibits proliferation and migration of human extravillous trophoblast derived HTR-8/SVneo cells via down-regulating cyclin D1. *Placenta.* 2012;33(10):824–829. https://doi.org/10.1016/j.placenta.2012.07.012.
188. Wang H, Zhang L, Guo X, et al. MiR-195 modulates oxidative stress-induced apoptosis and mitochondrial energy production in human trophoblasts via flavin adenine dinucleotide-dependent oxidoreductase domain-containing protein 1 and pyruvate dehydrogenase phosphatase regulatory subunit. *J Hypertens.* 2018;36(2):306–318. https://doi.org/10.1097/HJH.0000000000001529.

189. Wang Y, Zhang Y, Wang H, et al. Aberrantly up-regulated miR-20a in pre-eclampsic placenta compromised the proliferative and invasive behaviors of trophoblast cells by targeting forkhead box protein A1. *Int J Biol Sci.* 2014;10(9):973–982. https://doi.org/10.7150/ijbs.9088.
190. Gao WL, Liu M, Yang Y, et al. The imprinted H19 gene regulates human placental trophoblast cell proliferation via encoding miR-675 that targets Nodal Modulator 1 (NOMO1). *RNA Biol.* 2012;9(7):1002–1010. https://doi.org/10.4161/rna.20807.
191. Bai Y, Yang W, Yang Hx, et al. Downregulated miR-195 detected in Preeclamptic Placenta affects trophoblast cell invasion via modulating ActRIIA expression. *PLoS One.* 2012;7(6). https://doi.org/10.1371/journal.pone.0038875.
192. Hu TX, Wang G, Guo XJ, et al. MiR 20a,-20b and -200c are involved in hydrogen sulfide stimulation of VEGF production in human placental trophoblasts. *Placenta.* 2016; 39:101–110. https://doi.org/10.1016/j.placenta.2016.01.019.
193. Ji L, Zhang L, Li Y, et al. MiR-136 contributes to pre-eclampsia through its effects on apoptosis and angiogenesis of mesenchymal stem cells. *Placenta.* 2017;50:102–109. https://doi.org/10.1016/j.placenta.2017.01.102.
194. Niu ZR, Han T, Sun XL, Luan LX, Gou WL, Zhu XM. MicroRNA-30a-3p is overexpressed in the placentas of patients with preeclampsia and affects trophoblast invasion and apoptosis by its effects on IGF-1. *Am J Obstet Gynecol.* 2018;218(2). https://doi.org/10.1016/j.ajog.2017.11.568, 249-249.e12.
195. Pang RTK, Leung CON, Ye TM, et al. MicroRNA-34a suppresses invasion through downregulation of Notch1 and Jagged1 in cervical carcinoma and choriocarcinoma cells. *Carcinogenesis.* 2010;31(6):1037–1044. https://doi.org/10.1093/carcin/bgq066.
196. Su MT, Tsai PY, Tsai HL, Chen YC, Kuo PL. miR-346 and miR-582-3p-regulated EG-VEGF expression and trophoblast invasion via matrix metalloproteinases 2 and 9. *Biofactors.* 2017;43(2):210–219. https://doi.org/10.1002/biof.1325.
197. Sun M, Chen H, Liu J, Tong C, Meng T. MicroRNA-34a inhibits human trophoblast cell invasion by targeting MYC. *BMC Cell Biol.* 2015;16(1). https://doi.org/10.1186/s12860-015-0068-2.
198. Tamaru S, Mizuno Y, Tochigi H, et al. MicroRNA-135b suppresses extravillous trophoblast-derived HTR-8/SVneo cell invasion by directly down regulating CXCL12 under low oxygen conditions. *Biochem Biophys Res Commun.* 2015;461(2): 421–426. https://doi.org/10.1016/j.bbrc.2015.04.055.
199. Umemura K, Ishioka SI, Endo T, Ezaka Y, Takahashi M, Saito T. Roles of microRNA-34a in the pathogenesis of placenta accreta. *J Obstet Gynaecol Res.* 2013;39(1):67–74. https://doi.org/10.1111/j.1447-0756.2012.01898.x.
200. Wang Y, Fan H, Zhao G, et al. MiR-16 inhibits the proliferation and angiogenesis-regulating potential of mesenchymal stem cells in severe pre-eclampsia. *FEBS J.* 2012;279(24):4510–4524. https://doi.org/10.1111/febs.12037.
201. Xu P, Zhao Y, Liu M, et al. Variations of microRNAs in human placentas and plasma from preeclamptic pregnancy. *Hypertension.* 2014;63(6):1276–1284. https://doi.org/10.1161/HYPERTENSIONAHA.113.02647.
202. Yan T, Liu Y, Cui K, Hu B, Wang F, Zou L. MicroRNA-126 regulates EPCs function: implications for a role of miR-126 in preeclampsia. *J Cell Biochem.* 2013;114(9): 2148–2159. https://doi.org/10.1002/jcb.24563.
203. Yu Y, Wang L, Liu T, Guan H. MicroRNA-204 suppresses trophoblast-like cell invasion by targeting matrix metalloproteinase-9. *Biochem Biophys Res Commun.* 2015;463(3): 285–291. https://doi.org/10.1016/j.bbrc.2015.05.052.

204. Zhu L, Lv R, Kong L, Cheng H, Lan F, Li X. Genome-wide mapping of 5mC and 5hmC identified differentially modified genomic regions in late-onset severe preeclampsia: a pilot study. *PLoS One*. 2015;10(7). https://doi.org/10.1371/journal.pone.0134119.
205. Farrokhnia F, Aplin JD, Westwood M, Forbes K. MicroRNA regulation of mitogenic signaling networks in the human placenta. *J Biol Chem*. 2014;289(44): 30404–30416. https://doi.org/10.1074/jbc.M114.587295.
206. Zhang Y, Fei M, Xue G, et al. Elevated levels of hypoxia-inducible microRNA-210 in pre-eclampsia: new insights into molecular mechanisms for the disease. *J Cell Mol Med*. 2012;16(2):249–259. https://doi.org/10.1111/j.1582-4934.2011.01291.x.
207. He X, He Y, Xi B, et al. LncRNAs expression in preeclampsia placenta reveals the potential role of LncRNAs contributing to preeclampsia pathogenesis. *PLoS One*. 2013; 8(11). https://doi.org/10.1371/journal.pone.0081437.
208. McAninch D, Roberts CT, Bianco-Miotto T. Mechanistic insight into long noncoding RNAs and the placenta. *Int J Mol Sci*. 2017;18(7). https://doi.org/10.3390/ijms18071371.
209. Nordin M, Bergman D, Halje M, Engström W, Ward A. Epigenetic regulation of the Igf2/H19 gene cluster. *Cell Prolif*. 2014;47(3):189–199. https://doi.org/10.1111/cpr.12106.
210. Song X, Rui C, Meng L, et al. Long non-coding RNA RPAIN regulates the invasion and apoptosis of trophoblast cell lines via complement protein C1q. *Oncotarget*. 2017;8(5): 7637–7646. https://doi.org/10.18632/oncotarget.13826.
211. Xu Y, Ge Z, Zhang E, et al. The lncRNA TUG1 modulates proliferation in trophoblast cells via epigenetic suppression of RND3. *Cell Death Dis*. 2017;8(10). https://doi.org/10.1038/cddis.2017.503.
212. Xu Y, Lian Y, Zhang Y, et al. The long non-coding RNA PVT1 represses ANGPTL4 transcription through binding with EZH2 in trophoblast cell. *J Cell Mol Med*. 2018; 22(2):1272–1282. https://doi.org/10.1111/jcmm.13405.
213. Zhang Y, Zou Y, Wang W, et al. Down-regulated long non-coding RNA MEG3 and its effect on promoting apoptosis and suppressing migration of trophoblast cells. *J Cell Biochem*. 2015;116(4):542–550. https://doi.org/10.1002/jcb.25004.
214. Zou Y, Jiang Z, Yu X, et al. Upregulation of long noncoding RNA SPRY4-IT1 modulates proliferation, migration, apoptosis, and network formation in trophoblast cells HTR-8SV/neo. *PLoS One*. 2013;8(11). https://doi.org/10.1371/journal.pone.0079598.
215. Steegers E, Dadelszen, Duvekot J, Pijnenborg R. Pre-eclampsia. *Lancet*. 2010; 376(9741):60279–60286. https://doi.org/10.1016/S0140-6736.
216. Yeung KR, Chiu CL, Pidsley R, Makris A, Hennessy A, Lind JM. DNA methylation profiles in preeclampsia and healthy control placentas. *Am J Physiol Heart Circ Physiol*. 2016;310(10):H1295–H1303. https://doi.org/10.1152/ajpheart.00958.2015.
217. Albers RE, Kaufman MR, Natale BV, et al. Trophoblast-specific expression of Hif-1α results in preeclampsia-like Symptoms and fetal growth restriction. *Sci Rep*. 2019;9(1). https://doi.org/10.1038/s41598-019-39426-5.
218. Kulkarni A, Chavan-Gautam P, Mehendale S, Yadav H, Joshi S. Global DNA methylation patterns in placenta and its association with maternal hypertension in Pre-eclampsia. *DNA Cell Biol*. 2011;30(2):79–84. https://doi.org/10.1089/dna.2010.1084.
219. Leavey K, Wilson SL, Bainbridge SA, Robinson WP, Cox BJ. Epigenetic regulation of placental gene expression in transcriptional subtypes of preeclampsia. *Clin Epigenet*. 2018;10(1). https://doi.org/10.1186/s13148-018-0463-6.

220. Jia N, Li J. Role of circular RNAs in preeclampsia. *Dis Markers*. 2019;2019. https://doi.org/10.1155/2019/7237495.
221. Chakraborty D, Cui W, Rosario GX, et al. HIF-KDM3A-MMP12 regulatory circuit ensures trophoblast plasticity and placental adaptations to hypoxia. *Proc Natl Acad Sci USA*. 2016;113(46):E7212–E7221. https://doi.org/10.1073/pnas.1612626113.
222. Alahari S, Post M, Rolfo A, Weksberg R, Caniggia I. Compromised JMJD6 histone demethylase activity affects VHL gene repression in preeclampsia. *J Clin Endocrinol Metab*. 2018;103(4):1545–1557. https://doi.org/10.1210/jc.2017-02197.
223. Xie D, Zhu J, Liu Q, et al. Dysregulation of HDAC9 represses trophoblast cell migration and invasion through TIMP3 activation in preeclampsia. *Am J Hypertens*. 2019;32(5):515–523. https://doi.org/10.1093/ajh/hpz006.
224. Zadora J, Singh M, Herse F, et al. Disturbed placental imprinting in preeclampsia leads to altered expression of DLX5, a human-specific early trophoblast marker. *Circulation*. 2017;136(19):1824–1839. https://doi.org/10.1161/CIRCULATIONAHA.117.028110.
225. van Dijk M, van Bezu J, van Abel D, et al. The STOX1 genotype associated with preeclampsia leads to a reduction of trophoblast invasion by α-T-catenin upregulation. *Hum Mol Genet*. 2010;19(13):2658–2667. https://doi.org/10.1093/hmg/ddq152.
226. Jiang Y, Yu YC, Ding GL, Gao Q, Chen F, Luo Q. Intrauterine hyperglycemia induces intergenerational Dlk1-Gtl2 methylation changes in mouse placenta. *Oncotarget*. 2018;9(32):22398–22405. https://doi.org/10.18632/oncotarget.23976.
227. Reichetzeder C, Dwi Putra SE, Pfab T, et al. Increased global placental DNA methylation levels are associated with gestational diabetes. *Clin Epigenet*. 2016;8(1). https://doi.org/10.1186/s13148-016-0247-9.
228. Bouchard L, Thibault S, Guay SP, et al. Leptin gene epigenetic adaptation to impaired glucose metabolism during pregnancy. *Diabetes Care*. 2010;33(11):2436–2441. https://doi.org/10.2337/dc10-1024.
229. Bouchard L, Hivert MF, Guay SP, St-Pierre J, Perron P, Brisson D. Placental adiponectin gene DNA methylation levels are associated with mothers' blood glucose concentration. *Diabetes*. 2012;61(5):1272–1280. https://doi.org/10.2337/db11-1160.
230. El Hajj N, Pliushch G, Schneider E, et al. Metabolic programming of MEST DNA methylation by intrauterine exposure to gestational diabetes mellitus. *Diabetes*. 2013;62(4):1320–1328. https://doi.org/10.2337/db12-0289.
231. Houde AA, Guay SP, Desgagné V, et al. Adaptations of placental and cord blood ABCA1 DNA methylation profile to maternal metabolic status. *Epigenetics*. 2013;8(12):1289–1302. https://doi.org/10.4161/epi.26554.
232. Côté S, Gagné-Ouellet V, Guay SP, et al. PPARGC1α gene DNA methylation variations in human placenta mediate the link between maternal hyperglycemia and leptin levels in newborns. *Clin Epigenet*. 2016;8(1). https://doi.org/10.1186/s13148-016-0239-9.
233. Anju S, Vincenzo B. Intrauterine growth restriction (IUGR): etiology and diagnosis. *Curr Obst Gynecol Rep*. 2013:102–111. https://doi.org/10.1007/s13669-013-0041-z.
234. Deshpande SS, Balasinor NH. Placental defects: an epigenetic perspective. *Reprod Sci*. 2018;25(8):1143–1160. https://doi.org/10.1177/1933719118766265.
235. Hillman SL, Finer S, Smart MC, et al. Novel DNA methylation profiles associated with key gene regulation and transcription pathways in blood and placenta of growth-restricted neonates. *Epigenetics*. 2015;10(1):50–61. https://doi.org/10.4161/15592294.2014.989741.

236. Eggermann T, Gonzalez D, Spengler S, Arslan-Kirchner M, Binder G, Schönherr N. Broad clinical spectrum in Silver-Russell syndrome and consequences for genetic testing in growth retardation. *Pediatrics*. 2009;123(5):e929–e931. https://doi.org/10.1542/peds.2008-3228.
237. Eggermann T, Begemann M, Binder G, Spengler S. Silver-Russell syndrome: genetic basis and molecular genetic testing. *Orphanet J Rare Dis*. 2010;5(1). https://doi.org/10.1186/1750-1172-5-19.
238. Yamaguchi Y, Tayama C, Tomikawa J, et al. Placenta-specific epimutation at H19-DMR among common pregnancy complications: its frequency and effect on the expression patterns of H19 and IGF2. *Clin Epigenet*. 2019;11(1). https://doi.org/10.1186/s13148-019-0712-3.
239. Park JH, Stoffers DA, Nicholls RD, Simmons RA. Development of type 2 diabetes following intrauterine growth retardation in rats is associated with progressive epigenetic silencing of Pdx1. *J Clin Invest*. 2008;118(6):2316–2324. https://doi.org/10.1172/JCI33655.
240. Iglesias-Platas I, Martín Trujillo A, Court F, Monk D. Distinct promoter methylation and isoform-specific expression of RASFF1A in placental biopsies from complicated pregnancies. *Placenta*. 2015;36(4):397–402. https://doi.org/10.1016/j.placenta.2015.01.011.
241. Fu Q, Yu X, Callaway CW, Lane RH, McKnight RA. Epigenetics: intrauterine growth retardation (IUGR) modifies the histone code along the rat hepatic IGF-1 gene. *FASEB (Fed Am Soc Exp Biol) J*. 2009;23(8):2438–2449. https://doi.org/10.1096/fj.08-124768.
242. Roifman M, Choufani S, Turinsky AL, et al. Genome-wide placental DNA methylation analysis of severely growth-discordant monochorionic twins reveals novel epigenetic targets for intrauterine growth restriction. *Clin Epigenet*. 2016;8(1). https://doi.org/10.1186/s13148-016-0238-x.

CHAPTER

7

Role of placenta in developmental programming of sex-specific adult outcomes

Kirtan Kaur

Department of Medicine, Hematology, and Medical Oncology, Icahn School of Medicine at Mount Sinai, New York, NY, United States

Extensive research has demonstrated that the placenta can influence fetal growth and developmental trajectory during prenatal life.[1,2] Moreover, the placenta may even dictate postnatal and adult health outcomes later in life. Many specific changes are involved in the process of placentation and placentogenesis and in maintaining placental function during pregnancy. While there are many mechanisms that exist, this chapter will focus on discussing the main epigenetic mechanisms and how epigenetic programming may impact later life adult health outcomes. Epigenetic or "above" genetic modifications are those that influence and regulate gene transcription and expression but do not alter the genetic code.[3] Thus the environment may play a crucial role in epigenetic alterations that can be heritable.

Mechanisms of developmental programming

The three main epigenetic mechanisms that are involved in regulating placental function are DNA methylation, noncoding RNAs, and histone modifications.

DNA methylation is one of the most well-investigated epigenetic marks and understood epigenetic mechanisms. DNA methylation is an epigenetic process that regulates or controls gene expression and cellular differentiation.[4,5] DNA methyltransferases (DNMTs) are a family of enzymes that catalyze DNA methylation via the covalent transfer of a methyl group ($-CH_3$) to the fifth carbon of a cytosine residue (5′-CpG-3′), resulting in a 5-methylcytosine (5 mC).

Typically, DNA methylation leads to gene silencing as a form of regulating gene expression.[6] This occurs because the methylated promoter CpG site prevents transcription factors from binding to the gene promoter site. This is an important aspect of DNA methylation as it is a tightly orchestrated event in early embryogenesis and occurs within the trophoblast-derived placenta. The placenta, which derives from

Perinatal and Developmental Epigenetics. https://doi.org/10.1016/B978-0-12-821785-6.00007-4

trophoblast cells, is not subject to remethylation to the extent of the embryoblast and therefore remains in a global hypomethylated state.[7] Within the placenta, metastable epialleles and genomic imprinting are of value to consider under the context of DNA methylation and subsequent placental/fetal developmental trajectory.

Metastable epialleles are alleles that are susceptible to epigenetic modifications leading to variability in expression.[8] These alleles can maintain their epigenetic marks across generations but are also susceptible to changes due to endogenous and exogenous gestational exposures prior to gastrulation. Therefore, these epigenetic marks may contribute to programming susceptibility to later-life disease states.

Studies investigating maternal dietary exposures are elucidating the effects on the fetal epigenome using the Agouti Viable Yellow (A^{vy}) Mouse Model.[9,10] This model exploits the *Agouti* gene, which encodes for a molecule that induces the production of a yellow pheomelanin pigment from melanocytes instead of the wild-type black eumelanin. Once the *Agouti* allele transcription is initiated at the promoter site, the hair follicles express the wild-type brown, or agouti coat color. This gene is of value to study epigenetics due to the fact that when the upstream transposable element is hypermethylated, the *Agouti* gene is silenced. As a result, the mice can display coats ranging from heavily mottled, or a mix of predominantly brown and some yellow, to completely yellow. The yellow mice are also obese when compared with the wild-type. This model is also useful as an "epigenetic biosensor" with respect to maternal diets. Maternal dietary supplementation to boost methylation with nutrients such as folate (folic acid), methylcobalamin (vitamin B12), and trimethylglycine (betaine) have demonstrated changes in offspring coat color similar to the pseudo-agouti (brown coat) phenotype, indicating a greater degree of CpG methylation.[11]

Within the context of the placenta, CpG methylation of metastable epialleles was recently studied for its association with neonate growth potential through early childhood.[12] Since metastable epialleles are conserved across various tissues, the placenta included, using the placenta may offer insights into subsequent postnatal growth rate.

In this study by Clark et al., placental samples from the Extremely Low Gestational Age Newborns (ELGAN) cohort were assessed for CpG methylation in metastable epialleles in association with postnatal body mass index (BMI) during early childhood ages 1, 2, and 10 years (2019). These outcomes were also assessed for any placental sex-specific differences. Among the genes identified as being associated with BMI z-score were *PLAGL/HYMAI*, *HDAC4*, *MEST*, and *EN1* which relate to obesity and adipogenesis. The sex-stratified analyses demonstrated that BMI z-scores in females were associated with placental imprinted gene *PLAGL1/HYMAI*. These epigenetic marks may contribute to obesity programming in utero as evidenced by placental epigenetic signatures.

The other aspect of DNA methylation to consider is genomic imprinting. Genomic imprinting is an epigenetic phenomenon whereby one copy of a given gene is active depending on its parent-of-origin.[13] Some gene alleles may be active solely based on inheritance pattern, either maternal or paternal. Genes that are imprinted are marked via methylation within germ cells to identify which copy of a given gene was inherited from the mother or father. Not all genes are imprinted within an organism's genome.

The parental conflict theory, also known as the kinship theory, posits that fetal growth is driven by paternally expressed imprinted genes,[14] whereas the maternally expressed genes attempt to limit fetal growth and preserve her own survival, fitness, and resources. Thus we see that imprinting is typically evident in genes that determine the fate of metabolism, fetal, and placental growth.[15,16]

To exemplify this kinship theory, the imprinted genes *IGF2* and *H19* on chromosome 11p15 are noteworthy. In this case, the paternally expressed gene *IGF2* (maternally imprinted) is a growth promoter, whereas the maternally expressed *H19* (paternally imprinted) is a growth suppressor. The paternally expressed *IGF2* promotes nutrient transfer from the mother to the fetus to support placental and fetal growth. Meanwhile the maternal *H19* seeks to preserve maternal survival and resources. Due to the differential methylation of the imprint control region (ICR) of *H19* and *IGF2*, the expression of both genes is coregulated. The maternal unmethylated ICR diverts enhancers to the *H19* promoter, allowing for its expression. In contrast, the methylated paternal ICR allows for *IGF2* expression, and *H19* is silenced (Fig. 7.1).

The loss of imprinting (LOI) resulting from dysregulated monoallelic expression is commonly seen in tumors. Within the context of the placenta, LOI can result from various exposures in utero, ranging from maternal nutrition to infection and xenobiotic exposures. As a result, poor health outcomes may occur. A study by Kappil et al. [7] was one of the first to assess whether environmental pollutant exposures during pregnancy affected placental methylation genome-wide and the *H19/IGF2* imprinting and expression levels within the National Children's Study. The

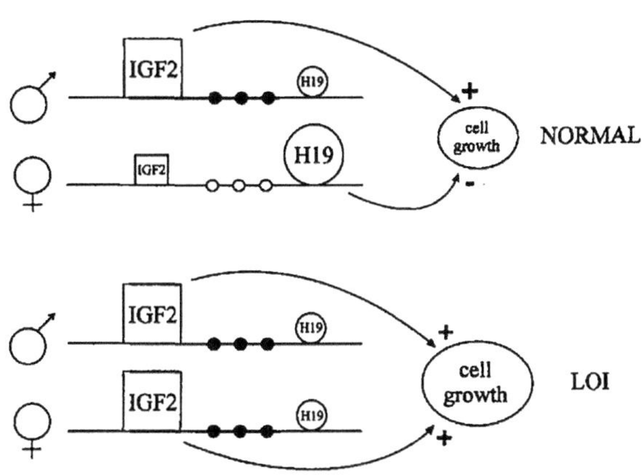

FIGURE 7.1

Genomic imprinting at the H19-IGF2 locus.

Adapted from: Steenman et al. Loss of imprinting of IGF2 is linked to reduced expression and abnormal methylation of H19 in Wilms' tumor. Nat Genet. *1994;7:437.*

study found that total PBDE levels in placental villous tissues were positively associated with global DNA methylation. Total PCB levels were positively associated with H19 expression, thus indicating that organic pollutants can dysregulate imprinting and other epigenetic marks that influence placental and fetal growth.

At present, we are aware of approximately 240 imprinted genes in humans.[17] Expression patterns of imprinted genes vary based on the human tissue in which it is expressed. For example, imprinted genes expressed within the brain or muscle may differ from those found to be expressed in tissues such as the uterus or placenta.[16] Some imprinted genes are placental tissue-specific (i.e., *PHLDA2*) and are highly expressed and can lead to placental overgrowth.[18,19] Other imprinted genes can play a distinct role in the development of intrauterine growth restriction (IUGR) when expression or methylation is disrupted.

Noncoding RNAs are another important epigenetic mechanism that impacts developmental programming. Nonencoding/coding RNAs (ncRNAs) are nucleotides that do not code for a protein product.[20] Despite not encoding functional proteins, ncRNAs are able to regulate gene expression.[20–22] ncRNAs can be microRNAs (miRNAs), Piwi-interacting RNA (piRNAs), endogenous small-interfering RNAs (-siRNAs), or long noncoding RNAs (lncRNAs). Typically, the smaller ncRNAs (siRNA and miRNA) tend to function in transcriptional gene silencing, whereas lncRNAs are involved in genomic imprinting or chromosomal inactivation. Within the placenta, ncRNAs are susceptible to dysregulation and can thus play a role in disease pathogenesis (i.e., preeclampsia). Overexpression and downregulation of lncRNAs such as *H19* or *PVT1* have been recently shown to be involved in regulating human trophoblast cells.[21,22] *H19* overexpression in the trophoblasts leads to increased invasion but reduced cell viability (Xu). lncRNAs can serve even as potential biomarkers to diagnose preeclampsia given their ability to regulate trophoblast migration and invasion.[23] *PVT1* expression is decreased in placentas from gestational diabetic and preeclamptic pregnancies compared with normal pregnancies.[22] The impact of ncRNAs on trophoblast function, placental development, and obstetric complications can then indirectly influence later-life health outcomes of the neonate.[20]

Lastly, histone modifications are another key aspect of epigenetic programming. Histones are proteins that condense and organize our DNA into chromatin.[24] Histones can be modified through various interactions with the proteins comprising of the histone. Therefore, histones are able to regulate and influence gene expression. The covalent posttranslational modifications (PTMs) to the histone subunits that impact gene expression are methylation, phosphorylation, acetylation, ubiquitylation, and SUMOylation. These modifications can either silence or activate transcription by altering the structure of chromatin. Methylation and acetylation are key events in histone modifications.[25] In early pregnancy, important occurrences such as uterine receptivity, maternal immune tolerance, and decidualization are all regulated in some regard by histone modifications. For example, both histone methylation and acetylation are critical in transforming the endometrial stromal cells and preparing it for successful implantation of the conceptus at the blastocyst stage.

Sex-specific effects and disease programming in utero

Though epigenetic mechanisms may be universal, our responses to them may be more/less sensitive and possibly dimorphic solely based on our biological sex. Even under normal circumstances, a placenta derived from a male fetus will differ from a female placenta at the level of placental function, adaptability, and susceptibility. It is important to consider sex differences when discussing development in utero and health outcomes, since biological sex plays a crucial role in how gene expression and disease risk arise.

The placenta demonstrates differences in many areas, such as hormone signaling, immune modulation (i.e., via cytokines), and glucocorticoid transporters.[26–28] These physiologic functions can then be altered epigenetically in the placenta by maternal environmental exposures to xenobiotics or maternal physiological changes (i.e., gestational diabetes mellitus, poor nutrition), and maternal stress. The placental epigenome may drive sexually dimorphic responses based on these various factors, such as prenatal stressors and in utero exposures.

There is unequal gene expression between male and female fetuses, based on X- and Y-sex chromosome–linked genes, especially within the placenta.[29] Many X-chromosome-linked genes are involved in the process of placentogenesis.[26,30,31] Thus, early-life epigenome changes can have diverse effects on gene expression, which will differ in placentas derived from female embryos (XX) versus male embryos (XY). The placenta can modulate the effects of prenatal exposures based on the sexually unique epigenome; this developmental programming may impact chronic health outcomes and disease susceptibility based on the sex of the offspring. For example, embryonic biological sex dictates placental and fetal size.[32–34] Male embryos exhibit faster rates of cell division when compared with female embryos at the same embryonic stage.[34] Similarly, at birth, males demonstrate greater placental efficiency as demonstrated by greater fetus-to-placenta weight ratio indices when compared with their female counterparts at the same gestational age.[35] This may provide some added protection to male fetuses to resist the effects of environmental stressors. Epigenetic processes can drive these sexually dimorphic patterns of growth and influence cell fate decisions and program fetal tissues.[29]

Epigenetic marks can remain stable in response to changes in utero. This developmental programming may then confer chronic disease risk in later life.[36] Due to its dynamic nature and plasticity, the placenta is susceptible to influences of the maternal environment and toxicant insult that can induce changes in methylation.[37–39] Some examples in the research literature include the impacts of maternal cigarette smoking,[40] prenatal air pollution exposure,[41] and even intimate partner violence during pregnancy[42] on DNA methylation changes and/or associated neurobehavioral outcomes in the neonate.

One study assessed whether sexually dimorphic patterns of placental DNA methylation are conserved across different cohorts.[43] In this study, two cohorts were examined for placental DNA methylation, the Extremely Low Gestational Age Newborns cohort (ELGAN, described earlier) and the New Hampshire Birth

Cohort Study (NHBCS), a cohort with normal gestational ages. Over 48,000 CpG sites in placentas taken from both male and female infants were first assessed. The CpG site differential methylation status was then validated in placentas from the NHBCS cohort. The authors reported that more than 99% of the differentially methylated CpG sites were conserved. These genes encoded for immune function, growth, and transcription factor signaling and cell membrane micronutrient transport. They also reported that 587 genes in both cohorts were differentially methylated in male placentas compared with female placentas. The majority of differentially methylated genes were present on X-chromosome-linked genes, possibly due to female X chromosome inactivation. A similar study assessed genome-wide methylation within the placenta and found that more than half of the genes present on X chromosome probes were hypermethylated in male placentas versus females.[44] Different sites along the X chromosome have varying levels of methylation in males compared with their female counterparts. These studies demonstrate that placental methylomes are inherently dimorphic in nature based on the biological sex of the fetus.

Other studies explored the effects of maternal environmental exposures to xenobiotics, air pollution, and cigarette smoking. In the study assessing the role of air pollution, specifically ambient ozone and nitrogen dioxide, exposure in placental epigenetic changes, placental and cord blood biospecimen samples were collected and sex-specific global DNA methylation, and differentially methylated regions (DMRs) were measured.[41] The study reported that prenatal ozone and nitrogen dioxide exposures had sex-specific effects on DNA methylation, in a sexually dimorphic manner. Six DMRs were associated with prenatal nitrogen dioxide exposure, three of which were sex-specific. Ozone exposure similarly demonstrated three sex-specific associated DMRs. An increase in prenatal nitrogen dioxide decreased DNA methylation within females at the *CYP2E1* locus. Increased ozone exposure was associated with a loss of methylation at the *PM20D1* promoter within females. In males by contrast, increased ozone exposure was associated with DMR at the *RNF39* locus. Additionally, nitrogen dioxide was associated with two male DMRs, and lower methylation was evident. The study exploring maternal smoking and altered patterns of placental methylation found that multiple loci within the *RUNX3* gene were affected.[40] These studies can help elucidate the perturbations in our epigenomes from early life that may manifest in later life in sexually dimorphic manners despite exposures remaining similar.

The genes that are identified to be under DNA methylation control can go on to impact toxicant transport and metabolism, which in turn can effect fetal exposure. The genes can also impact nutrient transport between mother and fetus and thus growth and development, and immune function is also influenced by gene expression based on epigenome regulation.[43] Studies have demonstrated this in the context of maternal exposure to plasticizers such as biaphenol-A (BPA) and phthalates such a Di(2-ethylhexyl) or DEHP during pregnancy.[45] One specific study demonstrated that prenatal DEHP exposures assessed in maternal urine in the third trimester were associated with disrupted methylation of the placental *IGF2/H19* domain.[46]

In this study, fetal growth restriction was evident, and two metabolites of DEHP were associated with the decrease in methylation of placental IGF2 when assessed at delivery in the maternal decidua. Taken together, these epigenetic changes may impact later-life metabolism, immune function, and disease risk.

To illustrate how epigenetic changes can result postnatal and adulthood disorders following perturbations within the in utero environment, we will address studies exploring neurobehavioral outcomes. The placenta is sometimes described as a "third brain" as it is the link between the fetus, which is undergoing neurodevelopment, and the mother. Epigenetic changes within the placenta that affect gene expression can impact fetal neurodevelopment. As described previously, DNA methylation is tightly orchestrated in early embryogenesis as it institutes genomic regulation at the tissue level. The placenta, through epigenetic mechanisms, can program the fetal hypothalamic–pituitary–adrenal (HPA) axis, which may lead to adverse neurobehavioral and neuroendocrine outcomes in the offspring.[47] Placental tissue has an active neuroendocrine system.[48,49] In the placenta, the HPA pathway is active in the placenta. Various peptides and hormones such as oxytocin, prostaglandins, corticotropin-releasing factor, and urocortin are involved in the stress response in addition to general maintenance of pregnancy.

Placental DNA methylation patterns have been reported to be consistent and conserved across various study populations; thus, using an established infant neurobehavioral assessment can help researchers investigate DNA methylation and associated neurobehavioral outcomes.[50–52] The standardized assessment employed for infant neurobehavior is known as the Neonatal Intensive Care Unit (NICU) Network Neurobehavioral Scale (NNNS). This tests a variety of scales in the neonate for scores of attention, handling, self-regulation, arousal, excitability, lethargy, hypertonicity, hypotonicity, nonoptimal reflexes, asymmetric reflexes, quality of movement, stress/abstinence, and habituation. A genome-wide study by Paquette et al. reported that 50 genes assessed for DNA methylation in the placenta were related to cell adhesion, and nervous system development was associated with scales of attention, quality of movement, lethargy, and arousal from the NNNS (2016). DNA methylation patterns in specific genes have been associated with specific NNNS scale summary scores. Low methylation of gene *NR3C1*, and high methylation of *HSD11B2* allude to fetal programming within stress regulation pathways that are demonstrated in NNNS outcome measures.[53] *NR3C1* or the Nuclear Receptor Subfamily 3 Group C Member 1 gene is a protein coding gene that encodes for the glucocorticoid receptor; it can act both as a transcription factor and as a regulator of other transcription factors.[54] The glucocorticoid receptor encoded by *NR3C1* binds cortisol and early-life stress (i.e., poor treatment in preschool of children from low socioeconomic areas) has been linked to hypermethylation of this gene.[54,55]

Cry acoustics are another important neonatal neurodevelopmental outcome of interest to study in relation to prenatal epigenetics. Characteristics and measures of cry acoustics include cry latency, threshold, utterances, short utterances, phonation, hyperphonation, dysphonation, cry mode changes, fundamental frequency, first formant, second formant, duration, duration of inspiration, amplitude, variability in f0, variability in F1 and F2, and variability in amplitude.[56,57] These measures of cry acoustics are insightful of infant neurodevelopment as they can demonstrate regulation of the HPA system, the stress response, and autonomic nervous system control involving cranial nerves as well as respiration and vibration in the vocal cords.[57] These measures are also associated with medical conditions, as they have specific biological mechanisms that drive them, which may be perturbed by aberrant epigenetic patterns.[51,52,58] Additionally, cry acoustics may be indicative of later-life speech and language development. The gene *NR3C1* again has been implicated in relation to its susceptibility to methylation.[59] Placental methylation of *NR3C1* has been associated with cry acoustic characteristic measures in relation to glucocorticoid receptor expression control and stress responses in the neonate, postnatally.

Epigenetic dysregulation of the HPA axis during the fetal stage may underlie neurobehavioral and neuroendocrine outcomes and mental health disorders in adulthood. This has been demonstrated in cases of child abuse, depression, and domestic violence during pregnancy.[42,60–66] Prenatal stress in the mother can adversely affect learning and memory in the neonate's hippocampus based on its biological sex and even the time point of exposure.[67] This can confer risks of developing cognitive and emotional disorders, such as schizophrenia and autism. Offspring that are exposed prenatally to stress demonstrate morphological differences within the hippocampus (lower volume) and cerebral cortex (asymmetries) at birth.[68–70] These structures in the developing brain may be vulnerable to epigenetic programming and may play a role in cognitive disorders later in life. Placental DNA methylation in the promoter region of *NR3C1* at 13 CpG sties was assessed at birth.[71] To gauge infant self-regulation, a modified Face-to-Face Still-Face (FFSF) Assessment was administered. The FFSF consists of a 2-min play interaction and still-face by the mother where she does not smile, talk, or touch the infant. Following this, she has a 2-min reunion, in which she resumes playing normally with the infant. Infant self-regulation was assessed in responses. Cortisol in the infants was also measured prior to the stress, after entering the laboratory and poststress after the FFSF assessment. Analysis of the placental *NR3C1* CpG sites in relation to infant self-regulation and cortisol reactivity revealed that greater DNA methylation of one CpG site was related to higher cortisol levels and infant self-regulation. This study demonstrates how early-life neuroendocrine stress responses and behavior can be predicted by placental epigenetic changes.

Sex-specific epigenetic effects on the newborn are evident in cases of prenatal maternal depression, where maternal glucocorticoid exposure can drive epigenetic programming in the neonate.[72] Male babies born to mothers that experienced depression during pregnancy demonstrated higher *NR3C1* methylation in buccal swab. Females by contrast were more hypomethylated for the same gene. This study highlights the postnatal epigenetic effects that remain following prenatal exposures, which may underlie neonatal stress responses and behavior. Another similar study found that maternal stress during the prenatal period releases higher levels of cortisol that the fetus is exposed to in utero.[73] In this study, the authors report that female children aged 7 demonstrated changes in the right amygdala of the brain. The females also demonstrated greater emotional problems following prenatal cortisol exposures via maternal stress experiences. In adulthood, we may see mental health disorders that may have been epigenetically programmed based during the prenatal period in a sex-specific manner. A study of Thai individuals with major depressive disorder (MDD) assessed levels of methylation in buccal swabs based on the patient's biological sex.[74] While MDD patients in this study demonstrated seven CpG sites with higher levels of methylation, when assessed for sex differences, only females displayed significantly different methylation compared with males with MDD.

This chapter predominantly dealt with the impacts of placental DNA methylation. As this is the most well-studied and easily assayed aspect of epigenetic modification, many studies focus on assessing it within placental tissues. Sex differences in this area of research are only just beginning to understand how DNA methylation can confer sexually dimorphic effects and possibly program disease in a sex-specific manner. Maternal stress in pregnancy and prenatal toxicant exposures can be drivers of epigenetic programming of the placenta. This can further impact the HPA axis and the developing fetal brain, especially at the well-studied *NR3C1* glucocorticoid receptor gene locus. These prenatal epigenomic modifications in the placenta are sensitive to the maternal environment and may program the fetus toward adverse neurobehavioral and growth-related outcomes in the postnatal and later-life disease outcomes.

References

1. Fowden AL, Ward JW, Woods FPB, Forhead AJ, Constancia M. Programming placental nutrient transport capacity. *J Physiol*. 2006;572:5–15.
2. Sandovici I, Hoelle K, Angiolini E, Constancia M. Placental adaptations to the maternal-fetal environment: implications for fetal growth and developmental programming. *Reprod Biomed Online*. 2012;25:68–89.
3. Bird A. Perceptions of epigenetics. *Nature*. 2007;447(7143):396–398.
4. Moore L, Le T, Fan G. DNA methylation and its basic function. *Neuropsychopharmacol*. 2013;38:23–38.
5. Holliday R, Pugh JE. DNA modification mechanisms and gene activity during development. *Science*. 1975;187(4173):226–232.
6. Watt F, Molloy PL. Cytosine methylation prevents binding to DNA of a HeLa cell transcription factor required for optimal expression of the adenovirus major late promoter. *Genes Dev*. 1988;2:1136–1143.
7. Kappil M, Wright RO, Sanders AP. Developmental origins of common disease: epigenetic contributions to obesity. *Annu Rev Genom Hum Genet*. 2016;17:177–192.
8. Rakyan VK, Blewitt ME, Druker R, Preis JI, Whitelaw E. Metastable epialleles in mammals. *Trends Genet*. 2002;18:348–351.
9. Duhl DM, Vrieling H, Miller KA, Wolff GL, Barsh GS. Neomorphic agouti mutations in obese yellow mice. *Nat Genet*. 1994;8:59–65.
10. Cooney CA, Dave AA, Wolff GL. Maternal methyl supplements in mice affect epigenetic variation and DNA methylation of offspring. *J Nutr*. 2002;132:2393S–2400S.

11. Waterland RA, Jirtle RL. Transposable elements: targets for early nutritional effects on epigenetic gene regulation. *Mol Cell Biol*. 2003;23:5293–5300.
12. Clark J, Martin E, Bulka CM, Smeester L, Santon H, O'Shea TM. Associations between placental CpG methylation of metabstable epialleles and childhood body mass index across ages one, two and ten in the Extremely Low Gestational Age Newborns (ELGAN) cohort. *Epigenetics*. 2019;14(11):1102–1111.
13. Jirtle RL, Skinner MK. Environmental epigenomics and disease susceptibility. *Nat Rev Genet*. 2007;8:253–262.
14. Moore T, Haig D. Genomic imprinting in mammalian development: a parental tug-or-war. *Trends Genet*. 1991;7(2):45–49.
15. Frost JM, Moore GE. The importance of imprinting in the human placenta. *PLoS Genet*. 2010;6:e1001015.
16. Piedrahita Jorge A. The role of imprinted genes in fetal growth abnormalities. *Birth Defects Res A Clin Mol Teratol*. 2011;91:682–692.
17. Jirtle RL. *Imprinted genes: by species*. Imprinted Gene Databases; 2012.
18. Tunster SJ, Tycko B, John RM. The imprinted Phlda2 gene regulates extraembryonic energy stores. *Mol Cell Biol*. 2010;30:295–306.
19. Lee MP, Feinberg AP. Genomic imprinting of a human apoptosis gene homologue, TSSC3. *Cancer Res*. 1998;58:1052–1056.
20. McAninch D, Roberts CT, Bianco-Miotto T. Mechanistic insight into long noncoding RNAs and the placenta. *Int J Mol Sci*. 2017;18(7):1371.
21. Xu J, Xia Y, Zhang H, Guo H, Feng K, Zhang C. Overexpression of long non-coding RNA H19 promotes invasion and autophagy via the PI3K/AKT/mTOR pathways in trophoblast cells. *Biomed Pharmacother*. 2018;101:691–697.
22. Wang Q, Lu X, Li C, et al. Down-regulated long non-coding RNA PVT1 contributes to gestational diabetes mellitus and preeclampsia via regulation of human trophoblast cells. *Biomed Pharmacother*. 2019;120:109501.
23. Luo X, Li X. Long non-coding RNAs serve as diagnostic biomarkers of preeclampsia and modulate migration and invasiveness of trophoblast cells. *Med Sci Monit*. 2018; 24:84–91.
24. Cacabelos R, Tellado I, Cacabelos P. Chapter 1 – The epigenetic machinery in the life cycle and pharmacoepigenetics. *Pharmacoepigenetics*. 2019;10:1–100.
25. Logan PC, Ponnampalam AP, Rahnama F, Lobie PE, Mitchell MD. The effect of DNA methylation inhibitor 5-Aza-2'-deoxycytidine on human endometrial stromal cells. *Hum Reprod*. 2010;25(11):2859–2869.
26. Gabory A, Roseboom TJ, Moore T, Moore LG, Junien C. Placental contribution to the origins of sexual dimorphism in health and diseases: sex chromosomes and epigenetics. *Biol Sex Differ*. 2013;4(1):5.
27. Januar V, Desoye G, Novakovic B, Cvitic S, Saffery R. Epigenetic regulation of human placental function and pregnancy outcome: considerations for causal inference. *Am J Obstet Gynecol*. 2015;213(Suppl. 4):S182–S196.
28. Tarrade A, Panchenko P, Junien C, Gabory A. Placental contribution to nutritional programming of health and diseases: epigenetics and sexual dimorphism. *J Exp Biol*. 2015; 218(Pt 1):50–58.
29. Mueller BR, Bale TL. Sex-specific programming of offspring emotionality after stress early in pregnancy. *J Neurosci*. 2008;28(36):9055–9065.

30. Mao J, Zhang X, Sieli PT, Falduto MT, Torres KE, Rosenfeld CS. Contrasting effects of different maternal diets on sexually dimorphic gene expression in the murine placenta. *Proc Natl Acad Sci USA*. 2010;107(12):5557–5562.
31. Clifton VL, Hodyl NA, Murphy VE, Giles WB, Baxter RC, Smith R. Effect of maternal asthma, inhaled glucocorticoids and cigarette use during pregnancy on the newborn insulin-like growth factor axis. *Growth Hormone IGF Res*. 2010;20(1):39–48.
32. Lampl M, Gotsch F, Kusanovic JP, et al. Sex differences in fetal growth responses to maternal height and weight. *Am J Hum Biol*. 2010;22(4):431–443.
33. Ishikawa H, Rattigan A, Fundele R, Burgoyne PS. Effects of sex chromosome dosage on placental size in mice. *Biol Reprod*. 2003;69(2):483–488.
34. Mittwoch U. Blastocysts prepare for the race to be male. *Hum Reprod*. 1993;8(10): 1550–1555.
35. Eriksson JG, Kajantie E, Osmond C, Thornburg K, Barker DJ. Boys live dangerously in the womb. *Am J Hum Biol*. 2010;22(3):330–335.
36. Skinner MK. Environmental epigenetic transgenerational inheritance and somatic epigenetic mitotic stability. *Epigenetics*. 2011;6(7):838–842.
37. Fuke C, Shimabukuro M, Petronis A, et al. Age related changes in 5-methylcytosine content in human peripheral leukocytes and placentas: an HPLC-based study. *Ann Hum Genet*. 2004;68:196–204.
38. Price EM, Cotton AM, Penaherrera MS, McFadden DE, Kobor MS, Robinson W. Different measures of "genome-wide" DNA methylation exhibit unique properties in placental and somatic tissues. *Epigenetics*. 2012;7:652–663.
39. Novakovic B, Yuen RK, Gordon L, et al. Evidence for widespread changes in promoter methylation profile in human placenta in response to increasing gestational age and environmental/stochastic factors. *BMC Genom*. 2011;12:529.
40. Maccani JZ, Koestler DC, Houseman EA, Marsit CJ, Kelsey KT. Placental DNA methylation alterations associated with maternal tobacco smoking at the RUNX3 gene are also associated with gestational age. *Epigenomics*. 2013;5(6):619–630.
41. Ladd-Acosta C, Feinberg JI, Brown SC, et al. Epigenetic marks of prenatal air pollution exposure found in multiple tissues relevant for child health. *Environ Int*. 2019;126: 363–376.
42. Radtke KM, Ruf M, Gunter HM, et al. Transgenerational impact of intimate partner violence on methylation in the promoter of the glucocorticoid receptor. *Transl Psychiatry*. 2011;1:e21.
43. Martin E, Smeester L, Bommarito PA, et al. Sexual epigenetic dimorphism in the human placenta: implications for susceptibility during the prenatal period. *Epigenomics*. 2017; 9(3):267–278.
44. Liu J, Morgan M, Hutchison K, Calhoun VD. A study of the influence of sex on genome wide methylation. *PLoS One*. 2010;5(4):e10028.
45. Strakovsky RS, Schantz SL. Impacts of bisphenol A (BPA) and phthalate exposures on epigenetic outcomes in the human placenta. *Environ Epigenet*. 2018;4(3):dvy022.
46. Zhao Y, Chen J, Wang X, Song Q, Xu HH, Zhang YH. Third trimester phthalate exposure is associated with DNA methylation of growth-related genes in human placenta. *Sci Rep*. 2016;6:33449.
47. Christensen BC, Houseman EA, Marsit CJ, et al. Aging and environmental exposures alter tissue-specific DNA methylation dependent upon CpG island context. *PLoS Genet*. 2009;5(8):e1000602.

48. Petraglia F, Imperatore A, Challis JRG. Neuroendocrine mechanisms in pregnancy and parturition. *Endocr Rev.* 2010;31(6):783–816.
49. O'Donnell K, O'Connor TG, Glover V. Placental tissue has an active neuroendocrine system and may be involved in fetal programming of cortisol and stress response, response and immune function. Prenatal stress and neurodevelopment of the child: focus on the HPA axis and role of the placenta. *Dev Neurosci.* 2009;31(4):285–292.
50. Lester B, Tronick E. The neonatal intensive care unit network neurobheavioral scale (NNNS). *Pediatrics.* 2004;113:631–699.
51. Lester B. A biosocial model of infant crying. In: Lipsitt L, Rovee-Collier C, eds. *Advances in infant research.* NY, USA: Ablex; 1984:167–212.
52. Lester B. Infant crying and the development of communication. In: Fox N, Davidson R, eds. *Affective development: a psychobiological perspective.* NJ, USA: Erlbaum Press; 1984:231–258.
53. Appleton AA, Lester BM, Armstrong DA, Lesseur C, Marsit CJ. Examining the joint contribution of placental NR3C1 and HSD11B2 methylation for infant neurobehavior. *Psychoneuroendocrinology.* 2015;52:32–42.
54. Cicchetti D, Handley ED. Methylation of the glucocorticoid receptor gene, nuclear receptor subfamily 3, group C, member 1 (NR3C1), in maltreated and nonmaltreated children: associations with behavioral undercontrol, emotional lability/negativity, and externalizing and internalizing symptoms. *Dev Psychopathol.* 2017;29(5): 1795–1806.
55. Tyrka AR, Ridout KK, Parade SH, Paquette A, Marsit CJ, Seifer R. Childhood maltreatment and methylation of FK506 binding protein 5 gene (FKBP5). *Dev Psychopathol.* 2015;27(4 Pt 2):1637–1645.
56. Lester BM, Marsit CJ. Epigenetic mechanisms in the placenta related to infant neurodevelopment. *Epigenomics.* 2018;10(3):321–333.
57. LaGasse LL, Neal AR, Lester BM. Assessment of infant cry: acoustic cry analysis and parental perception. *Ment Retard Dev Disabil Res Rev.* 2005;11(1):83–93.
58. Kheddache Y, Tadj C. Acoustic measures of the cry characteristics of healthy newborns and newborns with pathologies. *J Biomed Sci Eng.* 2013;6:796–804.
59. Sheinkopf SJ, Righi G, Marsit CJ, Lester BM. Methylation of the glucocorticoid receptor (NR3C1) in placenta is associated with infant cry acoustics. *Front Behav Neurosci.* 2016; 10:100.
60. Romens SE, McDonald J, Svaren J, Pollak SD. Associations between early life stress and gene methylation in children. *Child Dev.* 2015;86(1):303–309.
61. Perroud N, Dayer A, Piguet C, et al. Childhood maltreatment and methylation of the glucocorticoid receptor gene NR3C1 in bipolar disorder. *Br J Psychiatry.* 2014;204(1): 30–35.
62. Conradt E, Lester BM, Appleton AA, Armstrong DA, Marsit CJ. The roles of DNA methylation of NR3C1 and 11beta-HSD2 and exposure to maternal mood disorder in utero on newborn neurobehavior. *Epigenetics.* 2013;8(12):1321–1329.
63. McGowan PO, Sasaki A, D'Alessio AC, et al. Epigenetic regulation of the glucocorticoid receptor in human brain associates with childhood abuse. *Nat Neurosci.* 2009;12(3): 342–348.
64. Oberlander TF, Weinberg J, Papsdorf M, Grunau R, Misri S, Devlin AM. Prenatal exposure to maternal depression, neonatal methylation of human glucocorticoid receptor gene (NR3C1) and infant cortisol stress responses. *Epigenetics.* 2008;3(2):97–106.

65. Lester BM, Marsit CJ, Conradt E, Bromer C, Padbury JF. Behavioral epigenetics and the developmental origins of child mental health disorders. *J Dev Orig Health Dis*. 2012; 3(6):395–408.
66. Gunnar M, Quevedo K. The neurobiology of stress and development. *Annu Rev Psychol*. 2007;58:145–173.
67. Mueller BR, Bale TL. Early prenatal stress impact on coping strategies and learning performance is sex dependent. *Physiol Behav*. 2007;91(1):55–65.
68. Wadhwa PD. Psychoneuroendocrine processes in human pregnancy influence fetal development and health. *Psychoneuroendocrinology*. 2005;30(8):724–743.
69. Kofman O. The role of prenatal stress in the etiology of developmental behavioral disorders. *Neurosci Biobehav Rev*. 2002;26(4):457–470.
70. van Os K, Selten JP. Prenatal exposure to maternal stress and subsequent schizophrenia. The May 1940 invasion of The Netherlands. *Br J Psychiatry*. 1998;172:324–326.
71. Conradt E, Fei M, Lagasse L, et al. Prenatal predictors of infant self-regulation: the contributions of placental DNA methylation of NR3C1 and neuroendocrine activity. *Front Behav Neurosci*. 2015;9:130.
72. Braithwaite EC, Kundakovic M, Ramchandani PG, Murphy SE, Champagne F. Maternal prenatal depressive symptoms predict infant NR3C1 1F and BDNF IV DNA methylation. *Epigenetics*. 2015;10:408–417.
73. Buss C, Davis EP, Shahbaba B, Pruessner JC, Head K, Sandman CA. Maternal cortisol over the course of pregnancy and subsequent child amygdala and hippocampus volumes and affective problems. *Proc Natl Acad Sci USA*. 2012;109(20):E1312–E1319.
74. Nantharat M, Wanitchanon T, Amesbutr M, Tammachote R, Praphanphoj V. Glucocorticoid receptor gene (*NR3C1*) promoter is hypermethylated in Thai females with major depressive disorder. *Genet Mol Res*. 2015;14(4):19071–19079.

Further reading

1. Burton GJ, Fowden AL, Thornburg KL. Placental origins of chronic disease. *Physiol Rev*. 2016;96(4):1509–1565.
2. Compere SJ, Palmiter RD. DNA methylation controls the inducibility of the mouse metallothionein-I gene lymphoid cells. *Cell*. 1981;25(1):233–240.
3. Haley DW, Stansbury K. Infant stress and parent responsiveness: regulation of physiology and behavior during still-face and reunion. *Child Dev*. 2003;74(5):1534–1546.
4. Khashan AS, Abel KM, McNamee R, Pedersen MG, Webb RT, Baker PN, Kenny LC, Mortensen PB. Higher risk of offspring schizophrenia following antenatal maternal exposure to severe adverse life events. *Arch Gen Psychiatry*. 2008;65(2):146–152.
5. Paquette AG, Houseman EA, Green BB, et al. Regions of variable DNA methylation in human placenta associated with newborn neurobehavior. *Epigenetics*. 2016;11(8): 603–613.
6. Rosenfeld CS. Sex-specific placental response in fetal development. *Endocrinology*. 2015; 156(10):3422–3434.
7. Stroud LR, Papandonatos GD, Salisbury AL, et al. Epigenetic regulation of placental NR3C1: mechanism underlying prenatal programming of infant neurobehavior by maternal smoking? *Child Dev*. 2016;87(1):49–60.
8. Wei J, Huang K, Yang C, Kang C. Non-coding RNAs as regulators in epigenetics. *Oncol Rep*. 2017;37:3–9.

SECTION 3

Epigenetics, hormones, sex, and developmental programming

CHAPTER

8

Role of epigenetics in shaping sex differences in brain development and behavior

Laura R. Cortes, Nancy G. Forger

Neuroscience Institute, Georgia State University, Atlanta, GA, United States

Introduction

The human body comprises hundreds of diverse cell types that control different facets of physiology and govern our thoughts and behavior. Yet every cell stems from the same fertilized egg and contains the same genetic material: 22 pairs of autosomal chromosomes and 1 pair of sex chromosomes. This amazing diversity among cells carrying essentially the same DNA underscores the profound dependence of development on "epigenetics," literally, modifications "upon" the chromatin that cause lasting changes in gene expression without changing the underlying DNA nucleotide sequence.

Sexual differentiation (the differentiation of the initially bipotential embryo to become male or female) also depends on epigenetic mechanisms. This is most obvious when considering animal species without chromosomal sex. In some reptiles, for example, there are no sex chromosomes; genetically identical individuals may become either male or female.[1–3] Sex determination in painted turtles (*Chrysemys picta*), for instance, is the result of temperature differences during egg incubation, which cause epigenetic modifications to genes involved in hormone signaling.[4] Thus, *epi*genetic processes determine sex in this species. In mammals, males and females are not genetically identical: all female cells carry two X chromosomes, while male cells carry one X and one Y chromosome. As we discuss in the following, however, epigenetic processes nonetheless play an important role in mammalian sexual differentiation, including that of the central nervous system. In this chapter, we describe sex differences across multiple levels of epigenetic regulation and discuss how sex chromosomes, gonadal hormones, and epigenetic mechanisms may intersect to shape brain development and influence behavior.

Perinatal and Developmental Epigenetics. https://doi.org/10.1016/B978-0-12-821785-6.00009-8

Introduction to epigenetics

The word "epigenetics" was coined by Conrad Waddington to explain the observation that myriad phenotypes could arise from the same genotype.[5] The definition was later modified to include the *heritable* nature of the epigenome during cell division in developmental biology.[6] However, this definition excludes the role of epigenetics in nondividing cells, such as neurons, where epigenetic modifications to chromatin have been shown to be important for many processes. Here, we define epigenetics as long-lasting covalent modifications to chromatin that alter gene expression, without reference to heritability.

The backbone of DNA is composed of phosphate groups that carry a negative charge. Cells can pack an incredible amount of DNA into a single nucleus (each human cell contains about 6 feet of DNA) by spooling the DNA around positively charged histone proteins. DNA and its associated proteins are termed *chromatin*, and the basic unit of chromatin is the *nucleosome*: ~147 base pairs of DNA wrapped around a "core" of eight histone proteins.[7] The N-terminal region of histones contains a long, amino acid tail that can be covalently modified by the addition of methyl, acetyl, lysine, ubiquitin, or phosphate groups, among others.[8] The nomenclature of these modifications (e.g., H3K4me3) specifies the histone protein (e.g., histone 3), position of the targeted amino acid (e.g., fourth lysine (K) from the N-terminus), and type of modification (e.g., three methyl groups). Depending on the modification, these marks may either repress or activate the expression of neighboring genes.

The best characterized histone modification is acetylation, which is carried out by histone acetyltransferases (HATs). HATs add acetyl groups to lysine residues, thus neutralizing the attraction to DNA. Another type of histone modification is phosphorylation of serine or threonine residues, which is carried out by kinases and can similarly loosen histone–DNA interactions. In contrast, histone methylation does not rely on electrostatic interactions to regulate transcription. Depending on the amino acid modified and the number of methyl marks added, additional proteins are recruited that lead to gene activation or repression. Removing these marks entails a different set of enzymes: histone deacetylases (HDACs) remove acetyl marks, histone phosphatases remove phosphate groups, and histone demethylases remove methyl marks. New histone modifications have recently been discovered, such as dopaminylation and serotonylation,[9,10] but have not yet been investigated in the context of sexual differentiation of the brain and thus will not be discussed here.

In addition to histone modifications, DNA itself can be chemically modified, most frequently by the addition of methyl groups on the fifth carbon of cytosine. This generally correlates with gene repression and is catalyzed by DNA methyltransferases (DNMTs).[11,12] Although it was initially believed that methylation was restricted to cytosines followed by guanines (CpG, where p indicates the phosphate group linking two nucleotides), it is now appreciated that methylation can occur on cytosines followed by other nucleotides, especially adenine (A). In fact, in contrast to the rest of the body, DNA methylation in the brain occurs just as frequently at

CpH sites [H indicating A, C, or thymine (T)],[13] and recent work demonstrates that adenine can also be methylated in the mouse brain.[14] For the most part, the role of non-CpG methylation in sexual differentiation has not yet been explored.

DNA methylation marks can be quite stable but can also be removed through a series of steps catalyzed by ten–eleven translocation (TET) enzymes, which begins with the oxidation of methylcytosine (mC) to hydroxymethylcytosine (hmC).[15] An unmodified cytosine is then inserted in its place through the base excision repair pathway.[16] Recent work suggests that hmC may be a stable epigenetic mark in its own right, and not just an intermediary in demethylation.[17] In contrast to mC, hmC marks generally correlate with gene expression.[18] In support of this suggestion, hmC marks are primarily located in enhancers and gene bodies of actively transcribed genes, whereas mC is frequently found in intergenic repetitive elements,[18–20] suggesting that hmC is not simply a step in the process of demethylation. The role of hmC in sexual differentiation is just beginning to be explored,[21] as discussed in the following.

Posttranscriptional gene silencing is also an important mechanism for controlling gene expression. For example, microRNAs (miRNAs) are short, noncoding nucleotide fragments that bind to mRNA to generally prevent translation.[22] Although miRNAs do not fit our definition of covalent modifications to chromatin, they are increasingly recognized as epigenetic regulators of gene expression, and a few studies now link miRNAs to sexual differentiation of the brain.

As described in the following, there are sex differences across multiple levels of the epigenetic process (e.g., in expression or activity of the enzymes that place or remove epigenetic marks, in the global presence of particular epigenetic marks, and in "readers" of epigenetic marks), supporting the hypothesis that epigenetic mechanisms play an important role in sexual differentiation of the brain.

Gonadal steroid hormones drive sexual differentiation and are epigenetic modifiers

Although sex determination in mammals is genetic, the main role of the X and Y chromosomes in sexual differentiation is to determine what hormones are produced. The Y chromosome contains the *Sry* gene, which initiates a cascade of molecular events during embryonic development that leads to the development of testes. Decades of research supports the idea that sexual differentiation in mammals depends on the steroid hormones secreted from the testes (primarily testosterone and its estrogenic metabolites). Gonadal steroids bind to their intracellular receptors, which function as transcription factors to alter gene expression and cause masculinization of the body.[23,24] Circulating gonadal steroids also enter the brain[25] and orchestrate masculinization and defeminization of the nervous system by altering the expression of genes involved in cell death, synapse formation, and other functions.[26]

It can be argued that many, if not most, actions of steroid hormones involve epigenetic modifications. In classic gonadal steroid hormone action, the steroid binds to its intracellular receptor, which then binds to steroid response elements near the promoter regions of target genes. Coactivators and corepressors are then recruited that modify the chromatin to increase or (less often) decrease gene expression. For example, steroid receptor coactivator-1 (SRC1) is a histone acetyltransferase, which primarily acts on histone 3,[27–29] and female rats with genetic ablation of SRC1 are resistant to the defeminizing effects of testosterone on sexual behavior.[30] These "mini" epigenetic modifications in association with steroid hormone action occur both in peripheral cells[31] and in the brain.[32] It therefore may not be surprising that an increasing number of studies have offered direct demonstrations that (1) there are sex differences in the brain epigenome and (2) hormone-dependent brain sexual differentiation requires epigenetic mechanisms.[33–36]

In addition to the role of sex steroid hormones in brain sexual differentiation, more recent work has identified gonad-*independent* effects of sex chromosome complement on the developing mammalian brain.[37] Epigenetic mechanisms may underlie some of these hormone-independent effects[38] and, in fact, the actions of sex chromosomes, sex steroid hormones, and epigenetics are often intertwined. *Here, we explore how each factor and their interactions shape sexual differentiation of the brain.*

An extreme sex difference in epigenetic regulation: X chromosome inactivation

X chromosome inactivation is often overlooked when thinking about sex differences in the epigenome but is a process that may have repercussions for epigenetic control of gene expression in all female cells. Because female mammals have two X chromosomes, they have a double dosage of X chromosome genes. However, one X chromosome is randomly inactivated in each XX cell during embryogenesis, presumably to minimize sex differences in the expression of X-linked genes.[39] This process begins with the coating of one X chromosome with *Xist* long noncoding RNA, followed by recruitment of chromatin silencers (histone deacetylases and methyltransferases), and DNMT3b-directed DNA methylation (Fig. 8.1).[40]

Thus, the epigenetic silencing of nearly an entire chromosome (but note that a small number of genes escape X-inactivation) occurs in every cell in the body and brain of females, and in no cells of males. Moreover, the inactivation state must be actively maintained throughout the life of the cell.[41] If there were no other compensatory mechanisms, the inactivated X chromosome might create a "heterochromatic sink" that limits the availability of epigenetic machinery that would otherwise act on other chromosomes[42]; indeed, under carefully controlled experimental conditions, the expression of hundreds of autosomal genes is sensitive to sex chromosome complement.[43] To overcome this, XX cells may require increased

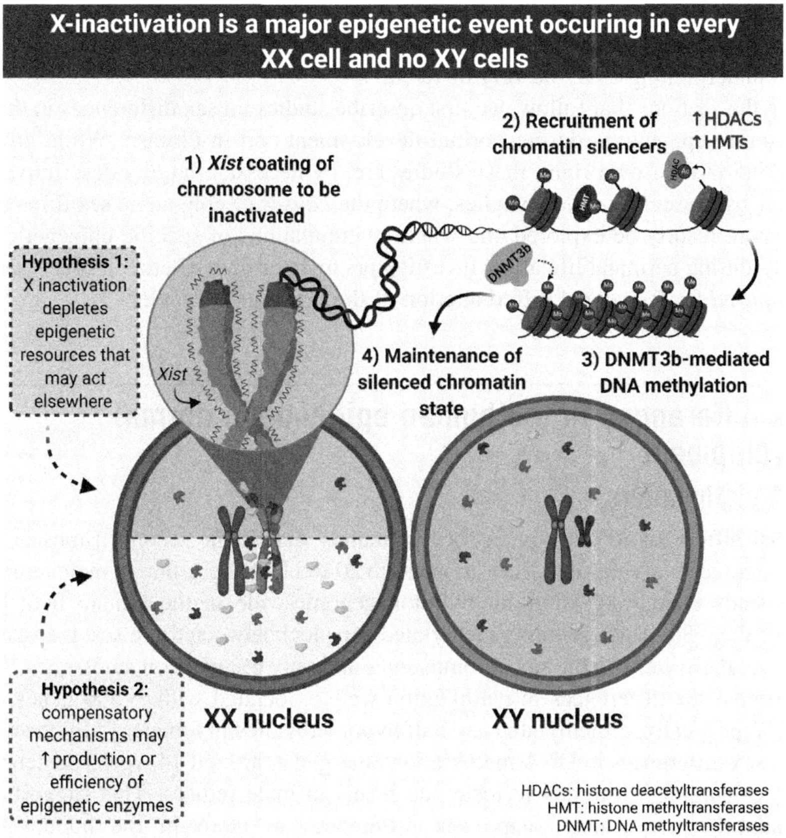

FIGURE 8.1 X-inactivation is a major epigenetic event occurring in every XX cell and no XY cells.

Early in development, one X chromosome is randomly chosen for inactivation in every XX cell of mammals. The process involves (1) coating of the chromosome to be inactivated with the noncoding RNA *Xist*; (2) the recruitment of chromatin silencers including, but not limited to, histone deacetylases (HDACs) and histone methyltransferases (HMTs); (3) DNA methylation, mediated primarily by DNA methyltransferase 3b (DNMT3b); and (4) maintenance of the inactivated state of the silenced X chromosome throughout life through epigenetic mechanisms. This major deployment of epigenetic machinery in every XX cell (and no XY cell) suggests two hypotheses (left side of figure). H1: X-inactivation may create an epigenetic "sink" that depletes epigenetic resources in XX cells that could act on other chromosomes or, H2: XX cells must compensate with increased production or efficiency of epigenetic enzymes. These two possibilities are not mutually exclusive.

Created with Biorender.com.

expression of DNA methylation enzymes, enzyme efficiency, or substrate material to meet the higher demand of epigenetic regulation in each and every cell. Either way, it is important to remember that the purpose of the largest known epigenetic sex difference is to make gene expression more equal in males and females. In other words, even when the sexes do not differ with respect to gene expression, the epigenetic underpinnings may be very different.

In the sections that follow, we first describe studies on sex differences in the human brain epigenome during normal development and in disease. While groundbreaking in their own right, these studies are, by necessity, largely descriptive. We follow by presenting rodent studies, where the *causes* of epigenetic sex differences can more readily be explored and where manipulations of specific epigenetic processes during perinatal life allow investigators to determine whether a given process is *required* for the sexual differentiation of the brain and behavior.

Sex differences in the human epigenome: normal development

DNA methylation

Sexual differentiation in humans occurs mainly during the second trimester, with testosterone levels elevated from about 10 to 20 weeks of gestation in male fetuses.[44] One study examining DNA methylation genome-wide in the human fetal brain found about 8000 differentially methylated gene loci between the sexes; the vast majority of these were on the sex chromosomes and only 6% were on autosomes.[45] The autosomal sex differences in methylation were associated with ~500 genes, with half of the loci hypermethylated and half hypomethylated in females.[45] Interestingly, most sex differences in DNA methylation emerged at around 14 weeks postconception, coinciding with peak testosterone levels in male fetuses. This suggests that gonadal hormones may shape sex differences in utero in the human brain methylome.

In another study, DNA methylation was examined in the dorsolateral prefrontal cortex of humans from 14 weeks gestation to adulthood. About 5% of CpG sites on autosomes differed between the sexes, and the majority of these sites had greater methylation in females.[46] Similarly, the majority of genetic loci with sex differences in DNA methylation in the frontal cortex of adults showed greater methylation in women.[47] Interestingly, about 20% of the genes that were differentially methylated in the adult frontal cortex had sex differences in the same direction in fetal cortical tissue, suggesting a subset of genes with stable sex differences across the life span.[45,47]

Given the barriers to obtaining large numbers of neural samples, many studies on the human epigenome have relied on peripheral tissues, such as blood cells. Although one should proceed with caution when using peripheral samples as a surrogate for the brain, there is some support for this approach based on correlations in DNA methylation between blood cells and brain samples.[48,49] At birth, umbilical cord blood cells contain a small number of sex differences in methylation of CpG sites on autosomes, and most of these sites have higher methylation in girls.[50,51]

A bias toward greater DNA methylation in females is also found when examining differentially methylated regions (as opposed to single CpG sites).[51] Many of the differentially methylated marks were found near genes associated with nervous system development, suggesting that—if the same pattern is seen in the brain—DNA methylation contributes sex-specifically to brain development. Saliva samples from adults reveal a higher number of sexually differentiated CpG sites on autosomes, relative to fetal brain and newborn blood—and similarly, most of these have increased methylation in females, even when accounting for sex differences in cellular composition.[52–55]

Thus, work to date illustrates that DNA methylation in the human brain differs between the sexes starting before birth and suggests a sex bias in DNA methylation, such that more differentially methylated genetic loci are hypermethylated in females than in males. A limitation of these studies (and several of the rodent studies mentioned below) is that the methods used do not distinguish between methylation and hydroxymethylation. This is important, given that mC marks are associated with gene repression, while hmC marks are associated with gene activation. Interestingly, in lymphocytes from umbilical cord blood of newborns, girls had twice the number of "methylated" CpG sites within CpG islands (regions with a high density of CpG sites, typically located near promoters), transcriptional start sites, and shores/shelves (regions located up to 4 kb from CpG islands).[56] Boys, on the other hand, had more "methylated" CpG sites in gene bodies and within the 3′ untranslated region. Given that these different gene locations are associated with mC and hmC, respectively, it is important for future experiments to use techniques that distinguish between these marks to investigate their independent role in shaping the male and female brain.

Histones and microRNAs

In an unusual study using living humans rather than postmortem brain samples, Gilbert and colleagues used positron emission tomography (PET) to examine HDAC expression and found sex-specific differences in brain regions associated with emotion and memory, including the amygdala and hippocampus.[57] The expression of several histone and chromatin regulatory factors is also sexually dimorphic within the brain. For example, SET nuclear oncogene, a histone chaperone partly responsible for regulating access of transcription machinery to chromatin, is expressed in a sexually dimorphic manner in the human and macaque brain.[58]

A recent analysis of publicly available transcriptomic data sets also discovered widespread sex differences in the expression of miRNAs in human tissues, including brains.[59] In addition, the functional roles of many male-biased miRNAs were associated with mechanisms underlying histone modifications, pointing toward a possible interplay between the two types of epigenetic processes in controlling gene expression.

Genomic imprinting

Genomic imprinting refers to the differential expression of an allele depending on whether it was inherited from the mother or the father, and is due to the placement of epigenetic marks during oogenesis or spermatogenesis. Genomic imprinting could cause neural sex differences, if it involves imprinting of genes on the X chromosome: While autosomal imprinted genes are inherited equally by both sexes, females inherit both a paternal and maternal X chromosome, whereas males inherit only the maternal X.

Genomic imprinting of X chromosome genes has been proposed as a possible mechanism underlying the large sex difference in the incidence of autism,[60,61] but this remains theoretical. Humans have relatively few imprinted genes (<100),[62] and none clearly confirmed on the X chromosome. Work on patients with Turner syndrome, a disorder caused by the absence of one X chromosome in girls, provides the best evidence to date for imprinting on the human X chromosome because cognitive and anatomical effects of the disorder differ depending on whether the single X chromosome is inherited from the mother or the father.[63,64] This suggests that gene expression depends on parent-of-origin, but falls short of a direct demonstration of imprinted genes.

A few X-linked imprinted genes are found in mice.[65] A pair of 2010 papers reported many newly identified imprinted genes in mice, and over 300 autosomal genes that showed sex-specific parent-of-origin expression in the brain.[66,67] A subsequent study called into question the majority of the newly identified imprinted genes,[68] however, and it is not clear if parent-of-origin expression biases of autosomal genes would necessarily be due to canonical germline imprinting.[69] Thus, while genomic imprinting of the X chromosome is an important potential contributor to sexual differentiation of the brain, no definitive links have so far been demonstrated.

Implications for neurological disorders and diseases

Many neurological disorders and diseases exhibit sex differences in incidence, age of onset, or severity. For example, boys are more likely than girls to be diagnosed with developmental disorders, such as autism or attention deficit hyperactivity disorder,[70] whereas women are at an increased risk to suffer from anxiety, depression, posttraumatic stress disorders, and Alzheimer's disease.[71–73] Sex biases in disorders and diseases reflect sex differences in risk and/or protective factors, some of which are likely to be epigenetic.

A growing body of literature ties epigenetic changes associated with specific genes to developmental and adult-onset neurological disorders.[74–78] Other neurological disorders have been linked to gene mutations in histone-modifying enzymes, DNMTs, or methyl-binding domain (MBD) proteins, which act as the "readers" of DNA methylation by binding to mC.[79] miRNAs may also underlie sex differences in disease susceptibility: Among schizophrenic patients, expression of an estrogen-responsive miRNA, miR-30b, in the cerebral cortex is reduced relative to controls in women, but not men,[80] hinting at a sex-specific epigenetic underpinning of this disorder. If one sex "relies" on a given epigenetic mechanism more than the other to control gene expression, then any disruption of that mechanism will affect the sexes differentially. For example, a possible conclusion from the studies referenced before is that females rely more on DNA methylation to control gene expression than males, because many genes with differential methylation between the sexes are hypermethylated in females.[46,47,50,51] If so, then any disorder affecting DNMTs, TETs, or MBD proteins would have a more profound impact on the regulation of gene expression in females.

Sex differences in the brain epigenome: rodent studies

Animal studies allow researchers to investigate sex differences in the neural epigenome in tightly controlled settings. This is especially important given that many epigenetic studies have demonstrated that the environment substantially alters the epigenome. By eliminating group differences in diet, stress, etc., these confounding variables can be ruled out, and the effects of sex on the epigenome can be studied in isolation. Rodent studies also allow for the manipulation of epigenetic processes during critical periods of development, which is not possible in human studies. Fig. 8.2 provides an illustrative summary of recent findings, which are detailed in the following.

DNA methylation

The preoptic area (POA) of the hypothalamus has been the target of intense scrutiny as a brain region where many morphological and neurochemical sex differences have been described in rodents. Most of these sex differences are due to differential exposure to perinatal testosterone, or an estrogenic metabolite of testosterone (such as estradiol), in males and females.[36] The first study to report sex differences in an epigenetic mark in the rodent brain used pyrosequencing to examine DNA methylation patterns associated with steroid hormone receptor genes in the POA of rats (see Fig. 8.2). The authors reported sex differences in the intron 1 promoter region of the gene for estrogen receptor alpha (ERα).[81] Importantly, the difference(s) were reversed by treating newborn females with estradiol, indicating that they are due to perinatal gonadal steroid exposure. In addition, sex differences in the epigenome were dynamic across life, as sex differences in CpG sites disappeared and reappeared at different locations with age.[81] Similarly, male rats have increased CpG methylation within the ERα 1b promoter in the POA on postnatal day (P) 8, and treating female pups with estradiol on the first 2 days of life masculinizes (increases) CpG methylation relative to control-treated females.[82]

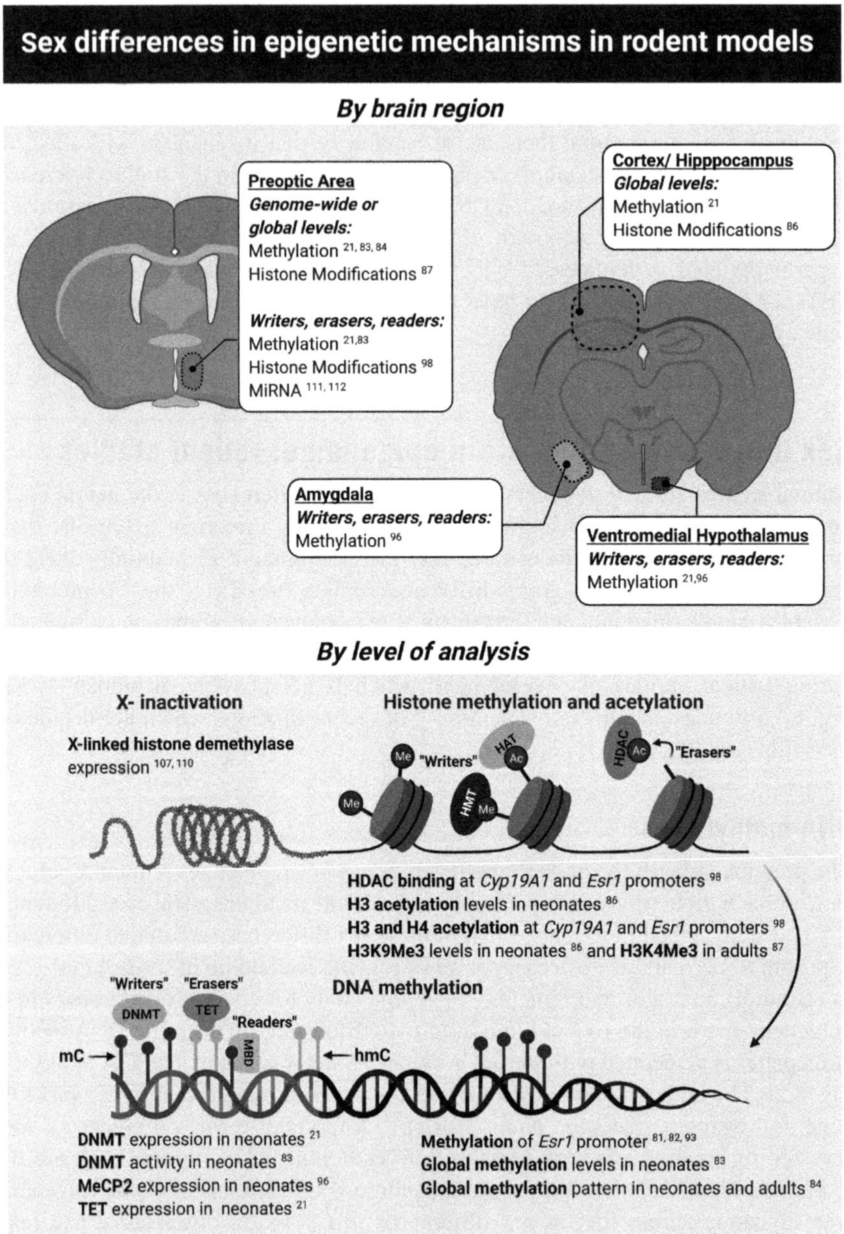

FIGURE 8.2 Sex differences in epigenetic mechanisms in rodent models.

Top: Several brain regions have been the focus of studies reporting sex differences in epigenetic marks, or in expression of epigenetic "writers," "erasers," or "readers." These include the preoptic area of the hypothalamus (in some studies combined with the neighboring bed nucleus of the stria terminalis), amygdala, ventromedial nucleus of the

A later study found that female rats have higher global levels of DNA methylation in the POA on the first days of life than do males.[83] Again, perinatal gonadal steroids are the cause of this difference because it is eliminated when females are treated with estradiol at birth.[83] Female mice also have higher global levels of DNA methylation in the POA during neonatal life, suggesting that this epigenetic sex difference is conserved across species (see Fig. 8.2).[21] Nugent et al. also showed that many sex differences in DNA methylation function to converge gene expression.[83] For example, while inhibition of DNA methylation early in life abolishes some sex differences in gene expression in the rat POA, it causes many more genes to become differentially expressed in males and females.[83] This surprising result implies that sex differences in DNA methylation may *primarily* function to equalize gene expression between male and female brains. Germane to the discussion of disease susceptibility, above, any global perturbation to DNA methylation might uncover a plethora of sex differences in neural circuit functioning where they did not previously exist.

Ghahramani and colleagues examined the effects of neonatal testosterone treatment on DNA methylation throughout the genome in the POA and neighboring bed nucleus of the stria terminalis (POA/BNST) as well as the striatum of male and female mice. Female mice treated with testosterone at birth had DNA methylation patterns that were significantly more masculine than control females, although not fully masculinized.[84] Interestingly, the large majority of sex differences and effects of neonatal testosterone on the methylome were not observed perinatally, but only emerged in adulthood.[84] This was seen despite the fact that all mice in the study were gonadectomized prior to puberty (i.e., effects could not be attributed to sex differences in adult levels of circulating gonadal steroids). Thus, perinatal exposure to gonadal steroids leads to widespread differences in the epigenome that emerge later in life, in what may be termed an "epigenetic echo."[85]

Histone modifications

There are also sex differences in histone modifications in the rodent brain, including in areas not traditionally regarded as highly sexually differentiated. For example, male mice have higher levels of H3 acetylation and methylation in the cortex and

hypothalamus, and cortex/hippocampus. *Bottom:* Levels of analysis include sex differences in gene dosage (e.g., X-linked histone demethylases that escape X chromosome inactivation); histone acetylation and histone methylation associated with specific genes or more globally within a brain region; DNA cytosine methylation (mC) and DNA cytosine hydroxymethylation (hmC) associated with specific genes or brain areas; and expression of the enzymes that place (DNA methyltransferase, DNMT), remove (ten–eleven translocases; TET), or "read" methyl marks (methyl-binding domain proteins, MBDs), such as methyl-CpG-binding protein 2 (MeCP2).

Created with Biorender.com.

hippocampus around the time of birth (i.e., at the time of the testosterone surge in rodents), which disappears by P6.[86] Perinatal testosterone treatment masculinizes H3 acetylation, but not methylation, in females demonstrating that gonadal hormones establish some sex differences in histone modifications. In the BNST/POA of mice, about 250 genes have significant sex differences in H3K4me3, a mark associated with active genes or genes poised for activation.[87] Most of the sex differences in H3K4me3 did not correlate with sex differences in gene expression, however.[87] This is not entirely surprising, if one considers that some sex differences in the brain epigenome function to make gene expression between the sexes more similar, as alluded to before in the section on X chromosome inactivation. The dual-function hypothesis for neural sexual differences popularized by De Vries[88] emphasizes that while some sex differences in brain structure and/or gene expression make males and females different, others perform a compensatory role and make the sexes more similar, and this is likely to apply to epigenetic sex differences, as well.

Although the vast majority of studies have focused on the contribution of gonadal hormones on sexual differentiation of the epigenome, sex chromosomes and gonadal hormones interact during development to augment or *reduce* sex differences, and this applies to histone modifications, as well. In one study, neural stem cells were isolated from XX and XY embryonic mice prior to the testosterone surge.[89] When testosterone was added to these cells, the expression of 2854 genes was altered in XX cells and only 792 in XY cells. At baseline, histone acetylation levels did not differ; however, testosterone treatment differentially altered acetylation levels of specific residues in XY and XX cells.[89] Thus, the effect of gonadal hormones on histone acetylation and on gene expression depends on sex chromosome complement.

Sex differences in the epigenetic writers, erasers, and readers

The previous section demonstrates that many sex differences in epigenetic marks in the brains of rodents are due to exposure to gonadal steroids early in life. The question then becomes: how do steroid hormones cause those changes? One important mechanism may be via hormonal regulation of the expression of epigenetic "writers" and "erasers," the enzymes that, respectively, place and remove epigenetic marks, or by altering the "readers" that interpret those marks. In addition, sex chromosome complement may influence the expression of epigenetic writers in a hormone-independent manner, and this may be one way that chromosomal sex causes the brain to develop differently in males and females.

DNA methylation

Three DNMTs, the "writers" of DNA methylation, are expressed in the brain (*Dnmt1*, *Dnmt3a*, *Dnmt3b*). In addition, the brain expresses *Dnmt3L*, which the protein product does not itself have enzymatic activity but increases the efficiency of

DNMT3a and DNMT3b.[90,91] Female mice have greater expression of *Dnmt1* and *Dnmt3L* than males within the neonatal POA (see Fig. 8.2).[21] Females administered testosterone on the first 2 days of life have intermediate expression, not significantly different from control males or females, demonstrating at least partial regulation by postnatal gonadal hormones. In neonatal rats, no sex difference in DNMT expression in the POA was found, but females had greater DNMT *catalytic activity* during the first few days of life than did males.[83] In this case, estradiol treatment completely masculinized DNMT activity. Future studies will need to test whether sex differences in DNMT3a (or DNMT3b) activity depend on DNMT3L and whether DNMT3L is required for establishing sex differences in the brain.

Sex differences in DNMT activity or expression presumably contribute to hypermethylation in the female POA after birth.[83] However, total DNA methylation depends on the balance between the placement and removal of methyl marks, so sex differences in DNA methylation could also be due to demethylation by the TET enzymes (the DNA methylation "erasers"). Indeed, compared with females, male mice have higher *Tet1* and *Tet2* expression in the ventromedial nucleus of the hypothalamus on P1 and higher *Tet2 and Tet3* expression in the POA on P7.[21] Treatment of females with testosterone at birth partially masculinized expression of *Tet2* and *Tet3*.[21] Similarly, in cancer cell lines, estradiol upregulates *Tet2* within 24 h and leads to demethylation of estrogen-regulated genes.[92] In a recent study, we tested the functional significance of sex differences in TET expression in the neonatal hypothalamus by giving intracerebroventricular injections of small interfering RNAs to *Tet2* and *Tet3* to male mice at birth. This treatment reduced TET expression in males to female-like levels and also prevented masculinization of ERα expression in the arcuate nucleus of the hypothalamus.[93] Thus, DNA demethylation is apparently important for the development of sex differences in ERα expression in the brain.

Epigenetic "readers" are recruited to epigenetically modified chromatin to regulate transcription, such as the MBD proteins described before.[94,95] Newborn female rats have higher levels of the MBD family member methyl-CpG-binding protein 2 (MeCP2) in the amygdala and ventromedial nucleus of the hypothalamus.[96] Mutations in the MeCP2 gene in humans cause Rett syndrome, a severe neurodevelopmental disorder, illustrating the importance of "reading" the epigenetic code.

Histone modifications

As mentioned before, sex differences in H3 acetylation in the cortex and hippocampus are shaped early in life by gonadal hormones.[86] These sex differences likely reflect the activity of aromatized testosterone bound to steroid receptors, which recruit epigenetic modifiers, i.e., SRC1.[27–29] Pfaff and colleagues hypothesize that early-life gonadal hormones create a sex-specific histone "code" that epigenetically programs the male and female brain to engage in sexually differentiated behaviors later in life.[97]

Sex differences in histone modifications may also regulate the production of estradiol and expression of the main estradiol receptor. On embryonic day 21 (one day before birth), male rats have increased H3 acetylation in a promoter of the gene for aromatase (the enzyme that converts testosterone to estradiol) and H4 acetylation in a promoter region of ERα in the medial POA.[98] Similarly, at birth, males also have 2.5 times more binding of HDAC2 on a promoter of the gene encoding aromatase and seven times more binding on the promoter region of the ERα gene in the medial POA.[98] These male biases disappear by P3,[98] coinciding with the end of the testosterone surge.[99] Thus, sex differences in histone modifications have, in several cases, been traced to perinatal levels of gonadal hormones.

There may also be epigenetic changes across the estrous cycle in the adult female brain.[100,101] For example, estradiol decreases histone deacetylase levels and increases global and gene-specific H3 acetylation in the brains of adult female mice.[101–103] Because levels of estradiol fluctuate across the estrous cycle, sex differences in histone acetylation may therefore wax and wane across the cycle. Indeed, two recent studies have reported changes in chromatin condensation in the brain in high versus low estrogen states, and linked chromatin dynamics to gene expression and behavioral changes.[104,105]

Other sex differences in histone-modifying enzymes may be gonad-independent and due to genetic sex. Some genes on the X chromosome escape X-inactivation and are expressed at higher levels in females than in males.[40,106] For example, *Kdm5c* is an X-linked gene that routinely escapes X chromosome inactivation and encodes a histone demethylase. *Kdm5c* is expressed at higher levels in female tissues, including in the brain,[87,107] and mutations are associated with altered neuronal development in mice and cognitive dysfunction in humans.[108,109] *Kdm6a* is another X-linked gene encoding a histone demethylase that also escapes X-inactivation and is expressed at higher levels in the female brain throughout development.[110] Given the important role of histone demethylation in the regulation of gene expression, it is easy to see how female-biased expression of *Kdm5c* and *Kdm6a* might lead to sex differences in transcription, if not offset by compensatory mechanisms in males.

MicroRNAs

Although only a handful of studies have so far been reported, there are also sex differences in miRNA expression in the rodent brain that are responsive to estrogen exposure early in life. Morgan and Bale found sex differences in about 160 miRNAs in the mouse hypothalamus on P2, with more than half disrupted by aromatase inhibition (which feminizes estrogen levels in males).[111,112] How these differences in miRNA affect the neural transcriptomes of males and females is currently unknown.

Manipulating epigenetic marks during development: effects on sexual differentiation of neural morphology and behavior

The studies described so far do not distinguish whether any particular epigenetic mechanisms are *required* for brain sexual differentiation. To address that question, it is necessary to disrupt an epigenetic process during the period of sexual differentiation (the first week or so of life in rodents) and then determine whether sex differences in the brain and behavior are disrupted later in life. Several studies that have done so are described in the following and are summarized schematically in Figs. 8.3 and 8.4.

The first such study looked at the role of histone acetylation on masculinization of the principal nucleus of the BNST (BNSTp) in mice. The BNSTp is larger and contains more neurons in males than in females of several species, including but

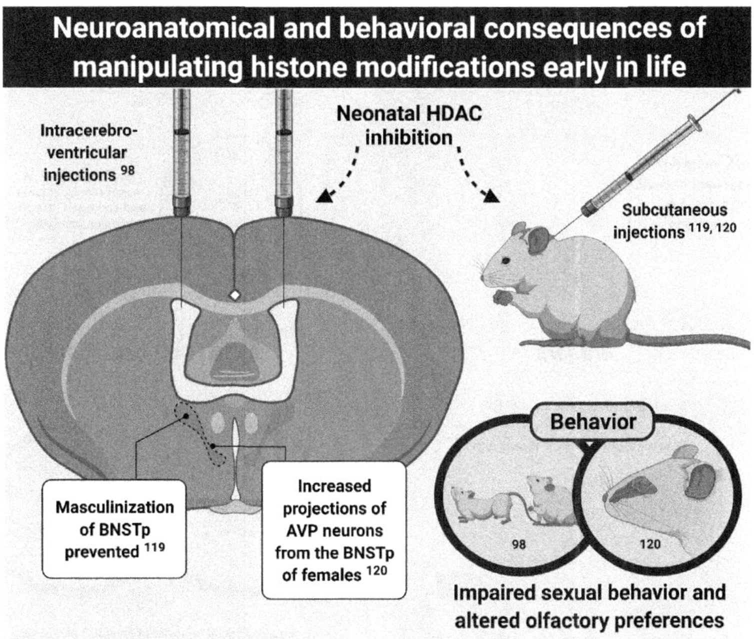

FIGURE 8.3 Neuroanatomical and behavioral consequences of manipulating epigenetic marks early in life.

Several studies have examined the consequences of inhibiting histone deacetylase (HDAC) activity in neonatal rodents. Newborns were given injections that were either subcutaneous (SCs) or directly into the brain (intracerebroventricular, ICV). Effects were examined later in life and include preventing masculinization of cell number in the principal nucleus of the bed nucleus of the stria terminalis (BNSTp); sex-specific changes in projections of neurons in the BNSTp expressing the neuropeptide arginine vasopressin (AVP); disrupting masculinization of sexual behavior; and effects on olfactory behavior.

Created with Biorender.com.

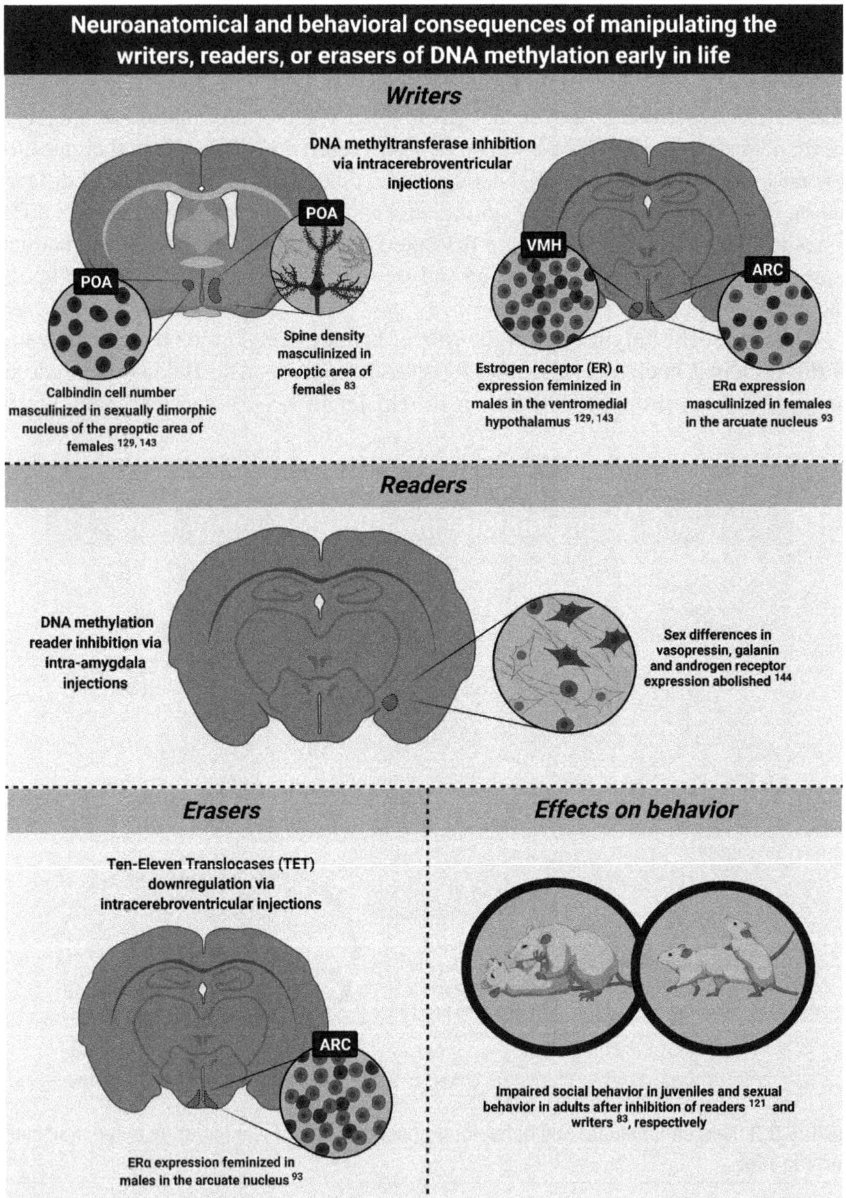

FIGURE 8.4 Neuroanatomical and behavioral consequences of manipulating the writers, readers, or erasers of DNA methylation early in life.

Several studies have examined the consequences of inhibiting DNA methylation writers (DNAs methyltransferases) or readers (methyl binding proteins) during neonatal life. Injections were made into the lateral ventricles or directly into the brain region of interest. Effects include a disruption of sex differences in dendritic spine density in the preoptic not limited to mice, rats, and humans.[113–118] Murray and colleagues injected newborn mice with the HDAC inhibitor valproic acid and found that masculinization of BNSTp volume and neuron number was completely prevented at weaning (see Fig. 8.3).[119] In a follow-up study, Murray et al. discovered that valproic acid also masculinized vasopressin projections from the BNST in adulthood and increased olfactory preference toward soiled bedding of the same sex in females.[120] Similarly, Matsuda and colleagues injected a different HDAC inhibitor (trichostatin A) or antisense oligonucleotides to HDAC2 and HDAC4 into the cerebral ventricles of male rats at birth and found impaired masculine sexual behaviors in adulthood.[98] Thus, the orchestration of histone acetylation around the time of birth apparently is required for the sexual differentiation of the brain and behavior.

In a study of similar design, Nugent et al. gave intracerebroventricular injections of the DNA methyltransferase inhibitor zebularine to newborn rats on the first 2 days of life.[83] The inhibition of DNA methylation masculinized dendritic spine density of neurons in the POA of females and sexual behavior in adulthood (see Fig. 8.4).[83] Interestingly, zebularine treatment or genetic downregulation of DNMTs outside of the critical period for sexual differentiation (which is generally thought to end after the first postnatal week) also partially masculinized the copulatory behavior of females, suggesting an extended period during which epigenetic programming may affect brain differentiation, and that may depend on the availability of epigenetic factors.

As discussed earlier, there is a sex difference in MeCP2 expression on P1 in the rat amygdala, a brain region that has been implicated in the control of social play behavior.[96] When Kurian and colleagues used small interfering (si)RNA to disrupt sex differences in MeCP2 expression during the first 3 days of life, sex differences in juvenile social play were abolished.[121] In addition, one type of MeCP2 mutation seen in Rett syndrome inhibits the ability of MeCP2 to bind to nuclear receptor corepressor (NCoR).[122] NCoR binds to Kaiso and other methyl-binding domain proteins and represses neural gene expression by recruiting HDACs.[123,124] Interestingly, there are also sex differences in NCoR expression in areas of the brain with high sex steroid receptor expression. Male and estradiol-treated females have lower levels of nCoR mRNA in the developing hypothalamus and amygdala than control females.[125] Moreover, a transient reduction of NCoR expression in the neonatal amygdala disrupts sex differences in juvenile play and anxiety-like behavior, by feminizing the behavior of males.[125]

area of the hypothalamus (POAs); calbindin cell number in the sexually dimorphic nucleus of the POA (SDN-POA); estrogen receptor (ER) α cell number in the ventromedial nucleus of the hypothalamus (VMHs); vasopressin (AVP), galanin, or androgen receptor (AR) expression; and social (play) behavior and sexual behavior.

Created with Biorender.com.

Thus, coordinated changes in epigenetic marks or the readers of those marks, often orchestrated by testosterone or its metabolites, are required for sexual differentiation of brain anatomy and behavior.

Manipulating epigenetic marks during development: sexual differentiation of neurochemistry

The most common, and most understudied, type of sex difference in the brain is sex differences in neurochemical phenotype. The chemical phenotype of a cell is defined based on its stable gene expression patterns. In the rodent brain, for example, females have more neurons expressing kisspeptin and tyrosine hydroxylase in the anteroventral periventricular nucleus of the hypothalamus (AVPV),[126] while males have a greater number of vasopressin-expressing neurons in the BNSTp.[127] These sex differences are determined by gonadal steroid exposure early in life[128–130] and regulate important physiological and behavioral processes, such as puberty[131] and social behavior.[132,133] Females also have more neurons expressing ERα in the ventromedial hypothalamus and arcuate nucleus of the hypothalamus, and these cells sex-specifically control energy homeostasis.[134–138]

Cellular diversity depends on the carefully orchestrated expression of specific transcription factors during development and culminates in unique epigenomic signatures across different tissue types, brain regions, and cell types within the same region.[139–141] Throughout the body, DNA methylation controls cellular differentiation by silencing cell fate regulators[142] and plays a significant role in maintaining cell phenotype throughout life. Therefore, we hypothesized that sex differences in neuronal phenotype are the result of differences in epigenetic mechanisms, including DNA methylation.

We tested this by administering a DNA methylation inhibitor at birth directly into the brains of male and female mice. Intracerebroventricular injections of zebularine transiently reduced DNMT activity by 54% in the POA 6 hours later, which returned to baseline after 24 hours.[129] Despite the transient reduction, early-life inhibition of DNMT activity had a lasting effect on the number of cells expressing ERα in the ventromedial nucleus of the hypothalamus and the POA. Females normally have more ERα cells in both regions, and neonatal zebularine treatment increased ERα cell number in males, thereby reducing or eliminating the sex differences.[129,143] Conversely, males have more cells that express calbindin in the POA, and zebularine treatment also affected this sex difference, by increasing calbindin cell number in females.[129,143] Therefore, inhibition of DNA methylation can cause both masculinization and feminization of neuronal phenotype, depending on the brain region or cell type. Zebularine did not alter cell death markers or the total number of surviving cells at weaning, demonstrating that DNA methylation determines the proportion of cells expressing these markers, rather than the number of remaining cells.[143]

Not all cell types are similarly sensitive to zebularine treatment, however, because the sex difference in kisspeptin cell number in the AVPV was not altered by our treatments.

Males have greater expression of the neuropeptide vasopressin in the amygdala than do females, and this sex difference also depends on DNA methylation. siRNA injections against MeCP2 during the first 3 days of life abolish sex differences in vasopressin expression in the amygdala of rats 2 weeks after birth by feminizing males (see Fig. 8.4).[144] Androgen receptor and galanin are similarly decreased to female levels, while ERα is unaffected. By adulthood, vasopressin fibers remain feminized in the amygdala, lateral septum, and BNST in males treated with siRNA, but androgen receptor and galanin return to baseline.

Total methylation depends on a balance between the placement and removal of methyl marks, and interfering with demethylation (such as with siRNAs to TET enzymes) also disrupts sex differences.[93] Thus, DNA methylation early in life is critical for establishing sex differences in neurochemical phenotype and mediates both masculinization (e.g., calbindin) or feminization (e.g., ERα, vasopressin) of cell type, depending on the marker and brain area.

Future directions

The emerging field of "neuroepigenetics" is rapidly advancing with unprecedented resolution of epigenetic sequencing. This, combined with an increasing number of studies that include sex as a biological factor in response to the 2015 NIH mandate (NIH Notice: NOT-OD-15−102), will bring us closer to understanding the development of sex differences in the neural epigenome. We conclude by identifying four areas where progress is needed—and possible—over the next several years.

1. **The role of hydroxymethyl cytosine**. Because sex differences in the expression of TET enzymes are more striking than those of DNMTs, at least in the neonatal mouse brain,[21] sex differences in DNA methylation might depend more on differences in demethylation than on methylation, per se. For this reason, an increased focus on hmC is needed. In addition, hmC is now recognized as an epigenetic mark in its own right. Hydroxymethyl marks can be quite stable, they accumulate with age, and they are found at much higher rates in the brain than in any somatic tissue.[145,146] Thus, hydroxymethylation is likely to play a role in neural function in general and in sex differences in neural function in particular. Until very recently, the large majority of studies interrogating DNA methylation used techniques (e.g., those depending on bisulfite sequencing) that do not distinguish mC from hmC, so a reevaluation of reported sex differences in DNA "methylation" is in order. With the development of TET-assisted bisulfite sequencing and other techniques,[147] it is now possible to distinguish mC from hmC, and this will be important going forward, given that these two marks are associated with opposite effects on gene expression.

2. **Overcoming cellular heterogeneity**. Studies of the neural epigenome have been hampered by the heterogenous nature of most brain samples. Even a small tissue punch of the brain contains multiple cells types, including several types of glia and many different neuronal phenotypes. Because epigenetic marks are cell type specific, the presence of many cell types will mask epigenetic changes to any one cell type, and this may be why the sex differences in epigenetic marks found in the brain thus far have been relatively subtle. The first studies to address this problem performed epigenetic analyses of brain tissue after separating neurons from glial cells, a relatively crude but important refinement that confirmed the divergent epigenetic landscape of these two cell types.[148,149] Other studies are just starting to isolate specific neuronal classes for epigenetic analyses.[141] An important limitation is that cell purification steps often result in too little starting material for subsequent epigenetic analyses. At this point, cell type–specific analyses are still the exception and have not yet been applied to sexual differentiation research. Until we overcome this obstacle, we are likely to find only small sex differences. Moreover, we will have only a vague idea of what it means for one sex to have more of a specific epigenetic mark than the other in a given brain region; what cell type, neural circuit, or function the difference relates to will remain obscure.
3. **Manipulating epigenetic marks**. Finding a sex difference in an epigenetic mark does not address whether it is functionally relevant. For example, a decrease in the expression of a given gene might be associated with increased DNA methylation of a particular cytosine reside of that gene, but this is simply a correlation. To demonstrate causation, it is necessary to manipulate specific marks. Epigenetic editing techniques were developed several years ago that depend on CRISPR (clustered regularly interspaced short palindromic repeats)-based tools. Normally, the Cas9 enzyme uses CRISPR sequences as a guide to recognize and cleave specific DNA sequences. By using "dead" Cas9 (dCas9; a mutant Cas9 without endonuclease activity) tethered to chromatin modifiers (e.g., writers or erasers of epigenetic marks) along with guide RNAs, it is possible to achieve site-specific epigenetic modifications. For example, Chen and colleagues[150] used CRISPR-dCas9 to target the histone acetyltransferase p300 specifically to enhancer regions of the *Fos* gene in neurons and showed that H3K27 acetylation is required for activity-regulated changes in *Fos* gene expression. Similarly, targeting of the active methyltransferase domain of DNMT proteins to dCas9 can be used to site-specifically increase DNA methylation,[151] and by directing dCas9-TET fusions to a specific gene promotor, Liu and colleagues were able to increase the expression of that gene in neurons.[152,153] This technology has not yet been applied to neural sex differences. In many cases, manipulation of a single epigenetic mark is likely to have no effect, simply because there are hundreds, if not thousands, of epigenetic marks associated with any gene and many are likely redundant. For this reason, it may be some time before we achieve the "gold standard" of proving that any

given sex difference in an epigenetic mark associated with a specific gene underlies a sex difference in gene expression, neural function, or behavior.

4. **Integrating across epigenetic marks**. Most studies on sex differences in brain epigenetics have focused on DNA methylation or histone acetylation. However, there are over a dozen different histone posttranslational modifications and hundreds of possible sites on histone tails that can be modified. When considered together with DNA methylation and hydroxymethylation, the result is a dizzying array of combinatorial possibilities for the epigenetic regulation of each gene. While it may not be possible to examine all epigenetic marks in the gene(s) of interest, it also may not be necessary, if the goal is to evaluate the integrated action of all epigenetic modifications on chromatin structure. ATAC-seq, for example, is a recently developed technique that uses a hyperactive transposase to insert sequencing tags into open regions of chromatin, genome-wide. The tagged DNA sequences are then isolated, amplified, and sequenced. Sex differences in chromatin accessibility have very recently been identified using ATAC-seq. For example, in glial cells of patients with late-onset Alzheimer's disease, many genes show sex-specific changes in chromatin accessibility.[154] In the mouse ventral hippocampus, about a third of chromatin is responsive to sex or estrous cycle, with high levels of estradiol associated with opening of regions containing *Egr1*-binding sites,[104] while in the mouse BNST, genes that are expressed more in males tend also to be associated with opening of chromatin conformation in response to estradiol.[155] In the next few years, we are likely to see many more studies identifying sex differences in chromatin accessibility of neural genes, which promises to provide much-needed insight into the combinatorial effect of all epigenetic mechanisms.

References

1. Charnier M. Action de la température sur la sex-ratio chez l'embryon d'*Agama agama* (Agamidae, Lacertilien). *Comptes Rendus Seances Soc Biol Ses Fil.* 1966;160(3): 620–622.
2. Deeming DC, Ferguson MW. Environmental regulation of sex determination in reptiles. *Philos Trans R Soc Lond Ser B Biol Sci.* 1988;322(1208):19–39. https://doi.org/10.1098/rstb.1988.0111.
3. Spotila JR, Standora EA, Morreale SJ, Ruiz GJ. Temperature dependent sex determination in the green turtle (*Chelonia mydas*): effects on the sex ratio on a natural nesting beach. *Herpetologica.* 1987;43(1):74–81.
4. Radhakrishnan S, Literman R, Neuwald JL, Valenzuela N. Thermal response of epigenetic genes informs turtle sex determination with and without sex chromosomes. *Sexual Dev.* 2018. https://doi.org/10.1159/000492188.
5. Waddington C. Genetic assimilation of the bithorax phenotype. *Evolution.* 1956;1. https://doi.org/10.2307/2406091.

6. Holliday R, Pugh JE. DNA modification mechanisms and gene activity during development. *Science*. 1975;187(4173):226–232. https://doi.org/10.1126/science.1111098.
7. Varshavsky AJ, Georgiev GP. Studies on chromatin—V. A model for the structure of chromatin subunit. *Mol Biol Rep*. 1975;2(3):255–262. https://doi.org/10.1007/BF00356996.
8. Bannister AJ, Kouzarides T. Regulation of chromatin by histone modifications. *Cell Res*. 2011;21(3):381–395. https://doi.org/10.1038/cr.2011.22.
9. Farrelly LA, Thompson RE, Zhao S, et al. Histone serotonylation is a permissive modification that enhances TFIID binding to H3K4me3. *Nature*. 2019;567(7749):535–539. https://doi.org/10.1038/s41586-019-1024-7.
10. Lepack A, Werner C, Stewart A, et al. Dopaminylation of histone H3 in ventral tegmental area regulates cocaine seeking. *Science*. 2020;368(6487):197–201. https://doi.org/10.1126/science.aaw8806.
11. Jeltsch A. Molecular enzymology of mammalian DNA methyltransferases. *Curr Top Microbiol Immunol*. 2006;301:203–225. https://doi.org/10.1007/3-540-31390-7_7.
12. Miranda TB, Jones PA. DNA methylation: the nuts and bolts of repression. *J Cell Physiol*. 2007;213(2):384–390. https://doi.org/10.1002/jcp.21224.
13. Lister R, Mukamel E, Nery J, et al. Global epigenomic reconfiguration during mammalian brain development. *Science*. 2013;341(6146). https://doi.org/10.1126/science.1237905.
14. Yao B, Cheng Y, Wang Z, et al. DNA N6-methyladenine is dynamically regulated in the mouse brain following environmental stress. *Nat Commun*. 2017;8(1). https://doi.org/10.1038/s41467-017-01195-y.
15. Tahiliani M, Koh KP, Shen Y, et al. Conversion of 5-methylcytosine to 5-hydroxymethylcytosine in mammalian DNA by MLL partner TET1. *Science*. 2009;324(5929):930–935. https://doi.org/10.1126/science.1170116.
16. Gong Z, Zhu JK. Active DNA demethylation by oxidation and repair. *Cell Res*. 2011;21(12):1649–1651. https://doi.org/10.1038/cr.2011.140.
17. Bachman M, Uribe-Lewis S, Yang X, Williams M, Murrell A, Balasubramanian S. 5-Hydroxymethylcytosine is a predominantly stable DNA modification. *Nat Chem*. 2014;6(12):1049–1055. https://doi.org/10.1038/nchem.2064.
18. Lin IH, Chen YF, Hsu MT. Correlated 5-hydroxymethylcytosine (5hmC) and gene expression profiles underpin gene and organ-specific epigenetic regulation in adult mouse brain and liver. *PLoS One*. 2017;12(1). https://doi.org/10.1371/journal.pone.0170779.
19. Wen L, Tang F. Genomic distribution and possible functions of DNA hydroxymethylation in the brain. *Genomics*. 2014;104(5):341–346. https://doi.org/10.1016/j.ygeno.2014.08.020.
20. Wen L, Li X, Yan L, et al. Whole-genome analysis of 5-hydroxymethylcytosine and 5-methylcytosine at base resolution in the human brain. *Genome Biol*. 2014;15(3). https://doi.org/10.1186/gb-2014-15-3-r49.
21. Cisternas CD, Cortes LR, Bruggeman EC, Yao B, Forger NG. Developmental changes and sex differences in DNA methylation and demethylation in hypothalamic regions of the mouse brain. *Epigenetics*. 2020;15(1–2):72–84. https://doi.org/10.1080/15592294.2019.1649528.

22. O'Carroll D, Schaefer A. General principals of miRNA biogenesis and regulation in the brain. *Neuropsychopharmacology.* 2013;38(1):39–54. https://doi.org/10.1038/npp.2012.87.
23. Tsai MJ, O'Malley BW. Molecular mechanisms of action of steroid/thyroid receptor superfamily members. *Annu Rev Biochem.* 1994;63:451–486. https://doi.org/10.1146/annurev.bi.63.070194.002315.
24. Wu MV, Tollkuhn J. Estrogen receptor alpha is required in GABAergic, but not glutamatergic, neurons to masculinize behavior. *Horm Behav.* 2017;95:3–12. https://doi.org/10.1016/j.yhbeh.2017.07.001.
25. McEwen BS, Plapinger L, Chaptal C, Gerlach J, Wallach G. Role of fetoneonatal estrogen binding proteins in the associations of estrogen with neonatal brain cell nuclear receptors. *Brain Res.* 1975;96(2):400–406. https://doi.org/10.1016/0006-8993(75)90755-6.
26. Lenz KM, McCarthy MM. Organized for sex—steroid hormones and the developing hypothalamus. *Eur J Neurosci.* 2010;32(12):2096–2104. https://doi.org/10.1111/j.1460-9568.2010.07511.x.
27. Molenda-Figueira HA, Murphy SD, Shea KL, et al. Steroid receptor coactivator-1 from brain physically interacts differentially with steroid receptor subtypes. *Endocrinology.* 2008;149(10):5272–5279. https://doi.org/10.1210/en.2008-0048.
28. Spencer TE, Jenster G, Burcin MM, et al. Steroid receptor coactivator-1 is a histone acetyltransferase. *Nature.* 1997;389(6647):194–198. https://doi.org/10.1038/38304.
29. Xu J, Li Q. Review of the in Vivo functions of the p160 steroid receptor coactivator family. *Mol Endocrinol.* 2003;17(9):1681–1692. https://doi.org/10.1210/me.2003-0116.
30. Auger AP, Tetel MJ, McCarthy MM. Steroid receptor coactivator-1 (SRC-1) mediates the development of sex-specific brain morphology and behavior. *Proc Natl Acad Sci U S A.* 2000;97(13):7551–7555. https://doi.org/10.1073/pnas.97.13.7551.
31. Yaşar P, Ayaz G, User SD, Güpür G, Muyan M. Molecular mechanism of estrogen–estrogen receptor signaling. *Reprod Med Biol.* 2017;16(1):4–20. https://doi.org/10.1002/rmb2.12006.
32. Tetel MJ, Auger AP, Charlier TD. Who's in charge? Nuclear receptor coactivator and corepressor function in brain and behavior. *Front Neuroendocrinol.* 2009;30(3):328–342. https://doi.org/10.1016/j.yfrne.2009.04.008.
33. Forger NG. Past, present and future of epigenetics in brain sexual differentiation. *J Neuroendocrinol.* 2018:e12492. https://doi.org/10.1111/jne.12492.
34. Forger NG, Strahan JA, Castillo-Ruiz A. Cellular and molecular mechanisms of sexual differentiation in the mammalian nervous system. *Front Neuroendocrinol.* 2016;40:67–86. https://doi.org/10.1016/j.yfrne.2016.01.001.
35. McCarthy MM. Is sexual differentiation of brain and behavior epigenetic? *Curr Opin Behav Sci.* 2019;25:83–88. https://doi.org/10.1016/j.cobeha.2018.10.005.
36. McCarthy MM, de Vries GJ, Forger NG. Sexual differentiation of the brain: mode, mechanisms, and meaning. In: *Hormones, Brain and Behavior Online.* United States: Elsevier Inc.; 2009:1707–1746. https://doi.org/10.1016/B978-008088783-8.00054-1.
37. Arnold AP, Chen X. What does the four core genotypes mouse model tell us about sex differences in the brain and other tissues? *Front Neuroendocrinol.* 2009;30(1):1–9. https://doi.org/10.1016/j.yfrne.2008.11.001.

38. Sekido R. The potential role of SRY in epigenetic gene regulation during brain sexual differentiation in mammals. In: *Advances in Genetics*. 86. United Kingdom: Academic Press Inc.; 2014:135–165. https://doi.org/10.1016/B978-0-12-800222-3.00007-3.
39. Lyon MF. Gene action in the X-chromosome of the mouse (mus musculus L.). *Nature*. 1961;190(4773):372–373. https://doi.org/10.1038/190372a0.
40. Fang H, Disteche CM, Berletch JB. X inactivation and escape: epigenetic and structural features. *Front Cell Dev Biol*. 2019;7. https://doi.org/10.3389/fcell.2019.00219.
41. Maxfield Boumil R, Lee J. Forty years of decoding the silence in X-chromosome inactivation. *Hum Mol Genet*. 2001:2225–2232. https://doi.org/10.1093/hmg/10.20.2225.
42. Zuckerkandl E. A possible role of inert heterochromatin in cell differentiation. Action of and competition for locking molecules. *Biochimie*. 1974;56(6–7):937–954. https://doi.org/10.1016/S0300-9084(74)80516-X.
43. Wijchers PJ, Yandim C, Panousopoulou E, et al. Sexual dimorphism in mammalian autosomal gene regulation is determined not only by sry but by sex chromosome complement as well. *Dev Cell*. 2010;19(3):477–484. https://doi.org/10.1016/j.devcel.2010.08.005.
44. Siiteri PK, Wilson JD. Testosterone formation and metabolism during male sexual difl-ferentiation in the human embryo. *J Clin Endocrinol Metab*. 1974;38(1):113–125. https://doi.org/10.1210/jcem-38-1-113.
45. Spiers H, Hannon E, Schalkwyk LC, et al. Methylomic trajectories across human fetal brain development. *Genome Res*. 2015;25(3):338–352. https://doi.org/10.1101/gr.180273.114.
46. Numata S, Ye T, Hyde TM, et al. DNA methylation signatures in development and aging of the human prefrontal cortex. *Am J Hum Genet*. 2012;90(2):260–272. https://doi.org/10.1016/j.ajhg.2011.12.020.
47. Xu H, Wang F, Liu Y, Yu Y, Gelernter J, Zhang H. Sex-biased methylome and transcriptome in human prefrontal cortex. *Hum Mol Genet*. 2014;23(5):1260–1270. https://doi.org/10.1093/hmg/ddt516.
48. Davies M, Volta M, Pidsley R, et al. Functional annotation of the human brain methylome identifies tissue-specific epigenetic variation across brain and blood. *Genome Biol*. 2012;13(6):R43. https://doi.org/10.1186/gb-2012-13-6-r43.
49. Horvath S, Zhang Y, Langfelder P, et al. Aging effects on DNA methylation modules in human brain and blood tissue. *Genome Biol*. 2012;13(10):R97. https://doi.org/10.1186/gb-2012-13-10-r97.
50. Adkins RM, Thomas F, Tylavsky FA, Krushkal J. Parental ages and levels of DNA methylation in the newborn are correlated. *BMC Med Genet*. 2011;12. https://doi.org/10.1186/1471-2350-12-47.
51. Yousefi P, Huen K, Davé V, Barcellos L, Eskenazi B, Holland N. Sex differences in DNA methylation assessed by 450K BeadChip in newborns. *BMC Genom*. 2015;16(1). https://doi.org/10.1186/s12864-015-2034-y.
52. Inoshita M, Numata S, Tajima A, et al. Sex differences of leukocytes DNA methylation adjusted for estimated cellular proportions. *Biol Sex Differ*. 2015;6(1). https://doi.org/10.1186/s13293-015-0029-7.
53. Liu J, Morgan M, Hutchison K, Calhoun VD. A study of the influence of sex on genome wide methylation. *PLoS One*. 2010;5(4). https://doi.org/10.1371/journal.pone.0010028.

54. Sun L, Lin J, Du H, et al. Gender-specific DNA methylome analysis of a Han Chinese longevity population. *BioMed Res Int*. 2014;2014. https://doi.org/10.1155/2014/396727.
55. Zaghlool SB, Al-Shafai M, Al Muftah WA, Kumar P, Falchi M, Suhre K. Association of dna methylation with age, gender, and smoking in an arab population. *Clin Epigenet*. 2015;7(1):6–7. https://doi.org/10.1186/s13148-014-0040-6.
56. Maschietto M, Bastos L, Tahira A, et al. Sex differences in DNA methylation of the cord blood are related to sex-bias psychiatric diseases. *Sci Rep*. 2017;7. https://doi.org/10.1038/srep44547.
57. Gilbert T, Zürcher N, Catanase M, et al. Neuroepigenetic signatures of age and sex in the living human brain. *Nat Commun*. 2019;10(1). https://doi.org/10.1038/s41467-019-11031-0.
58. Reinius B, Saetre P, Leonard JA, et al. An evolutionarily conserved sexual signature in the primate brain. *PLoS Genet*. 2008;4(6). https://doi.org/10.1371/journal.pgen.1000100.
59. Cui C, Yang W, Shi J, et al. Identification and analysis of human sex-biased MicroRNAs. *Dev Reprod Biol*. 2018;16(3):200–211. https://doi.org/10.1016/j.gpb.2018.03.004.
60. Badcock C. The imprinted brain: how genes set the balance between autism and psychosis. *Epigenomics*. 2011;3(3):345–359. https://doi.org/10.2217/epi.11.19.
61. Skuse DH. Imprinting, the X-chromosome, and the male brain: explaining sex differences in the liability to autism. *Pediatr Res*. 2000;47(1):9–16.
62. Babak T, DeVeale B, Tsang EK, et al. Genetic conflict reflected in tissue-specific maps of genomic imprinting in human and mouse. *Nat Genet*. 2015;47(5):544–549. https://doi.org/10.1038/ng.3274.
63. Lepage J-F, Hong DS, Mazaika PK, et al. Genomic imprinting effects of the X chromosome on brain morphology. *J Neurosci Official J Soc Neurosci*. 2013;33(19):8567–8574. https://doi.org/10.1523/JNEUROSCI.5810-12.2013.
64. Skuse DH, James RS, Bishop DV, et al. Evidence from Turner's syndrome of an imprinted X-linked locus affecting cognitive function. *Nature*. 1997;387(6634):705–708.
65. Davies W. Genomic imprinting on the X chromosome: implications for brain and behavioral phenotypes. *Ann N Y Acad Sci*. 2010;1204(Suppl):E14–E19. https://doi.org/10.1111/j.1749-6632.2010.05567.x.
66. Gregg C, Zhang J, Butler JE, Haig D, Dulac C. Sex-specific parent-of-origin allelic expression in the mouse brain. *Science (New York, NY)*. 2010;329(5992):682–685. https://doi.org/10.1126/science.1190831.
67. Gregg C, Zhang J, Weissbourd B, et al. High-resolution analysis of parent-of-origin allelic expression in the mouse brain. *Science (New York, NY)*. 2010;329(5992):643–648. https://doi.org/10.1126/science.1190830.
68. DeVeale B, van der Kooy D, Babak T. Critical evaluation of imprinted gene expression by RNA-Seq: a new perspective. *PLoS Genet*. 2012;8(3):e1002600. https://doi.org/10.1371/journal.pgen.1002600.
69. Kravitz SN, Gregg C. New subtypes of allele-specific epigenetic effects: implications for brain development, function and disease. *Curr Opin Neurobiol*. 2019;59:69–78. https://doi.org/10.1016/j.conb.2019.04.012.

70. Boyle CA, Boulet S, Schieve LA, et al. Trends in the prevalence of developmental disabilities in US children, 1997-2008. *Pediatrics*. 2011;127(6):1034–1042. https://doi.org/10.1542/peds.2010-2989.
71. Kessler RC, McGonagle KA, Swartz M, Blazer DG, Nelson CB. Sex and depression in the national comorbidity survey I: lifetime prevalence, chronicity and recurrence. *J Affect Disord*. 1993;29(2–3):85–96. https://doi.org/10.1016/0165-0327(93)90026-G.
72. Laws KR, Irvine K, Gale TM. Sex differences in Alzheimer's disease. *Curr Opin Psychiatr*. 2018;31(2):133–139. https://doi.org/10.1097/YCO.0000000000000401.
73. Tolin DF, Foa EB. Sex differences in trauma and posttraumatic stress disorder: a quantitative review of 25 years of research. *Psychol Bull*. 2006;132(6):959–992. https://doi.org/10.1037/0033-2909.132.6.959.
74. Chouliaras L, Mastroeni D, Delvaux E, et al. Consistent decrease in global DNA methylation and hydroxymethylation in the hippocampus of Alzheimer's disease patients. *Neurobiol Aging*. 2013;34(9):2091–2099. https://doi.org/10.1016/j.neurobiolaging.2013.02.021.
75. Gonzales ML, LaSalle JM. The role of MeCP2 in brain development and neurodevelopmental disorders. *Curr Psychiatr Rep*. 2010;12(2):127–134. https://doi.org/10.1007/s11920-010-0097-7.
76. Nagarajan RP, Hogart AR, Gwye Y, Martin MR, LaSalle JM. Reduced MeCP2 expression is frequent in autism frontal cortex and correlates with aberrant MECP2 promoter methylation. *Epigenetics*. 2006;1(4):172–182. https://doi.org/10.4161/epi.1.4.3514.
77. Tseng PT, Lin PY, Lee Y, et al. Age-associated decrease in global DNA methylation in patients with major depression. *Neuropsychiatric Dis Treat*. 2014;10:2105–2114. https://doi.org/10.2147/NDT.S71997.
78. Xu Y, Chen XT, Luo M, et al. Multiple epigenetic factors predict the attention deficit/hyperactivity disorder among the Chinese Han children. *J Psychiatr Res*. 2015;64:40–50. https://doi.org/10.1016/j.jpsychires.2015.03.006.
79. Jakovcevski M, Akbarian S. Epigenetic mechanisms in neurological disease. *Nat Med*. 2012;18(8):1194–1204. https://doi.org/10.1038/nm.2828.
80. Sharma S, Eghbali M. Influence of sex differences on microRNA gene regulation in disease. *Biol Sex Differ*. 2014;5(1). https://doi.org/10.1186/2042-6410-5-3.
81. Schwarz JM, Nugent BM, McCarthy MM. Developmental and hormone-induced epigenetic changes to estrogen and progesterone receptor genes in brain are dynamic across the life span. *Endocrinology*. 2010;151(10):4871–4881. https://doi.org/10.1210/en.2010-0142.
82. Kurian JR, Olesen KM, Auger AP. Sex differences in epigenetic regulation of the estrogen receptor-α promoter within the developing preoptic area. *Endocrinology*. 2010;151(5):2297–2305. https://doi.org/10.1210/en.2009-0649.
83. Nugent BM, Wright CL, Shetty AC, et al. Brain feminization requires active repression of masculinization via DNA methylation. *Nat Neurosci*. 2015;18(5):690–697. https://doi.org/10.1038/nn.3988.
84. Ghahramani NM, Ngun TC, Chen PY, et al. The effects of perinatal testosterone exposure on the DNA methylome of the mouse brain are late-emerging. *Biol Sex Differ*. 2014;5(1). https://doi.org/10.1186/2042-6410-5-8.
85. McCarthy MM, Nugent BM, Lenz KM. Neuroimmunology and neuroepigenetics in the establishment of sex differences in the brain. *Nat Rev Neurosci*. 2017;18(8):471–484. https://doi.org/10.1038/nrn.2017.61.

86. Tsai HW, Grant PA, Rissman EF. Sex differences in histone modifications in the neonatal mouse brain. *Epigenetics*. 2009;4(1):47–53. https://doi.org/10.4161/epi.4.1.7288.
87. Shen EY, Ahern TH, Cheung I, et al. Epigenetics and sex differences in the brain: a genome-wide comparison of histone-3 lysine-4 trimethylation (H3K4me3) in male and female mice. *Exp Neurol*. 2015;268:21–29. https://doi.org/10.1016/j.expneurol.2014.08.006.
88. De Vries GJ. Minireview: sex differences in adult and developing brains: compensation, compensation, compensation. *Endocrinology*. 2004;145(3):1063–1068. https://doi.org/10.1210/en.2003-1504.
89. Bramble MS, Roach L, Lipson A, et al. Sex-specific effects of testosterone on the sexually dimorphic transcriptome and epigenome of embryonic neural stem/progenitor cells. *Sci Rep*. 2016;6. https://doi.org/10.1038/srep36916.
90. Hata K, Okano M, Lei H, Li E. Dnmt3L cooperates with the Dnmt3 family of de novo DNA methyltransferases to establish maternal imprints in mice. *Development*. 2002;129(8):1983–1993.
91. Jia D, Jurkowska RZ, Zhang X, Jeltsch A, Cheng X. Structure of Dnmt3a bound to Dnmt3L suggests a model for de novo DNA methylation. *Nature*. 2007;449(7159):248–251. https://doi.org/10.1038/nature06146.
92. Wang L, Ozark PA, Smith ER, et al. TET2 coactivates gene expression through demethylation of enhancers. *Sci Adv*. 2018;4(11). https://doi.org/10.1126/sciadv.aau6986.
93. Cortes LR, Cisternas CD, Cabahug INKV, et al. DNA methylation and demethylation underlie the sex difference in estrogen receptor alpha in the arcuate nucleus. *Neuroendocrinology*. 2022. https://doi.org/10.1159/000519671.
94. Ballestar E, Wolffe AP. Methyl-CpG-binding proteins: targeting specific gene repression. *Eur J Biochem*. 2001;268(1):1–6. https://doi.org/10.1046/j.1432-1327.2001.01869.x.
95. Nan X, Ng HH, Johnson CA, et al. Transcriptional repression by the methyl-CpG-binding protein MeCP2 involves a histone deacetylase complex. *Nature*. 1998;393(6683):386–389. https://doi.org/10.1038/30764.
96. Kurian JR, Forbes-Lorman RM, Auger AP. Sex difference in Mecp2 expression during a critical period of rat brain development. *Epigenetics*. 2007;2(3):173–178. https://doi.org/10.4161/epi.2.3.4841.
97. Gagnidze K, Weil ZM, Pfaff DW. Histone modifications proposed to regulate sexual differentiation of brain and behavior. *Bioessays*. 2010;32(11):932–939. https://doi.org/10.1002/bies.201000064.
98. Matsuda KI, Mori H, Nugent BM, Pfaff DW, McCarthy MM, Kawata M. Histone deacetylation during brain development is essential for permanent masculinization of sexual behavior. *Endocrinology*. 2011;152(7):2760–2767. https://doi.org/10.1210/en.2011-0193.
99. Clarkson J, Herbison AE. Hypothalamic control of the male neonatal testosterone surge. *Phil Trans Biol Sci*. 2016;(1688):371. https://doi.org/10.1098/rstb.2015.0115.
100. Gagnidze K, Weil ZM, Faustino LC, Schaafsma SM, Pfaff DW. Early histone modifications in the ventromedial hypothalamus and preoptic area following oestradiol administration. *J Neuroendocrinol*. 2013;25(10):939–955. https://doi.org/10.1111/jne.12085.

101. Tomikawa J, Uenoyama Y, Ozawa M, et al. Epigenetic regulation of Kiss1 gene expression mediating estrogen-positive feedback action in the mouse brain. *Proc Natl Acad Sci*. 2012;109(20):E1294–E1301. https://doi.org/10.1073/pnas.1114245109.
102. Fortress AM, Kim J, Poole RL, Gould TJ, Frick KM. 17β-Estradiol regulates histone alterations associated with memory consolidation and increases Bdnf promoter acetylation in middle-aged female mice. *Learn Mem*. 2014;21(9):457–467. https://doi.org/10.1101/lm.034033.113.
103. Zhao Z, Fan L, Frick KM. Epigenetic alterations regulate estradiol-induced enhancement of memory consolidation. *Proc Natl Acad Sci*. 2010;107(12):5605–5610. https://doi.org/10.1073/pnas.0910578107.
104. Jaric I, Rocks D, Greally JM, Suzuki M, Kundakovic M. Chromatin organization in the female mouse brain fluctuates across the oestrous cycle. *Nat Commun*. 2019;10(1):2851. https://doi.org/10.1038/s41467-019-10704-0.
105. Rocks D, Shukla M, Ouldibbat L, et al. Sex-specific multi-level 3D genome dynamics in the mouse brain. *Nat Commun*. 2022;13:3438. https://doi.org/10.1038/s41467-022-30961-w.
106. Tukianen T, Villani AC, Yen A, et al. Landscape of X chromosome inactivation across human tissues. *Nature*. 2017;550(7675):244–248. https://doi.org/10.1038/nature24265.
107. Xu J, Burgoyne P, Arnold A. Sex differences in sex chromosome gene expression in mouse brain. *Hum Mol Genet*. 2002:1409–1419. https://doi.org/10.1093/hmg/11.12.1409.
108. Iwase S, Brookes E, Agarwal S, et al. A mouse model of X-linked intellectual disability associated with impaired removal of histone methylation. *Cell Rep*. 2016;14(5):1000–1009. https://doi.org/10.1016/j.celrep.2015.12.091.
109. Scandaglia M, Lopez-Atalaya JP, Medrano-Fernandez A, et al. Loss of Kdm5c causes spurious transcription and prevents the fine-tuning of activity-regulated enhancers in neurons. *Cell Rep*. 2017;21(1):47–59. https://doi.org/10.1016/j.celrep.2017.09.014.
110. Xu J, Deng X, Watkins R, Disteche CM. Sex-specific differences in expression of histone demethylases Utx and Uty in mouse brain and neurons. *J Neurosci*. 2008;28(17):4521–4527. https://doi.org/10.1523/JNEUROSCI.5382-07.2008.
111. Morgan CP, Bale TL. Early prenatal stress epigenetically programs dysmasculinization in second-generation offspring via the paternal lineage. *J Neurosci*. 2011;31(33):11748–11755. https://doi.org/10.1523/JNEUROSCI.1887-11.2011.
112. Morgan CP, Bale TL. Sex differences in microRNA-mRNA networks: examination of novel epigenetic programming mechanisms in the sexually dimorphic neonatal hypothalamus. *Biol Sex Differ*. 2017;8(1). https://doi.org/10.1186/s13293-017-0149-3.
113. Allen LS, Gorski RA. Sex difference in the bed nucleus of the stria terminalis of the human brain. *J Comp Neurol*. 1990;302(4):697–706. https://doi.org/10.1002/cne.903020402.
114. Forger NG, Rosen GJ, Waters EM, Jacob D, Simerly RB, De Vries GJ. Deletion of Bax eliminates sex differences in the mouse forebrain. *Proc Natl Acad Sci U S A*. 2004;101(37):13666–13671. https://doi.org/10.1073/pnas.0404644101.
115. Gotsiridze T, Kang N, Jacob D, Forger NG. Development of sex differences in the principal nucleus of the bed nucleus of the stria terminalis of mice: role of Bax-dependent cell death. *Dev Neurobiol*. 2007;67(3):355–362. https://doi.org/10.1002/dneu.20353.
116. Guillamón A, Segovia S, del Abril A. Early effects of gonadal steroids on the neuron number in the medial posterior region and the lateral division of the bed nucleus of

the stria terminalis in the rat. *Dev Brain Res.* 1988;44(2):281–290. https://doi.org/10.1016/0165-3806(88)90226-X.

117. Hines M, Davis FC, Coquelin A, Goy RW, Gorski RA. Sexually dimorphic regions in the medial preoptic area and the bed nucleus of the stria terminalis of the Guinea pig brain: a description and an investigation of their relationship to gonadal steroids in adulthood. *J Neurosci.* 1985;5(1):40–47. https://doi.org/10.1523/jneurosci.05-01-00040.1985.
118. del Abril A, Segovia S, Guillamón A. The bed nucleus of the stria terminalis in the rat: regional sex differences controlled by gonadal steroids early after birth. *Brain Res.* 1987;429:90110–90116. https://doi.org/10.1016/0165-3806(87.
119. Murray EK, Hien A, De Vries GJ, Forger NG. Epigenetic control of sexual differentiation of the bed nucleus of the stria terminalis. *Endocrinology.* 2009;150(9):4241–4247. https://doi.org/10.1210/en.2009-0458.
120. Murray EK, Varnum MM, Fernandez JL, De Vries GJ, Forger NG. Effects of neonatal treatment with valproic acid on vasopressin immunoreactivity and olfactory behaviour in mice. *J Neuroendocrinol.* 2011;23(10):906–914. https://doi.org/10.1111/j.1365-2826.2011.02196.x.
121. Kurian JR, Bychowski ME, Forbes-Lorman RM, Auger CJ, Auger AP. Mecp2 organizes juvenile social behavior in a sex-specific manner. *J Neurosci.* 2008;28(28):7137–7142. https://doi.org/10.1523/JNEUROSCI.1345-08.2008.
122. Lyst MJ, Ekiert R, Ebert DH, et al. Rett syndrome mutations abolish the interaction of MeCP2 with the NCoR/SMRT co-repressor. *Nat Neurosci.* 2013;16(7):898–902. https://doi.org/10.1038/nn.3434.
123. Ordentlich P, Downes M, Xie W, Genin A, Spinner NB, Evans RM. Unique forms of human and mouse nuclear receptor corepressor SMRT. *Proc Natl Acad Sci U S A.* 1999;96(6):2639–2644. https://doi.org/10.1073/pnas.96.6.2639.
124. Yoon HG, Chan DW, Reynolds AB, Qin J, Wong J. N-CoR mediates DNA methylation-dependent repression through a methyl CpG binding protein Kaiso. *Mol Cell.* 2003;12(3):723–734. https://doi.org/10.1016/j.molcel.2003.08.008.
125. Jessen HM, Kolodkin MH, Bychowski ME, Auger CJ, Auger AP. The nuclear receptor corepressor has organizational effects within the developing amygdala on juvenile social play and anxiety-like behavior. *Endocrinology.* 2010;151(3):1212–1220. https://doi.org/10.1210/en.2009-0594.
126. Semaan SJ, Kauffman AS. Sexual differentiation and development of forebrain reproductive circuits. *Curr Opin Neurobiol.* 2010;20(4):424–431. https://doi.org/10.1016/j.conb.2010.04.004.
127. De Vries GJ, Jardon M, Reza M, Rosen GJ, Immerman E, Forger NG. Sexual differentiation of vasopressin innervation of the brain: cell death versus phenotypic differentiation. *Endocrinology.* 2008;149(9):4632–4637. https://doi.org/10.1210/en.2008-0448.
128. Brock O, Bakker J. The two kisspeptin neuronal populations are differentially organized and activated by estradiol in mice. *Endocrinology.* 2013;154(8):2739–2749. https://doi.org/10.1210/en.2013-1120.
129. Cisternas* CD, Cortes* LR, Golynker I, Castillo-Ruiz A, Forger NG. Neonatal inhibition of DNA methylation disrupts testosterone-dependent masculinization of neurochemical phenotype. *Endocrinology.* 2020;161(1). https://doi.org/10.1210/endocr/bqz022.

130. De Vries GJ, Panzica GC. Sexual differentiation of central vasopressin and vasotocin systems in vertebrates: different mechanisms, similar endpoints. *Neuroscience*. 2006; 138(3):947–955. https://doi.org/10.1016/j.neuroscience.2005.07.050.
131. Terasawa E, Guerriero KA, Plant TM. Kisspeptin and puberty in mammals. *Adv Exp Med Biol*. 2013;784:253–273. https://doi.org/10.1007/978-1-4614-6199-9_12.
132. Koolhaas JM, Van Den Brink THC, Roozendaal B, Boorsma F. Medial amygdala and aggressive behavior: interaction between testosterone and vasopressin. *Aggress Behav*. 1990;16(3–4):223–229. https://doi.org/10.1002/1098-2337(1990)16:3/4<223::AID-AB2480160308>3.0.CO;2-#.
133. Veenema AH, Bredewold R, De Vries GJ. Sex-specific modulation of juvenile social play by vasopressin. *Psychoneuroendocrinology*. 2013;38(11):2554–2561. https://doi.org/10.1016/j.psyneuen.2013.06.002.
134. Correa SM, Newstrom DW, Warne JP, et al. An estrogen-responsive module in the ventromedial hypothalamus selectively drives sex-specific activity in females. *Cell Rep*. 2015;10(1):62–74. https://doi.org/10.1016/j.celrep.2014.12.011.
135. Frank A, Brown LM, Clegg DJ. The role of hypothalamic estrogen receptors in metabolic regulation. *Front Neuroendocrinol*. 2014;35(4):550–557. https://doi.org/10.1016/j.yfrne.2014.05.002.
136. Herber CB, Krause WC, Wang L, et al. Estrogen signaling in arcuate Kiss1 neurons suppresses a sex-dependent female circuit promoting dense strong bones. *Nat Commun*. 2019;10(1):163. https://doi.org/10.1038/s41467-018-08046-4.
137. Xu Y, Nedungadi TP, Zhu L, et al. Distinct hypothalamic neurons mediate estrogenic effects on energy homeostasis and reproduction. *Cell Metabol*. 2011;29(5):1232. https://doi.org/10.1016/j.cmet.2019.04.006.
138. van Veen JE, Kammel LG, Bunda PC, et al. Hypothalamic estrogen receptor alpha establishes a sexually dimorphic regulatory node of energy expenditure. *Nat Metabol*. 2020;2(4):351–363. https://doi.org/10.1038/s42255-020-0189-6.
139. Ghosh S, Yates AJ, Frühwald MC, Miecznikowski JC, Plass C, Smiraglia DJ. Tissue specific DNA methylation of CpG islands in normal human adult somatic tissues distinguishes neural from non-neural tissues. *Epigenetics*. 2010;5(6):527–538. https://doi.org/10.4161/epi.5.6.12228.
140. Ladd-Acosta C, Pevsner J, Sabunciyan S, et al. DNA methylation signatures within the human brain. *Am J Hum Genet*. 2007;81(6):1304–1315. https://doi.org/10.1086/524110.
141. Mo A, Mukamel E, Davis F, et al. Epigenomic signatures of neuronal diversity in the mammalian brain. *Neuron*. 2015;86(6):1369–1384. https://doi.org/10.1016/j.neuron.2015.05.018.
142. Reik W, Santos F, Mitsuya K, et al. Epigenetic asymmetry in the mammalian zygote and early embryo: relationship to lineage commitment? *Phil Trans Biol Sci*. 2003; 358(1436):1403–1409. https://doi.org/10.1098/rstb.2003.1326.
143. Mosley M, Weathington J, Cortes LR, et al. Neonatal inhibition of DNA methylation alters cell phenotype in sexually dimorphic regions of the mouse brain. *Endocrinology*. 2017;158(6):1838–1848. https://doi.org/10.1210/en.2017-00205.
144. Forbes-Lorman RM, Rautio JJ, Kurian JR, Auger AP, Auger CJ. Neonatal MeCP2 is important for the organization of sex differences in vasopressin expression. *Epigenetics*. 2012;7(3):230–238. https://doi.org/10.4161/epi.7.3.19265.

145. Globisch D, Münzel M, Müller M, et al. Tissue distribution of 5-hydroxymethylcytosine and search for active demethylation intermediates. *PLoS One*. 2010;5(12). https://doi.org/10.1371/journal.pone.0015367.
146. Penn NW, Suwalski R, O'Riley C, Bojanowski K, Yura R. The presence of 5-hydroxymethylcytosine in animal deoxyribonucleic acid. *Biochem J*. 1972;126(4): 781–790. https://doi.org/10.1042/bj1260781.
147. Yu M, Hon G, Szulwach K, et al. Base-resolution analysis of 5-hydroxymethylcytosine in the mammalian genome. *Cell*. 2012;149(6):1368–1380. https://doi.org/10.1016/j.cell.2012.04.027.
148. Li X, Baker-Andresen D, Zhao Q, Marshall V, Bredy TW. Methyl CpG Binding Domain Ultra-Sequencing: a novel method for identifying inter-individual and cell-type-specific variation in DNA methylation. *Gene Brain Behav*. 2014;13(7): 721–731. https://doi.org/10.1111/gbb.12150.
149. Shulha HP, Cheung I, Guo Y, Akbarian S, Weng Z. Coordinated cell type-specific epigenetic remodeling in prefrontal cortex begins before birth and continues into early adulthood. *PLoS Genet*. 2013;9(4). https://doi.org/10.1371/journal.pgen.1003433.
150. Chen LF, Lin YT, Gallegos DA, et al. Enhancer histone acetylation modulates transcriptional bursting dynamics of neuronal activity-inducible genes. *Cell Rep*. 2019;26(5). https://doi.org/10.1016/j.celrep.2019.01.032, 1174-1188.e5.
151. Vojta A, Dobrinic P, Tadic V, et al. Repurposing the CRISPR-Cas9 system for targeted DNA methylation. *Nucleic Acids Res*. 2016;44(12):5615–5628. https://doi.org/10.1093/nar/gkw159.
152. Liu XS, Wu H, Ji X, et al. Editing DNA methylation in the mammalian genome. *Cell*. 2016;167(1). https://doi.org/10.1016/j.cell.2016.08.056, 233-247.e17.
153. Liu XS, Wu H, Krzisch M, et al. Rescue of fragile X syndrome neurons by DNA methylation editing of the FMR1 gene. *Cell*. 2018;172(5). https://doi.org/10.1016/j.cell.2018.01.012, 979-992.e6.
154. Barrera J, Song L, Gamache JE, et al. Sex dependent glial-specific changes in the chromatin accessibility landscape in late-onset Alzheimer's disease brains. *Mol Neurodegener*. 2021;16(1):58. https://doi.org/10.1186/s13024-021-00481-0.
155. Gegenhuber B, Wu MV, Bronstein R, Tollkuhn J. Gene regulation by gonadal hormone receptors underlies brain sex differences. *Nature*. 2022;606:153–159. https://doi.org/10.1038/s41586-022-04686-1.

SECTION 4

Epigenetic programming by adverse perinatal influences

CHAPTER

9 Early-life stress exposure and epigenetic programming

Chloe M. Farrell

Institute of Psychiatry, Psychology & Neuroscience (IoPPN), King's College London, London, United Kingdom

Introduction to epigenetics

History of epigenetics

The British embryologist and geneticist, Conrad Hal Waddington, coined the term epigenetics in the first half of the 20th century to link the fields of developmental biology and genetics. Today, epigenetic events are defined as "*the structural adaptation of chromosomal regions so as to register, signal, or perpetuate altered activity states*"." This implies an alteration in the function of deoxyribonucleic acid (DNA) rather than in the base pair pattern itself. The "Central Dogma of Molecular Biology,"[1] although somewhat outdated today as a result of a number of major discoveries (including but not limited to noncoding ribonucleic acid [RNAs] and the importance of introns), remains an important concept in appreciating the importance of epigenetic mechanisms. The Central Dogma explains the flow of information from DNA to RNA: transcription, and from RNA to protein: translation. Epigenetic mechanisms are essential for the regulation of DNA structure and gene expression[2] as they can modulate events at the level of transcription and translation. In the mammalian system, epigenetic modifications can currently be placed into the following categories: DNA methylation, histone modifications, nucleosome positioning, and small noncoding RNAs.[3]

Epigenetic mechanisms

DNA methylation

Two of the four DNA bases, cytosine and adenine, can be methylated[4]; however, the evidence published to date is not sufficient to support the presence of methyladenine in mammals.[5] DNA methylation can be categorized as global or gene specific. The term DNA methylation generally refers to the addition of a methyl group, a central carbon atom bonded to three hydrogen atoms, to the 5-carbon of the cytosine ring resulting in 5-methylcytosine (5 mC), sometimes termed the "fifth base" of DNA. Methylation at gene promoter regions is typically associated with transcriptional silencing[6] and subsequent decreased gene expression as cellular chromatin becomes

Perinatal and Developmental Epigenetics. https://doi.org/10.1016/B978-0-12-821785-6.00001-3

tightly coiled, blocking access of transcriptional machinery to DNA. DNA methyltransferases are the enzymes which establish and maintain methylation patterns.[7] DNMT1 is largely responsible for the maintenance of DNA methylation, whereas DNMT3a and DNMT3b generally perform de novo methylation of unmethylated or hemimethylated DNA to assist in maintenance.[8] The majority of CG dinucleotides in the human genome are methylated. However, unmethylated CGs are grouped into clusters, known as CG islands, in the promoter regions of many genes.[9] DNA methylation often occurs at these CG islands; however, it can also occur in regions known as CG shores that are clusters of CGs that lie in close proximity to CG islands. DNA methylation of these CG shores is associated with transcriptional inactivation. It is important to mention that, although DNA methylation is typically associated with transcriptional repression, in some instances it can be coupled with transcriptional activation.[10] This has been observed at gene bodies, where the methylation positively correlated with gene expression.[11]

Histone modifications

Histones are subject to several different posttranslational modifications including, but not limited to, acetylation, methylation, phosphorylation, ubiquitination, and SUMOylation.[10] Histone acetylation (or deacetylation) typically refers to the addition (or removal) of acetyl groups to the highly conserved N-termini of the histones H3 and H4. Histone acetylation is facilitated by histone acetyltransferases (HATs) that catalyze the transfer of acetyl groups from acetyl-CoA to lysine residues, resulting in ε-*N*-acetyl lysine. Acetylation of histones H3 and H4 loosens nucleosome fibers, which leads to more accessible chromatin. This allows for transcriptional machinery to access promoters and initiate transcription more freely.[12] Deacetylation of histones is driven by histone deacetylases (HDACs). HDACs increase ionic interactions between positively charged histones and negatively charged DNA, which yields a more compact chromatin structure and represses gene transcription by limiting the accessibility of transcriptional machinery.[13] Histone methylation can occur on all basic residues: lysines,[14] arginines,[15] and histidines.[16] This methylation is catalyzed by enzymes belonging to three distinct families of proteins: protein arginine *N*-methyltransferase 1 (PRMT1) family, SET domain—containing protein family, and the non-SET domain proteins DOT 1 and DOT 1L (disruptor of telomeric silencing).[14] Histone methylation has been associated with both transcriptional activation and repression.[17] Histone phosphorylation can occur at all four nucleosomal histone tails. The tails contain acceptor sites that can be phosphorylated by numerous protein kinases and dephosphorylated by phosphatases. Histone phosphorylation can take place on serine, threonine, and tyrosine residues. Similar to methylation, the phosphorylation of histone tails has been associated with both transcriptional activation and repression.[18] Histone ubiquitination is most abundant at H2A and H2B. H2A ubiquitination is frequently correlated with gene silencing, while H2B is mostly associated with transcriptional activation.[19] Finally, histone SUMOylation has been shown to mediate transcriptional repression and gene silencing.[20]

Nucleosome positioning

The nucleosome is the basic repeating structural and functional unit of chromatin. It contains eight histone proteins and approximately 146 base pairs of DNA. Two of each of the histones H2A, H2B, H3, and H4 come together to form a histone octamer, which binds and wraps DNA.[21] Nucleosome acts as a physical barrier to transcription, as they block the access of transcriptional machinery to DNA and also inhibit the elongation of the transcripts by polymerases. The packaging of DNA into nucleosomes is thought to affect all stages of transcription. The precise position of nucleosomes and transcription start sites (TSS) has an important influence on transcription initiation.[10]

Small noncoding RNAs

Noncoding RNAs (ncRNAs) are RNAs that do not encode functional proteins and were actually originally considered to only regulate gene expression at the posttranscriptional level.[22] However, we now know that this is not the case and that many RNA classes play integral parts in gene expression from start to finish. Three well-characterized canonical small ncRNAs are microRNA (miRNA), short interfering RNA (siRNA), and Piwi-interacting RNA (piRNA).[23] Although they are best known for their role in regulating mRNA transcripts, they are also directly involved in chromatin-mediated gene silencing and DNA rearrangements.[24] Additionally, there are two classes of noncanonical small ncRNAs, transfer RNA (tRNA)–derived small RNA, and ribosomal RNA (rRNA)–derived small RNA, which are involved in transgenerational epigenetic inheritance.[23]

Mechanisms of stress

Stress and early-life stress

Stress is a physiological phenomenon, necessary for survival, defined as a process in which environmental demands strain an organism's adaptive capacity.[25,26] Stress triggers a number of physiological cascades that ultimately impact upon immune, cardiovascular, gastrointestinal, endocrine, and nervous system function.[27] In certain psychological and physiological circumstances, stress is necessary, and it helps that humans are good at handling short bouts of acute stress. However, problems arise with the presence of chronic stress as it is strongly associated with the prevalence of a number of disorders. Of particular note is stress experienced early in life. This is important as the brain continues to mature well into the 20s[28] resulting in infancy and childhood being periods of particularly high synaptic plasticity in the brain, during which experience(s) can have long-lasting effects on development.[26]

The term early-life stress (ELS) broadly refers to stress occurring during the childhood period (i.e., spanning prenatal, postnatal, childhood, and adolescence). ELS is a term widely used, colloquially and in the clinical setting, which describes many types of adverse experiences a child may encounter, including, but not limited to, exposure to toxins, nutritional restriction, abuse, neglect, maternal physical and/or mental health disorders, maltreatment, and limited family resources. Throughout this chapter, the term ELS will encompass these many different adverse environments. The developmental origins of adult health and disease (DOHaD) hypothesis state that early adverse experiences can result in individuals developing programmed risks for later chronic disorders including cardiovascular, neurodegenerative, and cognitive diseases in adult life.[29] The latency between exposure to adversity and development of a disorder can span years, and even decades in some instances.[30] This chapter will endeavor to include mention of the impact of different forms of stress experienced in childhood (prior to the age of 18) and their impact; however, it is important to highlight at this stage that the vast majority of the current literature and research conducted to date, within the context of epigenetics, has focused on DNA methylation changes at stress-regulatory genes and the subsequent impact(s) on adult stress systems, and as such, a significant proportion of the chapter will be dedicated to that. Additionally, I would like to highlight at this stage that although it is well documented that exposure to ELS increases the risk for psychopathology[31] and cognitive issues, there are also cases where ELS actually promotes the development of resilience. It has been suggested that exposure to moderate, but not mild or severe, ELS has an "inoculation-like" effect that actually boosts emotional regulation and resilience.[32] In any case, despite the variability observed in the occurrence of vulnerability versus resilience, the overarching consensus is that ELS is significantly undesirable for a myriad of reasons and that many adult disorders could be prevented or at least present in less severe forms by lowering the risk and occurrence of ELS.

The hypothalamic–pituitary–adrenal and sympathetic–adrenal–medullary axes

Developmental and physiological overview

In humans, stress is modulated by the sympathetic–adrenal–medullary (SAM) axis and the hypothalamic–pituitary–adrenal (HPA) axis. SAM axis activation is considered to regulate short-term effects with rapid responses, while HPA axis activation leads to short- and long-term effects. The SAM and HPA axes interact at several levels, working in tandem in most target tissues.[33] In fact, continuous stress is a significant contributing factor in the incidence of cancer, coronary heart disease, accidental injury, respiratory disorders, and suicide.[25] Although there is a significant body of work on the link between stress and its effects on many body systems, the most extensive research and the focus of this chapter will be the examination of stress and its link to the development of central nervous system complications, mainly psychiatric disorders and cognitive issues.

The SAM axis acts via the sympathetic nervous system (SNS) and the adrenal medulla leading to the so-called "*fight-or-flight*"" response. Physical and psychogenic stressors activate brainstem and/or forebrain limbic structures[34] and result in a rapid response through autonomic nervous system (ANS) processes, leading to sympathetic nerve stimulation of many target organs.[35] Preganglionic neurons make synapses directly with chromaffin cells of the adrenal medulla causing the synthesis and release of the catecholamines adrenaline and noradrenaline from the glandular tissue.[33] The effects of adrenaline and noradrenaline include vasoconstriction to the organs of the gastrointestinal tract, skin, and kidneys, to allow maximum blood flow to the heart, brain, and skeletal muscles[36] during a perceived stressor. Adrenaline and noradrenaline mediate their effects through interactions with adrenergic receptors. The effects of these receptors are transduced by G protein–coupled receptors (GPCRs): α-1 by G_q/11, α-2 by G_i/o, and β by G_s.[37]

The HPA axis is composed of the paraventricular nucleus (PVN) of the hypothalamus, the anterior lobe of the pituitary gland, and the adrenal cortex of the adrenal glands, located on the kidneys. The HPA axis is controlled by peptides released from the hypothalamus: corticotropin-releasing hormone (CRH) and arginine vasopressin (AVP). Various brain inputs, in particular from the hippocampus, amygdala and the raphe nuclei, converge on the PVN and the supraoptic nuclei to modify the release or inhibition of CRH or AVP. CRH, carried in a small blood vessel system to the anterior pituitary, stimulates the synthesis and release of adrenocorticotrophic hormone (ACTH) from the anterior pituitary gland. ACTH is released into the bloodstream and goes on to stimulate the production of glucocorticoids, such as cortisol, from the zona fasciculata of the adrenal cortex.[38]

It has been proposed that the aforementioned programming for disorders later in life, following exposure to adversity, can begin from as early as the prenatal period. The "fetal programming hypothesis" proposes that alterations in fetal nutrition and endocrine status result in developmental adaptations that permanently change structure, physiology, and metabolism, thereby predisposing to cardiovascular, metabolic, and endocrine diseases in adult life.[39] It is proposed that the HPA axis is programmed by circulating glucocorticoids at critical time points during gestation.[40,41] From approximately 25 weeks gestation, there is an exponential increase in maternal baseline cortisol levels as CRH is secreted from the placenta.[42] Rather than the negative feedback loop, characteristic of the HPA axis outside of pregnancy, placental CRH establishes a positive feedback loop over the course of gestation. This results in the simultaneous increase of CRH, ACTH, and cortisol within the maternal and fetal compartments.[43] Glucocorticoids play an essential role in the 'programming' of the developing fetal brain[44]; however, there is then the risk that exposure to excess glucocorticoids in utero, as a result of maternal hypercortisolemic state (e.g., pregnancy plus severe depression,[45] infection,[46] and/or Cushing's disease[47]), could then alter the development of the fetal HPA axis.[48] This mechanism is believed to act as a preemptive measure so the fetus is prepared for potential external challenges or a hostile environment once they are born. However, these preemptive measures are typically unnecessary and have been shown to be associated with numerous negative outcomes.[49–52] While extensive research has investigated the molecular mechanisms driving these, and many other, adversity-associated outcomes, much is still unknown about the underlying biology. Epigenetic mechanisms provide an etiological link with ELS exposure to psychological and physical outcomes observed in childhood and adulthood as it is known that psychological and physical responses are the result of complex interactions between genetics and environmental cues.

Cortisol is an important steroid hormone which functions to regulate a wide range of processes throughout the body. Upon release, it binds to both the high-affinity low-capacity mineralocorticoid receptors (MRs) and the low-affinity high-capacity glucocorticoid receptors (GRs) and mediates tissue-specific effects[53] resulting in immunomodulation and the mobilization of energy from endogenous stores. MR and GR can form heterodimers resulting in the regulation of target genes.[54] Glucocorticoids feedback to the hippocampus, the PVN of the hypothalamus, and the anterior pituitary gland resulting in receptor-mediated negative feedback inhibition of the HPA axis. This negative feedback loop ensures the maintenance of relatively low levels of cortisol during normal physiological conditions.[55] FK506-binding protein 5 (FKBP5) is a GR cochaperone protein that is crucial to effective HPA axis regulation. It regulates glucocorticoid function by binding to the receptor complex, and when so bound, cortisol binds with lower affinity.[56] FKBP5 expression is induced by cortisol, as part of an intracellular ultrashort negative feedback loop for glucocorticoid activity.[57] Alterations in HPA axis activity, both hyper- and hypoactivity, have been shown to produce anxiety- and depressive-like behaviors, altered internalizing and externalizing behaviors, suicidal ideation, posttraumatic stress disorder, and depression.[58] This HPA axis dysregulation has been measured in several ways. Firstly, cortisol levels in plasma have been used to assess HPA axis activity state with increased cortisol levels being found in the plasma of patients with depression,[59,60] with relative HPA axis hypersecretion more likely in severe depression, and the highest rates being found in those suffering with psychotic depression.[61] The dexamethasone suppression test (DST) is another way of measuring HPA axis activity and has long been used to assess cortisol feedback function in individuals diagnosed with depression.[62] In the DST, dexamethasone is given in various doses, and cortisol levels, usually 12 h later, are measured. Dexamethasone is a synthetic glucocorticoid that should provide negative feedback and suppress the secretion of ACTH, leading to reduced ACTH levels and decreased cortisol levels. However, reduced suppression of ACTH and cortisol has been reported in various studies examining the DST in depressed and suicidal cohorts.[63–65] The cortisol awakening response (CAR) refers to the brisk increase of cortisol levels within 20–30 min of awakening in the morning.[66] Several studies have demonstrated an increased CAR in currently depressed individuals[67,68] as well as recovered patients with depression.[69] This link between HPA axis hyperactivity and development of stress-related disorders has been hypothesized to result from defective regulatory feedback inhibition of glucocorticoids at the level of the hippocampus. It has been suggested that this dysfunctional feedback results from a reduced efficacy of central glucocorticoid and/or glucocorticoid receptor function,[70] which may be driven, in part, by epigenetic mechanisms.[71]

Stress and oxytocin

Oxytocin is a neuropeptide mainly synthesized in the paraventricular and supraoptic nuclei of the hypothalamus. It plays a key role in social cognition, social behaviors, mother–infant bonding, and fear conditioning.[72] Oxytocin mediates its effects through the oxytocin receptor (OXTR), a widely expressed $G_{\alpha q}$ protein–coupled receptor[73] encoded by the *OXTR* gene, located on chromosome 3p25.3.[74] Under normal physiological conditions, oxytocin works in tandem with the HPA axis by exerting stress-minimizing effects to regulate cortisol levels.[75] The dysregulation of oxytocin function has been hypothesized to result in psychiatric and physical disorders resulting from early relational trauma, a form of ELS. Moreover, it has been suggested that oxytocin dysregulation exacerbates HPA axis dysfunction and PTSD when ELS has occurred.[76]

Stress and the serotonergic system

Serotonin, or 5-hydroxytryptamine (5-HT), is a monoamine neurotransmitter primarily found in the gastrointestinal (GI) tract and central nervous system (CNS).[77] 5-HT plays key roles in neural activities and behaviors. Serotonin, secreted by serotonergic neurons, regulates mood, appetite, and sleep and is important for cognition, memory, and learning.[78] Once released, serotonin can bind to the numerous serotonin receptor subtypes, which act via inhibition of adenylyl cyclase (5-HT_{1A}, $5\text{-HT}_{1D\alpha}$, $5\text{-HT}_{1D\beta}$), stimulation of adenylyl cyclase (5-HT4, 5-HT7), opening of K^+ channels (5-HT1A), closing of K+ channels (5-HT2A), stimulation of phosphoinositide-specific phospholipase C (5-HT2A, 5-HT2B, 5-HT_{2C}), and as a ligand-gated cation channel (5-HT_3). The action of serotonin is terminated by reuptake via the 5-HT transporter (5-HTT), a protein encoded by the solute carrier family 6 member 4 (*SLC6A4*) gene and enzymatically by monoamine oxidase (MAO) A and B. Serotonin imbalances have been implicated in the pathophysiology of many disorders including depression, anxiety disorders, impulse control disorders,[79] serotonin syndrome, akathisia, and chronic dystonias.[80] HPA axis activity is inextricably linked with serotonergic system functioning[81]; therefore, overactivation of the HPA axis, due to chronic stress, has the potential to negatively impact upon serotonergic function and it is associated metabolic pathways. One example of this is seen in the tryptophan/kynurenine pathway. The essential amino acid tryptophan is essential for the production of serotonin; however, approximately 90% of tryptophan is used for the synthesis of kynurenine. The conversion of tryptophan to kynurenine requires the induction of either of the enzymes, indoleamine 2,3-dioxygenase (IDO), or tryptophan 2,3-dioxygenase (TDO). IDO can be induced in response to a proinflammatory environment, and TDO can be induced in response to increased cortisol levels. Whether converted via IDO or TDO, kynurenine can then be metabolized to either kynurenic acid (KYNA) by kynurenine aminotransferase (KAT) or to 3-hydroxykynurenine (3-HK) by kynurenine monooxygenase (KMO). 3-HK is further metabolized to 3-hydoxyanthranilic acid (3-HAA) by kynureninase (KYNU) or to xanthurenic acid (XA) by KAT. 3-HAA is further converted to quinolinic acid, the precursor for nicotinamide adenine dinucleotide (NAD+), by 3-hydroxyanthranilate 3,4-dioxygenase (HAOO). In the central nervous system, KYNA is reported to have neuroprotective effects. Other kynurenine pathway products give rise to metabolites with reactive oxidative properties including 3-HK, 3-HAA, and quinolinic acid that act as agonists at the NMDA receptor and may contribute to excitotoxicity and neurotoxicity.[82] It has been hypothesized that the decreased tryptophan concentrations observed in depression could result from an increase in the metabolism of tryptophan in the kynurenine pathway, induced by altered inflammatory processes and overactivation of the HPA axis.[83] Reduced peripheral availability of tryptophan and increased activation of the kynurenine pathway and cortisol have been demonstrated to correlate with major depression and suicide[84] confirming the inextricable link between the two pathways and the impact that they can have on mental health. A second example of the link between serotonergic system function and HPA axis activity is seen at the *SLC6A4* 5-HTTLPR polymorphism, which is located in the gene's promoter region. The short (*s*) and long (*l*) alleles in the 5-HTTLPR have been shown to impact upon transcriptional activity rates of *SLC6A4*. Gotlib and colleagues noted that girls homozygous for the *s* allele produced higher and more prolonged levels of cortisol in response to a stressor when compared with girls who had one or two copies of the *l* allele[85] illustrating the association between HPA axis activity and the serotonin system. Finally, it has been demonstrated that carriers of the low-expression *MAOA*, the gene responsible for encoding the MAO-A enzyme, variants are less able to achieve adequate HPA axis negative feedback in response to endocrine challenges,[86] further securing the link between HPA axis function and the state of the serotonergic system.

Stress and brain-derived neurotrophic factor

Brain-derived neurotrophic factor (BDNF) is a neurotrophic factor that supports the differentiation, maturation, and survival of neurons. Additionally, it serves as a neurotransmitter and is essential for learning and memory.[87] The BDNF gene, *BDNF*, is located at chromosome 11p14.1 and is comprised of 11 exons and 9 promoters that regulate the expression of numerous splice variants.[88] A single nucleotide polymorphism (SNP) of rs6265 in *BDNF* results in the substitution of valine (Val) to methionine (Met) at codon 66 (Val66Met), which affects synaptic targeting of BDNF-containing vesicles and the activity-dependent release of BDNF.[89] This loss-of-function SNP may actually alter HPA axis physiology with consequential changes in the response to stress.[90] A significant body of research has demonstrated that chronic stress decreases hippocampal BDNF expression resulting in cognitive and synaptic deficits.[91] More interestingly, long-term changes in the BDNF pathway are associated with exposure to childhood adversity and the development of a number of stress-related disorders in adulthood,[92] making *BDNF* an interesting candidate in the search for ELS-mediated epigenetic modifications and their functional consequences.

Candidate early-life stress—mediated epigenetic markers

The glucocorticoid receptor (NR3C1)

Within the context of ELS and epigenetics, the glucocorticoid receptor (GR), or nuclear receptor subfamily 3, group C, member 1 (NR3C1) gene is the most investigated marker. This is not surprising given the critical role of the GR in the stress response system. The human glucocorticoid receptor cDNA was isolated in 1985.[93] The gene is located on chromosome 5q31-32 and is composed of 9 exons. The 184 nucleotides of exon 1 represent exclusively the 5′/-untranslated region (5′ ′ UTR). *NR3C1* exon 1F has attracted much attention in the past decade for the potential role that it may play in the pathogenesis of psychiatric disorders via DNA methylation. The most investigated epigenetic modification to date is hypermethylation of a nerve growth factor—inducible protein A (NGFI-A) binding site located in the hippocampal *NR3C1* exon 1F (1_7) promoter.[94] NGFI-A belongs to a family of immediate-early gene-encoded transcription factors, which binds to and activates transcription at the DNA sequence GCGGGGGCGG.[95] The seminal paper by Weaver and colleagues[96] described that rat pups born to low maternal care mothers had more exaggerated HPA responses to stress and had corresponding reduced hippocampal GR protein expression and diminished glucocorticoid feedback sensitivity, compared with the offspring born to high care mothers. These results were mostly reversed by cross-fostering, suggesting a nongenomic transmission of the aforementioned differences and the possibility of an epigenetically driven mechanism. The authors concluded that these effects were the result of significant increases in the methylation of the NGFI-A-binding site at the exon 1_7 promoter of the *Nr3c1* gene resulting from maternal care. On the basis of these findings, many research groups began the task of translating this work into the human setting. The human *NR3C1* gene has a highly similar 5′ structure to that of the rat and, and the human exon 1F, discovered by Turner and Muller[97] is the ortholog to the rat 1_7 region.[98] McGowan[99] carried out one of the first translational studies on the *NR3C1* exon 1F promoter. The study findings reported that those who had a history of childhood abuse and who died by suicide had increased NGFI-A-binding site methylation at the exon 1F promoter, decreased levels of *NR3C1* mRNA and decreased mRNA transcripts that possessed the exon 1F variant. There were no differences in the epigenetic or other measures between those who were not abused and died by suicide and the controls, suggesting that the observed changes were related to exposure to childhood adversity, rather than death by suicide. These findings translated the previous results demonstrated in rats[96] to humans and indicated an important impact of adversity early in life on the epigenetic regulation of hippocampal *NR3C1* expression. Oberlander et al.[100] examined the link between prenatal exposure to maternal depression and levels of methylation at the NGFI-A-binding site of the *NR3C1* gene in newborns and their HPA stress reactivity at 3 months of age. They found increased methylation at the exon 1F promoter NGFI-A-binding site and increased salivary cortisol stress responses in the babies born to depressed women. The findings were associated with maternal third trimester depressed/anxious mood, again supporting the hypothesis of an epigenetic process linking antenatal mood to increased methylation at key regulatory sites located within the *NR3C1* gene. Perroud et al.[101] examined the correlation between the extent of methylation of the *NR3C1* exon 1F promoter and the severity of childhood maltreatment in adults with major depressive disorder (MDD), borderline personality disorder (BPD), and MDD plus comorbid posttraumatic stress disorder (PTSD). They found that childhood sexual abuse, its severity and the number of type of maltreatments, positively correlated with *NR3C1* methylation. Their findings further support the concept that methylation of the NR3C1 gene could result from the effects of childhood adversity and could be a mechanism that links childhood adversity to adult psychiatric illness. Radtke et al.[102] demonstrated that maternal exposure to intimate partner violence during pregnancy influenced the methylation status of the *NR3C1* exon 1F promoter in adolescent offspring. This was the first study to show that prenatal exposure to stress could have a long-term outcome on the methylation status in human offspring. The children in this study ranged in age from 10 to 19 years. It is worth noting that this study did not find that intimate partner violence impacted the mothers' methylation status, indicating that these epigenetic effects are possibly more likely to occur earlier in life, when presumably, systems are more plastic. Following on from this, Radtke and colleagues[103] more recently identified a positive association between specific methylation at the *NR3C1* exon 1F and childhood maltreatment. The methylation of this site was correlated to depressive symptoms as well as BPD symptoms. The authors remarked that the participants in their study were exposed to moderate levels of maltreatment, highlighting the risks that moderate stress, and not just severe maltreatment, can have on mental health status. Martín-Blanco and colleagues examined *NR3C1* exon 1F methylation in a cohort of patients with borderline personality disorder,[104] a disorder whose onset is strongly linked with exposure to ELS,[105] and found that positive correlations existed between *NR3C1* methylation levels and exposure to childhood trauma as well as with BPD symptom severity. Romens et al. and Cicchetti andand Handley have demonstrated associations with increased *NR3C1* exon 1F methylation and ELS, specifically physical maltreatment[106] and general maltreatment,[107] respectively. Cicchetti andand Handley additionally highlighted that the cohort that had experienced early life maltreatment had difficulties with emotional lability—negativity, more externalizing behaviors, and greater depressive symptoms providing even more evidence of the potential functional consequences that may result from ELS-induced epigenetic modifications at the *NR3C1* exon 1F promoter. Farrell et al. demonstrated that DNA methylation levels within the *NR3C1* exon 1F promoter were associated with emotional abuse severity during childhood. Moreover, their study also showed that the degree of *NR3C1* exon 1F methylation was positively correlated with increased HPA axis activity using morning salivary cortisol concentrations as a proxy measure.[108] Morning cortisol measurements are often used to assess HPA axis activity in the clinical research setting.[109] A number of separate research groups have replicated and reproduced similar findings in child[110–113] and adult[114–118] populations. Collectively, these findings indicate a probable role of ELS-mediated *NR3C1* exon 1F promoter hypermethylation in the pathogenesis of depression and other psychiatric disorders linked to chronic stress. This hypermethylation essentially blocks the binding of NGFI-A to its binding site at the promoter and subsequently affects the level of GR protein in the cytoplasm of the cell. The hypothesis underlying this is that the dysfunction in GR protein leads to a subsequent inhibition of glucocorticoid-GR function, resulting in dysregulation of HPAA negative feedback. uncontrolled cortisol release and ultimately depression.[94] It is necessary to highlight that presumably, hypermethylation at the *NR3C1* gene leading to decreased levels of GR protein and inhibition of the GC-GR complex does not exclusively occur at the level of the hippocampus/brain, as demonstrated by various studies that indicate similar epigenetic profiles in peripheral tissues.[100–104,106–108] If this is indeed the case, then hypermethylation at the *NR3C1* gene, as a result of early-life adversity, can potentially dysregulate glucocorticoid effects in other physiological systems, as the GR protein is expressed in diverse cells types and the GC–GR complex in involved in a myriad of functions, including cellular metabolism, immunosuppression, and glucose homeostasis.[119] This is potentially extremely interesting and an area deserving of further investigation as each of these functions has previously been implicated in the pathogenesis of depression.[120–122] This emphasizes the "domino effect" that results from dysregulation of GR protein expression. Further discussion of this goes beyond the scope of this chapter; however, this warrants further consideration from the scientific community.

Although many studies published to date do suggest a probable association between *NR3C1* exon 1F promoter hypermethylation and negative mental health outcomes, particularly in those who have been exposed to ELS, not all findings have been in agreement with this hypothesis. A study by Steiger et al.,[123] which assessed *NR3C1* methylation levels within the context of bulimia nervosa, found no relationship between methylation levels and exposure to childhood abuse. However, it is worth noting that increased methylation at *NR3C1* exon 1C was observed in the patients with bulimia, suggesting that additional examination of the other untranslated first exons (particularly D, J, E, B, F C, and H that are located within a CpG island upstream of exon 2[123]) is warranted. Steiger et al. are not the only research group to publish the finding of no association between *NR3C1* exon 1F promoter methylation and early-life adversity exposure.[124–128] A study by Alt and colleagues[129] investigating glucocorticoid receptor expression in postmortem brain tissues in individuals who had suffered from depression found that there was no significant difference in methylation patterns between both groups (controls and MDD) and between different brain regions for any of the CG sites that they examined (including the promoter region). However, it is worth highlighting that the authors did report that NGFI-A expression was significantly reduced in the hippocampus, the cingulate gyrus, and the nucleus accumbens of MDD patients. In their discussion, it is suggested that some of the study participants may have been victims of childhood abuse events and built a case that increased *NR3C1* methylation, sometimes seen in depression, may represent an MDD subtype following from exposure to ELS. Decreases in *NR3C1* methylation have also been observed in the literature. A series of studies from the Yehuda group have examined *NR3C1* methylation in posttraumatic stress disorder (PTSD), which can sometimes clinically present itself similarly to depression.[94] Lower methylation levels at the NR3C1 exon 1F promoter have been observed in individuals suffering from PTSD[130] as well as offspring with both maternal and paternal PTSD.[131] Tyrka et al. also reported decreased *NR3C1* methylation levels in their study, which retrospectively examined the effects of childhood trauma and the death or separation of a parent.[132] Additionally and interestingly, studies from Suderman et al. and Labonté et al. have revealed both increases and decreases in methylation patterns at different *NR3C1* loci within the same sample groups.[133,134] The differences noted in the literature between ELS-mediated increases and decreases in *NR3C1* methylation observed in various psychiatric and/or psychological disorders demonstrate that methylation at the 1F promoter can be increased or decreased in response to ones' adverse environment. Both conditions can develop after experiencing childhood trauma,[135,136] but it is yet to be fully understood how this adversity can result in psychopathologies with opposing biological characteristics. Further research is needed to better understand what factors are involved in instigating these opposing trajectories.

FK506-binding protein

FK506-binding protein 5 (FKBP5) is one of the immunophilin proteins that make up the inactive glucocorticoid receptor multiprotein complex in the cytoplasm. FKBP5 negatively regulates glucocorticoid function by two main mechanisms: (1) when FKBP5 is bound to the receptor complex, cortisol binds with lower affinity and the nuclear translocation of the receptor is less efficient,[56] and (2) *FKBP5* expression is induced by steroids, inducing glucocorticoids, as part of an intracellular ultrashort negative feedback loop for glucocorticoid activity.[57] Thus, when cortisol levels are relatively high, FKBP5 levels are increased, leading to reduced activation of cortisol-induced translational activity. Possessing an *FKBP5* risk allele may define a neuroendocrinologically distinct subtype of depression. In vivo experiments performed on the cells of individuals with the minor T-allele rs1360780 displayed impaired GC—GR sensitivity, whereas cells from individuals who did not possess the risk T-allele did not display this impaired sensitivity following stimulation.[56] It is interesting that in this study, the genotype-dependent differences were not observed at baseline. The authors pointed out that this mirrors previous animal studies performed that mice lacking the *FKBP5* gene only displayed alterations in endocrine measures following acute or chronic stress.[137] FKBP5 has attracted attention as a protein possibly implicated in the area of epigenetic-mediated gene—childhood trauma interaction, leading to adult mental health illnesses. Klengel et al. carried out the first human study looking specifically at demethylation of the *FKBP5* gene—ELSS interaction. They identified that individuals suffering from PTSD who were exposed to ELS and who carried a functional *FKBP5* risk T-allele (locus rs1360780) had demethylation at intron 7 of the *FKBP5* gene. This intronic demethylation was associated with an increase in FKBP5 induction by GR activation, leading to glucocorticoid receptor resistance, reduced binding of GCs to the GR, and ultimately dysregulation of the stress hormone system. Non et al. performed a study examining the effect of childhood institutionalization on *FKBP5* intron 7 methylation and found a negative correlation between the level of methylation and percent time in institutional care at baseline when the average age was 22 months.[138] Similar to the findings in the aforementioned research, in a longitudinal study by Parade and colleagues, children exposed to physical, sexual, and emotional abuse as well as supervisory and physical neglect, frequent change of residence or homelessness, inadequate food or clothing, and/or witnessing violence had decreased *FKBP5* intron 7 methylation levels.[139] Additionally, given the longitudinal nature of the study, the authors could demonstrate that these decreased methylation levels were maintained, on average, across the duration of the study. However, the results also highlighted that individual variability over time at both *FKBP5* sites was important. The findings of this study highlighted that ELS comes in many forms and so future research should endeavor to capture the effects of different stress types to evaluate a fuller picture of childhood trauma and its effects. Tyrka et al. performed a study on a cohort of children ranging in age from 3 to 5 years with child welfare documentation of moderate-to-severe maltreatment in the previous 6 months and levels of *FKBP5* intron 7 methylation.[58] In line with other studies, the authors reported decreased methylation levels at the CpG sites that were measured and that as FKBP5 alters glucocorticoid receptor responsiveness, these *FKBP5* gene methylation changes may be a mechanism of the effects of ELS exposure. Findings linking decreased *FKBP5* methylation and exposure to ELS have additionally been reported by Tozzi et al.[140] However, similar to the *NR3C1* methylation findings, not all studies have been in agreement with this trend/hypothesis, with one study reporting that increased *FKBP5* methylation levels were associated with exposure to childhood stress[141] and others reporting no associations between *FKBP5* methylation levels and ELS exposure whatsoever.[108,142–144]

The mineralocorticoid receptor (NR3C2)

The mineralocorticoid receptor (MR), or nuclear receptor subfamily 3 group C member 2 (NR3C2), is encoded by the *NR3C2* gene, located at chromosome 4q31.23 and spans 366,523bp. When considering the role of the HPA axis in depression and stress-related disorders, the majority of research to date has focused on the glucocorticoid receptor. The glucocorticoid receptor has been considered to be more important in the regulation of the stress response and its regulation when endogenous glucocorticoid levels are high. However, as negative feedback regulation of the HPA axis relies on both glucocorticoid receptors and mineralocorticoid receptors, some studies have shifted focus onto the potential role that mineralocorticoid receptors may have in the pathophysiology of depression and other stress-related disorders. It has been suggested that there may be an imbalance in the MR/GR ratio in depressed patients that could play a role in the serotonin alterations observed in MDD.[53] Brain MR expression has been shown to be reduced in the hippocampus, inferior frontal gyrus, and cingulate gyrus of individuals who had suffered with depression,[145] further supporting the hypothesis that MR differences may contribute to the depressive phenotype. Individuals with psychotic depression have diminished feedback inhibition of the HPA axis in response to the MR agonist fludrocortisone compared with healthy controls,[146] which highlights the possible role that MR dysregulation may have in the etiology of stress-related disorders. Unfortunately, there has been very little research directly focused on ELS exposure and its potential relationship with *NR3C2* methylation levels. Galbally and colleagues have demonstrated that maternal early pregnancy depression was associated with decreased *NR3C2* methylation in the placental tissue and alterations in infant cortisol reactivity.[147] These findings support the theory of specific fetal programming pathways linked to maternal ill health during pregnancy, mediated via epigenetic mechanisms that have the potential to result in long-term negative health outcomes. However, it is important for other research groups to independently replicate these results and to undertake additional research on *NR3C2* methylation within the context of ELS exposure, before a consensus on the relationship between the two can be determined.

The oxytocin receptor

The neuropeptide, oxytocin, which plays a key role in social cognition, social behaviors, mother–infant bonding, and fear conditioning,[72] mediates its effects via the oxytocin receptor (OXTR),[73] which is encoded by the *OXTR* gene, located on chromosome 3p25.3 and containing three introns and four exons.[74] As the dysregulation of oxytocin function has been hypothesized to result in psychiatric and physical disorders resulting from ELS, it is an interesting candidate to examine within the context of epigenetics. A number of studies have chosen to measure *OXTR* methylation as it is expressed ubiquitously, rather than *OXT*, which has a very fixed regional expression pattern.[74] When compared with some of the other candidate markers mentioned in this chapter, *OXTR* methylation has not been reported on to the same degree. However, the findings to date appear to be in agreement with the conclusion that *OXTR* methylation levels are not associated with exposure to ELS or early-life adversity.[114,148–151] One study, by Fujisawa et al., that measured *OXTR* methylation in children and young adults, aged 6–20, reported higher levels of methylation in children exposed to childhood maltreatment. Moreover, these increased methylation levels were negatively associated with gray matter volumes, and the cohort of children exposed to ELS showed lower gray matter volumes when compared with their non-ELS counterparts.[152] It is worth highlighting that the main difference between the numerous studies reporting no associations between *OXTR* methylation and ELS exposure and the Fujisawa study, which reported significant associations between the two, is the fact that the Fujisawa study was conducted in children with a mean age of 12.9 years. The studies referenced in this section, which reported no associations between *OXTR* methylation and ELS exposure, were conducted in adults with mean ages ranging from 20.18 to 41.0 years. Perhaps, *OXTR* methylation patterns are more responsive to ELS-mediated changes and revert to baseline levels once the source of stress has been removed. Significantly, more research will need to be undertaken to better understand the influence, if any, that ELS has on *OXTR* methylation and whether it is responsible for disorders linked to oxytocin dysregulation.

Solute carrier family 6 member 4 and monoamine oxidase-A

Serotonin (5-HT), a monoamine involved in the regulation of mood, sleep, appetite, motor activity, and cognitive function,[153] is the neurotransmitter most frequently implicated and investigated in the literature of affective disorders. As previously mentioned, there are important links between HPA axis activity and serotonergic functioning, with alterations in one system often resulting in knock-on effects on the other. 5-HT activity is regulated in the brain by the serotonin transporter (5-HTT) that effectively shunts serotonin back into the presynaptic neuron. Over 20 years ago, Heils and colleagues discovered a functional polymorphism in the 5-HTT gene, solute carrier family 6 member 4 (*SLC6A4*), referred to as 5-HTTLPR: a 44bp deletion/insertion that results in two different alleles; the short (*s*) variant having less transcriptional activity than the long variant (*l*),[154] which has since been associated with altered stress responses.[85] Additionally, 5-HT activity is regulated by monoamine oxidase-A, a mitochondrial enzyme encoded by the *MAOA* gene. MAOA is located on the X chromosome and contains a functional length polymorphism in its promoter region. The most frequently occurring short allele (*MAOA-L*) is transcribed less efficiently than the most frequent long allele (*MAOA-H*) and has been repeatedly associated with conduct behaviors in males exposed to childhood maltreatment.[115] This makes both *SLC6A4* and *MAOA* prime candidates for the investigation into the role of epigenetic alterations linked to ELS exposure and outcomes later on in life. Unsurprisingly, many research groups have investigated whether there are ELS-mediated epigenetic differences within these serotonergic genes, and similar to other candidate markers reported on in this chapter, the findings have been mixed. The majority of studies published to date have demonstrated that increased *SLC6A4* methylation levels are associated with individuals who have been exposed to ELS categorized under childhood adversity,[155,156] early stressful life events,[157] trauma,[116] and physical[158] and sexual abuse.[159] A smaller number of studies have reported alternative patterns of *SLC6A4* methylation in study participants who have experienced ELS including decreased methylation levels,[138] mixed methylation levels[160] (i.e., regions with both hyper- and hypomethylation), and instances where no associations[161] were observed between methylation levels and exposure to stress early in life. Currently, the majority of the available literature does indicate that ELS increases *SLC6A4* methylation levels in child and adult populations. The functional consequences of this marker require more detailed characterization; however, there have been studies that link this aberrant methylation finding with stress-related alterations in hippocampal volume[158] as well as effects on brain function during emotional processing.[162] Fewer studies have reported on *MAOA* methylation levels within the context of ELS exposure. The results that have been reported to date suggest that women with a history of ELS have higher *MAOA* methylation levels[163]; however, a lack of association between *MAOA* DNA methylation and ELS has also been reported in the literature.[115] These inconsistencies warrant further investigation by the scientific community so that a clearer understanding of the potential role of ELS-mediated effects at the *MAOA* gene can be obtained.

Brain-derived neurotrophic factor

Brain-derived neurotrophic factor (BDNF) is a neurotrophic factor that behaves as a neurotransmitter and is key in the support of neurogenesis, neuroprotection, and neuroplasticity. The BDNF gene, *BDNF*, has become an interesting candidate in the search for ELS-mediated epigenetic modifications as long-term changes in the BDNF pathway are associated with exposure to childhood adversity and the development of a number of stress-related disorders in adulthood.[92] Although the exact molecular mechanisms are not yet understood, as BDNF has a functional role in brain regions involved in the control of emotional and behavioral regulation,[164] it is conceivable that epigenetic changes at the BDNF gene, driven by stress, could result in maladaptive BDNF function. In a previously mentioned study by Barker and colleagues, which measured methylation levels at *NR3C1*, *OXTR*, and *BDNF* in a retrospective setting, it was documented that *BDNF* methylation at 1 of 73 loci was associated with childhood emotional abuse.[114] Other research groups have replicated similar findings of *BDNF* methylation being associated with exposure to early-life adversity[116,165,166]; however, as seen with the other candidate genes included in this chapter, research reporting the opposite relationship[167] (i.e., *BDNF* methylation being negatively associated with ELS exposure) has also been published. Similar to the other candidate markers mentioned throughout this chapter, additional research will need to be undertaken to better understand the potential role of ELS-mediated epigenetic effects at the *BDNF* gene and how they may influence the pathophysiology of disease.

Conclusion

Exposure to ELS is associated with an increased risk for the development of a wide range of negative health outcomes, both psychological and physiological, later on in life, with adults experiencing multiple adverse events in childhood dying almost 20 years earlier than those with no experience of adversity early in life.[168] Epigenetic regulation of gene expression can be viewed as the adaptation of the genome to environmental influences, positive and negative; a process that has been referred to as genomic plasticity. Therefore, it is possible to conceptualize the epigenome as the interface through which the environment can influence genetic processes and, as a result, regulate behaviors at least partially in response to environmental needs.[169] A series of pivotal studies by Weaver and colleagues identified that exposure to an adverse environment early in life negatively impacted upon HPA axis activity via DNA methylation at a key regulatory region at the gene encoding the glucocorticoid receptor.[96] Following on from this seminal finding, a large body of work has emerged investigating DNA methylation at the human glucocorticoid receptor gene (*NR3C1*) and its association with early-life stressors such as maltreatment, neglect, and abuse as well as the potential impact of stressors experienced in the prenatal period such as maternal depression, maternal exposure to intimate partner violence, and the impact of transgenerational traumas. A significant proportion of these studies have documented that *NR3C1* DNA methylation is increased in those exposed to adversity early in life; however, not all published research is in agreement with this. A number of studies have reported decreased *NR3C1* methylation in childhood adversity, and others have reported no association between the two at all. Similar trends are noted across a plethora of studies examining epigenetic alterations at candidate genes associated with the stress response (*FKBP5*, *SLC6A4*, *MAOA*, *OXTR*, *BDNF*, and others) and in studies utilizing an epigenome-wide approach. A recent systematic review by Parade et al. outlines a number of potential reasons for the disparities documented across the current research and makes a number of excellent recommendations for future research,[170] such as (11) including longitudinal timepoints, (22) standardizing definitions or tools used to measure childhood adversity, (33) focusing analyses on regions with greater variability to increase power (44) considering participant characteristics (i.e., sex, ancestry, age), and their interactions with candidate markers, (55) considering the impact of pharmacological and psychological therapies on outcomes, (66) ensuring the precise location of individual CG sites to facilitate meta-analysis efforts as well as guaranteeing consistency across different research groups, and (77) accounting for methodological differences across studies and their reported findings. It is also worth noting that the explosion of research examining ELS exposure and epigenetic programming has mainly focused on DNA methylation and not the other known epigenetic modifications. I would recommend that these modifications are given greater attention in future research to complement current findings. While there is no doubt that ELS does indeed influence the programming of epigenetic markers across the epigenome and particularly at genes associated with HPA axis function, given its complex nature, there are a number of additional considerations that need to be made and implemented across the landscape of epigenetic research before we truly understand the complexities of ELS exposure and its relationship with epigenetic programming.

References

1. Crick F. Central dogma of molecular biology. *Nature*. 1970;227(5258):561–563. https://doi.org/10.1038/227561a0.
2. Urdinguio RG, Sanchez-Mut JV, Esteller M. Epigenetic mechanisms in neurological diseases: genes, syndromes, and therapies. *Lancet Neurol*. 2009;8(11):1056–1072. https://doi.org/10.1016/S1474-4422(09)70262-5.
3. Arif M, Sadayappan S, Becker RC, Martin LJ, Urbina EM. Epigenetic modification: a regulatory mechanism in essential hypertension. *Hypertens Res*. 2019;42(8): 1099–1113. https://doi.org/10.1038/s41440-019-0248-0.
4. Suresh K, Viswanathan C, Trilochan M. Epigenetics of modified DNA bases: 5-methylcytosine and beyond. *Front Genet*. 2018. https://doi.org/10.3389/fgene.2018.00640.
5. Douvlataniotis K, Bensberg M, Lentini A, Gylemo B, Nestor CE. No evidence for DNA N6-methyladenine in mammals. *Sci Adv*. 2020;6(12). https://doi.org/10.1126/sciadv.aay3335.
6. Anastasiadi D, Esteve-Codina A, Piferrer F. Consistent inverse correlation between DNA methylation of the first intron and gene expression across tissues and species. *Epigenet Chromatin*. 2018;11(1):37. https://doi.org/10.1186/s13072-018-0205-1.
7. Kulis M, Esteller M. *DNA Methylation and Cancer*. Vol 70. 2010:27–56. https://doi.org/10.1016/B978-0-12-380866-0.60002-2. Spain.
8. Liao J, Karnik R, Gu H, et al. Targeted disruption of DNMT1, DNMT3A and DNMT3B in human embryonic stem cells. *Nat Genet*. 2015;47(5):469–478. https://doi.org/10.1038/ng.3258.
9. LDH K, ME R. DNA methylation: a form of epigenetic control of gene expression. *Obstet Gynaecol*. 2010:37–42. https://doi.org/10.1576/toag.12.1.037.27556.
10. Portela A, Esteller M. Epigenetic modifications and human disease. *Nat Biotechnol*. 2010;28(10):1057–1068. https://doi.org/10.1038/nbt.1685.
11. Hellman A, Chess A. Gene body-specific methylation on the active X chromosome. *Science*. 2007;315(5815):1141–1143. https://doi.org/10.1126/science.1136352.
12. Anton E, BP B. Histone acetylation: a switch between repressive and permissive chromatin. *EMBO Rep*. 2002:224–229. https://doi.org/10.1093/embo-reports/kvf053.
13. Ropero S, Esteller M. The role of histone deacetylases (HDACs) in human cancer. *Mol Oncol*. 2007;1(1). https://doi.org/10.1016/j.molonc.2007.01.001 (Epub).
14. Martin C, Zhang Y. The diverse functions of histone lysine methylation. *Nat Rev Mol Cell Biol*. 2005;6(11):838–849. https://doi.org/10.1038/nrm1761.
15. Lorenzo D, Bedford A,T. Histone arginine methylation. *FEBS Lett*. 2010;585(13). https://doi.org/10.1016/j.febslet.2010.11.010 (Epub).
16. Pattaroni C, Jacob. Histone methylation in the nervous system: functions and dysfunctions. *Mol Neurobiol*. 2012;47(2). https://doi.org/10.1007/s12035-012-8376-4 (Epub).
17. Hublitz P, Albert M, Peters AHFM. Mechanisms of transcriptional repression by histone lysine methylation. *Int J Dev Biol*. 2009;53(2–3):335–354. https://doi.org/10.1387/ijdb.082717ph.
18. Rossetto, Avvakumov N, Côté J. Histone phosphorylation: a chromatin modification involved in diverse nuclear events. *Epigenetics*. 2012;7(10). https://doi.org/10.4161/epi.21975 (Epub).

19. Cao J, Yan Q. Histone ubiquitination and deubiquitination in transcription, DNA damage response, and cancer. *Front Oncol*. 2012;2. https://doi.org/10.3389/fonc.2012.00026.
20. Shiio Y, Eisenman RN. Histone SUMOylation is associated with transcriptional repression. *Proc Natl Acad Sci USA*. 2003;100(23):13225. https://doi.org/10.1073/pnas.1735528100.
21. Annunziato A. DNA packaging: nucleosomes and chromatin. *Nat Educ*. 2008;1(1):26.
22. Wei JW, Huang K, Yang C, Kang CS. Non-coding RNAs as regulators in epigenetics (Review). *Oncol Rep*. 2017;37(1):3–9. https://doi.org/10.3892/or.2016.5236.
23. Zhou T, Ferguson BS. Chapter 3—small non-coding RNAs as epigenetic regulators. In: *Translational Epigenetics*. Vol 14. Academic Press; 2019:37–47. https://doi.org/10.1016/B978-0-12-816843-1.00003-5.
24. Collins LJ, Schönfeld B, Chen XS. The epigenetics of non-coding RNA. In: *Handbook of Epigenetics: The New Molecular and Medical Genetics*. Elsevier Inc.; 2011. https://doi.org/10.1016/B978-0-12-375709-8.00004-6.
25. Salleh MR. Life event, stress and illness. *Malays J Med Sci*. 2008;15(4):9–18. http://www.bioline.org.br/pdf?mj08034.
26. Smith KE, Pollak SD. Early life stress and development: potential mechanisms for adverse outcomes. *J Neurodev Disord*. 2020;12(1). https://doi.org/10.1186/s11689-020-09337-y.
27. Yaribeygi H, Panahi Y, Sahraei H, Johnston TP, Sahebkar A. The impact of stress on body function: a review. *EXCLI J*. 2017;16:1057–1072. https://doi.org/10.17179/excli2017-480.
28. Johnson SB, Blum RW, Giedd JN. Adolescent maturity and the brain: the promise and pitfalls of neuroscience research in adolescent health policy. *J Adolesc Health*. 2009;45(3):216–221. https://doi.org/10.1016/j.jadohealth.2009.05.016.
29. Vaiserman AM. Epigenetic programming by early-life stress: evidence from human populations. *Dev Dynam*. 2015;244(3):254–265. https://doi.org/10.1002/dvdy.24211.
30. Shonkoff JP, Boyce WT, McEwen BS. Neuroscience, molecular biology, and the childhood roots of health disparities: building a new framework for health promotion and disease prevention. *JAMA, J Am Med Assoc*. 2009;301(21):2252–2259. https://doi.org/10.1001/jama.2009.754.
31. Coley EJL, Demaestri C, Ganguly P, et al. Cross-generational transmission of early life stress effects on HPA regulators and bdnf are mediated by sex, lineage, and upbringing. *Front Behav Neurosci*. 2019;13:101. https://www.frontiersin.org/article/10.3389/fnbeh.2019.00101.
32. Parker KJ, Buckmaster CL, Hyde SA, Schatzberg AF, Lyons DM. Nonlinear relationship between early life stress exposure and subsequent resilience in monkeys. *Sci Rep*. 2019;9(1):16232. https://doi.org/10.1038/s41598-019-52810-5.
33. Dornela GL, Teixeira RM, Polianna D-P, Norberto G-C, Henriquede LUE. A comprehensive overview on stress neurobiology: basic concepts and clinical implications. *Front Behav Neurosci*. 2018. https://doi.org/10.3389/fnbeh.2018.00127.
34. Ulrich-Lai YM, Herman JP. Neural regulation of endocrine and autonomic stress responses. *Nat Rev Neurosci*. 2009;10(6):397–409. https://doi.org/10.1038/nrn2647.
35. Sharpley CF. Neurobiological pathways between chronic stress and depression: dysregulated adaptive mechanisms? *Clin Med Psychiatr*. 2009;2:33–45.

36. Piazza JR, Almeida DM, Dmitrieva NO, Klein LC. Frontiers in the use of biomarkers of health in research on stress and aging. *J Gerontol B Psychol Sci Soc Sci*. 2010;65(5): 513–525. https://doi.org/10.1093/geronb/gbq049.
37. Bylund D. In: Lennarz WJ, Lane MD, eds. *Encyclopedia of Biological Chemistry*. Second. Academic Press; 2013.
38. Bellavance MA, Rivest S. The HPA—immune axis and the immunomodulatory actions of glucocorticoids in the brain. *Front Immunol*. 2014;31(5):136. https://doi.org/10.3389/fimmu.2014.00136.
39. Barker D. In utero programming of chronic disease. *Clin Sci*. 1998;95:115–128.
40. Kapoor A, Dunn E, Kostaki A, Andrews MH, Matthews SG. Fetal programming of hypothalamo-pituitary-adrenal function: prenatal stress and glucocorticoids. *J Physiol*. 2006;572(1):31–44. https://doi.org/10.1016/j.poly.2005.06.060.
41. Duthie L, Reynolds RM. Changes in the maternal hypothalamic-pituitary-adrenal axis in pregnancy and postpartum: influences on maternal and fetal outcomes. *Neuroendocrinology*. 2013;98(2):106–115. https://doi.org/10.1159/000354702.
42. Allolio B, Hoffmann J, Linton EA, Winkelmann W, Kusche M, Schulte HM. Diurnal salivary cortisol patterns during pregnancy and after delivery: relationship to plasma corticotrophin-releasing-hormone. *Clin Endocrinol*. 1990;33(2):279–289. https://doi.org/10.1111/j.1365-2265.1990.tb00492.x.
43. King BR, Smith R, Nicholson RC. The regulation of human corticotrophin-releasing hormone gene expression in the placenta. *Peptides*. 2001;22(11):1941–1947. https://doi.org/10.1016/s0196-9781(01)00486-7.
44. O'Leary N. *An Investigation of the Role of Antenatal Depression, Maternal Cortisol and Postnatal Interactive Behaviour on Infant Neurodevelopment in the First Year of Life*; 2019. http://hdl.handle.net/2262/89745.
45. Gillespie CF, Nemeroff CB. Hypercortisolemia and depression. *Psychosom Med*. 2005; 67(1):S26–S28. https://doi.org/10.1097/01.psy.0000163456.22154.d2.
46. Fareau GG, Vassilopoulou-Sellin R. Hypercortisolemia and infection. *Infection in High-Risk Populations*. 2007;21(3):639–657. https://doi.org/10.1016/j.idc.2007.06.001.
47. Hinojosa-Amaya JM, Varlamov EV, McCartney S, Fleseriu M. Hypercortisolemia recurrence in cushing's disease; a diagnostic challenge. *Front Endocrinol*. 2019;10: 740. https://www.frontiersin.org/article/10.3389/fendo.2019.00740.
48. Davis EP, Glynn LM, Waffarn F, Sandman CA. Prenatal maternal stress programs infant stress regulation. *J Child Psychol Psychiatry Allied Discip*. 2011;52(2):119–129. https://doi.org/10.1111/j.1469-7610.2010.02314.x.
49. Tiffany F, Miguel D, John D, et al. Prenatal depression effects on the fetus and the newborn. *Infant Behav Dev*. 2004:216–229. https://doi.org/10.1016/j.infbeh.2003.09.010.
50. Keenan K, Gunthorpe D, Grace D. Parsing the relations between SES and stress reactivity: examining individual differences in neonatal stress response. *Infant Behav Dev*. 2007;30(1):134–145. https://doi.org/10.1016/j.infbeh.2006.08.001.
51. Brennan PA, Pargas R, Walker EF, Green P, Newport DJ, Stowe Z. Maternal depression and infant cortisol: influences of timing, comorbidity and treatment. *JCPP (J Child Psychol Psychiatry)*. 2008;49(10):1099–1107. https://doi.org/10.1111/j.1469-7610.2008.01914.x.
52. Gutteling BM, De Weerth C, Buitelaar JK. Prenatal stress and children's cortisol reaction to the first day of school. *Psychoneuroendocrinology*. 2005;30(6):541–549. https://doi.org/10.1016/j.psyneuen.2005.01.002.

53. Young EA, Lopez JF, Murphy-Weinberg V, Watson SJ, Akil H. Mineralocorticoid receptor function in major depression. *Arch Gen Psychiatr.* 2003;60(1):24–28. https://doi.org/10.1001/archpsyc.60.1.24.
54. Trapp T, Holsboer F. Heterodimerization between mineralocorticoid and glucocorticoid receptors increases the functional diversity of corticosteroid action. *Trends Pharmacol Sci.* 1996;17(4):145–149. https://doi.org/10.1016/0165-6147(96)81590-2.
55. Anacker, Zunszain P, Carvalho L, Pariante CM. *The Glucocorticoid Receptor: Pivot of Depression and of Antidepressant Treatment? Psychoneuroendocrinology.* 2011;36. https://doi.org/10.1016/j.psyneuen.2010.03.007. PMID: 20399565.
56. Menke A, Klengel T, Rubel J, et al. Genetic variation in FKBP5 associated with the extent of stress hormone dysregulation in major depression. *Gene Brain Behav.* 2013; 12(3):289–296. https://doi.org/10.1111/gbb.12026.
57. Binder EB. The role of FKBP5, a co-chaperone of the glucocorticoid receptor in the pathogenesis and therapy of affective and anxiety disorders. *Psychoneuroendocrinology.* 2009;34(1). https://doi.org/10.1016/j.psyneuen.2009.05.021 (PMID).
58. Tyrka AR, Ridout KK, Parade SH, Paquette A, Marsit CJ, Seifer R. Childhood maltreatment and methylation of FK506 binding protein 5 gene (FKBP5). *Dev Psychopathol.* 2015;27(4 Pt 2):1637–1645. https://doi.org/10.1017/S0954579415000991.
59. Young EA, Lopez JF, Murphy-Weinberg V, Watson SJ, Akil H. Hormonal evidence for altered responsiveness to social stress in major depression. *Neuropsychopharmacology.* 2000;23(4):411–418. https://doi.org/10.1016/S0893-133X(00)00129-9.
60. Gibbons JL, McHugh PR. Plasma cortisol in depressive illness. *J Psychiatr Res.* 1962; 2(1):162–171. https://doi.org/10.1016/0022-3956(62)90006-7.
61. Belanoff JK, Kalehzan M, Sund B, Fleming Ficek SK, Schatzberg AF. Cortisol activity and cognitive changes in psychotic major depression. *Am J Psychiatr.* 2001;158(10): 1612–1616. https://doi.org/10.1176/appi.ajp.158.10.1612.
62. Carroll BJ, Hall RCW, Beresford TP. *Dexamethasone Suppression Test.* Dordrecht: Springer Netherlands; 1984:3–28. https://doi.org/10.1007/978-94-011-6725-3_1.
63. Fountoulakis K, Iacovides A, Fotiou F, Karamouzis M, Demetriadou A, Kaprinis G. Relationship among dexamethasone suppression test, personality disorders and stressful life events in clinical subtypes of major depression: an exploratory study. *Ann Gen Hosp Psychiatr.* 2004;3(1). https://doi.org/10.1186/1475-2832-3-15, 15-15.
64. Nuller JL, Ostroumova MN. Resistance to inhibiting effect of dexamethasone in patients with endogenous depression. *Acta Psychiatr Scand.* 1980;61(2):169–177. https://doi.org/10.1111/j.1600-0447.1980.tb00576.x.
65. Yerevanian BI, Feusner JD, Koek RJ, Mintz J. The dexamethasone suppression test as a predictor of suicidal behavior in unipolar depression. *J Affect Disord.* 2004;83(2–3): 103–108. https://doi.org/10.1016/j.jad.2004.08.009.
66. Fries E, Dettenborn L, Kirschbaum C. The cortisol awakening response (CAR): facts and future directions. *Int J Psychophysiol.* 2009;72(1):67–73. https://doi.org/10.1016/j.ijpsycho.2008.03.014.
67. Vreeburg SA, Hoogendijk WJG, Van Pelt J, et al. Major depressive disorder and hypothalamic-pituitary-adrenal axis activity: results from a large cohort study. *Arch Gen Psychiatr.* 2009;66(6):617–626. https://doi.org/10.1001/archgenpsychiatry.2009.50.
68. Bhagwagar Z, Hafizi S, Cowen PJ. Increased salivary cortisol after waking in depression. *Psychopharmacology.* 2005;182(1):54–57. https://doi.org/10.1007/s00213-005-0062-z.

69. Bhagwagar Z, Hafizi S, Cowen PJ. Increase in concentration of waking salivary cortisol in recovered patients with depression. *Am J Psychiatr.* 2003;160(10):1890–1891. https://doi.org/10.1176/appi.ajp.160.10.1890.
70. Frodl, Dinan TG. A review of atypical depression in relation to the course of depression and changes in HPA axis organization. *Psychoneuroendocrinology.* 2012;37(10). https://doi.org/10.1016/j.psyneuen.2012.03.009 (Epub).
71. Palma-Gudiel H, Córdova-Palomera A, Leza JC, Fañanás L. Glucocorticoid receptor gene (NR3C1) methylation processes as mediators of early adversity in stress-related disorders causality: a critical review. *Neurosci Biobehav Rev.* 2015;55:520–535. https://doi.org/10.1016/j.neubiorev.2015.05.016.
72. Jones C, Barrera I, Brothers S, Ring R, Wahlestedt C. Oxytocin and social functioning. *Dialogues Clin Neurosci.* 2017;19(2):193–201. https://doi.org/10.31887/DCNS.2017.19.2/cjones.
73. McKay EC, Counts SE. Oxytocin receptor signaling in vascular function and stroke. *Front Neurosci.* 2020;14:1003. https://www.frontiersin.org/article/10.3389/fnins.2020.574499.
74. Maud C, Ryan J, McIntosh JE, Olsson CA. The role of oxytocin receptor gene (OXTR) DNA methylation (DNAm) in human social and emotional functioning: a systematic narrative review. *BMC Psychiatr.* 2018;18(1):154. https://doi.org/10.1186/s12888-018-1740-9.
75. Li Y, Hassett AL, Seng JS. Exploring the mutual regulation between oxytocin and cortisol as a marker of resilience. *Arch Psychiatr Nurs.* 2019;33(2):164–173. https://doi.org/10.1016/j.apnu.2018.11.008.
76. Seng JS. Posttraumatic oxytocin dysregulation: is it a link among posttraumatic self disorders, posttraumatic stress disorder, and pelvic visceral dysregulation conditions in women? *J Trauma & Dissociation.* 2010;11(4):387–406. https://doi.org/10.1080/15299732.2010.496075.
77. Ormsbee HS, Fondacaro JD. Action of serotonin on the gastrointestinal tract (42016). *PSEBM (Proc Soc Exp Biol Med).* 1985;178(3):333–338. https://doi.org/10.3181/00379727-178-42016.
78. Lv J, Liu F. The role of serotonin beyond the central nervous system during embryogenesis. *Front Cell Neurosci.* 2017;11:74. https://www.frontiersin.org/article/10.3389/fncel.2017.00074.
79. Marazziti D. Understanding the role of serotonin in psychiatric diseases. *F1000Research.* 2017;6. https://doi.org/10.12688/f1000research.10094.1, 180-180.
80. Ener RA, Meglathery SB, Decker WAV, Gallagher RM. Serotonin syndrome and other serotonergic disorders. *Pain Med.* 2003;4(1):63–74. https://doi.org/10.1046/j.1526-4637.2003.03005.x.
81. Leonard BE. HPA and immune axes in stress: involvement of the serotonergic system. *Neuroimmunomodulation.* 2006;13(5–6):268–276. https://doi.org/10.1159/000104854.
82. O'Farrell K, Fagan E, Connor TJ, Harkin A. Inhibition of the kynurenine pathway protects against reactive microglial-associated reductions in the complexity of primary cortical neurons. *Eur J Pharmacol.* 2017;810:163–173. https://doi.org/10.1016/j.ejphar.2017.07.008.
83. Müller N, Schwarz MJ. The immune-mediated alteration of serotonin and glutamate: towards an integrated view of depression. *Mol Psychiatr.* 2007;12(11):988–1000. https://doi.org/10.1038/sj.mp.4002006.

84. Messaoud A, Mensi R, Douki W, et al. Reduced peripheral availability of tryptophan and increased activation of the kynurenine pathway and cortisol correlate with major depression and suicide. *World J Biol Psychiatr.* 2018;20(9). https://doi.org/10.1080/15622975.2018.1468031. Epub).
85. GI H, Jutta J, MK L. Joachim H. HPA Axis reactivity: a mechanism underlying the associations among 5-HTTLPR, stress, and depression. *Biol Psychiatr.* 2008:847–851. https://doi.org/10.1016/j.biopsych.2007.10.008.
86. Jabbi M, Korf J, Kema IP, et al. Convergent genetic modulation of the endocrine stress response involves polymorphic variations of 5-HTT, COMT and MAOA. *Mol Psychiatr.* 2007;12(5):483–490. https://doi.org/10.1038/sj.mp.4001975.
87. Bathina S, Das UN. Brain-derived neurotrophic factor and its clinical Implications. *Arch Med Sci.* 2015;11(6):1164–1178. https://doi.org/10.5114/aoms.2015.56342.
88. Brigadski T, Leßmann V. The physiology of regulated BDNF release. *Cell Tissue Res.* 2020;382(1):15–45. https://doi.org/10.1007/s00441-020-03253-2.
89. Park C, Kim J, Namgung E, et al. The BDNF Val66Met polymorphism affects the vulnerability of the brain structural network. *Front Hum Neurosci.* 2017;11:400. https://www.frontiersin.org/article/10.3389/fnhum.2017.00400.
90. Notaras M, van den Buuse M. Neurobiology of BDNF in fear memory, sensitivity to stress, and stress-related disorders. *Mol Psychiatr.* 2020;25(10):2251–2274. https://doi.org/10.1038/s41380-019-0639-2.
91. Miranda M, Morici JF, Zanoni MB, Bekinschtein P. Brain-derived neurotrophic factor: a key molecule for memory in the healthy and the pathological brain. *Front Cell Neurosci.* 2019;13:363. https://www.frontiersin.org/article/10.3389/fncel.2019.00363.
92. Bondar NP, Merkulova TI. Brain-derived neurotrophic factor and early-life stress: multifaceted interplay. *J Biosci.* 2016;41(4):751–758. https://doi.org/10.1007/s12038-016-9648-3.
93. Hollenberg SM, Weinberger C, Ong ES, et al. Primary structure and expression of a functional human glucocorticoid receptor cDNA. *Nature.* 1985;318(6047):635–641. https://doi.org/10.1038/318635a0.
94. Farrell C, O'Keane V. Epigenetics and the glucocorticoid receptor: a review of the implications in depression. *Psychiatr Res.* 2016;242:349–356. https://doi.org/10.1016/j.psychres.2016.06.022.
95. Swirnoff AH, Milbrandt J. DNA-binding specificity of NGFI-A and related zinc finger transcription factors. *Mol Cell Biol.* 1995;15(4):2275–2287. https://doi.org/10.1128/MCB.15.4.2275.
96. WIC G, Nadia C, CF A, et al. Epigenetic programming by maternal behavior. *Nat Neurosci.* 2004:847–854. https://doi.org/10.1038/nn1276.
97. Turner JD, Muller CP. Structure of the glucocorticoid receptor (NR3C1) gene 5′ untranslated region: identification, and tissue distribution of multiple new human exon 1. *J Mol Endocrinol.* 2005;35(2):283–292. https://doi.org/10.1677/jme.1.01822.
98. Moser D, Molitor A, Kumsta R, Tatschner T, Riederer P, Meyer J. The glucocorticoid receptor gene exon 1-F promoter is not methylated at the NGFI-A binding site in human hippocampus. *World J Biol Psychiatr.* 2007;8:262–268. https://doi.org/10.1080/15622970701429862.
99. McGowan PO, Sasaki A, D'Alessio AC, et al. Epigenetic regulation of the glucocorticoid receptor in human brain associates with childhood abuse. *Nat Neurosci.* 2009;12(3):342–348. https://doi.org/10.1038/nn.2270.

100. Oberlander TF, Weinberg J, Papsdorf M, Grunau R, Misri S, Devlin AM. Prenatal exposure to maternal depression, neonatal methylation of human glucocorticoid receptor gene (NR3C1) and infant cortisol stress responses. *Epigenetics*. 2008;3(2):97–106. https://doi.org/10.4161/epi.3.2.6034.
101. Perroud N, Paoloni-Giacobino A, Prada P, et al. Increased methylation of glucocorticoid receptor gene (NR3C1) in adults with a history of childhood maltreatment: a link with the severity and type of trauma. *Transl Psychiatry*. 2011;1(12):e59. https://doi.org/10.1038/tp.2011.60.
102. Radtke KM, Ruf M, Gunter HM, et al. Transgenerational impact of intimate partner violence on methylation in the promoter of the glucocorticoid receptor. *Transl Psychiatry*. 2011;1(7). https://doi.org/10.1038/tp.2011.21. e21-e21.
103. Radtke K, Schauer M, Gunter H, et al. Epigenetic modifications of the glucocorticoid receptor gene are associated with the vulnerability to psychopathology in childhood maltreatment. *Transl Psychiatry*. 2015;5(5):e571. https://doi.org/10.1038/tp.2015.63.
104. Martín-Blanco A, Ferrer M, Soler J, et al. Association between methylation of the glucocorticoid receptor gene, childhood maltreatment, and clinical severity in borderline personality disorder. *J Psychiatr Res*. 2014;57(1):34–40. https://doi.org/10.1016/j.jpsychires.2014.06.011.
105. Cattane N, Rossi R, Lanfredi M, Cattaneo A. Borderline personality disorder and childhood trauma: exploring the affected biological systems and mechanisms. *BMC Psychiatr*. 2017;17(1). https://doi.org/10.1186/s12888-017-1383-2, 221-221.
106. Romens SE, McDonald J, Svaren J, Pollak SD. Associations between early life stress and gene methylation in children. *Child Dev*. 2015;86(1):303–309. https://doi.org/10.1111/cdev.12270.
107. Cicchetti D, Handley ED. Methylation of the glucocorticoid receptor gene, nuclear receptor subfamily 3, group C, member 1 (NR3C1), in maltreated and nonmaltreated children: associations with behavioral undercontrol, emotional lability/negativity, and externalizing and internalizing symptoms. *Dev Psychopathol*. 2017;29(5):1795–1806. https://doi.org/10.1017/S0954579417001407.
108. Farrell C, Doolin K, O' Leary N, et al. DNA methylation differences at the glucocorticoid receptor gene in depression are related to functional alterations in hypothalamic–pituitary–adrenal axis activity and to early life emotional abuse. *Psychiatr Res*. 2018;265:341–348. https://doi.org/10.1016/j.psychres.2018.04.064.
109. Golden SH, Wand GS, Malhotra S, Kamel I, Horton K. Reliability of hypothalamic-pituitary-adrenal axis assessment methods for use in population-based studies. *Eur J Epidemiol*. 2011;26(7):511–525. https://doi.org/10.1007/s10654-011-9585-2.
110. Parade SH, Ridout KK, Seifer R, et al. Methylation of the glucocorticoid receptor gene promoter in preschoolers: links with internalizing behavior problems. *Child Dev*. 2016;87(1):86–97. https://doi.org/10.1111/cdev.12484.
111. Parent J, Parade SH, Laumann LE, et al. Dynamic stress-related epigenetic regulation of the glucocorticoid receptor gene promoter during early development: the role of child maltreatment. *Dev Psychopathol*. 2017;29(5):1635–1648. https://doi.org/10.1017/S0954579417001298.
112. Van Der Knaap LJ, Riese H, Hudziak JJ, et al. Glucocorticoid receptor gene (NR3C1) methylation following stressful events between birth and adolescence. The TRAILS study. *Transl Psychiatry*. 2014;4. https://doi.org/10.1038/tp.2014.22.
113. Tyrka AR, Parade SH, Eslinger NM, et al. Methylation of exons 1D, 1F, and 1H of the glucocorticoid receptor gene promoter and exposure to adversity in preschool-aged

children. *Dev Psychopathol.* 2015;27(2):577–585. https://doi.org/10.1017/S0954579415000176.

114. Barker V, Walker, Evans K, Lawrie SM. Methylation of glucocorticoid receptor (NR3C1), BDNF and oxytocin receptor genes in association with childhood maltreatment in schizophrenia and schizoaffective disorder. *Schizophr Res.* 2019;216. https://doi.org/10.1016/j.schres.2019.11.050 (Epub).
115. MP A, Yabin W, WCC Y, et al. Genetic and epigenetic associations of MAOA and NR3C1 with depression and childhood adversities. *Int J Neuropsychopharmacol.* 2013:1513–1528. https://doi.org/10.1017/s1461145713000102.
116. Peng H, Zhu Y, Strachan E, et al. Childhood trauma, DNA methylation of stress-related genes, and depression: findings from two monozygotic twin studies. *Psychosom Med.* 2018;80(7):599–608. https://doi.org/10.1097/PSY.0000000000000604.
117. Shields AE, Wise LA, Ruiz-Narvaez EA, et al. Childhood abuse, promoter methylation of leukocyte NR3C1 and the potential modifying effect of emotional support. *Epigenomics.* 2016;8(11):1507–1517. https://doi.org/10.2217/epi-2016-0074.
118. Tyrka AR, Price LH, Marsit C, Walters OC, Carpenter LL. Childhood adversity and epigenetic modulation of the leukocyte glucocorticoid receptor: preliminary findings in healthy adults. *PLoS One.* 2012;7(1). https://doi.org/10.1371/journal.pone.0030148.
119. Necela B, Cidlowski J. Mechanisms of glucocorticoid receptor action in noninflammatory and inflammatory cells. *Proc Am Thorac Soc.* 2004;1(3):239–246. https://doi.org/10.1513/pats.200402-005MS.
120. Nagasawa M, Ogino Y, Kurata K, et al. Hypothesis with abnormal amino acid metabolism in depression and stress vulnerability in Wistar Kyoto rats. *Amino Acids.* 2012;43(5):2101–2111. https://doi.org/10.1007/s00726-012-1294-y.
121. Raison C, Miller A. The neuroimmunology of stress and depression. *Semin Clin Neuropsychiatry.* 2001;6:277–294.
122. Nikolaus S, Larisch R, Beu M, Vosberg H, Müller-Gärtner H-W. Diffuse cortical reduction of neuronal activity in unipolar major depression: a retrospective analysis of 337 patients and 321 controls. *Nucl Med Commun.* 2000;21(12). https://journals.lww.com/nuclearmedicinecomm/Fulltext/2000/12000/Diffuse_cortical_reduction_of_neuronal_activity_in.5.aspx.
123. Steiger H, Labonté B, Groleau P, Turecki G, Israel M. Methylation of the glucocorticoid receptor gene promoter in bulimic women: associations with borderline personality disorder, suicidality, and exposure to childhood abuse. *Int J Eat Disord.* 2013;46(3):246–255. https://doi.org/10.1002/eat.22113.
124. Hecker T, Radtke KM, Hermenau K, Papassotiropoulos A, Elbert T. Associations among child abuse, mental health, and epigenetic modifications in the proopiomelanocortin gene (POMC): a study with children in Tanzania. *Dev Psychopathol.* 2016;28(4):1401–1412. https://doi.org/10.1017/S0954579415001248.
125. Alexander N, Kirschbaum C, Wankerl M, et al. Glucocorticoid receptor gene methylation moderates the association of childhood trauma and cortisol stress reactivity. *Psychoneuroendocrinology.* 2018;90:68–75. https://doi.org/10.1016/j.psyneuen.2018.01.020.
126. Fiacco S, Gardini ES, Mernone L, Schick L, Ehlert U. DNA methylation in healthy older adults with a history of childhood adversity—findings from the women 40+ healthy aging study. *Front Psychiatr.* 2019;10. https://doi.org/10.3389/fpsyt.2019.00777.

127. Schür RR, van Leeuwen JMC, Houtepen LC, et al. Glucocorticoid receptor exon 1 F methylation and the cortisol stress response in health and disease. *Psychoneuroendocrinology*. 2018;97:182–189. https://doi.org/10.1016/j.psyneuen.2018.07.018.
128. Wang W, Feng J, Ji C, et al. Increased methylation of glucocorticoid receptor gene promoter 1F in peripheral blood of patients with generalized anxiety disorder. *J Psychiatr Res*. 2017;91:18–25. https://doi.org/10.1016/j.jpsychires.2017.01.019.
129. Alt SR, Turner JD, Klok MD, et al. Differential expression of glucocorticoid receptor transcripts in major depressive disorder is not epigenetically programmed. *Psychoneuroendocrinology*. 2010;35(4):544–556. https://doi.org/10.1016/j.psyneuen.2009.09.001.
130. Yehuda R, Flory, Bierer L, et al. Lower methylation of glucocorticoid receptor gene promoter 1F in peripheral blood of veterans with posttraumatic stress disorder. *Biol Psychiatr*. 2014;77(4). https://doi.org/10.1016/j.biopsych.2014.02.006 (Epub).
131. Yehuda R, Daskalakis NP, Lehrner A, et al. Influences of maternal and paternal PTSD on epigenetic regulation of the glucocorticoid receptor gene in Holocaust survivor offspring. *Am J Psychiatr*. 2014;171(8):872–880. https://doi.org/10.1176/appi.ajp.2014.13121571.
132. Tyrka AR, Parade SH, Welch ES, et al. Methylation of the leukocyte glucocorticoid receptor gene promoter in adults: associations with early adversity and depressive, anxiety and substance-use disorders. *Transl Psychiatry*. 2016;6(7). https://doi.org/10.1038/tp.2016.112.
133. Suderman M, McGowan PO, Sasaki A, et al. Conserved epigenetic sensitivity to early life experience in the rat and human hippocampus. *Proc Natl Acad Sci U S A*. 2012; 109(2):17266–17272. https://doi.org/10.1073/pnas.1121260109.
134. Labonte B, Yerko V, Gross J, et al. Differential glucocorticoid receptor exon 1 B, 1 C, and 1 H expression and methylation in suicide completers with a history of childhood abuse. *Biol Psychiatr*. 2012;72(1):41–48. https://doi.org/10.1016/j.biopsych.2012.01.034.
135. Kessler RC, Magee WJ. Childhood adversities and adult depression: basic patterns of association in a US national survey. *Psychol Med*. 1993;23(3):679–690. https://doi.org/10.1017/S0033291700025460.
136. Childhood physical abuse and combat-related posttraumatic stress disorder in Vietnam veterans. *Aust J Pharm*. 1993;150(2):235–239. https://doi.org/10.1176/ajp.150.2.235.
137. Hartmann, Wagner K, Liebl C, et al. The involvement of FK506-binding protein 51 (FKBP5) in the behavioral and neuroendocrine effects of chronic social defeat stress. *Neuropharmacology*. 2011;62(1). https://doi.org/10.1016/j.neuropharm.2011.07.041 (Epub).
138. Non AL, Hollister BM, Humphreys KL, et al. DNA methylation at stress-related genes is associated with exposure to early life institutionalization. *Am J Phys Anthropol*. 2016; 161(1):84–93. https://doi.org/10.1002/ajpa.23010.
139. Parade SH, Parent J, Rabemananjara K, et al. Change in FK506 binding protein 5 (FKBP5) methylation over time among preschoolers with adversity. *Dev Psychopathol*. 2017;29(5):1627–1634. https://doi.org/10.1017/S0954579417001286.
140. Tozzi L, Farrell C, Booij L, et al. Epigenetic changes of FKBP5 as a link connecting genetic and environmental risk factors with structural and functional brain changes in major depression. *Neuropsychopharmacology*. 2018;43(5):1138–1145. https://doi.org/10.1038/npp.2017.290.

141. Harms MB, Birn R, Provencal N, et al. Early life stress, FK506 binding protein 5 gene (FKBP5) methylation, and inhibition-related prefrontal function: a prospective longitudinal study. *Dev Psychopathol.* 2017;29(5):1895–1903. https://doi.org/10.1017/S095457941700147X.
142. Bustamante AC, Aiello AE, Guffanti G, Galea S, Wildman DE, Uddin M. FKBP5 DNA methylation does not mediate the association between childhood maltreatment and depression symptom severity in the Detroit Neighborhood Health Study. *J Psychiatr Res.* 2018;96:39–48. https://doi.org/10.1016/j.jpsychires.2017.09.016.
143. Klinger-König J, Hertel J, Van der Auwera S, et al. Methylation of the FKBP5 gene in association with FKBP5 genotypes, childhood maltreatment and depression. *Neuropsychopharmacology.* 2019;44(5):930–938. https://doi.org/10.1038/s41386-019-0319-6.
144. Yeo S, Enoch M-A, Gorodetsky E, et al. The influence of FKBP5 genotype on expression of FKBP5 and other glucocorticoid-regulated genes, dependent on trauma exposure. *Gene Brain Behav.* 2017;16(2):223–232. https://doi.org/10.1111/gbb.12342.
145. Klok MD, Giltay EJ, Van der Does AJW, et al. A common and functional mineralocorticoid receptor haplotype enhances optimism and protects against depression in females. *Transl Psychiatry.* 2011;1(12). https://doi.org/10.1038/tp.2011.59. e62-e62.
146. Lembke A, Gomez R, Tenakoon L, et al. The mineralocorticoid receptor agonist, fludrocortisone, differentially inhibits pituitary-adrenal activity in humans with psychotic major depression. *Psychoneuroendocrinology.* 2013;38(1):115–121. https://doi.org/10.1016/j.psyneuen.2012.05.006.
147. Galbally M, Watson SJ, van IJzendoorn M, et al. The role of glucocorticoid and mineralocorticoid receptor DNA methylation in antenatal depression and infant stress regulation. *Psychoneuroendocrinology.* 2020;115. https://doi.org/10.1016/j.psyneuen.2020.104611.
148. Gouin JP, Zhou QQ, Booij L, et al. Associations among oxytocin receptor gene (OXTR) DNA methylation in adulthood, exposure to early life adversity, and childhood trajectories of anxiousness. *Sci Rep.* 2017;7(1):7446. https://doi.org/10.1038/s41598-017-07950-x.
149. Kogan SM, Bae D, Cho J, Smith AK, Nishitani S. Childhood adversity, socioeconomic instability, oxytocin-receptor-gene methylation, and romantic-relationship support among young African American men. *Psychol Sci.* 2019;30(8):1234–1244. https://doi.org/10.1177/0956797619854735.
150. Smearman EL, Almli LM, Conneely KN, et al. Oxytocin receptor genetic and epigenetic variations: association with child abuse and adult psychiatric symptoms. *Child Dev.* 2016;87(1):122–134. https://doi.org/10.1111/cdev.12493.
151. Womersley, Hemmings S, Ziegler C, et al. Childhood emotional neglect and oxytocin receptor variants: association with limbic brain volumes. *World J Biol Psychiatr.* 2019; 21(7). https://doi.org/10.1080/15622975.2019.1584331. Epub).
152. Fujisawa TX, Nishitani S, Takiguchi S, Shimada K, Smith AK, Tomoda A. Oxytocin receptor DNA methylation and alterations of brain volumes in maltreated children. *Neuropsychopharmacology.* 2019;44(12):2045–2053. https://doi.org/10.1038/s41386-019-0414-8.
153. FN J, MH Y, GG A. Antagonism of serotonin receptor mediated neuroendocrine and temperature responses by atypical neuroleptics in the rat. *Eur J Pharmacol.* 1988: 463–469. https://doi.org/10.1016/0014-2999(88)90544-4.

154. Heils A, Teufel A, Petri S, et al. Allelic variation of human serotonin transporter gene expression. *J Neurochem.* 1996;66(6):2621–2624. https://doi.org/10.1046/j.1471-4159.1996.66062621.x.
155. Timothy A, Benegal V, Shankarappa B, Saxena S, Jain S, Purushottam M. Influence of early adversity on cortisol reactivity, SLC6A4 methylation and externalizing behavior in children of alcoholics. *Prog Neuro Psychopharmacol Biol Psychiatr.* 2019;94. https://doi.org/10.1016/j.pnpbp.2019.109649.
156. Kang HJ, Kim JM, Stewart R, et al. Association of SLC6A4 methylation with early adversity, characteristics and outcomes in depression. *Prog Neuro Psychopharmacol Biol Psychiatr.* 2013;44:23–28. https://doi.org/10.1016/j.pnpbp.2013.01.006.
157. Van Der Knaap LJ, Riese H, Hudziak JJ, et al. Adverse life events and allele-specific methylation of the serotonin transporter gene (SLC6A4) in adolescents. *Psychosom Med.* 2015;77(3):246–255. https://doi.org/10.1097/PSY.0000000000000159.
158. Booij L, Szyf M, Carballedo A, et al. DNA methylation of the serotonin transporter gene in peripheral cells and stress-related changes in hippocampal volume: a study in depressed patients and healthy controls. *PLoS One.* 2015;10(3). https://doi.org/10.1371/journal.pone.0119061.
159. Beach SRH, Brody GH, Todorov AA, Gunter TD, Philibert RA. Methylation at SLC6A4 is linked to family history of child abuse: an examination of the Iowa adoptee sample. *Am J Med Genet, Part B: Neuropsychiatric Genetics.* 2010;153(2):710–713. https://doi.org/10.1002/ajmg.b.31028.
160. Okada S, Morinobu S, Fuchikami M, et al. The potential of SLC6A4 gene methylation analysis for the diagnosis and treatment of major depression. *J Psychiatr Res.* 2014; 53(1):47–53. https://doi.org/10.1016/j.jpsychires.2014.02.002.
161. Wankerl M, Miller R, Kirschbaum C, Hennig J, Stalder T, Alexander N. Effects of genetic and early environmental risk factors for depression on serotonin transporter expression and methylation profiles. *Transl Psychiatry.* 2014;4. https://doi.org/10.1038/tp.2014.37.
162. Frodl T, Szyf M, Carballedo A, et al. DNA methylation of the serotonin transporter gene (SLC6A4) is associated with brain function involved in processing emotional stimuli. *J Psychiatr Neurosci.* 2015;40(5):296–305. https://doi.org/10.1503/jpn.140180.
163. Checknita D, Bendre M, Ekström TJ, et al. Monoamine oxidase A genotype and methylation moderate the association of maltreatment and aggressive behaviour. *Behav Brain Res.* 2020:382. https://doi.org/10.1016/j.bbr.2020.112476.
164. Fachim HA, Corsi-Zuelli F, Loureiro CM, et al. Early-life stress effects on BDNF DNA methylation in first-episode psychosis and in rats reared in isolation. *Prog Neuro Psychopharmacol Biol Psychiatr.* 2021;108:110188. https://doi.org/10.1016/j.pnpbp.2020.110188.
165. Perroud N, Salzmann A, Prada P, et al. Response to psychotherapy in borderline personality disorder and methylation status of the BDNF gene. *Transl Psychiatry.* 2013. https://doi.org/10.1038/tp.2012.140. e207-e207.
166. Thaler L, Gauvin L, Joober R, et al. Methylation of BDNF in women with bulimic eating syndromes: associations with childhood abuse and borderline personality disorder. *Prog Neuro Psychopharmacol Biol Psychiatr.* 2014;54:43–49. https://doi.org/10.1016/j.pnpbp.2014.04.010.
167. Chang L, Wang Y, Ji H, et al. Elevation of peripheral BDNF promoter methylation links to the risk of Alzheimer's disease. *PLoS One.* 2014;9(11). https://doi.org/10.1371/journal.pone.0110773.

168. Brown DW, Anda RF, Tiemeier H, et al. Adverse childhood experiences and the risk of premature mortality. *Am J Prev Med.* 2009;37(5):389–396. https://doi.org/10.1016/j.amepre.2009.06.021.
169. Turecki G. The molecular bases of the suicidal brain. *Nat Rev Neurosci.* 2014;15(12):802–816. https://doi.org/10.1038/nrn3839.
170. Parade SH, Huffhines L, Daniels TE, Stroud LR, Nugent NR, Tyrka AR. A systematic review of childhood maltreatment and DNA methylation: candidate gene and epigenome-wide approaches. *Transl Psychiatry.* 2021;11(1). https://doi.org/10.1038/s41398-021-01207-y.

CHAPTER

Early life substance abuse and epigenetic programming 10

Jonna M. Leyrer-Jackson[1], **Lauren E. Hood**[2], **M. Foster Olive**[2]

[1]*Department of Medical Education, School of Medicine, Creighton University, Phoenix, AZ, United States;* [2]*Department of Psychology, Arizona State University, Tempe, AZ, United States*

Substance abuse and substance use disorders

The use of legal and illegal substances of abuse represents a major problem across not only the United States, but also throughout the world. Misuse of psychoactive substances such as alcohol and nicotine-containing products, as well as illegal substances such as cocaine and other psychostimulants, marijuana, opiates, sedatives, and novel psychoactive substances (i.e., synthetic cannabinoids and cathinone derivatives), imposes a large economic and financial burden on society. Specifically, within the United States alone, substance abuse results in over 420 billion dollars annually, with at least 120 billion of these costs in healthcare alone.[1] Substance abuse not only represents a large financial burden but also has vast implications on life quality and health debilitation of the user, as well as negative impacts on social and education system within communities. For example, low socioeconomic status (SES),[2,3] early-life stress,[4] and childhood abuse[5–7] increase the likelihood of developing a substance abuse disorder (SUD) during adulthood. While it was previously thought that SUDs were simply a "lifestyle" choice or character weakness, more recent literature suggests that genetics and other biological factors play a significant role in the development of the substance abuse phenotype.[8] Specifically, patterns of gene expression and genetic modifications can be correlated to substance abuse, as well as the vulnerability to develop SUDs.[8–10] In the chapter herein, we discuss genetic alterations, through epigenetic modifications, which are either inherited or induced by various drugs of abuse, including nicotine, alcohol, cocaine, and tetrahydrocannabinol (THC), the main psychoactive component of marijuana.

Epigenetics and gene expression regulation

Every living eukaryotic cell contains a copy of a complete genome in the form of chromosomes. Long strands of nucleic acid wind around DNA-binding proteins called histones. Eight histones cross-link to form the core of a nucleosome, and nucleosomes cross-link to form a thick band of chromatin, effectively storing the code

Perinatal and Developmental Epigenetics. https://doi.org/10.1016/B978-0-12-821785-6.00008-6

for every gene in a highly dense package. Transcription, the process of assembling RNA nucleotides in a complementary order to DNA sequences, is the first step of gene expression. The identity of each cell and its defining form, function, and life span is a product of the intricate temporal regulation of gene expression. Gene regulation and expression is influenced by many different factors including DNA-altering and nonaltering processes. Epigenetics is defined as the control of gene expression or function by mechanisms other than changes in DNA sequences, which result in altered cellular phenotypes. Specifically, epigenetics describes the bridge between phenotype and genotype that alters the final molecular output of a chromosome without altering the underlying DNA sequence.[11] Epigenetic markers commonly act at the level of transcription with varying levels of duration, though some can work posttranscriptionally to alter the translation of mRNA transcripts and affect the splicing of isoforms of a particular gene.

Epigenetic changes are often due to environmental effects, which can give rise to diseases and disorders such as substance use and addiction phenotypes. Epigenetic gene modification results in gene activation or gene silencing, which can have vast downstream effects on both physiology and behavior. Through DNA methylation, histone modifications, and RNA-associated alterations, structural modifications of chromatin and gene expression are influenced and thus lead to alterations in phenotype or function. While epigenetics is still in its relative infancy, such chromatin modifications result in significant downstream changes that are often heritable and can affect multiple generations, a phenomena termed germline transfer. Epigenetic inheritance, through germline transfer, can allow adult onset of acquired traits and disease phenotypes in the absence of environmental stimuli that influence epigenetic modifications. In the context of substance abuse, maternal and paternal drug use before and during fetal development may have profound influences on adult drug seeking and addiction vulnerability, which will be discussed in detail in the following. Here, we provide an overview of each subtype of epigenetic modification that can occur due to environmental stimuli or germline transfer, and their ability to regulate gene expression.

Histone modification

Histones, protein complexes around which DNA winds, are primary components of chromatin. Chromatin can be modified to regulate DNA accessibility during processes such as DNA replication and repair and gene transcription through a series of covalent histone modifications. To date, covalent modifications have been characterized for more than 60 distinct modification sites, which include acetylation, methylation, phosphorylation, and ubiquitylation at various amino acids.[12] Ultimately, such histone modifications can lead to widespread DNA modifications that can ultimately give rise to a variety of diseases.[13] Histone modification occurs through acetylation and deacetylation and results in widening and relaxation of the chromatin structure, respectively, ultimately aiding in the facilitation or inhibition of DNA transcription.[14] Histone modifications are reversible and controlled by histone acetylases (HATs) and histone deacetylases (HDACs), which act as gene coactivators and corepressors, respectively.[14] HATs act by adding an acetyl group to the histone tails of the nucleosome, causing the relaxation of chromosomes. Chromosome relaxation promotes transcription by increasing the availability of DNA-binding sites to transcriptional machinery.[15] HDACs remove acetyl groups from histone tails of the nucleosome, condensing the chromosome, and prevents the binding of transcriptional machinery onto the DNA.[15] Histone acetylation can occur on the H2A, H2B, H3, and H4 histones, yet it is most commonly observed on the H3 histone within the brain.[16] In the following, we highlight histone acetylation mechanisms associated with drug abuse and vulnerability. Another common histone modification includes methylation, in which histone methyltransferases can add up to three methyl groups to histone tails at gene promoter sites, resulting in either repression or augmentation of gene transcription. Histone methylation can occur on all basic residues but is most extensively studied at the histone H3 lysine 4 (H3K4) site.[17] Histone modifications, while generally reversible and short-lasting, can provide the ability for long-term alterations to occur, such as DNA methylation.[15]

DNA methylation

DNA methylation is perhaps the most well-studied and characterized phenomenon associated with chromatin remodeling. The process of DNA methylation is characterized by the addition of methyl groups to DNA via DNA methyltransferases (DNMTs), which has two distinct effects—disruption of DNA-binding factors from binding to sequence targets, or recruitment of corepressors such as methyl-CpG-binding proteins (MeCPs), such as MeCP2.[18] DNA methylation is highly specific for regions rich in cytosine and guanine regions of the DNA, also known as CpG islands. Through either mechanism, gene silencing occurs due to condensation of the chromatin structure.[18] Covalent addition of methyl groups to the 5-carbon of the cytosine ring is perhaps the most widely characterized process of DNA methylation, which results in transcription inhibition.[19] Throughout development, DNA methylation and demethylation occur throughout the genome, resulting in unique and stabile tissue-specific gene transcription.[19] Interestingly, DNA methyltransferases, as well as other machinery responsible for DNA demethylation, can be found following development allowing for postdevelopmental modifications to occur. For example, neurons utilize these components to regulate DNA methylation in response to environmental and physiological stimuli, ultimately modifying brain circuitry. In the following, we have highlighted the effects of substance exposure and abuse on DNA methylation machinery and its proposed overarching effects.

RNA-associated modifications

Until the past decade or two, RNA was not considered to play a role in epigenetic processes. However, implications of noncoding RNAs and interference RNAs have recently gained interested for their potential role in epigenetic modifications due to their ability to form heterochromatin assemblies and specialized chromatin domains, and their ability to induce gene silencing.[20] Noncoding RNAs have the ability to bind to promoters and silence gene expression, which contributes to chromatin remodeling.[21] Furthermore, they have been shown to inhibit the replacement of histones by protamines,[22] contributing to long-lasting heritable epigenetic alterations.[23] The role of RNA in chromatin remodeling has also been shown for interference RNAs, where they mediate short antisense RNAs to inhibit translation and degrade cytoplasmic mRNA through posttranscriptional gene silencing.[24] Interestingly, due to their ability to regulate gene expression, interference RNAs are being used as a target for developing RNA therapeutics to elucidate the functions of genes and provide new avenues of treatment for diseases.[25] Thus, while previously overlooked, RNAs may play a pertinent role in mediating gene expression and may be partially responsible for contributing to heritable epigenetic changes.

Inherited predispositions (i.e., susceptibility of future generations)

Transgenerational epigenetic inheritance has recently gained traction as the ability of offspring to inherit epigenetic modifications that occurred in absence of the initial signal or environment that triggered the epigenetic change.[26] In addition to transgenerational inheritance, intergenerational epigenetic effects can be inherited when the offspring is exposed to drug substances and environmental effects in utero. Thus, the heritability of epigenetic modifications can be induced both transgenerationally and intergenerationally. One of the first studies to demonstrate transgenerational heritability of epigenetic alterations was found by Guerrero-Bosagna and colleagues,[27] who reported that vinclozolin, an environmental toxin, induced a pattern of sperm DNA methylation observed three generations after initial exposure in the F0 generation (i.e., in F3 males). These results were validated with other environmental toxins, reporting that exposure to pesticides, dioxin, and jet fuel can also induce DNA methylation patterns in sperm of the F3 generation.[28–31] Furthermore, in utero exposure to a variety of substances and environmental effects has led to a plethora of studies demonstrating intergenerational epigenetic inheritance. Specifically, an abundance of literature has shown maternal drug use to have long-lasting epigenetic effects on offspring in utero, which has the ability to influence adulthood drug use in the offspring.[32] In the following, we highlight relevant studies demonstrating epigenetic modifications induced by in utero exposure to drugs of abuse such as nicotine, alcohol, cocaine, and marijuana/THC. For additional information regarding transgenerational and intergenerational epigenetic inheritance.[33–35]

Early drug use and epigenetic modifications

Drug use and addiction, now formally referred to as substance use disorders, are similar to other psychiatric disorders such as schizophrenia and bipolar disorder in that they are heavily influenced by genetic susceptibility and environmental factors. Specifically, it is thought that epigenetic signatures are induced by early-life experience, including stressors and other environmental factors, which converge to alter gene expression patterns and confer a propensity toward the development of addictive phenotypes. Interestingly, while both genetic alterations and environmental stressors often work in tandem to contribute to drug-seeking and drug-addictive behaviors, environmental stressors during development such as a degraded home and low SES[2,36] play a substantial role predicting future drug use. For example, neighborhood poverty and low SES are significantly correlated with heroin and cocaine usage,[36] and SES predicts marijuana use among adolescents.[37] Thus, environmental effects can heavily influence drug use patterns among adolescents and adults, and may do so, in part, by promoting long-lasting epigenetic modifications.

In addition to environmental factors, more recent studies suggest that drug use and addictive phenotypes can be heavily influenced by heritable epigenetic modifications. In fact, the genetic heritability that influences substance abuse is estimated to be between 45% and 79%.[38–40] Intergenerational epigenetic inheritance caused by in utero drug exposure can heavily influence the vulnerability of offspring to developing addictive-like behaviors[23]; however, this heritability is likely influenced by drug class. Additional evidence suggesting transgenerational epigenetic inheritance has been reported, primarily with respect to paternal drug use that influences DNA methylation patterns of sperm.[23,41] Given the difficulty in studying the effects of drugs of abuse on epigenetic heritability, the field has many unanswered questions; however, recent advances in molecular biological technology have lent insight into genetic markers and epigenetic modifications that may act as predispositions to drug use. In the following sections, we aim to summarize major findings of both inherited and postnatal epigenetic modifications induced by five major drugs of abuse: nicotine, alcohol, cocaine, marijuana/THC, and opiates. Lastly, while we discuss some drug-induced germline alterations, we point the readers to a recent review by Zeid and Gould for a more thorough review on the effects of alcohol, nicotine, and cocaine.[42]

Nicotine

Nicotine, the primary addictive substance in tobacco-containing products, is the leading preventable cause of death within the United States.[43] Leading to over 400,000 premature deaths each year, nicotine decreases life expectancy by 14 years. Nicotine is among one of the most commonly abused substances, with usage among adolescents increasing in recent years due to nicotine-containing electronic cigarettes/vaping devices. According to the National Institute on Drug Abuse (NIDA), in 2019, combustible cigarette usage of any kind was reported by 10%, 14.2%, and 22.3% of 8th, 10th, and 12th graders, respectively, while the use of vaporizing devices, such as electronic cigarettes, was reported by 24.3%, 41%, and 45.6% of 8th, 10th, and 12th graders, respectively. Due to its addictive nature, cessation of nicotine usage is extremely difficult, with 60%–90% of individuals relapsing within the first year of intended abstinence. For this reason, prenatal exposure to nicotine by nicotine-using mothers during pregnancy is prevalent and has been shown to have long-lasting changes in offspring. Despite its negative health consequences for the fetus, including increased risk of miscarriage, stillborn, and premature birth, an average of 6%–11% of pregnant women continue to use nicotine products throughout pregnancy.[44] Maternal smoking during pregnancy has profound impacts on offspring (termed developmental nicotine exposure, or DNE), including behavioral predispositions as well as epigenetic alterations. In humans, up to 23 CpG sites can be altered in first-generation offspring due to DNE, including sites located within the aryl hydrocarbon receptor repressor (*Ahrr*), α-ketoglutarate-dependent dioxygenase FTO (*Fto*), myosin 1G (*Myo1g*), and FERM-domain containing 4A (*Frmd4a*) genes, which are associated with cancer development, obesity, cell signaling, and nicotine dependence, respectively.[45] Specifically, DNA methylation at these sites is long-lasting and contributes to high blood pressure as well as an increase in circulating high-density lipoprotein cholesterol and triacylglycerol levels.[45] Epigenetic modifications of brain-derived neurotrophic factor-6 exon (*Bdnf*) have also been observed in DNE offspring in humans. Using bisulfite sequencing of blood samples from adolescent whose mothers smoked during pregnancy, DNA methylation of the *Bdnf* promotor was found to be upregulated in individuals with DNE, which led to long-term suppression of *Bdnf* expression with potential effects on brain development and plasticity.[46] Behaviorally, DNE can enhance nicotine preference, risk-taking behavior, and hyperactivity in first- and second-generation adolescent offspring in rodents.[47] These behavioral alterations have been linked to epigenetic modifications including increases in global DNA methylation within the striatum and frontal cortex.[47] Furthermore, elevated levels of mRNA for absent, small, or homeotic discs two-like (*Ash2l*), which encodes a component of the histone methyltransferase complex, have been observed following DNE in mice and are linked to changes in cortical neuronal morphology and passive avoidance behavior.[48] While dendritic morphology and spine density within the cortex have been previously linked to DNE,[45] Jung et al. provided evidence of epigenetic markers that play a role in this process.[48] Using next-generation sequencing, these investigators found enriched trimethylation of the lysine 4 residue of histone 3 (*H3k4me3*) at promoter regions within multiple genes involved in developing and maintaining synaptic structure, specifically at excitatory synapses, including Ash2l and MADS box transcription enhancer factor 2 (*Mef2c*). Furthermore, *Mef2c* and *Ash2l* were shown to be regulated by nicotinic acetylcholine receptors (nAChRs), the primary target of nicotine.[48] Taken together, these findings indicate that maternal smoking and DNE can have profound effects on epigenetic markers in human offspring as well as in rodents, which can elicit powerful control over behavior and health-related abnormalities.

In addition to effects elicited by DNE, epigenetic modifications have been observed following nicotine exposure during development and adulthood. Specifically, acetylation of histone H4 within the dopamine D1 receptor (*Drd1*) gene has been reported within the rat prefrontal cortex following nicotine exposure produced by subcutaneous injections, which is hypothesized to promote enhanced expression of D1 receptor mRNA within this region.[49] Interestingly, these findings may lend insight into nicotine-induced elevations in synaptic dopamine levels. Additionally, DNA methylation of the monoamine oxidase B (*Maob*) promoter is upregulated in both humans with history of nicotine use and mice exposed to nicotine,[50] contributing to dysregulation in serotonin catabolism. These changes, in turn, may contribute to nicotine addiction, where increased *Maob* methylation has been previously associated with nicotine dependence in women.[51] Lastly, decreases in DNA methyltransferase 1 (*Dmnt1*) mRNA and protein, and increases in glutamic acid decarboxylase 67 (*Gad67*) expression within GABAergic neurons the prefrontal cortex, have been observed following subcutaneous nicotine exposure in adult mice (four injections/day for 4 days).[52] This dosage of nicotine was also found to decrease cytosine-5-methylation within the GAD67 promoter and prevented methionine-induced hypermethylation at the same promoter, which was blocked by pretreatment with the nAChR antagonist mecamylamine.[52] Taken together, it appears that nicotine exposure, pre- and postnatal, can induce long-lasting epigenetic changes, leading to increased propensity for dependence, further propagating a vicious cycle of inheritable alterations that may produce transgenerational tendencies toward nicotine abuse.

Alcohol

Alcohol use is heavily prevalent throughout the United States, as well as the rest of the world. In 2018, more than 55% of the US population over 18 years of age self-reported alcohol consumption within the past month. In addition, 28% of the population over 18 years old categorized their alcohol consumption as "binge drinking," defined as five or more drinks in the same occasion for men and four for women.[53] As the third leading cause of death in the United States, alcohol consumption leads to an estimated 88,000 premature deaths, and 3.3 million global deaths annually.[54] Like nicotine, maternal alcohol use during pregnancy has vast effects on the developing fetus, including epigenetic modifications. With estimates of two to seven cases per 1000 births, fetal alcohol spectrum disorders (FASDs) have become five times more prevalent in 2018 than the levels observed in 1996.[54] FASD results from prenatal alcohol exposure and often induces irreversible cognitive and behavioral disabilities as well as pre- and postnatal growth retardation and facial dysmorphisms.[55] Additionally, maternal and paternal alcohol abuse has profound effects on the likelihood of the child to abuse alcohol in adolescence or adulthood, where heritable epigenetic alterations have been determined to play a role.

Accumulating evidence has shown that prenatal alcohol exposure results in DNA methylation and histone modifications leading to the development of a wide array of FASDs, varying in severity. In humans, 5%–10% of fetuses exposed to alcohol develop some form of FASD, dependent on the severity and duration of alcohol exposure. Rodent models have proved beneficial for exploring epigenetic modifications associated with prenatal alcohol exposure and provide insight to key epigenetic mechanisms giving rise to FASD in humans. Specifically, mice exposed to alcohol in utero during days 9–11 of gestation display decreased global levels of methylation throughout the genome.[56] Furthermore, the hippocampus appears to be particularly sensitive to epigenetic changes induced by prenatal alcohol exposure, where changes in DNA methylation enzymes and binding proteins are observed.[57] In addition to DNA methylation, prenatal alcohol exposure also induces histone methylation, which may also play a role in the development of FASDs. Specifically, alcohol exposure in utero increases histone methyltransferase abundance, leading to alcohol-induced neurodegeneration in specific brain regions such as the hypothalamus.[58] In addition, alterations in histone H3 lysine 4 trimethylation (H3K4me3) and histone H3 lysine 27 trimethylation (H3K27me3) are observed following exposure to alcohol during critical neurodevelopmental periods.[59] Together, these studies heavily implicate epigenetic alterations induced by prenatal alcohol exposure that induce FASDs.

During adolescence, modifications to synaptic connections are often observed in animals exposed to alcohol. As observed with DNE, intermittent binge-like alcohol consumption in adolescent rats alters BDNF, FosB, and FosB cyclic-dependent kinase 5 (Cdk5) signaling within the prefrontal cortex and induce long-lasting changes in synaptic plasticity.[60] Interestingly, while most commonly explored for their role in neuroinflammatory processes, Toll-like 4 receptors (TLR4) are involved in this epigenetic modification, as rodents lacking TLR4 show reduced alcohol consumption and fail to elicit epigenetic synaptic alterations as compared with control animals.[61]

The anxiolytic effects of alcohol exposure are linked to epigenetic modifications within the amygdala. Specifically, a combination of increases in histone acetylation, levels of cyclic AMP response element-binding protein (CREB) levels, and inhibition of histone deacetylase within the amygdala are linked to chromatin relaxation following acute alcohol exposure.[62] Global acetylation of histone complexes H3 and H4 has also been observed in the amygdala of rats following acute alcohol exposure.[63] Furthermore, both studies also showed decreases in histone deacetylase levels within the amygdala, suggesting alcohol-induced epigenetic remodeling following acute exposure. Interestingly, innate high expression levels of histone deacetylase 2 (HDAC2) lead to excessive alcohol intake and are often identified as a heritable genetic predisposition to alcohol abuse.[64,65] HDAC2 expression is increased through adulthood in animals that experienced intermittent alcohol exposure during adolescence.[66] Kyzar and colleagues[66] also report a decrease in histone H3 lysine 9 (H3K9) acetylation as well as changes in expression of multiple genes associated with synaptic plasticity, including BDNF and activity-regulated cytoskeletal protein (Arg/Arg3.1) animals with intermittent alcohol exposure. Such alterations have been implicated in synaptic density changes within the central and medial nuclei of the amygdala of the amygdala in adulthood.[66,67] Thus, given the vast role of alcohol on histone modifications in adolescent exposed rodents, it is not surprising that these epigenetic modifications have the ability to influence adult psychopathology.

Cocaine

Cocaine is a commonly abused psychostimulant derived from the coca plant that can be snorted, smoked, or injected intravenously to produce feelings of euphoria and mental acuity. The psychostimulant effects of cocaine are mediated through enhanced dopaminergic activity throughout mesocorticolimbic circuitry via inhibition of the dopamine transporter.[68] Over 2.2 million Americans regularly use cocaine, at least 1 million of which suffer from cocaine use disorder.[69] 5%–6% of users develop cocaine dependence in the first year of use, a quicker rate as compared with those for marijuana and alcohol, with peak age of dependence in early adulthood.[70] Earlier use in adolescence is correlated to higher risk of developing cocaine-related problems. A national survey on drug use and health from 2017 estimates 26,000 adolescents between the ages of 12–17 use some form of cocaine.[70] Adolescents boys were more likely to report increased lifetime use and repeated use of cocaine over girls, though young women tended to initiate use at younger ages and have quicker transitions into a state a dependence.[71,72]

Environmental exposure to cocaine alters the epigenome, and some of the cocaine-induced changes in DNA methylation and histone modifications can be inherited by future generations.[42,73,74] On the paternal side, cocaine can accumulate in human semen and remain bound to sperm for up to 24 h, where it has been to shown reduce sperm count, motility, and the ability to infiltrate oocytes.[42,75] Chromosomes involved in spermatogenesis are particularly susceptible to heritable epigenetic remodeling,[73] and increased cocaine-induced deficits in spermatogenesis have been reported for adolescent mice compared with adults.[76]

Male rats that self-administered cocaine showed increased acetylation of the H3 histone at the BDNF promoter in sperm.[77] Consistent with multigenerational inheritance, these marks were passed to the F1 offspring to enhance expression levels of BDNF mRNA and protein within the medial prefrontal cortex and lowered their motivation for cocaine self-administration. DNA methylation patterns in the sperm of cocaine-dependent rats have also been shown to persist into the F1 generation.[78] In addition to histone acetylation and DNA methylation, mouse sperm has been shown to pass along mRNA transcripts for HDACs and DNMTs. Paternal cocaine exposure is also associated with anxiety- and depression-like phenotypes in offspring, which are dysfunctional mental states that are inherently vulnerable to substance abuse.[78–80]

In females, cocaine is known to stimulate luteinizing hormone, follicle-stimulating hormone, and estradiol release, leaving the endocrine system vulnerable to dysregulation.[42] Long-term dysregulation of reproductive hormones can affect germline integrity (i.e., number and morphology of oocytes).[42] Interestingly, maternal exposure to cocaine prior to pregnancy can induce behavioral sensitization to cocaine in male F1 offspring,[81] which is different than other outcomes described for paternal cocaine exposure. Sensitization to the effects cocaine is associated with elevated levels of the dopamine D1 receptor expressed in the medial prefrontal cortex (PFC).[81] Maternal exposure to cocaine is also associated with decreases in global

DNA methylation in hippocampal neurons of neonatal male offspring, with DNA methylation levels increasing in adolescent cocaine-exposed animals.[82]

Cocaine use alters DNA methylation in both a brain region and gene-specific manner.[73,83,84] Acute and chronic exposure to cocaine has been shown to increase global DNA methylation in the nucleus accumbens (NAc), accompanied by increased expression of DNMT3a, at multiple stages throughout the drug abuse cycle including withdrawal and cue-induced reinstatement (a model of relapse).[83,84] Global increases in DNA methylation do not equate to uniform hypermethylation of the entire genome; rather, DNA hypomethylation was also reported in the NAc in promoter regions for genes such as *FosB* following acute exposure.[83] In the PFC, cocaine exposure has been shown to induce global DNA hypomethylation and a reduction in DNMT3b mRNA and protein. Restored DNMT function prevents the acquisition of cocaine-conditioned place preference (CPP), but not CPP for food, suggesting DNA hypomethylation in the PFC contributes to the rewarding effects of cocaine.[85] Repeated cocaine exposure has also been shown to downregulate the expression of ten—eleven translocation methylcytosine dioxygenase 1 (*Tet1*) in the NAc, a key regulator of DNA demethylation. Cocaine- and withdrawal-induced changes in DNA methylation levels have been inversely correlated to some changes in gene expression.[83,86]

Pharmacological manipulation of DNMTs can influence the development of cocaine-related behaviors. Microinjections of the DNMT inhibitor RG108 ((2S)-2-(1,3-dioxoisoindol-2-yl)-3-(1H-indol-3-yl)) propanoic acid (RS108) into the NAc attenuated cue-induced reinstatement of cocaine seeking in rats, while administration of the methyl supplement S-adenosylmethionine to promote DNMT activity enhanced cocaine seeking.[86] Both effects from a single treatment of DMNT modulator persisted 1-month posttreatment, suggesting DNA methylation remains dynamic and amenable to epigenetic reprogramming despite drug-induced alterations.[86] Similarly, DNMT activity regulates the progression of cocaine behavioral sensitization, with inhibitors such as zebularine reducing sensitization and reducing DNA methylation at the MeCP2 promoter in the NAc.[83] DNMT inhibitors also prevent the acquisition of CPP for cocaine after local injections into regions required for learning and memory, such as the hippocampus and prelimbic cortex.[87] DMNTs can influence other forms of plasticity, as DNMT3a is crucial for cocaine-induced increases in dendritic spine density, and is therefore implicated in regulating structural and synaptic plasticity in the brain.[73]

Cocaine use enhances HAT activity while reducing the removal of acetyl marks on histones by HDACs to increase transcription levels and gene expression across mesolimbic regions of the brain, likely contributing to the neuroadaptations that underlie cocaine use disorder.[73] For example, mice genetically depleted of the HAT CBP (CREB-binding protein) exhibited lower levels of histone acetylation at the *FosB* promoter, lower expression of FosB protein, and decreased sensitivity to cocaine compared with controls.[88] Acute cocaine exposure also results in the global enhancement of histone H3 acetylation throughout the striatum and NAc.[88,89] After chronic cocaine self-administration in rats, the NAc, ventral tegmental area (VTA) and medial PFC all showed increases in histone H3 acetylation at the *Bdnf* promoter, though the exons within the promoter were differentially acetylated depending on the region.[73,90,91] These changes in histone acetylation persisted after withdrawal, suggesting stability in cocaine-driven histone modifications,[92] were correlated with elevated levels of BDNF mRNA and protein in the VTA and PFC.[73,91] Chronic cocaine exposure was also reported to reduce HDAC5 activity in the NAc, while upregulating HDAC5 target genes and increasing motivation for cocaine seeking.[16]

However, studies on alterations of histone acetylation in cocaine CPP and self-administration paradigms are not always in agreement. Systemic administration of HDAC inhibitors enhances cocaine CPP and locomotor activity in mice, which was reversed by the overexpression of HDAC4 or HDAC5 in the NAc shell and core subregions.[16] However, HDAC inhibitors also reduce the motivation for cocaine reinforcement in self-administration procedures,[93] suggesting that the relationship between HDACs and cocaine-seeking behavior is complex and requires more study.

In rats, cocaine exposure during adolescence resulted in decreased histone H3 methylation in the medial PFC that persisted into adulthood and led to cognitive impairments.[94] Repeated cocaine exposure also induced the downregulation of one enzyme responsible for histone 3 lysine 9 (H3K9) methylation, the lysine methyltransferase G9a. Reduced G9a activity led to subsequent reductions in H3K9 methylation in the NAc.[95] Cocaine-induced FosB expression was found to regulate the repression of G9a to promote dendritic spine plasticity in the NAc and cocaine CPP.[95] Later, this same group of investigators would find similar cocaine-induced repression of H3K9 methylation via G9a downregulation directed medium spiny neurons to adopt a *Drd1* subtype identity.[93] Reduced G9a levels also mediated susceptibility to social defeat stress in mice and, when paired with a history of cocaine self-administration, promoted the manifestation of depression-like phenotypes.[96]

Noncoding RNA levels are also impacted by cocaine exposure.[73] Of note, micro-RNA 212 (miR-212) has been observed to be upregulated in the dorsal striatum of rats after extended access cocaine self-administration.[97] Increased miR-212 levels negatively regulate self-administration to reduce cocaine intake, a behavioral outcome that can be reversed through the knockdown of the transcriptional repressor MeCP2 in the dorsal striatum.[97] Striatal BDNF protein levels also increased in response to cocaine exposure, echoing the expression patterns of MeCP2 and negatively correlating to miR-212 expression, indicating a homeostatic relationship between miR-212 and MeCP2 in the modulation of cocaine-induced BDNF expression.[97]

Marijuana/THC

Cannabis, or marijuana, derived from the plant *Cannabis sativa* contains over 440 biologically active compounds, at least 120 of which are cannabinoids that function within the body's endogenous endocannabinoid (eCB) system.[98] Cannabis is one of the most widely used psychoactive drugs in the United States, with an estimated

45% of Americans over the age of 12 having used cannabis in their lifetime. Of the 44,500, 8th-, 10th-, and 12th-grade students surveyed in 2018 across the country, and 11.8%, 28.8%, and 35.7% had used cannabis in the past year, respectively.[99] While cannabis overdose is almost never lethal (synthetic cannabinoids excluded), and responsible use is gaining social acceptance,[99,100] it remains a mind-altering drug with the potential for abuse. Cannabis use disorder (CUD), characterized by cravings and withdrawal, can develop due to the neuroadaptations induced by chronic THC exposure in an estimated 30% of users, and risk of developing CUD is positively correlated with increased use.[98,101] CUD is a polygenic disorder, with a heritability estimate ranging from 40% to 49%.[32] The risk for developing CUD peaks from late adolescence to early adulthood, and the onset of cannabis use in adolescents, is correlated to increased risk for developing CUD in adulthood, with worse treatment outcomes and high relapse rates.[98] For all age groups, men are more likely to develop CUD than women, though women typically develop more severe cases.[98] Cannabis use in adolescence has also been shown to increase the risk for use of cocaine and other psychostimulants later in life and can alter cocaine-induced reprogramming of the epigenome.[102]

Use of synthetic cannabinoids engineered to mimic the effects of THC is also on the rise in adolescent populations, with 3%–10% of American high school students having tried them.[99,103] These compounds, over 150 of which have been identified to date, have higher affinities for the CB1 receptor than natural cannabinoids, inducing stronger psychoactive reactions and therefore have higher addictive properties.[104] These novel chemicals are potentially dangerous due to dangerous side effects such as paranoia, delusions, coma, and death.[104] Criminalization of synthetic cannabinoids in the United States commenced in 2011, but the speed at which newer analogs are manufactured and the evolution of their chemical structures make them difficult to regulate.[99,104] While the legal status of cannabis use has been changing dramatically over the past few decades, with many states relaxing older restrictions to decriminalize medicinal and recreational use for adults, this does not necessarily translate to increased availability of cannabis to the teenage population.[99,105] High-school students surveyed in 2018 throughout the United States reported no perceived change in the ability to access marijuana compared with availability rates reported from previous years,[99] nor was any difference found in the reported amount of cannabis used in adolescents across states with legalized or criminalized recreational uses.[106] This trend is supported by the reported numbers of adolescents seeking treatment for CUDs, which has not increased following the legalization of cannabis across the United States in 2012.[107]

When smoked, vaporized, or ingested as food, cannabis induces a state of "high" intoxication with psychoactive qualities that are mediated primarily by the cannabinoid delta-9-tetrahydrocannabinol (THC), which activates G protein–coupled cannabinoid receptors.[104] THC can activate both the type 1 (CB1) and type 2 (CB2) cannabinoid receptors that are found throughout central and peripheral nervous systems, respectively, to impact everything from memory and executive functioning to reproductive systems and gut microbiota.[104,108]

The eCB system heavily influences many aspects of male and female reproductive systems,[109,110] and some epigenetic repercussions of cannabis exposure can be transmitted from parental gamete to offspring. Sperm quality is declined in human cannabis users, with clinical studies reporting deficits in sperm count, morphology, motility, ability to penetrate, and changes in DNA methylation patterns.[111,112] Activation of CB2 signaling using the CB2 agonist JWH-133 in young male mice led to a 30% reduction in sperm count, while decreases in sperm motility were found to be mediated through CB1 activation.[110] The sperm from these males also exhibited increased levels of DNA methylation on paternally imprinted genes, and these marks were found to be conserved in the placentas of their mating partners.[110] The offspring were often underdeveloped, with lasting deficits in weight and growth rate, suggesting paternal DNA methylation patterns are important for proper neonatal development. It is unknown if these epigenetic changes would persist to stunt the growth of subsequent generations. In rats, moderate doses of THC (2 mg/kg/day) administered for 12 days were sufficient exposure to increase DNA methylation in sperm that was passed on to the F1 generation, which also demonstrated attentional deficits.[111] Germline exposure to THC has also led to alterations in DNA methylation for genes important in glutamatergic neurotransmission and synaptic plasticity, i.e., the postsynaptic density protein 95 (Psd-95) encoding gene discs large homolog 4 (*Dlg4*), in the NAc of offspring.[113,114] Among its many functions, Psd-95 stabilizes NMDA receptors at excitatory synapses, and its epigenetic upregulation has been linked to the experience of drug reward.[114,115]

Less information is available on the maternal contributions to the multigenerational effects of cannabis, though evidence suggests cannabis exposure prior to motherhood can lead to heritable changes in behavior.[116] When female rats were treated with CB1 receptor agonists during adolescence and prior to mating, their male and female offspring experienced increased sensitization to morphine.[74,117] Adolescent females repeatedly exposed to THC prior to pregnancy also produced male offspring with diminished motivation for THC and D-amphetamine.[116] Another study subjected male and female rats to chronic THC exposure during adolescence, and in addition to a 40% reduction in the rate of pregnancy compared with controls, male offspring exhibited increased levels of heroin intake in adulthood and altered expression levels of the genes for CB1, glutamate, and dopamine receptors in the striatum.[118] However, effects of parental sex on behavioral outcomes of offspring were not reported. These studies suggest a history of THC exposure in females can trigger an atypical response to drug reward in offspring, and while hypothesized as potentially epigenetic in origin, no specific epigenetic modifications were identified in the manifestation of this phenotype.

Cannabis is one of the most commonly used drugs during pregnancy, and exposure to cannabis in utero can lead to epigenetic modifications in the developing embryo.[119] The expression level of the gene for the D2 dopamine receptor (*Drd2*) decreased in the NAc of rats exposed to cannabis prenatally, due to several epigenetic modifications including increases in the histone methylation associated with gene repression (dimethylated lysine 9) and decreases in the histone methylation associated with active transcription (trimethylated lysine 4).[120] The same reductions in *Drd2* expression were also found in human fetal brain samples from cannabis-using mothers.[120] The culmination of these changes suggests an increased sensitivity to opiates and other abused substances in THC exposed offspring, indicating that prenatal cannabis exposure may enhance susceptibility to future substance abuse.

Supporting this theory, adolescent rats exposed to THC experienced an upregulation of the opioid peptide precursor proenkephalin that directly mediated increased susceptibility to later heroin use.[121] These THC-induced increases in proenkephalin were mediated by reductions in histone H3 lysine 9 (H3K9) methylation in the NAc. Adolescent THC exposure was also found to, possibly as a result of these changes in histone methylation, reduce the dendritic complexity of prefrontal pyramidal neurons, an alteration in synaptic plasticity that is characteristic of psychiatric disorders including substance use disorders.[37,122] The expression of genes important for synaptic plasticity in the prefrontal cortex was impacted by adolescent cannabis exposure in female rats via fluctuations in histone modifications, primarily increased H3K9me3.[123] When exposed to cannabis during adolescence, humans carrying a homozygous genetic polymorphism that increases the activity of the dopamine-degrading enzyme catechol-O-methyl-transferase (COMT) exhibited increased methylation of COMT and were less likely to develop dependence or psychosis compared with cannabis users lacking both copies of the COMT polymorphism allele.[124] Chronic exposure to THC in adolescent mice also leads to DNA hypermethylation of the gene encoding regulator of G protein signaling 7 (*Rgs7*) in the hippocampus, resulting in memory impairments for fear conditioning and spatial navigation tasks, as well as impairments in coping responses to stressors.[103]

The CB2 receptor is primarily expressed in immune cells, and as a result, the eCB system promotes antiinflammatory responses and immunosuppression.[119] Embryonic exposure to cannabis can prevent proper development of T cells, leading to functional impairments later in life that negatively impact the efficacy of the immune system.[125] Exposure to cannabis in adulthood can result in dysregulation of the miRNAs that target mRNAs for proinflammatory cytokines in rodents and nonhuman primates, which in the latter ultimately slows the disease progression of simian immunodeficiency virus (SIV).[126,127] Chronic THC administration to SIV-infected nonhuman primates induces the upregulation of miRNAs such as miR-24 and miR-145, which target inflammatory signals, as well as increases in DNA methylation.[126] Histone methylation and acetylation are also altered by THC exposure in T cells of mice, promoting differentiation into type 2 versus type 1 helper T cells.[128] Neuroimmune activation can influence the development and progression of substance abuse disorders, highlighting the potential impact of THC exposure on CB receptor signaling in the immune system function and its contribution to future development of polydrug use disorders.[129]

Opioids

Opioid use within the United States has become a highly detrimental health issue over the past two decades. Roughly 8%–12% of prescription opioid users develop an opioid use disorder, and 80% of heroin users first misused prescription opioids. In 2018, 1.7 million Americans suffered from SUDs related to prescription opioids, and over 45,000 Americans died due to opioid overdose.[130] Like other substances discussed in this chapter, opiates have profound epigenetic effects, which can originate from paternal and maternal drug use prior to conception, in utero exposure as well as postnatal drug consumption.

Around 40%–60% of the vulnerability for developing opiate use disorders is heritable, which can be impacted by maternal and paternal opiate usage prior to conception.[131] Like each of the drugs discussed thus far, there is much evidence linking opiate-induced epigenetic alterations to profound effects in offspring, through gamete formation and genetic modification. Specifically, Byrnes and colleagues have reported that in rats, the offspring, both male and female, of mothers exposed to morphine prior to pregnancy, exhibit enhanced locomotor activity relative to the offspring of mothers not exposed to morphine.[132] Furthermore, neurobiological changes in offspring from male and female opioid users, even with use prior to conception, have been shown. For example, in rats, male offspring of mothers exposed to heroin during adolescence exhibit enhanced kappa opioid receptor mRNA expression within the NAc.[133] Additionally, morphine-exposed male rats produce F1 generation female offspring with enhanced β-endorphin levels within the hypothalamus, as well as increased corticosterone levels within blood serum.[134] The sperm of opioid-dependent male rats also exhibit enhanced mu opioid receptor (*Oprm1*) promotor methylation,[135] further demonstrating the potent effects of parental drug use on genetic alterations of future offspring.

Neonatal abstinence syndrome (NAS), also known as neonatal opioid withdrawal syndrome (NOWS), is a growing issue in which infants develop opioid dependence due to in utero opioid exposure, and experience signs of withdrawal within 2–3 days after birth.[136] NAS has increased by a factor of five within the past two decades, and it is predicted that genetic factors likely play a large role in the variability of the disorder.[136] Some have reported that genetic variations in prepronociceptin (*Pnoc*) and *COMT* genes can indicate the severity of NAS and may serve as important markers for predicting future childhood outcomes such as long-term effects on cognitive abilities and motor performance.[137,138] Additionally, decreased expression of the mu-opioid receptor *Oprm1* gene has also been correlated with the severity of NAS expression.[139] Infants with at least one copy of the G allele in *rs1799971*, a gene associated with *Oprm1*, exhibit less severe NAS requiring shorter hospital stays and reduced need for pharmacological treatment.[139] While NAS is becoming increasingly prevalent in light of the current opioid epidemic, less is known about the epigenetic alterations contributing to the syndrome. In fact, most research regarding epigenetics and NAS has been conducted by Wachman and colleagues, as reviewed before. Thus, further studies exploring epigenetic alterations and their specific contributions to NAS occurrence and severity are warranted.

Adult opiate use has also been shown to have epigenetic effects in the user, independent from those inherited. For example, chronic opiate exposure during adulthood induces global DNA methylation[140,141] and downregulates BDNF protein expression within the VTA. *Oprm1* promotor methylation is also observed in opiate-dependent humans,[142] and morphine has been found to suppress the binding of phospho-CREB to *Bdnf* promotors within the VTA of both adult rats and mice. For a more thorough discussion of the epigenetic mechanisms involved in opioid addiction, we refer readers to the recent review by Browne and colleagues.[143]

Molecular strategies targeting epigenetic modifications as novel therapeutics for treatment

To date, treatments for substance use disorders have relied on behavioral and psychosocial interventions as well as pharmacotherapies. However, due to the vast effects of drugs of abuse on epigenetic alterations, pharmacotherapies and molecular strategies targeting epigenetic-induced modifications may prove to be viable treatment options for individuals with substance abuse disorders. To date, preclinical studies on epigenetic therapies targeting small interfering RNAs (siRNA) and HDACs have yielded promising results as possible treatments for substance use disorders. Specifically, treatment with a gold nanorod nanoplex conjugated to an siRNA directed against the dopamine- and cAMP-regulated neuronal phosphoprotein 32 (DARPP-32, also known as protein phosphatase 1 regulatory inhibitor subunit 1B) in dopaminergic neurons of the VTA not only was sufficient to cross the blood–brain barrier but also successfully inhibited DARPP-32 expression,[135] which acts as a potent activator of the dopamine reward pathway. These results suggest that using siRNA nanoplexes for inhibiting DARPP-32, as well as other possible genes, could be a promising therapy for patients with substance use disorders.[18]

In addition, HDAC inhibitors have showed some promising results for inhibition of drug seeking, although results are controversial. Specifically, HDAC inhibitors have been shown to increase drug intake in animals self-administering cocaine; however, when given prior to acquisition of drug self-administration, a reduction in intake was observed.[93,144,145] Thus, these results suggest that administration timing and circumstances may alter efficacy of HDAC inhibitor treatment, a phenomenon that should be further explored. However, it should be noted that administration of an HDAC inhibitor prior to commencement of drug use would likely have limited translational value, since this would occur well before clinical presentation. While epigenetic targeting for therapy has been generally sparse in regard to substance abuse, application of these types of treatments has gained traction in other fields such as cancer research and treatment. Thus, therapeutic efficacy of epigenetic therapeutic targets may yield potential insight into the treatment of substance use disorders. However, a major challenge of treating substance use disorders with such therapies is that they often target ubiquitously expressed proteins that have additional off-target effects in the context of drug addiction.[146] Thus, the development of epigenetic therapies with direct actions is warranted.

Conclusions

Understanding the genetic underpinnings, heritability and epigenetic profiles contributing to substance abuse disorders are essential for effective therapeutic development and a complete understanding of disease states. Here, we have summarized recent evidence suggesting that substances of abuse can have profound effects on gene expression, which can be inherited both transgenerationally and intergenerationally (i.e., in utero exposure). While the effects of these substances on the genetic code are far from fully understood, recent advances in technology and understanding of epigenetics have greatly enhanced our understanding as a field within the past decade. Furthermore, a more complete understanding may advance treatment options for patients with substance abuse disorders by targeting known epigenetic alterations caused by these substances. Together, a more complete understanding of the link between substances of abuse and genetics will greatly improve our understanding, diagnosis, and treatment of substance abuse disorders.

Acknowledgments

This work was supported by National Institutes of Health Public Health Service grants R01 AA025590 and R01 DA042172 to MFO, and a postdoctoral National Research Service Award Postdoctoral F32 AA027962 to JMLJ.

References

1. McLellan AT. Substance misuse and substance use disorders: Why do they matter in healthcare? *Trans Am Clin Climatol Assoc.* 2017;128:112–130. PMID: 28790493.
2. Hanson MD, Chen E. Socioeconomic status and substance use behaviors in adolescents: the role of family resources versus family social status. *J Health Psychol.* 2007;12(1): 32–35. https://doi.org/10.1177/1359105306069073.
3. Boardman JD, Finch BK, Ellison CG, Williams DR, Jackson JS. Neighborhood disadvantage, stress, and drug use among adults. *J Health Soc Behav.* 2001;42(2):151–165. https://doi.org/10.2307/3090175.
4. Enoch MA. The role of early life stress as a predictor for alcohol and drug dependence. *Psychopharmacology.* 2011;214(1):17–31. https://doi.org/10.1007/s00213-010-1916-6.
5. Schückher F, Sellin T, Engström I, Berglund K. History of childhood abuse is associated with less positive treatment outcomes in socially stable women with alcohol use disorder. *BMC Wom Health.* 2019;19(1):159. https://doi.org/10.1186/s12905-019-0857-4.

6. Ai AL, Lee J. Childhood abuse, religious involvement, and lifetime substance use disorders among latinas nationwide. *Subst Use Misuse*. 2018;53(13):2099–2111. https://doi.org/10.1080/10826084.2018.1455701.
7. Austin AE, Shanahan ME, Zvara BJ. Association of childhood abuse and prescription opioid use in early adulthood. *Addict Behav*. 2018;76:265–269. https://doi.org/10.1016/j.addbeh.2017.08.033.
8. Uhl GR. Molecular genetics of substance abuse vulnerability: remarkable recent convergence of genome scan results. *Ann N Y Acad Sci*. 2004;1025:1–13. https://doi.org/10.1196/annals.1316.001.
9. Dick DM, Agrawal A. The genetics of alcohol and other drug dependence. *Alcohol Res Health*. 2008;31(2):111–118. http://pubs.niaaa.nih.gov/publications/arh312/111-118.pdf.
10. Kendler KS, Chen X, Dick D, et al. Recent advances in the genetic epidemiology and molecular genetics of substance use disorders. *Nat Neurosci*. 2012;15(2):181–189. https://doi.org/10.1038/nn.3018.
11. Goldberg AD, Allis CD, Bernstein E. Epigenetics: a landscape takes shape. *Cell*. 2007;128(4):635–638. https://doi.org/10.1016/j.cell.2007.02.006.
12. Wang G, Allis CD, Chi P. Chromatin remodeling and cancer, Part I: covalent histone modifications. *Trends Mol Med*. 2007;13(9):363–372.
13. Hake S, Xiao A, Allis CD. Linking the epigenetic "language" of covalent histone modifications to cancer. *Br J Cancer*. 2004;90:761–769. https://doi.org/10.1038/sj.bjc.6601575.
14. Vaissière T, Sawan C, Herceg Z. Epigenetic interplay between histone modifications and DNA methylation in gene silencing. *Mutat Res Rev Mutat Res*. 2008;659(1–2):40–48. https://doi.org/10.1016/j.mrrev.2008.02.004.
15. Cedar H, Bergman Y. Linking DNA methylation and histone modification: patterns and paradigms. *Nat Rev Genet*. 2009;10(5):295–304. https://doi.org/10.1038/nrg2540.
16. Renthal W, Nestler EJ. Histone acetylation in drug addiction. *Semin Cell Dev Biol*. 2009;20(4):387–394.
17. Greer EL, Shi Y. Histone methylation: a dynamic mark in health, disease and inheritance. *Nat Rev Genet*. 2012;13:343–357.
18. Ajonijebu DC, Abboussi O, Russell VA, Mabandla MV, Daniels WMU. Epigenetics: a link between addiction and social environment. *Cell Mol Life Sci*. 2017;74(15):2735–2747. https://doi.org/10.1007/s00018-017-2493-1.
19. Moore LD, Le T, Fan G. DNA methylation and its basic function. *Neuropsychopharmacology*. 2013;38:23–28.
20. Bernstein E, Allis CD. RNA meets chromatin. *Genes Dev*. 2005;19(14):1635–1655. https://doi.org/10.1101/gad.1324305.
21. Kim JB, Zaehres H, Wu G, et al. Pluripotent stem cells induced from adult neural stem cells by reprogramming with two factors. *Nature*. 2008;454:646–650.
22. Johnson GD, Lalancette C, Linnemann AK, Leduc F, Boissonneault G, Krawetz SA. The sperm nucleus: chromatin, RNA, and the nuclear matrix. *Reproduction*. 2011;141(1):21–36. https://doi.org/10.1530/REP-10-0322.
23. Vassoler FM, Sadri-Vakili G. Mechanisms of transgenerational inheritance of addictive-like behaviors. *Neuroscience*. 2014;264:198–206. https://doi.org/10.1016/j.neuroscience.2013.07.064.
24. Castel SE, Martienssen RA. RNA interference in the nucleus: roles for small RNAs in transcription, epigenetics and beyond. *Nat Rev Genet*. 2013;14(2):100–112. https://doi.org/10.1038/nrg3355.

25. Jessica C. RNA therapeutics: RNAi and antisense mechanisms and clinical applications. *Postdoc J.* 2016. https://doi.org/10.14304/SURYA.JPR.V4N7.5.
26. Heard E, Martienssen RA. Transgenerational epigenetic inheritance: myths and mechanisms. *Cell.* 2014;157(1):95–109. https://doi.org/10.1016/j.cell.2014.02.045.
27. Guerrero-Bosagna C, Settles M, Lucker B, Skinner MK. Epigenetic transgenerational actions of vinclozolin on promoter regions of the sperm epigenome. *PLoS One.* 2010;5(9):1–17. https://doi.org/10.1371/journal.pone.0013100.
28. Manikkam M, Guerrero-Bosagna C, Tracey R, Haque MM, Skinner MK. Transgenerational actions of environmental compounds on reproductive disease and identification of epigenetic biomarkers of ancestral exposures. *PLoS One.* 2012;7(2). https://doi.org/10.1371/journal.pone.0031901.
29. Manikkam M, Tracey R, Guerrero-Bosagna C, Skinner MK. Dioxin (TCDD) induces epigenetic transgenerational inheritance of adult onset disease and sperm epimutations. *PLoS One.* 2012;7(9). https://doi.org/10.1371/journal.pone.0046249.
30. Manikkam M, Tracey R, Guerrero-Bosagna C, Skinner MK. Pesticide and insect repellent mixture (permethrin and DEET) induces epigenetic transgenerational inheritance of disease and sperm epimutations. *Reprod Toxicol.* 2012;34(4):708–719. https://doi.org/10.1016/j.reprotox.2012.08.010.
31. Tracey R, Manikkam M, Guerrero-Bosagna C, Skinner MK. Hydrocarbons (jet fuel JP-8) induce epigenetic transgenerational inheritance of obesity, reproductive disease and sperm epimutations. *Reprod Toxicol.* 2013;36:104–116. https://doi.org/10.1016/j.reprotox.2012.11.011.
32. Yohn NL, Bartolomei MS, Blendy JA. Multigenerational and transgenerational inheritance of drug exposure: the effects of alcohol, opiates, cocaine, marijuana, andnicotine. *Prog Biophys Mol Biol.* 2015;118(1–2):21–33. https://doi.org/10.1016/j.pbiomolbio.2015.03.002.
33. Fernandez-Twinn DS, Constância M, Ozanne SE. Intergenerational epigenetic inheritance in models of developmental programming of adult disease. *Semin Cell Dev Biol.* 2015;43:85–95. https://doi.org/10.1016/j.semcdb.2015.06.006.
34. Martos SN, Tang Wy, Wang Z. Elusive inheritance: transgenerational effects and epigenetic inheritance in human environmental disease. *Prog Biophys Mol Biol.* 2015; 118(1–2):44–54. https://doi.org/10.1016/j.pbiomolbio.2015.02.011.
35. Perez MF, Lehner B. Intergenerational and transgenerational epigenetic inheritance in animals. *Nat Cell Biol.* 2019;21(2):143–151. https://doi.org/10.1038/s41556-018-0242-9.
36. Williams CT, Latkin CA. Neighborhood socioeconomic status, personal network attributes, and use of heroin and cocaine. *Am J Prev Med.* 2007;32(6):S203–S210. https://doi.org/10.1016/j.amepre.2007.02.006.
37. Miller DS, Miller TQ. A test of socioeconomic status as a predictor of initial marijuana use. *Addict Behav.* 1997;22(4):479–489. https://doi.org/10.1016/S0306-4603(96)00059-7.
38. Agrawal A, Lynskey MT. The genetic epidemiology of cannabis use, abuse and dependence. *Addiction.* 2006;101(6):801–812. https://doi.org/10.1111/j.1360-0443.2006.01399.x.
39. Kendler KS, Prescott CA, Myers J, Neale MC. The structure of genetic and environmental risk factors for common psychiatric and substance use disorders in men and women. *Arch Gen Psychiatr.* 2003;60(9):929–937. https://doi.org/10.1001/archpsyc.60.9.929.

40. Tsuang MT, Bar JL, Harley RM, Lyons MJ. The Harvard twin study of substance abuse: what we have learned. *Harv Rev Psychiatr.* 2001;9(6):267–279. https://doi.org/10.1080/10673220127912.
41. Yuan TF, Li A, Sun X, et al. Transgenerational inheritance of paternal neurobehavioral phenotypes: stress, addiction, ageing and metabolism. *Mol Neurobiol.* 2016;53(9): 6367–6376. https://doi.org/10.1007/s12035-015-9526-2.
42. Zeid D, Gould TJ. Impact of nicotine, alcohol, and cocaine exposure on germline integrity and epigenome. *Neuropharmacology.* 2020;173. https://doi.org/10.1016/j.neuropharm.2020.108127.
43. Prochaska JJ, Benowitz NL. The past, present, and future of nicotine addiction therapy. *Annu Rev Med.* 2016;67:467–486. https://doi.org/10.1146/annurev-med-111314-033712.
44. Drake P, Driscoll AK, Mathews TJ. Cigarette smoking during pregnancy: United States. *NCHS Data Brief.* 2016;305:1–8. PMID: 29528282.
45. Rauschert S, Melton PE, Burdge G, et al. Maternal smoking during pregnancy induces persistent epigenetic changes into adolescence, independent of postnatal smoke exposure and is associated with cardiometabolic risk. *Front Genet.* 2019;5(10):770.
46. Toledo-Rodriguez M, Lotfipour S, Leonard G, et al. Maternal smoking during pregnancy is associated with epigenetic modifications of the brain-derived neurotrophic factor-6 exon in adolescent offspring. *Am J Med Genet, Part B: Neuropsychiatric Genetics.* 2010;153(7):1350–1354. https://doi.org/10.1002/ajmg.b.31109.
47. Buck JM, Sanders KN, Wageman CR, Knopik VS, Stitzel JA, O'Neill HC. Developmental nicotine exposure precipitates multigenerational maternal transmission of nicotine preference and ADHD-like behavioral, rhythmometric, neuropharmacological, and epigenetic anomalies in adolescent mice. *Neuropharmacology.* 2019;149:66–82. https://doi.org/10.1016/j.neuropharm.2019.02.006.
48. Jung Y, Hsieh L, Lee AM, et al. An epigenetic mechanism mediates developmental nicotine effects on neuronal structure and behavior. *Nat Neurosci.* 2016;19(7):905–914.
49. Gozen O, Balkan B, Yildirim E, Koylu EO, Pogun S. The epigenetic effect of nicotine on dopamine D1 receptor expression in rat prefrontal cortex. *Synapse.* 2013;67(9): 545–552. https://doi.org/10.1002/syn.21659.
50. Launay JM, Del Pino M, Chironi G, et al. Smoking induces long-lasting effects through a monoamine-oxidase epigenetic regulation. *PLoS One.* 2009;4(11). https://doi.org/10.1371/journal.pone.0007959.
51. Philibert RA, Gunter TD, Beach SRH, Brody GH, Madan A. MAOA methylation is associated with nicotine and alcohol dependence in women. *Am J Med Genet B Neuropsychiatr Genet.* 2008:565–570. https://doi.org/10.1002/ajmg.b.30778.
52. Satta R, Maloku E, Zhubi A, et al. Nicotine decreases DNA methyltransferase 1 expression and glutamic acid decarboxylase 67 promoter methylation in GABAergic interneurons. *Proc Natl Acad Sci U S A.* 2008;105(42):16356–16361. https://doi.org/10.1073/pnas.0808699105.
53. National Institute on alcohol abuse and alcoholism. In: *Alcohol Facts and Statistics*; 2019. http://www.niaaa.nih.gov.
54. National Institute on alcohol abuse and alcoholism. In: *Alcohol Facts and Statistics*; 2018. http://www.niaaa.nih.gov.
55. Ramsay M. Genetic and epigenetic insights into fetal alcohol spectrum disorders. *Genome Med.* 2010;2(4). https://doi.org/10.1186/gm148.

56. Garro AJ, McBeth DL, Lima V, Lieber CS. Ethanol consumption inhibits fetal DNA methylation in mice: implications for the fetal alcohol syndrome. *Alcohol Clin Exp Res.* 1991;15(3):395–398. https://doi.org/10.1111/j.1530-0277.1991.tb00536.x.
57. Chen Y, Ozturk NC, Zhou FC. DNA methylation program in developing Hippocampus and its alteration by alcohol. *PLoS One.* 2013;8(3). https://doi.org/10.1371/journal.pone.0060503.
58. Bekdash RA, Zhang C, Sarkar DK. Gestational choline supplementation normalized fetal alcohol-induced alterations in histone modifications, DNA methylation, and proopiomelanocortin (POMC) gene expression in β-endorphin-producing POMC neurons of the hypothalamus. *Alcohol Clin Exp Res.* 2013;37(7):1133–1142. https://doi.org/10.1111/acer.12082.
59. Veazey KJ, Carnahan MN, Muller D, Miranda RC, Golding MC. Alcohol-induced epigenetic alterations to developmentally crucial genes regulating neural stemness and differentiation. *Alcohol Clin Exp Res.* 2013;37(7):1111–1122. https://doi.org/10.1111/acer.12080.
60. Montesinos J, Pascual M, Millán-Esteban D, Guerri C. Binge-like ethanol treatment in adolescence impairs autophagy and hinders synaptic maturation: role of TLR4. *Neurosci Lett.* 2018;682:85–91. https://doi.org/10.1016/j.neulet.2018.05.049.
61. Montesinos J, Pascual M, Pla A, et al. TLR4 elimination prevents synaptic and myelin alterations and long-term cognitive dysfunctions in adolescent mice with intermittent ethanol treatment. *Brain Behav Immun.* 2015;45:233–244. https://doi.org/10.1016/j.bbi.2014.11.015.
62. Pandey SC, Zhang H, Ugale R, Prakash A, Xu T, Misra K. Effector immediate-early gene arc in the amygdala plays a critical role in alcoholism. *J Neurosci.* 2008;28(10): 2589–2600. https://doi.org/10.1523/JNEUROSCI.4752-07.2008.
63. Sakharkar AJ, Zhang H, Tang L, Shi G, Pandey SC. Histone deacetylases (HDAC)-induced histone modifications in the amygdala: a role in rapid tolerance to the anxiolytic effects of ethanol. *Alcohol Clin Exp Res.* 2012;36(1):61–71. https://doi.org/10.1111/j.1530-0277.2011.01581.x.
64. Moonat S, Sakharkar AJ, Zhang H, Tang L, Pandey SC. Aberrant histone deacetylase2-mediated histone modifications and synaptic plasticity in the amygdala predisposes to anxiety and alcoholism. *Biol Psychiatr.* 2013;73(8):763–773. https://doi.org/10.1016/j.biopsych.2013.01.012.
65. Sakharkar AJ, Tang L, Zhang H, Chen Y, Grayson DR, Pandey SC. Effects of acute ethanol exposure on anxiety measures and epigenetic modifiers in the extended amygdala of adolescent rats. *Int J Neuropsychopharmacol.* 2014;17(12):2057–2067. https://doi.org/10.1017/S1461145714001047.
66. Kyzar EJ, Floreani C, Teppen TL, Pandey SC. Adolescent alcohol exposure: burden of epigenetic reprogramming, synaptic remodeling, and adult psychopathology. *Front Neurosci.* 2016;10. https://doi.org/10.3389/fnins.2016.00222. MAY.
67. Kyzar EJ, Pandey SC. Molecular mechanisms of synaptic remodeling in alcoholism. *Neurosci Lett.* 2015:11–19. https://doi.org/10.1016/j.neulet.2015.01.051.
68. Hummel M, Unterwald EM. D1 dopamine receptor: a putative neurochemical and behavioral link to cocaine action. *J Cell Physiol.* 2002;191(1):17–27. https://doi.org/10.1002/jcp.10078.
69. Kampman KM. The treatment of cocaine use disorder. *Sci Adv.* 2019;5.
70. Ryan SA. Cocaine use in adolescents and young adults. *Pediatr Clin.* 2019;66(6): 1135–1147. https://doi.org/10.1016/j.pcl.2019.08.014.

71. Schneider KE, Krawczyk N, Xuan Z, Johnson RM. Past 15-year trends in lifetime cocaine use among US high school students. *Drug Alcohol Depend.* 2018;183: 69–72. https://doi.org/10.1016/j.drugalcdep.2017.10.028.
72. John WS, Wu LT. Trends and correlates of cocaine use and cocaine use disorder in the United States from 2011 to 2015. *Drug Alcohol Depend.* 2017;180:376–384. https://doi.org/10.1016/j.drugalcdep.2017.08.031.
73. Sadri-Vakili G. Cocaine triggers epigenetic alterations in the corticostriatal circuit. *Brain Res.* 2015;1628:50–59. https://doi.org/10.1016/j.brainres.2014.09.069.
74. Vassoler FM, Johnson NL, Byrnes EM. Female adolescent exposure to cannabinoids causes transgenerational effects on morphine sensitization in female offspring in the absence of in utero exposure. *J Psychopharmacol.* 2013;27(11):1015–1022. https://doi.org/10.1177/0269881113503504.
75. Killinger CE, Robinson S, Stanwood GD. Subtle biobehavioral effects produced by paternal cocaine exposure. *Synapse.* 2012;66(10):902–908. https://doi.org/10.1002/syn.21582.
76. Pires A, Pieri P, Hage M, et al. Repeated inhalation of crack-cocaine affects spermatogenesis in young and adult mice. *Inhal Toxicol.* 2012;24(7):439–446.
77. Vassoler FM, White SL, Schmidt HD, Sadri-Vakili G, Christopher Pierce R. Epigenetic inheritance of a cocaine-resistance phenotype. *Nat Neurosci.* 2013;16(1):42–47. https://doi.org/10.1038/nn.3280.
78. Le Q, Li Y, Hou W, et al. Binge-like sucrose self-administration experience inhibits cocaine and sucrose seeking behavior in offspring. *Front Behav Neurosci.* 2017;11:184.
79. Smith JP, Book SW. Anxiety and substance use disorders: a review. *Psychiatr Times.* 2008;25(10):19–23.
80. White SL, Vassoler FM, Schmidt HD, Pierce RC, Wimmer ME. Enhanced anxiety in the male offspring of sires that self-administered cocaine. *Addiction Biol.* 2016;21(4): 802–810. https://doi.org/10.1111/adb.12258.
81. Sasaki A, Constantinof A, Pan P, Kupferschmidt DA, McGowan PO, Erb S. Cocaine exposure prior to pregnancy alters the psychomotor response to cocaine and transcriptional regulation of the dopamine D1 receptor in adult male offspring. *Behav Brain Res.* 2014;265:163–170. https://doi.org/10.1016/j.bbr.2014.02.017.
82. Novikova SI, He F, Bai J, Cutrufello NJ, Lidow MS, Undieh AS. Maternal cocaine administration in mice alters DNA methylation and gene expression in hippocampal neurons of neonatal and prepubertal offspring. *PLoS One.* 2008;3(4). https://doi.org/10.1371/journal.pone.0001919.
83. Anier K, Malinovskaja K, Aonurm-Helm A, Zharkovsky A, Kalda A. DNA methylation regulates cocaine-induced behavioral sensitization in mice. *Neuropsychopharmacology.* 2010;35(12):2450–2461. https://doi.org/10.1038/npp.2010.128.
84. Vaillancourt K, Ernst C, Mash D, Turecki G. DNA methylation dynamics and cocaine in the brain: progress and prospects. *Genes.* 2017;8(5). https://doi.org/10.3390/genes8050138.
85. Tian W, Zhao M, Li M, et al. Reversal of cocaine-conditioned place preference through methyl supplementation in mice: altering global DNA methylation in the prefrontal cortex. *PLoS One.* 2012;7(3). https://doi.org/10.1371/journal.pone.0033435.
86. Massart R, Barnea R, Dikshtein Y, et al. Role of DNA methylation in the nucleus accumbens in incubation of cocaine craving. *J Neurosci.* 2015;35(21):8042–8058. https://doi.org/10.1523/JNEUROSCI.3053-14.2015.

87. Han J, Li Y, Wang D, Wei C, Yang X, Sui N. Effect of 5-aza-2-deoxycytidine microinjecting into hippocampus and prelimbic cortex on acquisition and retrieval of cocaine-induced place preference in C57BL/6 mice. *Eur J Pharmacol.* 2010;642(1−3):93−98. https://doi.org/10.1016/j.ejphar.2010.05.050.
88. Levine AA, Guan Z, Barco A, Xu S, Kandel ER, Schwartz JH. CREB-binding protein controls response to cocaine by acetylating histones at the fosB promoter in the mouse striatum. *Proc Natl Acad Sci U S A.* 2005;102(52):19186−19191. https://doi.org/10.1073/pnas.0509735102.
89. Brami-Cherrier K, Roze E, Girault JA, Betuing S, Caboche J. Role of the ERK/MSK1 signalling pathway in chromatin remodelling and brain responses to drugs of abuse. *J Neurochem.* 2009;108(6):1323−1335. https://doi.org/10.1111/j.1471-4159.2009.05879.x.
90. Cleck JN, Ecke LE, Blendy JA. Endocrine and gene expression changes following forced swim stress exposure during cocaine abstinence in mice. *Psychopharmacology.* 2008;201(1):15−28. https://doi.org/10.1007/s00213-008-1243-3.
91. Schmidt HD, Sangrey GR, Darnell SB, et al. Increased brain-derived neurotrophic factor (BDNF) expression in the ventral tegmental area during cocaine abstinence is associated with increased histone acetylation at BDNF exon I-containing promoters. *J Neurochem.* 2012;120(2):202−209. https://doi.org/10.1111/j.1471-4159.2011.07571.x.
92. Freeman WM, Patel KM, Brucklacher RM, et al. Persistent alterations in mesolimbic gene expression with abstinence from cocaine self-administration. *Neuropsychopharmacology.* 2008;33(8):1807−1817. https://doi.org/10.1038/sj.npp.1301577.
93. Romieu P, Host L, Gobaille S, Sandner G, Aunis D, Zwiller J. Histone deacetylase inhibitors decrease cocaine but not sucrose self-administration in rats. *J Neurosci.* 2008; 28(38):9342−9348. https://doi.org/10.1523/JNEUROSCI.0379-08.2008.
94. Black YD, Maclaren FR, Naydenov AV, Carlezon WA, Baxter MG, Konradi C. Altered attention and prefrontal cortex gene expression in rats after binge-like exposure to cocaine during adolescence. *J Neurosci.* 2006;26(38):9656−9665. https://doi.org/10.1523/JNEUROSCI.2391-06.2006.
95. Maze I, Covington HE, Dietz DM, et al. Essential role of the histone methyltransferase G9a in cocaine-induced plasticity. *Science.* 2010;327(5962):213.
96. Covington HE, Maze I, Sun H, et al. A role for repressive histone methylation in cocaine-induced vulnerability to stress. *Neuron.* 2012;71(4):656−670.
97. Im HI, Hollander JA, Bali P, Kenny PJ. MeCP2 controls BDNF expression and cocaine intake through homeostatic interactions with microRNA-212. *Nat Neurosci.* 2010; 13(9):1120−1127. https://doi.org/10.1038/nn.2615.
98. Ferland JMN, Hurd YL. Deconstructing the neurobiology of cannabis use disorder. *Nat Neurosci.* 2020;23(5):600−610. https://doi.org/10.1038/s41593-020-0611-0.
99. Johnston LD. *Monitoring the Future National Survey Results on Drug Use 1975−2018: Overview, Key Findings on Adolescent Drug Use.* Ann Arbor: Institute for Social Research; 2019.
100. Hasin DS. US epidemiology of cannabis use and associated problems. *Neuropsychopharmacology.* 2018;43(1):195−212. https://doi.org/10.1038/npp.2017.198.
101. Panlilio LV, Justinova Z. Preclinical studies of cannabinoid reward, treatments for cannabis use disorder, and addiction-related effects of cannabinoid exposure.

Neuropsychopharmacology. 2018;43(1):116–141. https://doi.org/10.1038/npp.2017.193.

102. Scherma M, Qvist JS, Asok A, et al. Cannabinoid exposure in rat adolescence reprograms the initial behavioral, molecular, and epigenetic response to cocaine. *Proc Natl Acad Sci U S A.* 2020;117(18):9991–10002.
103. Tomas-Roig J, Benito E, Agis-Balboa RC, et al. Chronic exposure to cannabinoids during adolescence causes long-lasting behavioral deficits in adult mice. *Addiction Biol.* 2017;22(6):1778–1789. https://doi.org/10.1111/adb.12446.
104. Di Marzo V. New approaches and challenges to targeting the endocannabinoid system. *Nat Rev Drug Discov.* 2018;17(9):623–639.
105. Ta M, Greto L, Bolt K. Trends and characteristics in marijuana use among public school students—king county, Washington, 2004–2016. *MMWR Morb Mortal Wkly Rep.* 2019;68(39):845–850. https://doi.org/10.15585/mmwr.mm6839a3.
106. Midgette G, Reuter P. Has cannabis use among youth increased after changes in its legal status? A commentary on use of monitoring the future for analyses of changes in state cannabis laws. *Prev Sci.* 2020;21(1):137–145. https://doi.org/10.1007/s11121-019-01068-4.
107. Mennis J, Stahler GJ. Adolescent treatment admissions for marijuana following recreational legalization in Colorado and Washington. *Drug Alcohol Depend.* 2020;210. https://doi.org/10.1016/j.drugalcdep.2020.107960.
108. Walker OLS, Holloway AC, Raha S. The role of the endocannabinoid system in female reproductive tissues. *J Ovarian Res.* 2019;12(1).
109. Meccariello R. The epigenetics of the endocannabinoid system. *Int J Mol Sci.* 2020; 21(3).
110. Innocenzi E. Paternal activation of CB(2) cannabinoid receptor impairs placental and embryonic growth via an epigenetic mechanism. 2019;9.
111. Murphy SK, Hawkey AB, Hall BJ, et al. Paternal THC exposure in rats causes long-lasting neurobehavioral effects in the offspring. *Neurotoxicol Teratol.* 2019;74. https://doi.org/10.1016/j.ntt.2019.04.003.
112. Sansone A, Di Dato C, de Angelis C, et al. Smoke, alcohol and drug addiction and male fertility. *Reprod Biol Endocrinol.* 2018;16(1). https://doi.org/10.1186/s12958-018-0320-7.
113. Szutorisz H, Hurd YL. Epigenetic effects of cannabis exposure. *Biol Psychiatr.* 2016; 79(7):586–594. https://doi.org/10.1016/j.biopsych.2015.09.014.
114. Watson CT, Szutorisz H, Garg P, et al. Genome-wide DNA methylation profiling reveals epigenetic changes in the rat nucleus accumbens associated with cross-generational effects of adolescent THC exposure. *Neuropsychopharmacology.* 2015;40(13): 2993–3005. https://doi.org/10.1038/npp.2015.155.
115. Wang Z, Yan P, Hui T, Zhang J. Epigenetic upregulation of PSD-95 contributes to the rewarding behavior by morphine conditioning. *Eur J Pharmacol.* 2014;732(1): 123–129. https://doi.org/10.1016/j.ejphar.2014.03.040.
116. Pitsilis G, Spyridakos D, Nomikos GG, Panagis G. Adolescent female cannabinoid exposure diminishes the reward-facilitating effects of Δ9-tetrahydrocannabinol and d-amphetamine in the adult male offspring. *Front Pharmacol.* 2017;8. https://doi.org/10.3389/fphar.2017.00225.
117. Byrnes JJ, Johnson NL, Schenk ME, Byrnes EM. Cannabinoid exposure in adolescent female rats induces transgenerational effects on morphine conditioned place preference

in male offspring. *J Psychopharmacol.* 2012;26(10):1348–1354. https://doi.org/10.1177/0269881112443745.

118. Szutorisz H, DiNieri JA, Sweet E, et al. Parental THC exposure leads to compulsive heroin-seeking and altered striatal synaptic plasticity in the subsequent generation. *Neuropsychopharmacology.* 2014;39(6):1315–1323. https://doi.org/10.1038/npp.2013.352.
119. Zumbrun EE, Sido JM, Nagarkatti PS, Nagarkatti M. Epigenetic regulation of immunological alterations following prenatal exposure to marijuana cannabinoids and its long term consequences in offspring. *J Neuroimmune Pharmacol.* 2015;10(2):245–254. https://doi.org/10.1007/s11481-015-9586-0.
120. Dinieri JA, Wang X, Szutorisz H, et al. Maternal cannabis use alters ventral striatal dopamine D2 gene regulation in the offspring. *Biol Psychiatr.* 2011;70(8):763–769. https://doi.org/10.1016/j.biopsych.2011.06.027.
121. Tomasiewicz HC, Jacobs MM, Wilkinson MB, Wilson SP, Nestler EJ, Hurd YL. Proenkephalin mediates the enduring effects of adolescent cannabis exposure associated with adult opiate vulnerability. *Biol Psychiatr.* 2012;72(10):803–810. https://doi.org/10.1016/j.biopsych.2012.04.026.
122. Lachman HM, Papolos DF, Saito T, Yu YM, Szumlanski CL, Weinshilboum RM. Human catechol-O-methyltransferase pharmacogenetics: description of a functional polymorphism and its potential application to neuropsychiatric disorders. *Pharmacogenetics.* 1996;6(3):243–250. https://doi.org/10.1097/00008571-199606000-00007.
123. Prini P, Rusconi F, Zamberletti E, et al. Adolescent THC exposure in female rats leads to cognitive deficits through a mechanism involving chromatin modifications in the prefrontal cortex. *J Psychiatry Neurosci : JPN.* 2017;42(6):170082. https://doi.org/10.1503/jpn.170082.
124. van der Knaap LJ, Schaefer JM, Franken IHA, Verhulst FC, van Oort FVA, Riese H. Catechol-O-methyltransferase gene methylation and substance use in adolescents: the TRAILS study. *Gene Brain Behav.* 2014;13(7):618–625. https://doi.org/10.1111/gbb.12147.
125. Lombard C, Hegde VL, Nagarkatti M, Nagarkatti PS. Perinatal exposure to Δ 9-tetrahydrocannabinol triggers profound defects in T cell differentiation and function in fetal and postnatal stages of life, including decreased responsiveness to HIV antigens. *J Pharmacol Exp Therapeut.* 2011;339(2):607–617. https://doi.org/10.1124/jpet.111.181206.
126. Chandra LC, Kumar V, Torben W, et al. Chronic administration of Δ9-tetrahydrocannabinol induces intestinal anti-inflammatory microRNA expression during acute simian immunodeficiency virus infection of rhesus macaques. *J Virol.* 2015;89(2):1168–1181. https://doi.org/10.1128/JVI.01754-14.
127. Molina PE, Amedee A, Lecapitaine NJ, et al. Cannabinoid neuroimmune modulation of SIV disease. *J Neuroimmune Pharmacol.* 2011;6(4):516–527. https://doi.org/10.1007/s11481-011-9301-8.
128. Yang X, Hegde VL, Rao R, Zhang J, Nagarkatti PS, Nagarkatti M. Histone modifications are associated with 9-tetrahydrocannabinol-mediated alterations in antigenspecific t cell responses. *J Biol Chem.* 2014;289(27):18707–18718. https://doi.org/10.1074/jbc.M113.545210.
129. Crews FT. Immune function genes, genetics, and the neurobiology of addiction. *Alcohol Res Curr Rev.* 2012;34(3):355–361.
130. Blanco C. America's opioid crisis: the need for an integrated public health approach. *Transl Psychiatry.* 2020;10(1).

131. Ho MK, Goldman D, Heinz A, et al. Breaking barriers in the genomics and pharmacogenetics of drug addiction. *Clin Pharmacol Therapeut.* 2010;88(6):779–791. https://doi.org/10.1038/clpt.2010.175.
132. Byrnes EM. Transgenerational consequences of adolescent morphine exposure in female rats: effects on anxiety-like behaviors and morphine sensitization in adult offspring. *Psychopharmacology.* 2005;182(4):537–544. https://doi.org/10.1007/s00213-005-0122-4.
133. Byrnes JJ, Johnson NL, Carini LM, Byrnes EM. Multigenerational effects of adolescent morphine exposure on dopamine D2 receptor function. *Psychopharmacology.* 2013;227(2):263–272. https://doi.org/10.1007/s00213-012-2960-1.
134. Cicero TJ, Adams ML, Giordano A, Miller BT, O'Connor L, Nock B. Influence of morphine exposure during adolescence on the sexual maturation of male rats and the development of their offspring. *J Pharmacol Exp Therapeut.* 1991;256(3):1086–1093.
135. Bonoiu AC, Mahajan SD, Ding H, et al. Nanotechnology approach for drug addiction therapy: gene silencing using delivery of gold nanorod-siRNA nanoplex in dopaminergic neurons. *Proc Natl Acad Sci U S A.* 2009;106(14):5546–5550. https://doi.org/10.1073/pnas.0901715106.
136. Wachman EM, Farrer LA. The genetics and epigenetics of neonatal abstinence syndrome. *Semin Fetal Neonatal Med.* 2019;24(2):105–110. https://doi.org/10.1016/j.siny.2019.01.002.
137. Logan BA, Brown MS, Hayes MJ. Neonatal abstinence syndrome: treatment and pediatric outcomes. *Clin Obstet Gynecol.* 2013;56(1):186–192. https://doi.org/10.1097/GRF.0b013e31827feea4.
138. Wachman EM, Hayes MJ, Sherva R, et al. Association of maternal and infant variants in PNOC and COMT genes with neonatal abstinence syndrome severity. *Am J Addict.* 2017;26(1):42–49. https://doi.org/10.1111/ajad.12483.
139. Wachman EM, Hayes MJ, Sherva R, et al. Variations in opioid receptor genes in neonatal abstinence syndrome. *Drug Alcohol Depend.* 2015;155:253–259. https://doi.org/10.1016/j.drugalcdep.2015.07.001.
140. Doehring A, Oertel BG, Sittl R, Lötsch J. Chronic opioid use is associated with increased DNA methylation correlating with increased clinical pain. *Pain.* 2013;154(1):15–23. https://doi.org/10.1016/j.pain.2012.06.011.
141. Nielsen DA, Utrankar A, Reyes JA, Simons DD, Kosten TR. Epigenetics of drug abuse: predisposition or response. *Pharmacogenomics.* 2012;13(10):1149–1160. https://doi.org/10.2217/pgs.12.94.
142. Chorbov VM, Todorov AA, Lynskey MT, Cicero TJ. Elevated levels of DNA methylation at the OPRM1 promoter in blood and sperm from male opioid addicts. *J Opioid Manag.* 2011;7(4):258–264. https://doi.org/10.5055/jom.2011.0067.
143. Browne CJ, Godino A, Salery M, Nestler EJ. Epigenetic mechanisms of opioid addiction. *Biol Psychiatr.* 2020;87(1):22–33. https://doi.org/10.1016/j.biopsych.2019.06.027.
144. Hui B, Wang W, Li J. Biphasic modulation of cocaine-induced conditioned place preference through inhibition of histone acetyltransferase and histone deacetylase. *Saudi Med J.* 2010;31(4):389–393.
145. Romieu P, Deschatrettes E, Host L, Gobaille S, Sandner G, Zwiller J. The inhibition of histone deacetylases reduces the reinstatement of cocaine-seeking behavior in rats. *Curr Neuropharmacol.* 2011;9(1):21–25. https://doi.org/10.2174/157015911795017317.
146. Nestler EJ, Luscher C. The molecular basis of drug addiction: linking epigenetic to synaptic and circuit mechanisms. *Neuron.* 2019;102(1):48–59.

CHAPTER

11 Paternal and maternal environmental influences on offspring health: inter- and transgenerational epigenetic inheritance

Simeiyun Liu[1], Joshua F. Shaffer[1], Upasna Sharma

Department of Molecular, Cell and Developmental Biology, University of California, Santa Cruz, CA, United States

Introduction

Can our life experiences influence health and longevity in our future generations? Indeed, there is mounting evidence, from worms to humans, that parental environment can influence phenotypes in descendants. Such across-generational inheritance of parental environmental effects is sometimes also referred to as Lamarckian inheritance or inheritance of acquired characters. While there are many documented cases of transgenerational inheritance, the mechanism of such inheritance remains poorly understood. Historically, the inheritance of acquired traits was discredited by many scientists as there was no known mechanism by which the environment can change genetic material (DNA) to transmit environmental information to offspring. With advances in the field of epigenetics—inheritance that is independent of the underlying DNA sequence—there is a resurgent interest in transgenerational inheritance. Consequently, the past decade has seen a sharp surge in the number of studies examining parental exposure regimens, including but not limited to undernutrition, overnutrition, traumatic stress, and exposure to toxicants, and their influence on future generations.

Mechanistically, it is proposed that the environment can modulate epigenetic marks, including histone modifications, DNA cytosine methylation, and small noncoding RNAs, in gametes—sperm in males and eggs in females—and thus, influence phenotypes in offspring.[1] While epigenetic inheritance is well documented in model organisms such as worms and plants, such inheritance is considered challenging in mammals as mammalian epigenome undergoes global erasure and reprogramming

[1] Equal contribution.

Perinatal and Developmental Epigenetics. https://doi.org/10.1016/B978-0-12-821785-6.00005-0

during germ cell development and soon after fertilization. Recent studies reveal that certain epigenetic marks escape global reprogramming and, therefore, have the potential to serve as epigenetic information carriers to future generations.

The studies on parental exposure mostly fall under two categores (Fig. 11.1).[2] 1) intergenerational inheritance: inheritance mediated via direct exposure of either the germ cells and/or the fetus to the parental environment. For example, if the female (F0) is pregnant with male offspring, both its fetus (F1) and germ cells of the fetus (F2) will be directly exposed to the parental environment. In cases where a specific environmental exposure is introduced after birth, the germ cells (F1) will be directly exposed. Thus, intergenerational inheritance is mediated by epigenetic changes occurring in the gametes that can shape early embryonic development and phenotypes in the next generation. 2) transgenerational inheritance: inheritance of parental environmental effects in the generations that are not directly exposed to the particular environment. Depending on the timing of exposure, transgenerational inheritance is mostly observed in F3 and onward generations in the case of maternal inheritance, and F2 and onwards in the case of paternal inheritance. In mammals, there are fewer examples of pure transgenerational inheritance (discussed below), as compared to model organisms such as worms, where RNAi-mediated transgenerational silencing is well characterized.[3]

Another major disapproval of across-generational epigenetic inheritance in mammals comes from the so-called Weismann Barrier, which is proposed to prevent any flow of information from somatic cells to germ cells, and hence, the transmission of environmental information to the germ cells and offspring. We discuss recent

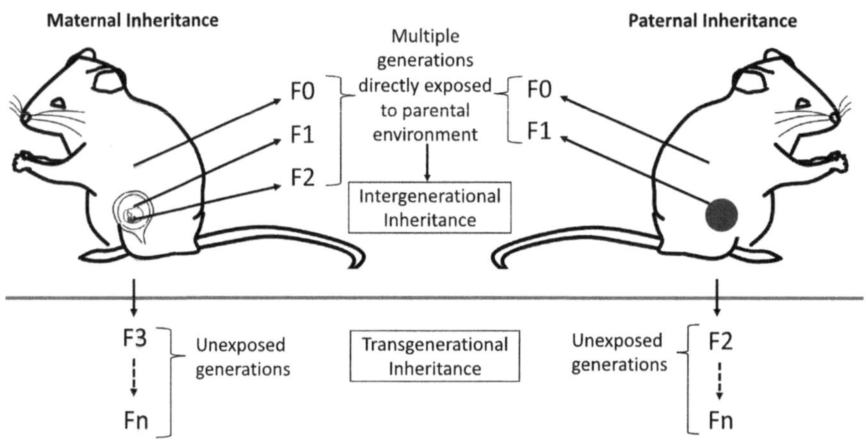

FIGURE 11.1 Intergenerational and Transgenerational Inheritance.

In the case of exposure of a pregnant female (F0), both its fetus (F1) and germ cells of the fetus (F2) are directly exposed to the parental environment. Such inheritance is mediated by epigenetic changes occurring in the gametes and constitutes intergenerational inheritance. Thus, inheritance of phenotypes in F3 generation will be considered true transgenerational inheritance (no exposure to parental environment). In males, exposure of adult (F0) to a specific environment will also expose the germline (F1), and therefore, inheritance in F2 and onward generations will be considered transgenerational inheritance.

studies on paternal epigenetic inheritance that demonstrate that small RNAs are potentially delivered from somatic cells to sperm via small extracellular vesicles. These studies suggest that epigenetic information (in the form of small RNAs) can be transmitted to the germline and provide a potential mechanism of how specific environmental information can be signaled to the germ cells to influence offspring health.

While recent studies shed some light on the mechanism of epigenetic inheritance of parental environment, many fundamental questions remain unanswered. In this chapter, we discuss the current understanding and explore potential mechanisms of across-generational inheritance by the paternal and maternal germline in mammals. Specifically, we discuss (1) various examples of inter- and transgenerational inheritance, (2) epigenetic marks in gametes and how the environment influences those epigenetic marks, and (3) how gametic epigenetic marks influence offspring phenotypes.

Germline epigenome and reprogramming

The three most well-characterized epigenetic marks in sperm and eggs include DNA cytosine methylation, histone modifications, and small noncoding RNAs. Establishment of proper epigenetic marks is essential for cell state commitment and development. Similarly, reprogramming of epigenetic marks is vital for the generation of a totipotent embryo, which has the potential to give rise to all cell types of the body. The germline epigenetic marks undergo global erasure during two waves: during germ cell differentiation and soon after fertilization. However, a fraction of the genome escapes global remodeling, providing an opportunity for transmission of environment-induced epigenetic changes to offspring.[4]

DNA methylation

Methylation of DNA at the 5th carbon on cytosine results in 5-methyl cytosine (5mC), sometimes also known as the fifth base. DNA cytosine methylation is one of the most well-characterized epigenetic marks in mammals and is associated with gene repression.[5] DNA methylation is globally erased and reset during two rounds of reprogramming, one during gametogenesis in primordial germ cells and second soon after fertilization in preimplantation embryos.[6] However, specific genomic regions escape global erasure of 5mC during these rounds of reprogramming. These regions include repeat elements enriched regions, such as Alu elements in humans and intracisternal A particle (IAP) elements in mice. Also, certain single copy regions escape erasure of 5mC.[4] Thus, DNA methylation can also serve as a potential epigenetic information carrier of paternal environmental information.

Histone modifications

In eukaryotes, DNA is wrapped around an octamer of histone proteins to form a nucleoprotein complex known as chromatin.[7] Chromatin formation alters the accessibility of the underlying DNA sequence to transcription machinery, thereby regulating gene expression. Posttranslational modifications of histones can further regulate gene expression by recruiting chromatin-modifying enzymes, activators, or repressors. During mammalian spermatogenesis, small basic proteins known as protamines replace the canonical histones.[8] Recent studies reveal that a subset of genomic loci escape this replacement. The specific loci that retain histones remain unclear to some extent, as some studies report histone retention on developmentally important gene promoters.[9,10] In contrast, others observed histone retention in the gene-poor regions of the genome.[11,12] Although the reason behind these conflicting results remains to be fully understood, it is clear that at least a subset of genomic loci retain histones in mature sperm and can, therefore, serve as carriers of paternal epigenetic information to the next generation.

Small RNAs

Recent studies provide evidence of the role of small RNAs in mammalian epigenetic inheritance. The major classes of small RNAs include microRNAs (miRNAs), endogenous small interfering RNAs (endo-siRNAs), and germline-enriched Piwi-interacting RNAs. Recent advances in RNA sequencing have also identified cleavage products of noncoding RNAs such as rRNAs and tRNAs. Small noncoding RNAs can regulate gene expression by regulating (1) transcription, (2) posttranscriptional gene expression by target mRNA degradation or deadenylation, and (3) translational repression.

As mature mammalian sperm shed most of their cytoplasm during spermiogenesis and they are transcriptionally silent, they were considered to be devoid of any functional RNA. Recent advances in sequencing technology revealed that sperm carry abundant RNAs, and at least a subset of these RNAs have been reported to function at fertilization. Early studies on human sperm revealed that sperm carry various RNA species,[13] which are potentially delivered to the oocyte at the time of fertilization.[14] Recent studies revealed that mature mammalian sperm have a unique RNA payload, comprising chiefly of tRNA-derived fragments (tRFs), with a lower abundance of miRNAs and piRNAs.[15–17] A comprehensive database for sperm RNAs has been generated as SpermBase, which curates both long (>200 nt) and small RNAs (<200 nt) in sperm of various species.[18] Interestingly, abundant tRFs in mature sperm have been reported in humans, mouse, rat, and bull, demonstrating that sperm-specific high abundance of tRFs is a widespread phenomenon. Recent studies discovered that sperm small RNA payload undergoes dramatic changes during posttesticular sperm maturation. In mice, the testicular germ cells are highly abundant in piRNAs.[19–23] After completing spermatogenesis in the testis, spermatozoa enter a tubule known as the epididymis, where they undergo maturation and become motile. Sperm small RNA payload is dramatically altered during epididymal sperm maturation, resulting in the mature epididymal sperm chiefly comprised of tRFs.[15,16,23,24] Mature sperm are also abundant in miRNAs, and it has been recently reported that sperm miRNAs regulate embryonic development in a strain-specific manner.[25,26]

Mus musculus oocytes also contain a range of different types of small RNAs, including tRNA and rRNA fragments, piRNAs, miRNAs, snoRNAs, and endo-siRNAs.[27] Endo-siRNAs appear to be the most abundant small RNA species in *M. musculus* oocytes. The oocytes of *Homo sapiens* also contain a variety of types of small RNAs though by far, the most abundant are oocyte short-Piwi-interacting small RNAs (os-piRNAs), which are not present in mouse oocytes but are present in monkey (*Macaca fascicularis*) oocytes.[28]

It should be noted that small RNA sequencing technology is still not advanced enough to capture all classes of small RNAs (especially those with modified ends or bases), and therefore, there very well could be other abundant classes of small RNAs in sperm and oocytes that are yet to be discovered.

Germline epigenetic inheritance in mammals

Genomic imprinting is one of the clearest examples of epigenetic inheritance in mammals, wherein certain genes are expressed in a parent-of-origin specific manner with the other allele being silenced epigenetically. After fertilization, the paternal genome is actively demethylated, while the maternal genome undergoes passive demethylation in a replication-dependent manner. This round of reprogramming is essential to generate a totipotent embryo from differentiated germ cells. The imprinting control regions escape this reprogramming to maintain parent-of-origin-specific gene expression.[29]

Epigenetic inheritance has also been observed at the *agouti viable yellow* (A^{vy}) allele locus. In the viable yellow (A^{vy}/a) mice, transcription originating in the IAP retrotransposon inserted upstream of the agouti gene (A) regulates ectopic expression of agouti protein, which results in yellow fur, obesity, diabetes, and increased susceptibility to tumors.[30] DNA methylation at the IAP elements regulates expression of the agouti gene; methylation results in wild-type coat color, whereas loss of methylation leads to yellow coat color. However, the degree of DNA methylation at the IAP element varies between individuals, resulting in coat color that varies in the spectrum from entirely yellow to full agouti and in between. Intriguingly, the coat color phenotype in offspring was related to mother's phenotype; if the mother had a yellow coat color, then the offspring were more likely to have yellow coat color.[31] These studies suggest that the A^{vy} locus either escapes or undergoes incomplete erasure of epigenetic marks in the female germline, and thus, allows epigenetic inheritance of the maternal phenotype.

Another earlier example of epigenetic inheritance in mammals comes from paramutation in the *kit* gene in mice. Paramutation is an epigenetic phenomenon where one silenced allele (paramutagenic allele) can silence the other homologous allele (paramutable allele) in trans, and these gene expression states are stably inherited. The insertion of LacZ in the *kit* locus results in white spots on the belly, feet, and tail. The offspring of male and female heterozygous for the mutation (*Kit tm1Alf*), despite a homozygous wild-type genotype, display to a variable extent the white spot phenotype characteristic of *Kit* mutant animals. Importantly, injection of RNA isolated from *kit* mutant mice into control zygotes resulted in the transmission of white-tailed phenotype in offspring generated from those zygotes, demonstrating a role of RNA in the epigenetic inheritance in mammals.[32]

Recent studies also provide evidence that changes in germline histone modifications can affect offspring phenotypes. Overexpression of a histone H3K4 demethylase (KDM1A) during spermatogenesis leads to a global loss of H3K4 methylation in sperm and altered development in offspring, including those that do not inherit the transgene *per se*[33] Similarly, the deletion of histone H3K27 demethylase KDM6A in paternal germline resulted in increased tumor incidence in wild-type offspring.[34] Importantly, altered histone modifications observed in sperm of these mice are also accompanied by small RNA[33] and cytosine methylation[34] changes, suggesting a cross-talk between different epigenetic information carriers.

Inheritance of paternal environmental exposure effects

Numerous studies suggest that the paternal environment can impact offspring's development and vulnerability to various pathological conditions. For example, human epidemiological studies indicate that ancestral nutrient availability can affect metabolic phenotypes in descendants. A study conducted on three generations of families from Overkalix in northern Sweden suggested that grand paternal food availability was linked to mortality rates in grandsons.[35–37] The Overkalix cohort study was aimed at investigating if overeating during a child's slow growth period (SGP) (prepubertal period) can increase the risk of death due to cardiovascular diseases in descendants. The data were collected from three cohorts born in 1890, 1905, and 1920. Historical records of harvests, food prices, and any other historical facts available were used to determine parental and grandparental food availability during their SGP. The risk of death due to diabetes increased if the grandfather was exposed to a surfeit of food during his SGP. These studies suggest that paternal diet can influence metabolic phenotypes in future generations. In the past decade, a multitude of rodent studies have focused on environmental effects transmitted via the paternal lineage to elucidate the mechanism of such inheritance. As with human epidemiological studies, rodent studies also demonstrate that paternal environment can influence phenotypes in offspring. For example, one of the early studies on preconception fasting in male mice revealed a consistent decrease in average serum glucose levels in both male and female offspring.[38]

As discussed before, paternal epigenetic inheritance can be categorized into intergenerational and transgenerational inheritance depending on the developmental stage of exposure (Fig. 11.1). In the case of exposure to the gestating female (F0), the male embryo/fetus (F1) and germline of the fetus (F2) are directly exposed. Hence, any paternal effects observed in F1 and F2 generations will be defined as intergenerational inheritance. F3 generation will be the first generation not directly exposed to the environmental insult, and thus, phenotypes seen in F3 or later generations will be characterized as transgenerational inheritance. In contrast, postnatal or adult exposure involves direct exposure of the male (F0) and its germline (F1), in which case inheritance of phenotypes in the F2 and later generations will be defined as transgenerational inheritance.

In the case of maternal inheritance (discussed in later sections), due to long-term interaction of the fetus with mother, it is challenging to distinguish between direct exposure of the environment to the embryo/fetus and molecular changes brought about by the environment in the oocytes. Although other factors in the seminal fluid could play a role in hereditary (such as proteins, vesicles, lipids, ions, etc.), fathers mostly contribute just sperm at the time of fertilization. Therefore, studying paternal inheritance is more straightforward, and consequently, a vast majority of studies on across-generational epigenetic inheritance are focused on the male germline. Nonetheless, there are some exciting studies on maternal inheritance, and those studies will be discussed in the later sections of this chapter.

Paternal environmental influences on offspring health

Several paternal environmental exposures, including diet, stress, and toxicant exposure, have been reported to influence offspring health. In most of the studies, males are exposed to a specific environment (undernutrition, overnutrition, stress, toxicants, etc.) either during fetal development, from weaning to sexual maturity or adulthood. Exposed males are next mated with unexposed females, and their offspring are examined for phenotypic changes such as changes in metabolism, behavior, etc. While there is an ever-increasing list of studies focusing on inheritance of paternal environmental effects, here we discuss a subset of studies in detail to describe general experimental setup and major conclusions.

Paternal dietary exposure

The influence of paternal diet or nutrient availability on F1 and F2 offspring has been studied extensively in mice and rats. These studies include dietary alterations, such as high fat, low protein, or western diet.[15,24,38–46] For instance, various studies have reported that paternal high-fat diet is correlated with glucose intolerance, obesity, weight gain, and reproductive problems in offspring.[40,41,43,47–50] Males fed a low protein diet sire offspring with altered lipid and cholesterol metabolism and glucose tolerance.[39,44,51] Preconception caloric restriction alters anxiety-like behavior in adult rat offspring.[52] Similarly, intermittent fasting in males resulted in increased glucose serum levels in offspring.[38]

In the studies mentioned before, the males were exposed from weaning onward; numerous studies have also examined the effects of in utero exposure of male offspring. Exposure during gestation is crucial to study as some of the major epigenetic marks, such as DNA methylation, are established in primordial germ cells during fetal development. In these studies, pregnant females were exposed to undernutrition in the form of starvation or low-protein diet[46,53,54] and overnutrition with a high-fat diet.[55–57] The exposed male offspring (F0) were then mated with control females to study inheritance in F1 offspring. The offspring of in utero exposed males (F1) also displayed altered metabolism and behavioral phenotypes.

Paternal exposure to traumatic stress

Paternal environmental exposures that can impact offspring health and development are not restricted to diet; paternal stress, such as social defeat stress, early-life trauma, and chronic variable stress, have also been reported to have a penetrant effect on offspring phenotypes.[58–62] These studies report that paternal stress is associated with reduced stress sensitivity in offspring. In some cases, altered cortisol levels and glucose metabolism are also observed in offspring. For example, one study investigated the transmission of paternal social defeat stress in F1 offspring.[62] Adult male mice were exposed to an aggressive male mouse for 10 min/day, after which they were separated by a plexiglass barrier that prevents physical contact but allows sensory stimulation. The control males were housed the same way but in the absence of exposure to an aggressive male. After 10 days of social defeat exposure and 1 month of recovery, the males were mated with control females. The male offspring sired by defeated fathers displayed depressive and anxiety-like behavior, such as social avoidance and increased plasma levels of stress-responsive hormone corticosterone.[62] However, in this study, transmission was not observed when offspring were generated by in vitro fertilization (IVF),[62] suggesting a nonepigenetic mechanism of inheritance (see later).

Nonetheless, there are more recent studies that implicate a role of epigenetic information carriers in sperm in transmission of paternal stress effects and will be discussed in the next section on mechanisms of inheritance of paternal environmental effects. One such study examined maternal separation combined with maternal stress (MSUS) paradigm, wherein male mice were exposed to MSUS in childhood and then mated with control females in adulthood. Offspring of MSUS fathers produced lower levels of corticosterone and displayed fewer anxiety-like behavior compared with control offspring.[58]

Paternal toxicant exposure

Other environmental stimuli investigated in paternal inheritance include toxicants and drugs, such as fungicide vinclozolin, alcohol, and nicotine. One of the earliest studies on epigenetic inheritance focused on the transmission of effects of exposure of pregnant female rats to endocrine disruptor vinclozolin. Exposure to vinclozolin resulted in decreased sperm number and motility in male offspring, and this defect was transmitted to males until F4 generation,[63] suggesting transgenerational inheritance of paternal environmental effects.

Recent studies have also documented transmission of altered behavioral and physiological phenotypes when fathers are exposed to toxicants such as carbon tetrachloride and drugs such as alcohol, nicotine, and cocaine.[64–67] For example, paternal preconception exposure to nicotine induces a higher metabolic tolerance for nicotine.[64] Interestingly, it was reported that the offspring of nicotine-exposed males are also tolerant to higher levels of cocaine.[64] Mechanistically, offspring of nicotine-exposed males displayed increased expression of genes involved in xenobiotic clearance, thus allowing better tolerance to both nicotine and cocaine. Therefore, exposure to nicotine seems adaptive as it provides a protective response in the next generation. It is important to note that the adaptive response was only observed in offspring that were preexposed to nicotine, suggesting that the phenotypes observed in offspring of exposed fathers are also influenced by the environment of the offspring.[64]

Mechanisms of inheritance of paternal environmental effects

As discussed before, numerous studies demonstrate that paternal environmental conditions can influence phenotypes in the future generations. However, the mechanism of such inheritance remains poorly understood. In this section, we discuss some of the recent studies that shed light on potential mechanisms of inheritance of paternal environmental effects and discuss outstanding questions in the field.

As fathers mainly contribute sperm at the time of fertilization, sperm is the most likely carrier of paternal environmental information. However, other factors such as the maternal judgment of male quality, factors in the seminal fluid (lipids, proteins, extracellular vesicles), or even the microbiome could have an effect on offspring phenotypes.[68,69] Use of assisted reproduction techniques for generation of F1 offspring, such as IVF or intracytoplasmic sperm injections (ICSI) where parents only contribute their gametes, allows a direct investigation of the role of sperm (or oocytes in the case of female germline inheritance) in the transmission of paternal environmental information. Indeed, recent studies report transmission of paternal environmental effects to offspring generated using IVF or ICSI, similar to those transmitted by natural mating.[15,24,41,69] These studies indicate that at least a subset of paternal environmental information can be transmitted to offspring via sperm. These studies bring the focus to the three most well-characterized epigenetic information carriers in sperm including, DNA cytosine methylation, histone modifications, and small noncoding RNAs. In the following, we discuss the potential mechanisms of epigenetic inheritance of paternal environmental effects via sperm epigenome.

DNA methylation

DNA methylation is the most well-established epigenetic information carrier in mammals. As a result, numerous studies have examined changes in DNA methylation in sperm in response to altered environmental conditions. For example, the mechanistic basis of intergenerational inheritance of paternal exposure to in utero undernutrition has been examined by a few studies. F1 offspring of females exposed to undernutrition display low birth weight, early-life adiposity, reduced muscle stem cell number and function, impaired pancreatic function, and progressive glucose intolerance.[54,70,71] Importantly, reduced birth weight and glucose intolerance were inherited in F2 generation through the paternal line, in the absence of any further environmental perturbation.[54] As the period of nutritional restriction coincides with the reacquisition of methylation in the male germ cells, Radford et al. focused on DNA methylation as the potential carrier of this paternal environmental information. Using immunoprecipitation of methylated DNA combined with high-throughput sequencing, they assessed the whole-genome distribution of methylation in F1 sperm. Overall, 111 hypomethylated regions were reported in undernutrition sperm, with 10%–30% absolute changes in methylation levels occurring mostly in the intergenic, nonrepetitive regions.[72] DNA methylation changes have also been examined in sperm of mice exposed to low-protein or high-fat diet,[39,49,50,73,74] folate-deficient diet,[75,76] undernutrition,[46,72] and toxicants.[77]

While these studies indicate that environmental conditions can modulate DNA methylation, the changes in DNA methylation in sperm are subtle (10%–30%) and occur at only a few loci (mostly intergenic). Given that DNA methylation is binary—methylation or no methylation—such small differences in DNA methylation imply that only a fraction of sperm carry differential methylation marks, and thus, cannot fully explain the altered phenotypes in the majority of offspring.[78] Studies have also reported changes in DNA methylation at ribosomal DNA in response to low-protein diet[46] and at satellite repeats in sperm of high-fat diet–exposed rats.[79] As repeats can differ in their copy number, a role of DNA methylation at repeat-rich regions in epigenetic inheritance is more plausible. Furthermore, it was recently reported that DNA methylation differences within individuals are more pronounced than the effect of paternal diet.[73] Overall, these studies suggest that the environment can modulate DNA methylation, but the extent to which DNA methylation plays a role in intergenerational inheritance remains to be determined.

Histone modifications

As mentioned before, while protamines replace most of the histones in mature sperm, specific loci retain histones and genetic studies demonstrate that global changes in histone modifications can influence phenotypes in offspring. However, there are few reports of environmental effects being transmitted by histone modifications, emphasizing a need for further exploration of role of histone modifications in intergenerational inheritance. Exposure to carbon tetrachloride can lead to altered levels of histone acetylation at specific promoters in mouse sperm.[66] Coadministration of bleomycin, etoposide, and cisplatin was reported to induce changes in histone modifications and chromatin accessibility in testicular germ cells.[80] A high-fat diet leads to altered histone H3 occupancy and histone H3 lysine 4 methylation (H3K4me) at some transcription regulatory genes in rat sperm.[81] A recent study reported that paternal low-protein diet regulates phosphorylation of a chromatin regulator (cyclic AMP-dependent transcription factor 7, ATF7), which in turn leads to a decrease in histone H3 lysine 9 methylation (H3K9me) on many genes and biogenesis of tRNA fragments in sperm.[82] While histone modifications can be altered by diet, histone retention changes were not detected in obese versus lean men.[74] Collectively, histones and histone modifications are potential carriers of paternal environmental effects. However, as in the case of DNA methylation, since only a subset of genomic loci retain histones, it is still unclear whether such changes can fully explain the altered phenotypes in offspring. There is a need for in-depth investigations in the role of histones in intergenerational inheritance.

Small RNAs

While a role of small RNAs in transgenerational inheritance has been well documented in model organisms, such as RNA interference (RNAi) in worms and paramutation in plants, recent studies implicate small RNAs in the transmission of paternal environmental effects in mammals.[15,24,42,43,50] As opposed to studies with DNA methylation and histone modifications where complicated genetic models are required to test their role in epigenetic inheritance, a function of sperm RNAs in transmitting paternal environmental information to offspring can be directly tested by injecting small RNAs purified from the sperm of exposed males into control zygotes. As discussed before, the MSUS mice display depressive-life behavior, and this phenotype is transmitted to their offspring. In one of the first studies to demonstrate a role of sperm RNAs in intergenerational inheritance, RNAs purified from sperm of MSUS mice were injected into one-cell control embryos. The adult offspring generated from those embryos displayed a subset of the behavioral and metabolic traits of MSUS mice. A specific set of miRNAs was reported to change in MSUS sperm in response to stress,[60] suggesting a potential role of these miRNAs in the transmission of paternal effects. In another study on paternal chronic stress, levels of nine specific miRNAs were significantly upregulated in sperm of exposed mice compared with control sperm.[58] Importantly, injection of all nine miRNAs combinatorically, not individually, in control zygotes resulted in transmission of paternal chronic stress effects in the offspring.[59,60]

In the case of paternal dietary exposure, levels of specific small RNAs were altered in sperm of mice exposed to low-protein or high-fat diets. Specifically, 5′ fragments of multiple isoacceptors of tRNA-Glycine, tRNA-LysCTT, and tRNA-HisGTT were upregulated, and a miRNA Let7c was downregulated in low-protein sperm.[15] In rats, high-fat diet also led to changes in specific small RNAs including, piRNAs, miRNAs, and tRFs.[50] Sperm of mice exposed to a high-fat and high-sugar diet (also called a Western diet) display altered levels of specific miRNAs.[42] Sperm of mice exposed to a high-fat diet were reported to have an ~11% increase in the abundance of all tRFs.[24] Injection of purified tRFs from high-fat sperm into control zygotes resulted in metabolic phenotypes in adult offspring (similar to those transmitted by natural mating), suggesting that tRFs act as epigenetic information carriers of paternal dietary information.[24] Finally, exposure to toxicants, such as vinclozolin and ethanol, also seems to affect the levels of tRFs and miRNAs in rodent sperm.[67,83] A recent study also reported a role of long RNAs in sperm in intergenerational inheritance of paternal early-life trauma.[84]

Overall, there is strong evidence that small RNAs are environmentally responsive epigenetic molecules in sperm and potential carriers of paternal environmental information.[85] However, there are many mechanistic questions that remain unanswered regarding the role of sperm RNAs in epigenetic inheritance. The oocyte has vast excess of RNAs compared with the amount of RNA delivered by sperm at the time of fertilization. It is estimated that the most abundant tRFs in mature sperm are around ~100–100,000 molecules per sperm.[15] Yet, zygotic injections of tRFs and miRNAs (mentioned before) demonstrate that sperm RNAs can regulate gene expression in early embryos.[15,24] This raises the question of how do sperm small RNAs regulate offspring phenotypes? Perhaps sperm RNAs have special features that make them more potent even at lower abundance. For example, posttranscriptional modifications on tRFs have been reported to enhance their stability.[24,86] Moreover, small-RNA loading onto effector complexes can provide target specificity.[87] Secondly, as mammals do not have a known RNA-dependent RNA polymerase, any RNA molecules delivered to the oocyte at fertilization will be diluted away with every cell division. How do sperm RNAs exert long-lasting effects to influence phenotypes in offspring? Small RNAs potentially regulate some key developmental events during early embryonic stages, which have secondary effects later in development.[1,85]

Inheritance of maternal environmental exposure effects

It is becoming ever more apparent that a range of environmental agents, including nutrients, drugs, and environmental toxicants, can have effects on the health of not only offspring exposed in utero but also on offspring multiple generations downstream of the environmental exposure. Research focused on the transmission of environmental effects through the female germline is, as aforementioned, complicated by several factors. For instance, human epidemiological studies of the Dutch Hunger Winter of 1944–45 report that children born to mothers that experienced severe undernutrition (consumption of lower than 1000 calories per day) during their pregnancy have higher risk of metabolic disorders in adulthood.[88] However, in such inheritance, it is difficult to distinguish between the possibility of maternal environment impacting offspring via environmentally directed molecular changes in the oocyte versus direct effects of maternal factors on the developing fetus. In addition, as with paternal inheritance, most research on environmental exposure effects in the offspring of mammals is limited to intergenerational inheritance when concerning the female germline. The exposure of a pregnant female to an environmental agent may expose not only the offspring as the fetus developing in utero but also the grand offspring as germ cells within the developing fetus (Fig. 11.1).

Maternal environmental influence on offspring health

As with paternal inheritance, various environmental agents have been examined for their ability to induce changes in phenotypes in future generations through the female germline, including dietary inputs, various toxicants, and drugs. Here, we discuss some examples and implications of the effects of some of these agents.

Maternal toxicant exposure

Interestingly, much of the work that exists to date regarding maternal inheritance of acquired environmental effects concerns exposure to synthetic organic molecules used widely in the industrial production of materials and those commonly used, either presently or in the recent past, as pesticides. Many of these chemicals are endocrine disruptors. One such chemical is di(2-ethylhexyl) phthalate, otherwise known as DEHP. DEHP is an organic compound commonly used to make materials more flexible and can be found in household items such as garden hoses and shower curtains. DEHP exposure in pregnant mice from conception until the end of lactation results in reproductive dysfunction in offspring, including lower weight, lower gonad weight, low sperm quality, and number, and altered expression of genes involved in gonad steroidogenesis.[89,90] In this case, the F1 and F2 generations were directly exposed to DEHP as embryos/fetuses and as germ cells, respectively, and the F3 generation would be the first generation to have not been directly exposed to DEHP. Any persistent effects of ancestral DEHP exposure seen in the F3 generation could be considered transgenerational. The F2 and F3 generations were generated by mating of F1 and F2 females, respectively, to male mice with no family history of DEHP exposure.[89] As with the F1 generation, both F2 and F3 generations had a lower number of primordial follicles and a higher number of preantral follicles in their ovaries compared with those of control animals. This suggests that this DEHP-associated phenotype was transmitted transgenerationally. Mechanistically, while F2 and F3 offspring did not show altered expression of steroidogenesis genes, as seen in F1, all three generations showed higher levels of *Gdf9* mRNA in ovaries. As *Gdf9* plays a role in folliculogenesis, this study suggests that impaired folliculogenesis phenotype is inherited transgenerationally. Whether this is a result of epigenetic changes in the germ cells was not explored in the study. Moreover, only a few genes were examined, and future research with global gene expression analysis will be able to shed light on the broader effects of DHEP. Finally, while there are limitations to extrapolating these findings to human DEHP exposure, they certainly provide impetus for more research into whether this compound can affect ovarian health in humans for multiple generations. Similarly, maternal exposure to TCDD (2,3,7,8-tetrachlorodibenzo-p-dioxin)[91] and the pesticide methoxychlor[92] led to altered phenotypes that were transmitted transgenerationally.

Maternal dietary effects

There is a well-documented link between poor gestational nutrition[93] or intrauterine growth retardation[94] and metabolic phenotypes in the offspring. In addition, studies have also reported that maternal obesity or high-fat diet consumption during gestation can increase the risk of cardiovascular disorders and other metabolic diseases in offspring.[41,95] However, most of these studies did not explore the effects of maternal nutrition on oocyte epigenome and its effects on offspring metabolism. To focus specifically on epigenetic inheritance via the germline, Huypens et al. generated embryos using IVF, wherein the only contribution from parents is their gametes, to study inheritance of obesity phenotype in offspring of parents exposed to a high-fat diet. Sperm, oocytes, or both were obtained from high-fat diet or control diet—fed mice (F0), and transferred to healthy foster mothers maintained on a normal diet to generate adult offspring (F1). Offspring were next challenged with high-fat diet and phenotypically analyzed for metabolic disorder, including weight gain, glucose intolerance, and insulin resistance.[41] The offspring resulting from normal-diet fathers and high-fat diet mothers showed significantly higher increases in body fat during the high-fat diet challenge than their counterparts from control parents. When challenged with an intraperitoneal injection of glucose, the offspring of high-fat diet mothers showed a significant increase in the area under the curve of blood glucose concentration over a 120-min interval relative to controls. This difference was accompanied by no significant changes in insulin levels over the interval, suggesting the development of an insulin resistance phenotype.[41] Huypens et al. thus showed that obesity and a type II diabetes-like insulin resistance phenotype, which were also seen in high-fat diet parents, could be inherited intergenerationally through the oocyte. For obvious reasons, these findings have profound implications if they can be applied to humans.

Transgenerational inheritance of maternal diet-induced changes has also been documented.[96] It has been reported that high-fat diet in conjunction with toxicant exposure can have synergistic effects. Chamorro-García et al. found that F4 offspring of male mice perinatally (in utero and during lactation) exposed to obesogen tributyltin (TBT) in their mothers drinking water in conjunction with a maternal high-fat diet showed impaired glucose tolerance and had significantly higher body fat relative to controls.[97] Exposure to a low-protein diet during pregnancy and throughout lactation has also been shown to affect metabolic phenotypes in the F3 generation, suggesting transgenerational inheritance of maternal undernutrition effects.[96]

Mechanisms of inheritance of maternal environmental effects

Based on the evidence including but not limited to what has been described before, it can reasonably be deduced that environmentally induced epigenetic changes can be inherited through the mammalian maternal germline. However, these works were largely empty of attestation with respect to how this transmission occurs mechanistically. As with inheritance via male germline, inheritance through the female germline is potentially mediated by environment-induced changes in one or more of the following epigenetic information carriers: small RNAs, histone modifications, and DNA methylation. In addition, a role of the mitochondria and/or the metabolome has also been implied. In the following sections, some of these possible mechanisms will be explored as it pertains to epigenetic inheritance of maternal environmental effects.

DNA methylation

One potential mechanism by which epigenetic inheritance of environmentally acquired traits occurs is through changes in methylation patterns of cytosine residues in genomic DNA. As discussed in previous sections, DNA methylation at the IAP retrotransposons inserted upstream of the *Agouti* locus regulates expression of the *Agouti* gene, resulting in a range of coat color in animals. A methyl donor-rich diet during pregnancy in *Avy* female mice is associated with a larger percentage of offspring with a wild-type coat color compared with offspring of mothers on control diet. Studies report that these changes in phenotype are associated with changes in the levels of DNA methylation at the IAP elements in response to a methyl-rich diet.[98,99] However, as the exposure was given during pregnancy, it is not clear whether these changes are due to a direct effect on the fetus or through oocyte-mediated changes.

Changes in DNA methylation at the promoter of two metabolic genes were reported in the oocytes of females exposed to a high-fat diet, as well as in the livers and oocytes of offspring that were raised on a normal diet.[100] As mentioned earlier, a widespread erasure of DNA methylation marks occurs in primordial germ cells, and the maternally imprinted genes (expressed from the paternal allele) are remethylated during oogenesis, while the paternally imprinted (expressed from the maternal allele) loci in the genome are left unmethylated during oogenesis. One of the imprinted genes in mice is *IGF2R*, which encodes insulin-like growth factor 2 receptor. *IGF2R* is maternally expressed and reciprocally imprinted with a long noncoding RNA called *Airn*.[101] While DNA methylation at this gene was not affected in the oocytes of females exposed to a high-fat diet,[100] changes were overserved in the oocytes of females given cyclophosphamide (CPM), a drug used in the treatment of breast cancer.[102] Female mice (F0) were given a single dose of CPM 12 weeks prior to conception of the F1 generation. Female F0 and F1 mice were mated to males with no history of CPM exposure to create the F1 and F2 generations, respectively. At 21 days of age, the oocytes of F1 and F2 females (including those of age-matched, unexposed females) were harvested, and the methylation status of *IGF2R's* promoter was queried via bisulfite sequencing. In both the F1 and F2 oocytes of the

CPM exposure group, *IGF2R's* promoter was significantly less methylated than that of control oocytes. While previous studies have focused on an environmental exposure in F0 pregnant females (direct exposure to F1 generation and the F2 generation in the form of the F1 germ cells), Di Emidio et al. focused on an environmental exposure that occurred prior to pregnancy with the F1 generation. Thus, the F2 generation was not directly exposed. Overall, this study demonstrates a transgenerational transmission of differential methylation status at the promoters of three imprinted genes in mice as a result of exposure to a drug (*IGF2R, PEG3,* and *H19*).[102]

For this differential methylation pattern phenotype to have been passed on from the F1 oocytes to ultimately be seen again in the F2 oocytes, the changes would have had to have survived the second wave of demethylation and resetting that occurs in the embryo.[103] They then would have had to survive the near global demethylation that occurs in primordial germ cells and resetting of methylation marks during oogenesis to see these patterns in the F2 oocytes. This implies that some sort of mechanistic memory is responsible for the transgenerational inheritance of this change in methylation levels that it either directly avoided the two waves of reprogramming or it was "remembered" and reset by an indirect mechanism. It is known that imprinted loci avoid the second wave of demethylation that occurs during preimplantation embryonic development in a manner that is dependent on the DNA methyltransferase 1 (DNMT1) that remains in embryonic nuclei during this process.[103] However, in the specific case of *IGF2R*, both maternal and paternal alleles are robustly expressed during murine preimplantation embryonic development.[104] Repression of the paternal allele is seen at E6.5 and onward though a bit of residual paternal expression persists in tissues that highly express *IGF2R*.[104] Proper imprinting of *IGF2R* appears to require DNA methylation, although it seems to be less important than for other imprinted genes such as *IGF2* and *H19*.[105] Yet, the reasoning for this requirement seems to be that DNA methylation blocks expression of the reciprocally imprinted *Airn* lncRNA on the maternal chromosome.[106] *Airn's* promoter is within the second intron of the *IGF2R* gene, and it drives the transcription of this 108-kb lncRNA in the direction opposite of *IGF2R* transcription.[106] Transcription of *Airn* through the *IGF2R* promoter region represses transcription of *IGF2R* from the paternal chromosome.[107] In differentiated embryonic stem cells and differentiating embryoids, paternal *IGF2R* promoter methylation is associated temporally with the onset of *Airn* expression, though *IGF2R* transcripts continue to be produced from the paternal allele at a constant level while transcription from the maternal allele increases substantially with differentiation.[108] *Airn* levels are high during early oogenesis and slowly taper off as oogenesis progresses to low levels in MII oocytes. Meanwhile, *IGF2R* mRNA levels are relatively consistent throughout oogenesis.[109] Perhaps, *IGF2R* promoter methylation established during oogenesis after primordial germ cell erasure is also correlated with *Airn* expression. If this is the case, then maybe modulation of *Airn* expression due to ancestral CPM exposure is responsible for the transgenerationally inherited changes in oocyte *IGF2R* promoter methylation seen by Di Emidio et al. Regardless, the mechanism by which these changes occur and their functional significance remains unclear.

Histone modifications

In supplement to those mechanisms already listed, another putative way for environmentally acquired epigenetic information to be transmitted through the mammalian female germline is by modification of histones. Alteration of oocyte histone modifications as a result of exposure to toxicants has been documented. It was found that porcine oocytes exposed to 250 μM of bisphenol A (BPA) when cultured as cumulus cell oocyte complexes (COCs) during in vitro maturation led to a significant decrease in the amount of H3K4me2 (histone H3 lysine 4 dimethylation) in the oocytes relative to controls as detected by immunostaining for that particular histone mark.[110] Interestingly, exposure of porcine COCs to MEHP (or mono-(2-ethylhexyl) phthalate, a metabolite of the previously discussed DEHP) led to an increase in oocyte H3K4me2 levels relative to controls, and this effect was reversed by exposure of the COCs to melatonin.[111] MEHP interacts with human estrogen receptor (expressed in *Saccharomyces cerevisiae*) in an antagonistic manner,[111] while BPA appears to interact with human and rat estrogen receptor in a partially agonistic manner.[112] Estrogen receptors are nuclear hormone receptors, as they translocate into the nucleus after ligand binding and alter gene expression upon binding to motifs in DNA. Melatonin receptors are G protein–coupled receptors[113] that are expressed in both cumulus cells and oocytes of cows[114] and sheep[115] during in vitro maturation. Melatonin receptor mRNA is expressed in porcine granulosa and cumulus cells but not oocytes.[116] This information suggests that at least some environmental toxicants can alter mammalian oocyte histone modifications via cell signaling pathways in either cumulus cells or the oocytes themselves.

Whether these toxicant-induced changes in oocyte histone modifications also occur during natural folliculogenesis and oocyte maturation is unclear as the results discussed before were in the context of in vitro maturation. In addition, whether these mechanisms can lead to intergenerational or transgenerational inheritance of environmentally acquired epigenetic changes is also unclear. While, to our knowledge, there are no examples in the literature of oocyte histone modifications or epigenetic changes dependent on particular histone marks inherited intergenerationally or transgenerationally through the mammalian female germline, histone modifications make for a likely mechanistic suspect in these processes. Indeed, histone modifications seem to take part in these processes in nonmammals. For example, H3K9me3 appears to be associated with temperature-induced gene expression changes that can be transmitted transgenerationally through the oocytes of *Caenorhabditis elegans.*[117]

Small RNAs

The section of this chapter that discusses paternal environmental effects upholds small RNAs in sperm as a vehicle for epigenetic inheritance of acquired traits. There is reason to believe that small RNAs could play a similar role in oocytes. The volume of an oocyte is also hundreds of times larger than that of a spermatozoan,[27] perhaps suggesting a vastly greater potential for oocyte small RNAs to mediate epigenetic inheritance of environmentally acquired epigenetic traits assuming that amount of small RNAs in gametes is proportional to gametic cytoplasmic volume. At least in some cases, oocyte small RNAs seem to vastly outnumber those contributed by sperm in early embryos. For example, a fragment derived from tRNA-Glu-CTC as well as several select miRNAs from a single oocyte were shown by qRT-PCR to outnumber their counterparts in a single sperm by many thousand-fold.[27]

Another clue that advocates for the possibility of small RNAs being involved in epigenetic inheritance of environmentally induced changes in mammals through the female germline comes from paramutation at the *Kit* locus. A paramutagenic allele was generated by inserting the copGFP (green fluorescent protein cloned from copepod *Pontellina plumata*) at the *Kit* locus.[118] This led to a significantly higher incidence of white-tail tips in the genetically wild-type offspring (paramutants) of mice heterozygous for the paramutagenic allele than what is seen in the regular C57BL/6J mouse colonies. This trend persisted through to the F3 generation when the paramutants were interbred. Crossing males heterozygous for the paramutagenic allele with either *Mov10l*$^{-/-}$ (expected to disrupt piRNAs production in all tissues, including oocytes) or *Zp3-Cre/Droshalox/-* (predicted to prevent the generation of miRNAs in oocytes) black-tipped tailed female mice led to approximately 90% of the offspring having white-tipped tails. This contrasted with crossing paramutagenic heterozygote males with genetically wild-type black-tipped tailed females, which led to only 55% of the offspring having white-tipped tails. Overall, this work showed the transgenerational transmission of a paramutation phenotype in mice and suggested that oocyte miRNAs and piRNAs potentially suppress the transmission of that phenotype.[118]

To our knowledge, what seems to be lacking in the literature is evidence implicating small RNAs in the intergenerational or transgenerational inheritance of an environmentally induced phenotypic change through the mammalian maternal lineage. Considering recent advances in sequencing of small RNAs in mammalian oocytes,[27,28] it is likely that much more information surrounding these topics will be elucidated in the coming years.

Mitochondrial inheritance and the metabolome

While small RNAs contributed to the zygote by both the sperm and oocyte appear to be important for embryonic development,[26] mitochondria are exclusively maternally inherited in animals. Therefore, mitochondria provide another potential mechanism through which maternal environmentally induced changes can be inherited. It has recently been suggested that mitochondrial morphology and molecular physiology can be altered by a high-fat and high-sugar (HF/HS) diet and that these changes can be transmitted transgenerationally to the F3 generation in mice.[119] Female C57BL/6J (F0) were fed an HF/HS diet for 6 weeks from conception to weaning of F1 offspring and mated to males fed a normal diet. Control diet was fed to all of the F1, F2, and F3 offspring, and F1 and F2 females were mated to control diet—fed males with no ancestral HF/HS exposure to generate the F2 and F3 offspring, respectively. F1, F2, and F3 offspring in the HF/HS group all showed increased mitochondrial size, altered inner mitochondrial membrane (IMM) morphology, and reduced levels of OPA1 protein, which is involved in cristae formation and IMM fusion, in skeletal muscle mitochondria relative to those of control skeletal muscle mitochondria.[119] As these changes were seen in the F1, F2, and F3 generations, they constitute an example of transgenerational inheritance. In addition, F1 oocytes had significantly larger mitochondria and lower OPA1 protein levels than those of controls and F2 oocytes had lower OPA1 protein levels.[119] While this work shows an example of transgenerational inheritance of altered mitochondrial phenotype through the maternal lineage (as males used for mating had no ancestral HF/HS exposure), it should be noted that OPA1 is a nuclear-encoded protein. This suggests that the changes seen are potentially wholly or partially a result of some underlying epigenetic changes, such as alterations in histone modifications or nuclear DNA methylation at the OPA1 gene. The study, however, did not examine mitochondrial size or IMM morphology in F2 oocytes of the HF/HS group. If these were not altered, that would give more credence to the supposition that other mechanisms were involved.

The idea that variations in other substrates for epigenetic modification could be linked to metabolism has been explored in other contexts. Disruption of folate metabolism via a hypomorphic mutation in the *Mtrr* gene (encodes methionine synthase reductase that activates methionine synthase, thereby allowing conversion of 5-methyltetrahydrofolate to methionine) leads to transgenerational epigenetic inheritance of developmental defects.[120] Genetically wild-type mouse embryos with genetically wild-type mothers and maternal grandfathers heterozygous for the hypomorphic *Mtrr* allele showed significant decreases at E10.5 in the numbers of normal embryos and in the length of the embryos as well as a significant increase in the occurrence of severe abnormalities present relative to control embryos.[120] Embryos of that same lineage and at the same developmental stage showed significant hypermethylation of DNA in the placenta across numerous differentially methylated regions (DMRs) and a corresponding reduction in mRNA expression from the associated imprinted genes. This would be considered a case of transgenerational epigenetic inheritance through the maternal germline as the epigenetic information was passed to the mother through the grandfather's sperm, and thus, subsequent generations were not exposed in utero. Epigenetic inheritance was, in this case, the result of an ancestral mutation. However, these findings may provide insight into mechanistic details surrounding the effects of disrupting the folate cycle via folate deficiency during pregnancy as the folate cycle ultimately provides a source of methyl groups for various cellular processes such as DNA methylation.[120]

Signaling environmental information to germline

Finally, a major challenge in understanding how parental environment can influence phenotypes in the future generations is that it is still unclear how parental environment, such as a traumatic experience or undernutrition, affects gametic epigenome. The environment can potentially induce changes in somatic cells, which then communicate environmental information to the germline. Studies from model organisms suggest that epigenetic information, in the form of small RNAs, can be transmitted from gonadal somatic cells to germ cells.[121] For instance, during gametogenesis in worms and plants, RNAs generated in the somatic support cells are transported to germ cells, and this process is proposed to be essential for proper germline maturation and maintenance of genomic integrity.[121–123] Such communication between soma and germline is possible due to direct contact between these cell types. For example, the sperm and vegetative nuclei are present in the same cytoplasm in plants. However, unlike these model organisms (ciliates, plants, and worms), there is no physical contact between the somatic accessory cells of the reproductive system and mature sperm in mammals. For this reason, as aforementioned, soma to germline flow of information is challenging in mammals and proposed to be inhibited by the so-called Weismann barrier.[124]

Some recent studies provide evidence of soma–germline communication in mammals.[15,23] Sperm small RNA payload changes drastically during sperm maturation in the epididymis; sperm acquire new tRFs and miRNAs during epididymal maturation.[15,23,125] Given that epididymal sperm are transcriptionally inert (no new transcripts synthesized in epididymal sperm), recent studies report that small RNAs generated in the somatic epididymis epithelium are shipped to mature sperm via extracellular vesicles (EVs) known as epididymosomes.[15,23,126] These studies suggest that paternal environmental information is also potentially transmitted from the epididymis epithelial to sperm in the form of small RNAs. Indeed, exposure to a low-protein diet results in similar changes in small RNA levels in sperm and the epididymis epithelial cells.[15] Moreover, chronic ethanol exposure in mice leads to changes in small RNA levels in the epididymis, epididymosomes, and sperm, implying that paternal environmental information is transmitted from epididymis to sperm via epididymosomes in the form of small RNAs.[67] Furthermore, when epididymosomes from ethanol-exposed males are incubated with control sperm and the resulting sperm is used for generating offspring, a subset of paternal ethanol exposure phenotypes are transmitted to offspring.[127] Together, these studies provide evidence of RNA-mediated communication between somatic cells and sperm and suggest a potential mechanism by which paternal environmental information can be transmitted to sperm, and thereby, to offspring.

Can maternal environment affect mammalian oocyte small RNA dynamics through soma–germline communication? It has been shown that EVs in human follicular fluid contain small RNAs and that the levels of some of these small RNAs are different between women with and without polycystic ovarian disease (PCOS).[128] Whether or not these EVs penetrate the zona pellucida to reach the oocyte is unclear. The more probable scenario is that these EVs modulate transcription in cumulus cells, which thereby may affect the loading of different RNAs into the oocyte. Indeed, EVs within the ovarian follicle have been shown to possess different miRNA profiles correlated with different phases of the estrous cycle, and these can alter transcript levels in cumulus cells accordingly.[129] In addition, alterations in the DNA methylation patterns as well as small and long noncoding RNA profiles of rat granulosa cells have been shown to occur in response to exposure to the pesticides vinclozolin and dichlorodiphenyltrichloroethane (DDT).[130] Soma–germline contact is known to be made through structures called transzonal projections (TZPs), which extend from cumulus cells and penetrate through the zona pellucida to the growing oocyte.[131] RNA molecules appear to be transported from the cumulus cells to growing bovine oocytes via TZPs.[132] Taking all of this information into account, it may also not be surprising if in the coming years evidence comes along demonstrating that environmental effects on somatic cells can result in epigenetic changes in the mammalian female germline that can be transmitted intergenerationally or transgenerationally.

Conclusions and future perspective

There is mounting evidence that epigenetic information is transmitted to future generations and that ancestral environment can influence phenotypes in descendants. Understanding the mechanisms of such inheritance is important not only for basic biology but also for making informed decisions about public health and policy. As epigenetic marks can be modulated by environmental conditions, intergenerational inheritance is potentially mediated via alterations to the gamete epigenome. Indeed, numerous recent studies report alteration of DNA methylation, histone modifications, and small RNA levels in gametes in response to changes in environmental conditions, and influence of such changes on offspring health.

While there is increasing evidence of parental environmental influence on offspring health, the critical steps of the process remain poorly understood in mammals: (1) Which epigenetic marks transmit parental environmental information? (2) How is environmental information signaled to influence epigenetic marks in germ cells? (3) How do those epigenetic signals influence embryonic development? (4) How do early changes in embryonic gene expression and development lead to altered phenotypes in adult offspring? (5) How epigenetic changes in one generation are transmitted transgenerationally? With growing evidence of epigenetic inheritance in mammals and advancement in technologies for profiling gametic and embryonic epigenomes, the coming years hold promise in solving some of these challenging, yet crucial questions in the field.

References

1. Sharma U, Rando OJ. Metabolic inputs into the epigenome. *Cell Metab.* 2017;25(3):544–558.
2. Heard E, Martienssen RA. Transgenerational epigenetic inheritance: myths and mechanisms. *Cell.* 2014;157(1):95–109.
3. Fire A, Xu S, Montgomery MK, et al. Potent and specific genetic interference by double-stranded RNA in *Caenorhabditis elegans. Nature.* 1998;391(6669):806–811.

4. Messerschmidt DM, Knowles BB, Solter D. DNA methylation dynamics during epigenetic reprogramming in the germline and preimplantation embryos. *Genes Dev.* 2014;28(8):812–828.
5. Siegfried Z, Simon I. DNA methylation and gene expression. *Wiley Interdiscip Rev Syst Biol Med.* 2010;2(3):362–371.
6. Hackett JA, Surani MA. DNA methylation dynamics during the mammalian life cycle. *Philos Trans R Soc Lond B Biol Sci.* 2013;368(1609):20110328.
7. Kornberg RD, Lorch Y. Twenty-five years of the nucleosome, fundamental particle of the eukaryote chromosome. *Cell.* 1999;98(3):285–294.
8. Balhorn R. The protamine family of sperm nuclear proteins. *Genome Biol.* 2007;8(9):227.
9. Hammoud S, Liu L, Carrell DT. Protamine ratio and the level of histone retention in sperm selected from a density gradient preparation. *Andrologia.* 2009;41(2):88–94.
10. Hammoud SS, Low DH, Yi C, Carrell DT, Guccione E, Cairns BR. Chromatin and transcription transitions of mammalian adult germline stem cells and spermatogenesis. *Cell Stem Cell.* 2014;15(2):239–253.
11. Yamaguchi K, Hada M, Fukuda Y, et al. Re-Evaluating the localization of sperm-retained histones revealed the modification-dependent accumulation in specific genome regions. *Cell Rep.* 2018;23(13):3920–3932.
12. Carone BR, Hung JH, Hainer SJ, et al. High-resolution mapping of chromatin packaging in mouse embryonic stem cells and sperm. *Dev Cell.* 2014;30(1):11–22.
13. Ostermeier GC, Dix DJ, Miller D, Khatri P, Krawetz SA. Spermatozoal RNA profiles of normal fertile men. *Lancet.* 2002;360(9335):772–777.
14. Ostermeier GC, Miller D, Huntriss JD, Diamond MP, Krawetz SA. Reproductive biology: delivering spermatozoan RNA to the oocyte. *Nature.* 2004;429(6988):154.
15. Sharma U, Conine CC, Shea JM, et al. Biogenesis and function of tRNA fragments during sperm maturation and fertilization in mammals. *Science.* 2016;351(6271):391–396.
16. Peng H, Shi J, Zhang Y, et al. A novel class of tRNA-derived small RNAs extremely enriched in mature mouse sperm. *Cell Res.* 2012;22(11):1609–1612.
17. Hutcheon K, McLaughlin EA, Stanger SJ, et al. Analysis of the small non-protein-coding RNA profile of mouse spermatozoa reveals specific enrichment of piRNAs within mature spermatozoa. *RNA Biol.* 2017;14(12):1776–1790.
18. Schuster A, Tang C, Xie Y, Ortogero N, Yuan S, Yan W. SpermBase: A database for sperm-borne RNA contents. *Biol Reprod.* 2016;95(5):99.
19. Kuramochi-Miyagawa S, Kimura T, Ijiri TW, et al. Mili, a mammalian member of piwi family gene, is essential for spermatogenesis. *Development.* 2004;131(4):839–849.
20. Aravin A, Gaidatzis D, Pfeffer S, et al. A novel class of small RNAs bind to MILI protein in mouse testes. *Nature.* 2006;442(7099):203–207.
21. Aravin AA, Sachidanandam R, Girard A, Fejes-Toth K, Hannon GJ. Developmentally regulated piRNA clusters implicate MILI in transposon control. *Science.* 2007;316(5825):744–747.
22. Li XZ, Roy CK, Dong X, et al. An ancient transcription factor initiates the burst of piRNA production during early meiosis in mouse testes. *Mol Cell.* 2013;50(1):67–81.
23. Sharma U, Sun F, Conine CC, et al. Small RNAs are trafficked from the epididymis to developing mammalian sperm. *Dev Cell.* 2018;46(4):481–494.e6.
24. Chen Q, Yan M, Cao Z, et al. Sperm tsRNAs contribute to intergenerational inheritance of an acquired metabolic disorder. *Science.* 2016;351(6271):397–400.
25. Conine CC, Sun F, Song L, Rivera-Perez JA, Rando OJ. Small RNAs gained during epididymal transit of sperm are essential for embryonic development in mice. *Dev Cell.* 2018;46(4):470–480.e3.

26. Yuan S, Schuster A, Tang C, et al. Sperm-borne miRNAs and endo-siRNAs are important for fertilization and preimplantation embryonic development. *Development*. 2016; 143(4):635–647.
27. Yang Q, Lin J, Liu M, et al. Highly sensitive sequencing reveals dynamic modifications and activities of small RNAs in mouse oocytes and early embryos. *Sci Adv*. 2016;2(6): e1501482.
28. Yang Q, Li R, Lyu Q, et al. Single-cell CAS-seq reveals a class of short PIWI-interacting RNAs in human oocytes. *Nat Commun*. 2019;10(1):3389.
29. Tucci V, Isles AR, Kelsey G, Ferguson-Smith AC, Erice Imprinting G. Genomic imprinting and physiological processes in mammals. *Cell*. 2019;176(5):952–965.
30. Duhl DM, Vrieling H, Miller KA, Wolff GL, Barsh GS. Neomorphic agouti mutations in obese yellow mice. *Nat Genet*. 1994;8(1):59–65.
31. Morgan HD, Sutherland HGE, Martin DIK, Whitelaw E. Epigenetic inheritance at the agouti locus in the mouse. *Nat Genet*. 1999;23(3):314–318.
32. Rassoulzadegan M, Grandjean V, Gounon P, Vincent S, Gillot I, Cuzin F. RNA-mediated non-mendelian inheritance of an epigenetic change in the mouse. *Nature*. 2006; 441(7092):469–474.
33. Siklenka K, Erkek S, Godmann M, et al. Disruption of histone methylation in developing sperm impairs offspring health transgenerationally. *Science*. 2015;350(6261): aab2006.
34. Lesch BJ, Tothova Z, Morgan EA, et al. Intergenerational epigenetic inheritance of cancer susceptibility in mammals. *Elife*. 2019;8:e39380.
35. Pembrey ME, Bygren LO, Kaati G, et al. Sex-specific, male-line transgenerational responses in humans. *Eur J Hum Genet*. 2006;14(2):159–166.
36. Kaati G, Bygren LO, Edvinsson S. Cardiovascular and diabetes mortality determined by nutrition during parents' and grandparents' slow growth period. *Eur J Hum Genet*. 2002; 10(11):682–688.
37. Bygren LO, Kaati G, Edvinsson S. Longevity determined by paternal ancestors' nutrition during their slow growth period. *Acta Biotheor*. 2001;49(1):53–59.
38. Anderson LM, Riffle L, Wilson R, Travlos GS, Lubomirski MS, Alvord WG. Preconceptional fasting of fathers alters serum glucose in offspring of mice. *Nutrition*. 2006; 22(3):327–331.
39. Carone BR, Fauquier L, Habib N, et al. Paternally induced transgenerational environmental reprogramming of metabolic gene expression in mammals. *Cell*. 2010;143(7): 1084–1096.
40. Ng SF, Lin RC, Laybutt DR, Barres R, Owens JA, Morris MJ. Chronic high-fat diet in fathers programs beta-cell dysfunction in female rat offspring. *Nature*. 2010;467(7318): 963–966.
41. Huypens P, Sass S, Wu M, et al. Epigenetic germline inheritance of diet-induced obesity and insulin resistance. *Nat Genet*. 2016;48(5):497–499.
42. Grandjean V, Fourre S, De Abreu DA, Derieppe MA, Remy JJ, Rassoulzadegan M. RNA-mediated paternal heredity of diet-induced obesity and metabolic disorders. *Sci Rep*. 2015;5:18193.
43. Fullston T, Ohlsson Teague EM, Palmer NO, et al. Paternal obesity initiates metabolic disturbances in two generations of mice with incomplete penetrance to the F2 generation and alters the transcriptional profile of testis and sperm microRNA content. *FASEB J*. 2013;27(10):4226–4243.
44. Watkins AJ, Sinclair KD. Paternal low protein diet affects adult offspring cardiovascular and metabolic function in mice. *Am J Physiol Heart Circ Physiol*. 2014;306(10): H1444–H1452.

45. Ly NH, Maekawa T, Yoshida K, Liu Y, Muratani M, Ishii S. RNA-sequencing Analysis of paternal low-protein diet-induced gene expression change in mouse offspring adipocytes. *G3 (Bethesda)*. 2019;9(7):2161–2170.
46. Holland ML, Lowe R, Caton PW, et al. Early-life nutrition modulates the epigenetic state of specific rDNA genetic variants in mice. *Science*. 2016;353(6298):495–498.
47. Su L, Patti ME. Paternal nongenetic intergenerational transmission of metabolic disease risk. *Curr Diab Rep*. 2019;19(7):38.
48. Stanford KI, Rasmussen M, Baer LA, et al. Paternal exercise improves glucose metabolism in adult offspring. *Diabetes*. 2018;67(12):2530–2540.
49. Wei Y, Yang CR, Wei YP, et al. Paternally induced transgenerational inheritance of susceptibility to diabetes in mammals. *Proc Natl Acad Sci USA*. 2014;111(5):1873–1878.
50. de Castro Barbosa T, Ingerslev LR, Alm PS, et al. High-fat diet reprograms the epigenome of rat spermatozoa and transgenerationally affects metabolism of the offspring. *Mol Metab*. 2016;5(3):184–197.
51. Watkins AJ, Dias I, Tsuro H, et al. Paternal diet programs offspring health through sperm- and seminal plasma-specific pathways in mice. *Proc Natl Acad Sci USA*. 2018;115(40):10064–10069.
52. Govic A, Penman J, Tammer AH, Paolini AG. Paternal calorie restriction prior to conception alters anxiety-like behavior of the adult rat progeny. *Psychoneuroendocrinology*. 2016;64:1–11.
53. Radford EJ, Isganaitis E, Jimenez-Chillaron J, et al. An unbiased assessment of the role of imprinted genes in an intergenerational model of developmental programming. *PLoS Genet*. 2012;8(4):e1002605.
54. Jimenez-Chillaron JC, Isganaitis E, Charalambous M, et al. Intergenerational transmission of glucose intolerance and obesity by in utero undernutrition in mice. *Diabetes*. 2009;58(2):460–468.
55. Dunn GA, Bale TL. Maternal high-fat diet promotes body length increases and insulin insensitivity in second-generation mice. *Endocrinology*. 2009;150(11):4999–5009.
56. Dunn GA, Bale TL. Maternal high-fat diet effects on third-generation female body size via the paternal lineage. *Endocrinology*. 2011;152(6):2228–2236.
57. Sarker G, Sun W, Rosenkranz D, et al. Maternal overnutrition programs hedonic and metabolic phenotypes across generations through sperm tsRNAs. *Proc Natl Acad Sci USA*. 2019;116(21):10547–10556.
58. Rodgers AB, Morgan CP, Bronson SL, Revello S, Bale TL. Paternal stress exposure alters sperm microRNA content and reprograms offspring HPA stress axis regulation. *J Neurosci*. 2013;33(21):9003–9012.
59. Rodgers AB, Morgan CP, Leu NA, Bale TL. Transgenerational epigenetic programming via sperm microRNA recapitulates effects of paternal stress. *Proc Natl Acad Sci USA*. 2015;112(44):13699–13704.
60. Gapp K, Jawaid A, Sarkies P, et al. Implication of sperm RNAs in transgenerational inheritance of the effects of early trauma in mice. *Nat Neurosci*. 2014;17(5):667–669.
61. Wu L, Lu Y, Jiao Y, et al. Paternal psychological stress reprograms hepatic gluconeogenesis in offspring. *Cell Metab*. 2016;23(4):735–743.
62. Dietz DM, Laplant Q, Watts EL, et al. Paternal transmission of stress-induced pathologies. *Biol Psychiatry*. 2011;70(5):408–414.
63. Anway MD, Cupp AS, Uzumcu M, Skinner MK. Epigenetic transgenerational actions of endocrine disruptors and male fertility. *Science*. 2005;308(5727):1466–1469.
64. Vallaster MP, Kukreja S, Bing XY, et al. Paternal nicotine exposure alters hepatic xenobiotic metabolism in offspring. *Elife*. 2017;6:e24771.
65. Vassoler FM, White SL, Schmidt HD, Sadri-Vakili G, Pierce RC. Epigenetic inheritance of a cocaine-resistance phenotype. *Nat Neurosci*. 2013;16(1):42–47.

66. Zeybel M, Hardy T, Wong YK, et al. Multigenerational epigenetic adaptation of the hepatic wound-healing response. *Nat Med.* 2012;18(9):1369–1377.
67. Rompala GR, Simons A, Kihle B, Homanics GE. Paternal preconception chronic variable stress confers attenuated ethanol drinking behavior selectively to male offspring in a pre-stress environment dependent manner. *Front Behav Neurosci.* 2018;12:257.
68. Rando OJ. Daddy issues: paternal effects on phenotype. *Cell.* 2012;151(4):702–708.
69. Dias BG, Ressler KJ. Parental olfactory experience influences behavior and neural structure in subsequent generations. *Nat Neurosci.* 2014;17(1):89–96.
70. Isganaitis E, Jimenez-Chillaron J, Woo M, et al. Accelerated postnatal growth increases lipogenic gene expression and adipocyte size in low-birth weight mice. *Diabetes.* 2009; 58(5):1192–1200.
71. Woo M, Isganaitis E, Cerletti M, et al. Early life nutrition modulates muscle stem cell number: implications for muscle mass and repair. *Stem Cells Dev.* 2011;20(10): 1763–1769.
72. Radford EJ, Ito M, Shi H, et al. In utero effects. In utero undernourishment perturbs the adult sperm methylome and intergenerational metabolism. *Science.* 2014;345(6198): 1255903.
73. Shea JM, Serra RW, Carone BR, et al. Genetic and epigenetic variation, but not diet, shape the sperm methylome. *Dev Cell.* 2015;35(6):750–758.
74. Donkin I, Versteyhe S, Ingerslev LR, et al. Obesity and bariatric surgery drive epigenetic variation of spermatozoa in humans. *Cell Metabol.* 2016;23(2):369–378.
75. Lambrot R, Xu C, Saint-Phar S, et al. Low paternal dietary folate alters the mouse sperm epigenome and is associated with negative pregnancy outcomes. *Nat Commun.* 2013;4: 1–13.
76. Ly L, Chan D, Aarabi M, et al. Intergenerational impact of paternal lifetime exposures to both folic acid deficiency and supplementation on reproductive outcomes and imprinted gene methylation. *Mol Hum Reprod.* 2017;23(7):461–477.
77. Skinner MK, Ben Maamar M, Sadler-Riggleman I, et al. Alterations in sperm DNA methylation, non-coding RNA and histone retention associate with DDT-induced epigenetic transgenerational inheritance of disease. *Epigenet Chromatin.* 2018;11(1):8.
78. Rando OJ. Intergenerational transfer of epigenetic information in sperm. *Cold Spring Harb Perspect Med.* 2016;6(5).
79. Youngson NA, Lecomte V, Maloney CA, et al. Obesity-induced sperm DNA methylation changes at satellite repeats are reprogrammed in rat offspring. *Asian J Androl.* 2016;18(6):930–936.
80. Bagheri-Sereshki N, Hales BF, Robaire B. The effects of chemotherapeutic agents, bleomycin, etoposide, and cisplatin, on chromatin remodeling in male rat germ cells. *Biol Reprod.* 2016;94(4):81.
81. Terashima M, Barbour S, Ren J, Yu W, Han Y, Muegge K. Effect of high fat diet on paternal sperm histone distribution and male offspring liver gene expression. *Epigenetics.* 2015;10(9):861–871.
82. Yoshida K, Maekawa T, Ly NH, et al. ATF7-Dependent epigenetic changes are required for the intergenerational effect of a paternal low-protein diet. *Mol Cell.* 2020;78(3): 445–458.e6.
83. Schuster A, Skinner MK, Yan W. Ancestral vinclozolin exposure alters the epigenetic transgenerational inheritance of sperm small noncoding RNAs. *Environ Epigenet.* 2016;2(1).
84. Gapp K, van Steenwyk G, Germain PL, et al. Alterations in sperm long RNA contribute to the epigenetic inheritance of the effects of postnatal trauma. *Mol Psychiatry.* 2018; 25(9):2162–2174.

85. Sharma U. Paternal contributions to offspring health: role of sperm small RNAs in intergenerational transmission of epigenetic information. *Front Cell Dev Biol.* 2019;7:215.
86. Chen Q, Yan W, Duan E. Epigenetic inheritance of acquired traits through sperm RNAs and sperm RNA modifications. *Nat Rev Genet.* 2016;17(12):733–743.
87. Czech B, Hannon GJ. Small RNA sorting: matchmaking for Argonautes. *Nat Rev Genet.* 2011;12(1):19–31.
88. Lumey LH, Stein AD, Kahn HS, et al. Cohort profile: the Dutch Hunger Winter families study. *Int J Epidemiol.* 2007;36(6):1196–1204.
89. Pocar P, Fiandanese N, Berrini A, Secchi C, Borromeo V. Maternal exposure to di(2-ethylhexyl)phthalate (DEHP) promotes the transgenerational inheritance of adult-onset reproductive dysfunctions through the female germline in mice. *Toxicol Appl Pharmacol.* 2017;322:113–121.
90. Pocar P, Fiandanese N, Secchi C, et al. Exposure to di(2-ethyl-hexyl) phthalate (DEHP) in utero and during lactation causes long-term pituitary-gonadal axis disruption in male and female mouse offspring. *Endocrinology.* 2012;153(2):937–948.
91. Bruner-Tran KL, Osteen KG. Developmental exposure to TCDD reduces fertility and negatively affects pregnancy outcomes across multiple generations. *Reprod Toxicol.* 2011;31(3):344–350.
92. Manikkam M, Haque MM, Guerrero-Bosagna C, Nilsson EE, Skinner MK. Pesticide methoxychlor promotes the epigenetic transgenerational inheritance of adult-onset disease through the female germline. *PLoS One.* 2014;9(7):e102091.
93. Petry CJ, Dorling MW, Pawlak DB, Ozanne SE, Hales CN. Diabetes in old male offspring of rat dams fed a reduced protein diet. *Int J Exp Diabetes Res.* 2001;2(2):139–143.
94. Simmons RA, Templeton LJ, Gertz SJ. Intrauterine growth retardation leads to the development of type 2 diabetes in the rat. *Diabetes.* 2001;50(10):2279–2286.
95. Sun B, Purcell RH, Terrillion CE, Yan J, Moran TH, Tamashiro KL. Maternal high-fat diet during gestation or suckling differentially affects offspring leptin sensitivity and obesity. *Diabetes.* 2012;61(11):2833–2841.
96. Benyshek DC, Johnston CS, Martin JF. Glucose metabolism is altered in the adequately-nourished grand-offspring (F3 generation) of rats malnourished during gestation and perinatal life. *Diabetologia.* 2006;49(5):1117–1119.
97. Chamorro-Garcia R, Diaz-Castillo C, Shoucri BM, et al. Ancestral perinatal obesogen exposure results in a transgenerational thrifty phenotype in mice. *Nat Commun.* 2017;8(1):2012.
98. Cooney CA, Dave AA, Wolff GL. Maternal methyl supplements in mice affect epigenetic variation and DNA methylation of offspring. *J Nutr.* 2002;132(suppl 8):2393S–2400S.
99. Dolinoy DC, Weidman JR, Waterland RA, Jirtle RL. Maternal genistein alters coat color and protects Avy mouse offspring from obesity by modifying the fetal epigenome. *Environ Health Perspect.* 2006;114(4):567–572.
100. Ge ZJ, Luo SM, Lin F, et al. DNA methylation in oocytes and liver of female mice and their offspring: effects of high-fat-diet-induced obesity. *Environ Health Perspect.* 2014;122(2):159–164.
101. Regha K, Sloane MA, Huang R, et al. Active and repressive chromatin are interspersed without spreading in an imprinted gene cluster in the mammalian genome. *Mol Cell.* 2007;27(3):353–366.
102. Di Emidio G, D'Aurora M, Placidi M, et al. Pre-conceptional maternal exposure to cyclophosphamide results in modifications of DNA methylation in F1 and F2 mouse oocytes: evidence for transgenerational effects. *Epigenetics.* 2019;14(11):1057–1064.

103. Zeng Y, Chen T. DNA methylation reprogramming during mammalian development. *Genes (Basel)*. 2019;10(4).
104. Lerchner W, Barlow DP. Paternal repression of the imprinted mouse Igf2r locus occurs during implantation and is stable in all tissues of the post-implantation mouse embryo. *Mech Dev*. 1997;61(1–2):141–149.
105. Li E, Beard C, Jaenisch R. Role for DNA methylation in genomic imprinting. *Nature*. 1993;366(6453):362–365.
106. Seidl CIM, Stricker SH, Barlow DP. The imprinted Air ncRNA is an atypical RNAPII transcript that evades splicing and escapes nuclear export. *Embo J*. 2006;25(15): 3565–3575.
107. Latos PA, Pauler FM, Koerner MV, et al. Airn transcriptional overlap, but not its lncRNA products, induces imprinted Igf2r silencing. *Science*. 2012;338(6113): 1469–1472.
108. Latos PA, Stricker SH, Steenpass L, et al. An in vitro ES cell imprinting model shows that imprinted expression of the Igf2r gene arises from an allele-specific expression bias. *Development*. 2009;136(3):437–448.
109. Deuve JL, Bonnet-Garnier A, Beaujean N, Avner P, Morey C. Antagonist Xist and Tsix co-transcription during mouse oogenesis and maternal Xist expression during pre-implantation development calls into question the nature of the maternal imprint on the X chromosome. *Epigenetics*. 2015;10(10):931–942.
110. Wang T, Han J, Duan X, et al. The toxic effects and possible mechanisms of Bisphenol A on oocyte maturation of porcine in vitro. *Oncotarget*. 2016;7(22):32554–32565.
111. Kim DH, Park CG, Kim SH, Kim YJ. The effects of mono-(2-ethylhexyl) phthalate (MEHP) on human estrogen receptor (hER) and androgen receptor (hAR) by YES/YAS in vitro assay. *Molecules*. 2019;24(8).
112. Gould JC, Leonard LS, Maness SC, et al. Bisphenol A interacts with the estrogen receptor alpha in a distinct manner from estradiol. *Mol Cell Endocrinol*. 1998; 142(1–2):203–214.
113. Reppert SM. Melatonin receptors: molecular biology of a new family of G protein-coupled receptors. *J Biol Rhythms*. 1997;12(6):528–531.
114. El-Raey M, Geshi M, Somfai T, et al. Evidence of melatonin synthesis in the cumulus oocyte complexes and its role in enhancing oocyte maturation in vitro in cattle. *Mol Reprod Dev*. 2011;78(4):250–262.
115. Tian X, Wang F, Zhang L, et al. Beneficial effects of melatonin on the in vitro maturation of sheep oocytes and its relation to melatonin receptors. *Int J Mol Sci*. 2017;18(4).
116. Kang JT, Koo OJ, Kwon DK, et al. Effects of melatonin on in vitro maturation of porcine oocyte and expression of melatonin receptor RNA in cumulus and granulosa cells. *J Pineal Res*. 2009;46(1):22–28.
117. Klosin A, Casas E, Hidalgo-Carcedo C, Vavouri T, Lehner B. Transgenerational transmission of environmental information in *C. elegans*. *Science*. 2017;356(6335): 320–323.
118. Yuan S, Oliver D, Schuster A, Zheng H, Yan W. Breeding scheme and maternal small RNAs affect the efficiency of transgenerational inheritance of a paramutation in mice. *Sci Rep*. 2015;5:9266.
119. Saben JL, Boudoures AL, Asghar Z, et al. Maternal metabolic syndrome programs mitochondrial dysfunction via germline changes across three generations. *Cell Rep*. 2016;16(1):1–8.
120. Padmanabhan N, Jia D, Geary-Joo C, et al. Mutation in folate metabolism causes epigenetic instability and transgenerational effects on development. *Cell*. 2013;155(1): 81–93.

121. Bourc'his D, Voinnet O. A small-RNA perspective on gametogenesis, fertilization, and early zygotic development. *Science*. 2010;330(6004):617–622.
122. Slotkin RK, Vaughn M, Borges F, et al. Epigenetic reprogramming and small RNA silencing of transposable elements in pollen. *Cell*. 2009;136(3):461–472.
123. Martinez G, Panda K, Kohler C, Slotkin RK. Silencing in sperm cells is directed by RNA movement from the surrounding nurse cell. *Nat Plants*. 2016;2:16030.
124. Liu YS. Inheritance of acquired characters in animals: a historical overview, further evidence and mechanistic explanations. *Ital J Zool*. 2011;78(4):410–417.
125. Nixon B, Stanger SJ, Mihalas BP, et al. The microRNA signature of mouse spermatozoa is substantially modified during epididymal maturation. *Biol Reprod*. 2015;93(4):91.
126. Reilly JN, McLaughlin EA, Stanger SJ, et al. Characterisation of mouse epididymosomes reveals a complex profile of microRNAs and a potential mechanism for modification of the sperm epigenome. *Sci Rep*. 2016;6:31794.
127. Rompala GR, Ferguson C, Homanics GE. Coincubation of sperm with epididymal extracellular vesicle preparations from chronic intermittent ethanol-treated mice is sufficient to impart anxiety-like and ethanol-induced behaviors to adult progeny. *Alcohol*. 2020;87:111–120.
128. Hu J, Tang T, Zeng Z, Wu J, Tan X, Yan J. The expression of small RNAs in exosomes of follicular fluid altered in human polycystic ovarian syndrome. *PeerJ*. 2020;8:e8640.
129. de Avila A, Bridi A, Andrade GM, et al. Estrous cycle impacts microRNA content in extracellular vesicles that modulate bovine cumulus cell transcripts during in vitro maturationdagger. *Biol Reprod*. 2020;102(2):362–375.
130. Nilsson E, Klukovich R, Sadler-Riggleman I, et al. Environmental toxicant induced epigenetic transgenerational inheritance of ovarian pathology and granulosa cell epigenome and transcriptome alterations: ancestral origins of polycystic ovarian syndrome and primary ovarian insufiency. *Epigenetics*. 2018;13(8):875–895.
131. El-Hayek S, Yang Q, Abbassi L, FitzHarris G, Clarke HJ. Mammalian oocytes locally remodel follicular architecture to provide the foundation for germline-soma communication. *Curr Biol*. 2018;28(7):1124–1231.e3.
132. Macaulay AD, Gilbert I, Scantland S, et al. Cumulus cell transcripts transit to the bovine oocyte in preparation for maturation. *Biol Reprod*. 2016;94(1):16.

CHAPTER

12 Modeling early-life adversity in the laboratory: animal models, their advantages, and future directions in extrapolating findings to humans

Taylor S. Campbell, Tania L. Roth

University of Delaware, Psychological and Brain Sciences, Newark, DE, United States

Introduction

State officials record child abuse into the National Child Abuse and Neglect Data System (NCANDS), which defines a victim as any child who is determined to have experienced at least one event of maltreatment.[1] In the NCANDS, maltreatment is defined as neglect, physical abuse, sexual abuse, or "other," with "other" being defined as maltreatment that does not fit into the first three categories, including threats of abuse/neglect.[1] The US Department of Health and Human Services 2020 Child Maltreatment report showed there are 678,000 abused or neglected child victims in the United States as of 2018, a 0.4% increase since 2014. Substantially more reports of abuse or neglect were obtained from the under 1 year of age group than any other age group.

Central nervous system development during the perinatal period is especially vulnerable to the consequences of stress as this is a time marked by an increase in changes at synapses across the brain (reviewed in Ref. 2). Children who experience early-life adversity (ELA) are more likely to develop depression[3] (reviewed in Refs. 2,4,5); posttraumatic stress disorder (reviewed in Refs. 2,4); bipolar disorder (reviewed in Ref. 4); substance use disorders (reviewed in Refs. 2,6); and schizophrenia (reviewed in Ref. 7). Furthermore, ELA may have implications in the development and progression of Alzheimer's disease (AD).[8,9]

With such a large number of young victims and far-reaching consequences, research aimed at understanding the impact ELA, from behavioral to molecular levels, is necessary for intervention and prevention strategies to best serve this population. Using nonhuman animals is essential in these efforts. Animal research represents an investment not just in the advancement of this science but also in human health.

Perinatal and Developmental Epigenetics. https://doi.org/10.1016/B978-0-12-821785-6.00013-X

The advantage of animal models

The use of animal models in biological research dates to Ancient Greece, with animal models still very much necessary for scientific advancement today.[10] Modeling ELA in the laboratory and using a multitude of behavioral and biological assays allows researchers to empirically parse out its consequences on development. Researchers can observe behavioral outcomes, such as the way an animal socializes with others, how the animal interacts with offspring, and how the animal performs on tests of anxiety- and depressive-like behaviors. Multiple biological assays can be utilized as well, to discover structural changes of organs, such as the brain, and underlying changes in protein expression and alterations of the genome.

Logistically speaking, animal models have many advantages. In fact, the rat genome contains roughly the same number of genes as the human genome and analogous genes for almost every gene associated with human behavior and disease[11] (Fig. 12.1). Animal models allow researchers to accrue large sample sizes in shorter time. The gestation period for rats and mice is roughly 20 days, producing litter sizes between 3 and 16 pups. Inbred rodents provide a unique opportunity for researchers to study experimental variables without the confound of genetic variation.[12] Outbred rodents are equally as advantageous because they afford genetic variability more indicative of human genetic diversity.[10] Long-term and cross-generational effects of ELA can more readily be observed using longitudinal research designs. Human lives are complex, making it impossible for researchers to control for all possible extraneous variables that can naturally occur during a person's life (reviewed in Ref. 13). Researchers can provide laboratory animals with identical diets and living quarters, to name a few additional controls. We know adversity experienced in development can lead to sex-specific long-term consequences (reviewed in

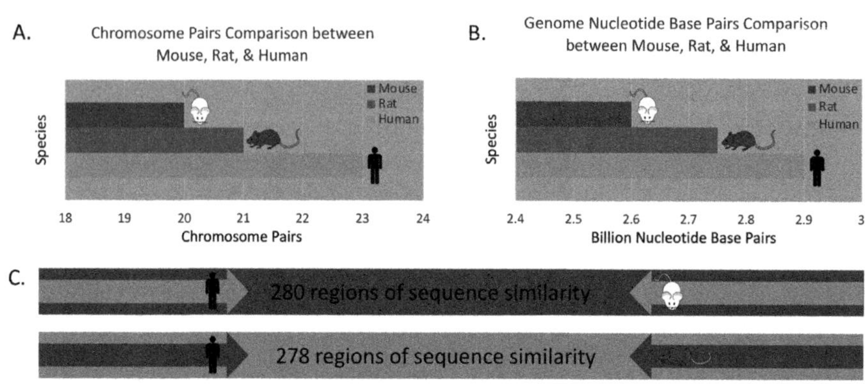

FIGURE 12.1 Human and rodent genetic similarity.

(A) Chromosome pairs comparison between mouse (20), rat (21), and human (23). (B) Nucleotide base pairs (bp) comparison between mouse (2.6 billion bp), rat (2.75 billion bp), and human (2.9 billion bp). (C) The human genome shares 280 ortholog segments with the mouse genome and 278 with the rat genome.[11]

Graphics by Brandon Campbell.

Ref. 2). Animal models provide an excellent ability to study many of the nuances of sex-specific differences. Altogether, such considerations land animal models in a prime position to experimentally parse out specific consequences of ELA.

Animal models used to study the consequences of early adversity

In the laboratory, researchers strive to achieve parallels of various aspects of ELA experienced by humans. For example, exposure to stress during gestation is associated with an increased susceptibility to develop psychological disorders (reviewed in Refs. 14,15). Prenatal stress models involve inducing stress in the mother during various gestational periods and analyzing the developmental outcomes of her pups. We know that parents who report higher levels of poverty and low resources are more likely to physically abuse their children.[16–19] A resource-poor environment for caregiving that disrupts parenting can be modeled by providing rodent mothers with inadequate nesting material.[20] Secure attachment to a caregiver, particularly the mother, is a strong moderator of susceptibility to juvenile delinquency and psychosocial deficits throughout development.[21–23] Laboratory researchers are also able to disrupt attachment by implementing a maternal separation paradigm in which the rodent mother is separated from her pups for a set amount of time over a predetermined number of days.[24] Together, these animal models (described in more detail in the following) along with others have revolutionized our understanding of just how far reaching the effects of adverse early experiences can be, ranging from the manifestation of cognitive/behavioral aberrations to physical brain changes to intergenerational consequences, with concomitant alterations of various neurochemical/endocrine systems and gene activity[25–30] (reviewed in Refs. 31,32).

Prenatal stressors can take many different forms. Some examples of stressor choices include constant white light, predator odor, novel objects placed in the home cage, restraint, unfamiliar noise for an extended time period, numerous home cage changes, and wet bedding.[33] Temporal delivery and variability of stressors require great consideration as offspring outcomes depend largely on the predictability of the stressors and the gestational time point[33,34] (Fig. 12.2).

Unpredictable stress can equate to the same stressor occurring at random intervals[34] or take the form of a chronic variable stress (CVS) model wherein a variety of random stressors are delivered at set or unpredictable time intervals.[33,35] Stress models that are unpredictable induce more aversive effects for the offspring, including motor development delays, likely due to the dam's inability to habituate or adapt to the stressor.[34] For example, pups exposed to CVS stress late in gestation have been documented to have higher birth weights,[33] whereas predictable stress in likewise late gestation results in normal birth weights.[34] In Mueller and Bale's[33] study, offspring stressed during mid- and late gestation born from stress-sensitive dams had higher birth weights that persisted into adulthood. Such data are particularly important considering the known link between high birth weight and likelihood for childhood obesity.[36] They argue for the importance of teaching stress coping skills for mothers who are more sensitive to emotional or psychological stress.

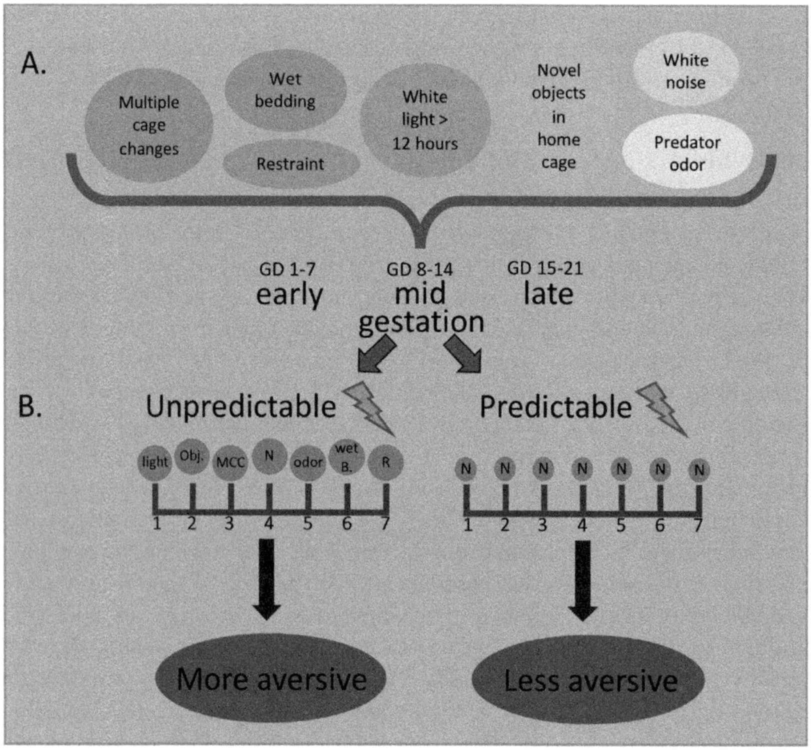

FIGURE 12.2 Prenatal stress model variation.

Diagram of prenatal stress methodology. (A) Prenatal stressor choices (from *left* to *right*) and gestation time points wherein the experimenter could target. (B) Examples of an unpredictable (*left*) and predicable prenatal stress (*right*). In the unpredictable stress experiment, the dam was exposed to a different stressor each day for 7 days. This results in more severe consequences in the pups due to the greater aversive nature of the stress. In contrast, the unpredictable stress experiment exposed the dam to the same stressor each day for 7 days. This results in less severe consequences in the pups as the dam likely habituates to the predicable stress.

MCCs, multiple cage changes; *N*, noise; *Obj.*, novel object; *R*, restraint; *wet B.*, wet bedding.

One of the longest-standing models of ELA is the maternal separation model wherein the mother is simply separated from the offspring for varying time intervals occurring anytime between the birth of the pups and weaning.[20,37] One advantage of this model is that it can be extended to any animal species that employs maternal care to young offspring. Though this model is informative in terms of offspring neurobiological and behavior development deviations caused by disrupting maternal care, divergent results on offspring outcomes occur with unstandardized methodology. Some examples of methodological cross-roads are deciding what age and duration to carry out the maternal separation (typically between 1 and 6 h), keeping pups together in the home cage while removing the dam to a novel location, separating pups from littermates, the duration of the experiment (ex. postnatal day (PN) 1–14 versus PN1-7), keeping control pups or an age-matched foster litter with the dam during the separation procedure, keeping pups on a warming pad, and placing pups near the dam such that they can still smell her and hear her vocalizations[20,37–39] (Fig. 12.3). Additionally, pup outcomes may be influenced by maternal behavior upon reunion. The dam's behavior is significantly altered after reuniting with her offspring following prolonged separation, thus influencing the quality of mother–pup interactions, a phenomenon abolished by housing a foster litter with the mother during the separation.[38]

In a limited bedding and nesting (LBN) model of ELA, the rodent mother is provided with scarce resources such that the stressful conditions of her environment elicit erratic treatment of her pups. This model to date has only been implemented in rats and mice. The LBN model was originally developed by Tallie Baram and is still a widely popular ELA model, used nationally in several laboratories.[20] In the original LBN model, also called the Baram model, the rodent mother and pups are housed continuously from PN 2–9 in a cage where they are atop a raised wire mesh platform (reviewed in Ref. 20; Fig. 3). The wire mesh adds to the severity of the stress by blocking contact with the scarce bedding material and inducing a level of discomfort to the rodent mother and pups. On PN 10, the mother and pups are moved to normal care housing cages with ample bedding, and the wire mesh is excluded from the environment. In the absence of the environmental stressors, the mother's behavior quickly normalizes, and the pups' stress is reduced (reviewed in Ref. 20). The LBN model also increases the unpredictability of the mother's behavior and results in erratic and sudden changes in the way she handles the pups.[20] In this environment, the mother may also contribute to stress in the pups due to the erratic nature of her behavior. This finding is paralleled in the human literature as fragmented sensory stimulation in early development is highly indicative of ELA consequences[22] (reviewed in Ref. 21).

The LBN model has since been modified into a milder form of ELA termed the scarcity–adversity model utilized by the Sullivan and Roth Labs (Fig. 12.3). The scarcity–adversity models include experiments wherein the wire mesh is omitted from the home cage and the mother and pups remain continually in the impoverished environment (i.e., PN 1–7 or PN 8–12), or experiments where pups are periodically exposed to aversive caregiving in a resource-poor environment outside of the home cage from PN 1–7 (reviewed in Ref. 20). Both of these variants result in a decrease of nurturing behaviors, such as licking, actively nursing, and hovering over the pups, and an increase in aversive behaviors, such as dropping and dragging, stepping on, and actively avoiding the pups. Another recent adaptation of the LBN model has included a Western diet.[40] A Western diet is one mainly comprised of little amounts of fiber and large quantities of sugar and saturated fats (reviewed in Ref. 40) and is indicative of food insecurity in the human population (reviewed in Refs. 40,41). Food insecurity is highly comorbid with poverty, both of which are risk factors

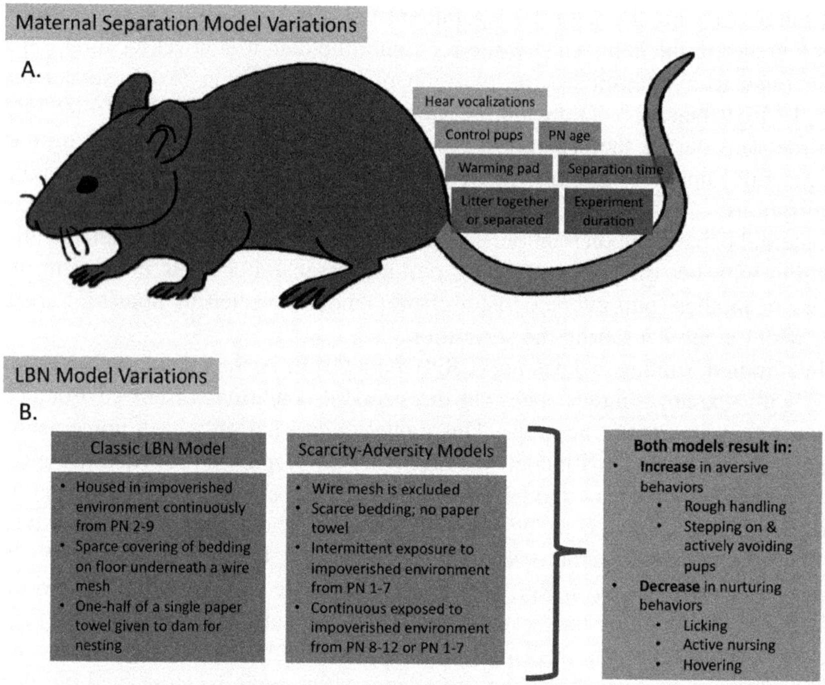

FIGURE 12.3 Postnatal stress models variations.

(A) Methodological choices when adopting a maternal separation model. Examples include keeping the pups close enough to hear the dam's vocalizations, the age of the pups (PN 1–7 or PN 1–14), placing the pups on warming pads, keeping the pups together or in isolation, the duration of the experiment, duration of the separation, and keeping control pups with the dam to control her behavior. (B) Limited bedding and nesting (LBN) model variations. In the classic LBN model, the dam and her pups are housed continuously in an impoverished environment atop wire mesh bedding from PN 2–9. A scarce amount of bedding is underneath the wire mesh to soak up excrement and the dam is given half of one paper towel as bedding material. The scarcity–adversity models describe a milder form of early-life adversity. The wire mesh is omitted, and the pups and dam are exposed intermittently to the impoverished environment (PN 1–7) or continuously (PN 8–12 or PN 1–7).[20]

Graphic by Brandon Campbell.

for the development of psychological and behavioral disorders (reviewed in Refs. 40,41).

Consequences of early adversity

Behavioral analyses

The Barker hypothesis states that the perinatal environment sets in motion neural programming that determines adult cognitive function and emotional health.[42] Consistent with this hypothesis, the perinatal period is marked by great neural plasticity that is sensitive to environmental regulation,[43] and consequently, people who experience various forms of ELA are more likely to develop cognitive dysfunction and psychiatric disorders.[44] Rodent pups that endure ELA exhibit phenotypes later in life including cognitive deficits, depressive-like behaviors, and greater anxiety levels.[20] The specific emotional and cognitive consequences of ELA are measured in the rodent by using standardized tests that measure cognitive ability and behavior. See Table 12.1 for a nonexhaustive list of documented behavioral and neurodevelopmental outcomes in rodents after experiencing ELA.

The forced swim test (FST) is a measure of a depressive-like or despair phenotype in rats and mice. Increases in the rodent's lack of movement (immobility), besides what is required to remain above water, or taking less time to become immobile, are coded as despair or depressive-like behaviors.[62] Numerous studies have demonstrated that ELA-exposed animals engage in increased immobility during the FST.[45–47] Results on the FST regarding ELA models, however, have been

Table 12.1 Summary of documented rodent behavioral and neurodevelopment consequences of ELA.

Rodent behavior	**Citations**
Cognitive deficits, depressive-like behaviors, and increased anxiety	Reviewed in Ref. 20
Increased immobility on the FST (depressive-like behavior)	Refs. 45–47
Increased latency to immobility on the FST (altered stress coping skills)	Ref. 48
No significant performance difference on FST compared with controls	Ref. 49
Increased anxiety-like behaviors on the EMP	Ref. 50
No significant performance difference on the EMP or OFT compared with controls	Refs. 48,51,52
Decreased performance on the NOR test (cognitive deficit)	Refs. 49,51,52
Decreased performance on the OLT (cognitive deficit)	Refs. 49,53
Decreased performance on the MWM (spatial and long-term memory deficit)	Refs. 52,54–56
Decreased social interaction (social behavior deficit)	Refs. 45,46,51,57
Aberrant maternal care behavior	Ref. 58
Rodent neurodevelopment	**Citation**
Aberrant amygdala development	Ref. 59; reviewed in Ref. 2
Aberrant hippocampal development	Reviewed in Refs. 2,60
HPA-axis dysfunction	Ref. 61

EMP, *elevated plus maze;* FST, *forced swim test;* MWM, *Morris water maze;* NOR, *novel object recognition;* OFT, *open field test;* OLT, *object location test.*

mixed. For example, some studies have shown rodents reared in the LBN model do not perform differently on the test compared with controls,[49] while others have observed an increased latency to immobility.[48] One way to interpret these mixed results is that ELA reared animals may also have altered stress coping skills.

The open field task (OFT) is a sensorimotor test used to measure explorative and locomotor behaviors in various rodent models. In the context of ELA animal models, the OFT can provide information on supposed anxiety levels of the animal, as less locomotor and exploratory behavior in the center of the arena would indicate greater anxiety. Additionally, anxiety can also be measured via the elevated plus maze (EPM). The maze is made up of an enclosed arm set perpendicular to an open arm. The number and duration of arm entries are recorded, with more time spent in the open arm areas corresponding to less anxiety.[63] Rats exposed to ELA have more anxiety-like behaviors and cognitive deficits as adult in the EPM and MWM.[50,52,55] However, some studies have not detected anxiety-like behaviors from the EPM or OFT.[48,51,52]

Cognitive deficits elicited by ELA exposure have been observed on a variety of tests. The novel object recognition test (NOR) and object location test (OLT) are favorable tests to measure memory in rodents because they appeal to the rodent's innate behavior of interacting with unique objects in the environment (reviewed in Ref. 64). The NOR is a memory task in which rodents are exposed to two identical objects in a familiar arena for several minutes during a training phase and the amount of time the rodent spends interacting with each object is recorded as a matter of seconds.[49] Naturally, rodents will spend more time with the novel object as they have a proclivity for novel objects and will have a memory of previously interacting with the training phase object. Mice and rats exposed to ELA have shown deficits on this task, as they did not interact more with the novel object.[49,51,52] Though these results provide evidence for ELA having an adverse effect on nonspatial memory without differences in sex, other studies have shown more variability and sex differences in deficits.[48] The OLT is similarly based on the notion that rodents will naturally spend less time with a familiar object than a novel one and discriminate between an object in a familiar location vs. a novel location. Naninck et al. 84 used the OLT to measure hippocampal-dependent cognition in ELA model mice. Results showed ELA reared male mice did not show a significant difference in the amount of time spent exploring the object in the novel location, whereas control mice and ELA female mice did demonstrate discrimination in object exploration. A similar relationship has also been documented in male rats such that their performance is impaired on this test as early as 2 months of age.[65] The Morris water maze (MWM) is a commonly used test of both spatial and long-term memory, and mice exposed to ELA have performance deficits in adulthood on hippocampus-dependent tasks such as the MWM.[54–56]

Numerous social interaction tests are also available to measure ELA-induced changes in social behavior. Testing can take place before puberty (around PN 25) to lessen the possible effects of hormonal changes in the two sexes (reviewed in Ref. 66) or later in adulthood. While results regarding changes in social behavior (i.e., social interaction) are mixed[67] and possibly sex-specific,[57,66] social deficits in ELA animals both during adolescence and adulthood have been documented.[45,46,51,57] Additionally, one can observe the effects of ELA on maternal behavior. Previous findings have shown ELA female rats grow up to exhibit significantly more abusive behaviors and less nurturing behaviors toward their pups.[58]

The timing of the behavioral testing (i.e., adolescence vs. adulthood) may explain some of the divergent results obtained between labs. For example, ELA adolescent rats can perform comparable with control rats on tests of cognitive ability or anxiety behavior[53]; however, behavioral and cognitive disruptions are evident in adulthood.[48] The exact timing and severity of ELA may also have an effect on the behavioral outcome of the animal. For example, rats showed a shorter latency to immobility on the FST when the LBN model was induced later in infancy development (PN 8–12). 45 This finding was not observed when a milder version of LNB model (scarcity–adversity) was implemented in early postnatal development, and instead, animals showed a longer latency to immobility (PN 1–7). 48 Likewise, evidence suggests that in the human, the behavioral consequences of ELA are influenced by the duration and severity of the stressors (review in Refs. 68,69). For example, using the Baram model, Molet et al. 79 found male adolescent rats to display depressive-like behaviors, whereas depressive-like behaviors in adolescence have not been observed using the milder scarcity–adversity ELA model.[48]

Consequences of early adversity

Biological analyses

Analyses of the various possible underlying biological mechanisms behind the vast cognitive and behavioral consequences of ELA are of great interest. A more complete understanding of mechanisms drives prevention and interventions strategies that could ameliorate the cognitive and emotional sequelae brought forth by aversive experiences early in development. Animal models provide the opportunity to study consequences at many levels, from structural abnormalities down to DNA alterations.

ELA has been linked to significant, though often transient, changes in body weight. After implementing an LBN model, mice pups reared in the LBN environment had a significant decrease in body weight relative to controls[49,70] that continued to PN 21,[70] but normalized relative to controls by PN 120–150.[49,70] Additionally, rats reared in a maternal separation model had lower body weights than controls discernible as early as PN 4 and still maintained at PN 51.[54] Weight lower than the standardized norms for children has been used as an identifier of malnutrition in institutionally reared children (reviewed in Ref. 71).

Internal effects of ELA can be understood by assessing organs that play a role in stress management. The thymus is involved in immune system responses and can be negatively affected by chronic stress (reviewed in Ref. 72). Previous research has shown a decrease in thymus weight due to mild chronic stress.[72] After rearing in an ELA environment, pups had a significant decrease in thymus weight at PN 9 compared with control mouse pups.[49] Male human infants who manifest a greater cortisol response to stress have been shown to exhibit smaller thymus size, while both male and female infants with a greater cortisol response threshold show disrupted immune system development.[73] This highlights possible individual and sex differences in the vulnerability of the thymus to stress. Rats exposed to chronic stress can show adrenal weight changes.[72] With the advancement of sonogram technologies, medical doctors have begun turning to the fetal adrenal gland as a possible biomarker of preterm birth risk.[16] Increased fetal adrenal volume has been associated with increased risk of preterm parturition, indicating that the fetal adrenal gland is likely particularly vulnerable to intrauterine stressors.[74] The severity ELA has been negatively correlated with lower heart rate variability; a reliable physiological measurement related to psychological disorders in humans.[75] Experiencing at least four bouts of ELA is additionally associated with cardiovascular disease later in life.[76] The pituitary gland, an integral player in an organism's stress response, shows accelerated growth in females who experienced ELA.[77]

ELA affects the development of many brain regions, including the HPA axis[61] (reviewed in Ref. 78), hippocampus[49] (reviewed in Ref. 78), prefrontal cortex,[58] amygdala[78] (reviewed in Ref. 2), and locus coeruleus (reviewed in Ref. 78). ELA causes early activation of the HPA axis, leading to a cascade of neuronal and behavioral changes, including deficits in attachment learning and elevated stress hormone levels in early adolescence.[78–80] Females, being particularly vulnerable to stress, have been shown to exhibit a blunted HPA axis response to stressors in adulthood after exposure to ELA.[61] Consistent with animal model observations, institutionally reared children show HPA axis dysregulation, with chronic activation of the HPA axis in early life leading to a blunted HPA axis response to stress.[81–83] In addition to disrupting the stress hormone system, ELA in humans is known to decrease basal levels of the attachment hormone oxytocin.[84] This finding has been recapitulated in rodent models of ELA such that these animals have altered oxytocin and arginine vasopressin levels later in life that are related to several social behavior deficits (reviewed in Ref. 85).

Though sensitivity to stress is necessary and adaptive in adulthood, early activation of the HPA axis can cause functionality changes in the brain (review in Ref. 32). ELA has been shown to quicken the maturation of the amygdala 59 and lead to an abnormal strengthening of the amygdala's connections to areas like the nucleus accumbens and thalamus, possibility taking away from frontal lobe connections as the rodent matures (reviewed in Ref. 2). Children typically show a decrease in amygdala activity when presented with parental cues (i.e., photograph of parents) as the parents act as an emotional buffer (reviewed in Ref. 86). Children adopted from institutions, however, are more likely to not show this decrease in amygdala activity and as such score higher on measures of anxiety.[86] Formerly institutionalized children show accelerated amygdala–PFC connectivity such that they exhibit a mature connectivity pattern instead of the immature connectivity that is typical of their age group.[87] This could be viewed as a possible adaptive neurologic response to stress, being that mature connections are typically related to lower anxiety.

However, the previously institutionalized children still exhibited greater anxiety levels than normally reared children.[87]

The hippocampus is highly implicated in cognitive performance and is especially vulnerable to ELA as it continues to develop after birth (reviewed in Ref. 49), with its critical period of development taking place during the first 2 weeks of life (reviewed in Ref. 2). ELA leads to disruptions in the development of hippocampal structure (i.e., dendritic spine density), synaptic connections, and function (i.e., memory deficits; reviewed in Refs. 2,60). Studies of children reared in institutions have further shed light on white matter connectivity deficits in areas that are heavily implicated in higher cognitive processing and emotional regulation (reviewed in Ref. 68). Moreover, children raised in poverty have significantly less gray matter in the frontal lobe, temporal lobe, and hippocampus.[88] This finding indicates children could be vulnerable to limitations in academic success beginning at a young age. Furthermore, structural brain region malformations might emerge differently as a function of the category or severity of the ELA experienced. For example, Hanson et al. 52 reported decreased amygdala volume in children who experienced low socioeconomic status (SES), physical abuse, and neglect but only decreased hippocampus volume in children who experienced physical abuse or low SES. The volume of these brain regions was also negatively correlated with both the frequency of ELA and increased behavioral problems.[89]

The effects of ELA can even be traced down to the level of DNA. One gene this is particularly true for is the brain-derived neurotrophic factor (BDNF) gene (reviewed in Ref. 90). Animal research has linked ELA to an increase in cortical Bdnf gene methylation.[58,91–93] DNA methylation refers to the addition of methyl groups to cytosines, with most data showing that DNA methylation is associated with inactive genes and compact chromatin structure (Fig. 12.4).

There exists a highly significant, positive correlation between the total number of childhood maltreatment encounters and average percentage of methylation of BDNF.[69] Alterations in BDNF methylation have been linked to multiple psychiatric disorders associated with ELA, such as PTSD, major depressive Disorder (MDD), 94 and borderline personality disorder.[69] More generally, research has also linked ELA to aberrations in genome-wide methylation.[95,96] See Table 12.2 for a nonexhaustive list of documented relationships between ELA and epigenetic alternations of this gene in rodents and humans.

Recovery interventions

ELA can have long-lasting consequences on later behavior. This is driven by underlying changes in brain architecture and function. Changes in environments, both early and later in life, however, can help reverse or ameliorate the consequences.

Physical exercise is known to lessen the consequences of stress, demonstrating even antidepressant-like effects (reviewed in Ref. 104). Therefore, it is plausible that exercise could serve as a valuable intervention or prevention strategy for the

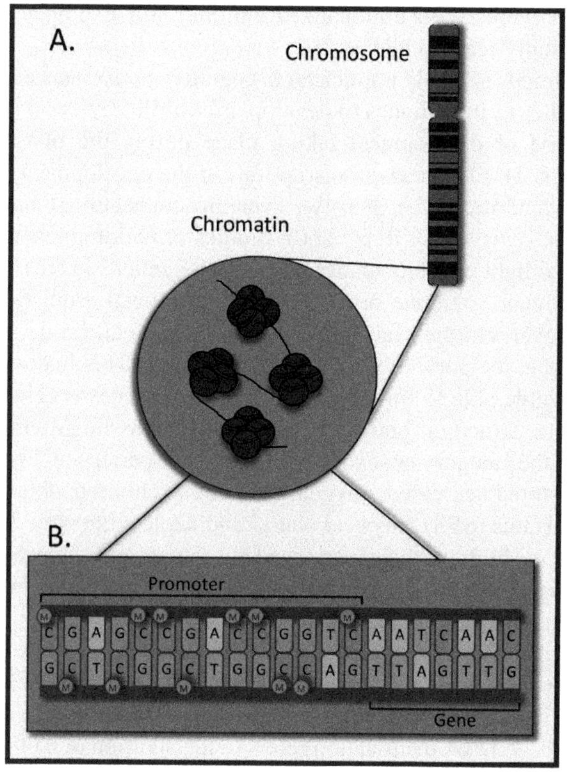

FIGURE 12.4 Gene methylation.

(A) Chromosomes are made up of DNA and proteins. Zooming in on the chromosome, we see that DNA is wrapped around structural proteins called histones organized in a tread-like strand (chromatin). (B) Zooming in further on the chromatin, we see the nucleic acid sequence. CpG sites are highly potent surrounding the promoter region of most genes. Methylation occurs when a methyl group is added to the cytosine of a CG site.

outcomes of ELA, being as ELA is a risk factor for developing depression later in life. Furthermore, voluntary exercise has been shown to contribute to neuroplasticity in both humans and nonhuman animals.[105] Generally, in an exercise intervention using rodents, the animals are housed in same sex groups including 2–3 animals with free access to a running wheel in the cage for a predetermined amount of time.[104,106] Wheel running as a form of physical exercise has been shown to rescue behavioral phenotypes suggestive of depression and abnormal gene expression (including Bdnf) in the hippocampus in rats that experienced maternal separation.[104] In humans, a strong correlation between increased exercise and psychological health exists (reviewed in Ref. 107). Additionally, acute exercise in humans has been shown to increase peripheral BDNF peripheral,[108] consistent with the notion that

Table 12.2 Summary of documented BDNF aberrations due to ELA in rodents and humans.

Brain region or peripheral cell type	Sex	Gene methylation/ expression finding	Species	Source
Prefrontal cortex	Females	Increased Bdnf methylation at the promotor region of exon IV	Long-Evans rats	Ref. 91
Prefrontal cortex	Both	Increased Bdnf methylation at the promotor region of exons IV and IX; decreased Bdnf mRNA (exon IX)	Long-Evans rats	Ref. 58
Amygdala	Females	Increased Bdnf methylation at the promotor region of exon IV	Long-Evans rats	Ref. 97
Ventral hippocampus	Both	Increased Bdnf methylation at the promotor region of exon IV	Long-Evans rats	Ref. 97
Dorsal hippocampus	Both	Increased Bdnf methylation at the promotor region of exon IV	Sprague–Dawley rats	Ref. 98
Dorsal hippocampus	Females	Decreased Bdnf expression	Wistar rats	Ref. 50
Peripheral whole-blood cells	Both	Increased BDNF methylation at one CG locus (exon not specified), additionally related to schizophrenia	*Homo sapiens*	Ref. 99
Peripheral blood leukocytes	Both	Increased BDNF methylation at the promotor region of exons I and IV, additionally related to bipolar disorder	*Homo sapiens*	Ref. 69
Peripheral whole-blood cells	Both	Increased BDNF methylation at the 3′ end of exon VI	*Homo sapiens*	Ref. 100
Peripheral blood lymphocytes	Females	Increased BDNF methylation at several CG sites on and near exon IV, additionally related to bulimia nervosa	*Homo sapiens*	Ref. 101
Peripheral whole-blood cells from umbilical cord	Both	Decreased BDNF expression in female babies and increased BDNF expression in male babies born to mothers who experienced ELA (transgenerational)	*Homo sapiens*	Ref. 102
Peripheral whole-blood cells	Both	Multiple CG site methylation differences at the BDNF promotor region of exon IV, additionally related to mood disorders	*Homo sapiens*	Ref. 103

BDNF upregulation could be a mechanism by which exercise therapies might help to manage depression.

Environmental enrichment in rodents is a facilitator of neurogenesis,[109] which can likewise help mitigate and reduce ELA consequences. Enrichment is a neuroprotectant from stress in general, aging, and brain injury; is known to enhance hippocampal molecular functioning; and also bolsters learning and memory (reviewed in Ref. 50). Environmental enrichment models in rodent experiments generally refer to the rodents being socially housed in a cage that includes a variety of toys, tunnels, ramps, and climbing mesh.[61] The complexity of the environment allows for the possibility of neuroprotection, positive epigenetic regulation, and behavioral deficit attenuation. Enrichment paradigms are also flexible regarding the number of animals in the cages, age of enrichment, duration of enrichment, and same/different sex housing. Depending on the size of the cage, rodents can be housed anywhere from 3 to 12 per cage. Enrichment can take place in adolescence or adulthood, with the duration at the discretion of the experimenter. Male and female rodents can even be housed together if enrichment occurs before sexual maturation. For example, three male and three female rats can be housed together in a large environmental enrichment cage from PN 21–36 and then moved to smaller enrichment cages in same-sex triplicates from PN 36–54.[110] In this case, enrichment spans the early adolescent and adult phases of life.

Rodent literature has shown environmental enrichment can function as a buffer to maladaptive effects of stress, as environmental enrichment that takes place concurrently with chronic stress during late adolescence ameliorates behavioral and HPA axis changes exhibited in adulthood during instances of stress in female rodents.[61] Environmental enrichment has also been shown to ameliorate memory deficits observed in working memory behavioral tasks induced by maternal separation.[110] Rats that experienced environmental enrichment after experiencing maternal separation displayed less anxiety-like behavior on the EPM and completed the MWM task in significantly less time.[50] Additionally, environmental enrichment ameliorated abnormally high plasma corticosterone levels after exposure to a stressor, returning corticosterone levels to that of control animals.[50] Epigenetically, environmental enrichment increases the amount of BDNF protein expression in the hippocampus of rats that experienced maternal separation.[50]

Consideration of timing of interventions is of course critical. For example, there is an apparent window of opportunity during pubertal development to recalibrate the HPA axis of institutionally reared children.[111] Current evidence-based interventions for solving the blunted HPA axis problem also lie in support for caregiver-based interventions.[82] The Multidimensional Treatment Foster Care for Preschoolers (MTFC-P), developed by the Oregon Social Learning Center, is an intensive intervention that provides robust resources and support to both the caregiver and the child to support the formation of an adaptive child–caregiver relationship.[83,112] The MTFC-P has been successful at lessening the increasingly blunted HPA axis response seen in institutionally reared children following adoption.[83] Another noteworthy course of action for shoring up proper emotional and behavioral development in these children is the Attachment and Biobehavioral Catch-up (ABC) intervention. The ABC intervention, developed by doctor Mary Dozier and colleagues, facilitates the caregiver–child relationship by teaching foster parents the importance of being guided by the child's lead, nurturing physical contact (i.e., hugging), and constructing an environment where the child can process emotions to stimulate self-regulation.[113] Children whose foster parents received the ABC intervention have been reported to have HPA axis regulation comparable with typically reared children and less behavioral problems postintervention compared with foster children whose foster parents did not receive the ABC training.[113] ABC training has also recently been shown to be associated with differential methylation in gene pathways associated with neuronal differentiation, neuronal development, and cell signaling.[95] Moreover, attachment training designed to increase parental sensitivity, or greater engagement in nurturing behaviors, appears to be advantageous in correctly the blunted HPA axis response to stressful situations. Children adopted internationally from institutions to parents who received attachment training showed a greater HPA axis response to stressors compared with children whose parents exhibited lower sensitivity.[81] This finding tracks with the animal literature, being that the fragmentation and inconsistent delivery of nurturing behaviors by the dam in the LBN model is proposed to be a major contributor to maladaptive rodent pup outcomes. Hence, caregiver-based relationship training likely has lasting therapeutic effects on the state of the child's epigenome, endocrine system, and behavior.

Additional work provides further evidence that the epigenome retains its responsivity and is a plausible mechanism for intervention outside of sensitive periods of development. For example, adult rats treated with a DNA methylation inhibitor showed recovery of ELA-induced aberrant DNA methylation[58] and behavior.[114] Furthermore, BPD patients with a high number of childhood trauma, after being treated with intensive-dialectical behavior therapy (I-DBT), the main treatment available for BPD, showed a significant decrease in BDNF methylation and BPD symptomatology.[69]

Summary

Animal models strive to elucidate the etiology and life course of behavior and diseases that would be impossible to discern using human studies alone.[115] Behavioral assessments measure the rodent's actions and then aim to liken these actions to human development and psychological states. We must though be critical in our interpretations and aware of limitations in extrapolation. For example, the forced swim test (FST) has historically been used as a measure of depressive-like behavior. Recently, its construct validity in measuring depression and despair has been called into question,[62,116] with more recent interpretations suggesting the test may be better suited for measuring the development of aberrant stress coping skills.[116] As with making assessments for any realm of behavior, employing a battery of behavioral tests designed to target the same or similar constructs is the best procedure.

Pharmacological treatments identified in preclinical animal models have often failed in clinical trials.[115,117] Rodents of different species, and even strains, can develop divergent behavioral and disease trajectories, challenging translation.[118] Variability in animal models of ELA can additionally lead to disparate outcomes.[118] It is thus worthwhile to examine ELA treatment prevention and intervention strategies in a variety of ELA models, ensuring sample sizes be utilized that meet appropriate power constraints,[118] detailing nuances in procedures, and examining the same outcome or treatment in multiple species/strains and model types.

Animal models nonetheless remain essential for determining the causality between ELA and epigenetic variation through rigorous experimentation. Using animal models, researchers can experimentally manipulate variables, such as stress exposure, and strictly control the environment in ways that are impossible to do in human studies. Being that ethical science practices prohibit scientists from intentionally exposing human infants to stress or abuse, most human studies on ELA are post hoc and correlational by nature. These studies are a valuable start to understanding the epigenetic and far-reaching effects of ELA, but they fail to provide a full picture. Animal models fill in the scientific gaps by determining the direct effect of ELA on the epigenome. In particular, they allow us to directly investigate neural epigenetic activity in vivo and compare it to peripheral (blood, saliva, hair, etc.) epigenetic activity that is measurable in humans, thereby extrapolating the effect of ELA on the human brain throughout development. Findings in animal models inform human scientists on the underlying mechanisms by which ELA exerts both short- and long-term consequences and thereby allows them to narrow in their focus on potential therapeutic targets.

As reviewed in this chapter, we observe behavioral and biological outcomes of ELA in animal models with direct parallels to humans. It is staggering to think such slight differences in early caregiving environments can have such a profound effect on behavioral and biological development. With the help of numerous research studies over the past several decades, it has become clear that environmental influences can have tremendous effects on genes and their end products, with organism–environment interactions now recognized as substantial contributors of individual symptomology and etiology of mental illness.[119] Scientific research is imperative to further elucidate the biological mechanisms behind the behavioral and biological aberrations caused by exposure to ELA. At this time, we can surmise that future treatments for those affected by ELA need to be tailored to the individual. Continued research of ELA prevention and treatment strategies in rodents sets the stage for empirically tested community resources and intervention programs for the best practices and resources to assist individuals in need.

Acknowledgments

Work on this chapter was supported by grants from The National Institute of Health and The Eunice Kennedy Shriver National Institute of Child Health and Human Development (R01HD087509).

References

1. U.S. Department of Health & Human Services, Administration for Children and Families, Administration on Children, Youth and Families, Children's Bureau. *Child Maltreatment 2018*; 2020. Available from: https://www.acf.hhs.gov/cb/research-data-technology/statistics-research/child-maltreatment.
2. Van Bodegom M, Homberg JR, Henckens MJAG. Modulation of the hypothalamic-pituitary-adrenal axis by early life stress exposure. *Front Cell Neurosci.* 2017;11. https://doi.org/10.3389/fncel.2017.00087.
3. Cicchetti D, Rogosch FA. Genetic moderation of child maltreatment effects on depression and internalizing symptoms by serotonin transporter linked polymorphic region (5-HTTLPR), brain-derived neurotrophic factor (BDNF), norepinephrine transporter (NET), and corticotropin releasing hormone receptor 1 (CRHR1) genes in African American children. *Dev Psychopathol.* 2014;26:1219–1239. https://doi.org/10.1017/S0954579414000984.
4. Syed S, Nemeroff C. Early life stress, mood, and anxiety disorders. *Chronic Stress.* 2017;247054701769446. https://doi.org/10.1177/2470547017694461.
5. Fumagalli F, Molteni R, Racagni G, Riva MA. Stress during development: impact on neuroplasticity and relevance to psychopathology. *Prog Neurobiol.* 2007;81(4): 197–217. https://doi.org/10.1016/j.pneurobio.2007.01.002.
6. Cicchetti D, Handley ED. Child maltreatment and the development of substance use and disorder. *Neurobiol Stress.* 2019;10. https://doi.org/10.1016/j.ynstr.2018.100144.
7. Bahari-Javan S, Varbanov H, Halder R, et al. HDAC1 links early life stress to schizophrenialike phenotypes. *Proc Natl Acad Sci USA.* 2017;114(23):E4686–E4694. https://doi.org/10.1073/pnas.1613842114.
8. Hoeijmakers L, Ruigrok SR, Amelianchik A, et al. Early-life stress lastingly alters the neuroinflammatory response to amyloid pathology in an Alzheimer's disease mouse model. *Brain Behav Immun.* 2017;63:160–175. https://doi.org/10.1016/j.bbi.2016.12.023.
9. Lesuis SL, Maurin H, Borghgraef P, Lucassen PJ, Van Leuven F, Krugers HJ. Positive and negative early life experiences differentially modulate long term survival and amyloid protein levels in a mouse model of Alzheimer's disease. *Oncotarget.* 2016; 7(26):39118–39135. https://doi.org/10.18632/oncotarget.9776.
10. Ericsson AC, Crim MJ, Franklin CL. A brief history of animal modeling. *Mo Med.* 2013;110(3):201–205.
11. Gibbs RA, Weinstock GM, Metzker ML, et al. Genome sequence of the brown Norway rat yields insights into mammalian evolution. *Nature.* 2004;428(6982):493–520. https://doi.org/10.1038/nature02426.
12. *National Academy of Sciences (US) and Institute of Medicine (US) Committee on the Use of Animals in*; 1991. https://www.ncbi.nlm.nih.gov/books/NBK223354/.
13. Phillips NLH, Roth TL. Animal models and their contribution to our understanding of the relationship between environments, epigenetic modifications, and behavior. *Genes.* 2019;10(1). https://doi.org/10.3390/genes10010047.
14. Faa G, Manchia M, Pintus R, Gerosa C, Marcialis MA, Fanos V. Fetal programming of neuropsychiatric disorders. *Birth Defects Res C Embryo Today Rev.* 2016;108(3): 207–223. https://doi.org/10.1002/bdrc.21139.

15. Glover V, O'Donnell KJ, O'Connor TG, Fisher J. Prenatal maternal stress, fetal programming, and mechanisms underlying later psychopathology—a global perspective. *Dev Psychopathol.* 2018;30(3):843–854. https://doi.org/10.1017/S095457941800038X.
16. Kim D, Epperson CN, Ewing G, Appleby D, Sammel MD, Wang E. Methodology for using 3-dimensional sonography to measure fetal adrenal gland volumes in pregnant women with and without early life stress. *J Ultrasound Med.* 2016;35(9):2029–2037. https://doi.org/10.7863/ultra.15.10046.
17. Maguire-Jack K, Font SA. Community and individual risk factors for physical child abuse and child neglect: variations by poverty status. *Child Maltreat.* 2017;22(3): 215–226. https://doi.org/10.1177/1077559517711806.
18. Prendergast S, MacPhee D. Trajectories of maternal aggression in early childhood: associations with parenting stress, family resources, and neighborhood cohesion. *Child Abuse Neglect.* 2020;99. https://doi.org/10.1016/j.chiabu.2019.104315.
19. Whipple EE, Webster-Stratton C. The role of parental stress in physically abusive families. *Child Abuse Neglect.* 1991;15(3):279–291. https://doi.org/10.1016/0145-2134(91)90072-L.
20. Walker CD, Bath KG, Joels M, et al. Chronic early life stress induced by limited bedding and nesting (LBN) material in rodents: critical considerations of methodology, outcomes and translational potential. *Stress.* 2017;20(5):421–448. https://doi.org/10.1080/10253890.2017.1343296.
21. Baram TZ, Davis EP, Obenaus A, et al. Fragmentation and unpredictability of early-life experience in mental disorders. *Am J Psychiatry.* 2012;169(9):907–915. https://doi.org/10.1176/appi.ajp.2012.11091347.
22. Bowlby J. Research into the origins of delinquent behaviour. *Br Med J.* 1950;1(4653): 570–573. https://doi.org/10.1136/bmj.1.4653.570.
23. Schroeder RD, Higgins GE, Mowen TJ. Maternal attachment trajectories and criminal offending by race. *Am J Crim Justice.* 2014;39(1):155–171. https://doi.org/10.1007/s12103-012-9192-0.
24. Banqueri M, Méndez M, Gómez-Lázaro E, Arias JL. Early life stress by repeated maternal separation induces long-term neuroinflammatory response in glial cells of male rats. *Stress.* 2019;22(5):563–570. https://doi.org/10.1080/10253890.2019.1604666.
25. Esposito EA, Jones MJ, Doom JR, MacIsaac JL, Gunnar MR, Kobor MS. Differential DNA methylation in peripheral blood mononuclear cells in adolescents exposed to significant early but not later childhood adversity. *Dev Psychopathol.* 2016;28(4): 1385–1399. https://doi.org/10.1017/S0954579416000055.
26. Hodel AS, Hunt RH, Cowell RA, Van Den Heuvel SE, Gunnar MR, Thomas KM. Duration of early adversity and structural brain development in post-institutionalized adolescents. *NeuroImage.* 2015;105:112–119. https://doi.org/10.1016/j.neuroimage.2014.10.020.
27. Loman MM, Johnson AE, Westerlund A, Pollak SD, Nelson CA, Gunnar MR. The effect of early deprivation on executive attention in middle childhood. *J Child Psychol Psychiatry Allied Discip.* 2013;54(1):37–45. https://doi.org/10.1111/j.1469-7610.2012.02602.x.
28. Loman MM, Wiik KL, Frenn KA, Pollak SD, Gunnar MR. Postinstitutionalized children's development: growth, cognitive, and language outcomes. *JDBP (J Dev Behav Pediatr).* 2009;30(5):426–434. https://doi.org/10.1097/DBP.0b013e3181b1fd08.

29. Lupien SJ, McEwen BS, Gunnar MR, Heim C. Effects of stress throughout the lifespan on the brain, behaviour and cognition. *Nat Rev Neurosci*. 2009;10(6):434–445. https://doi.org/10.1038/nrn2639.
30. Pitula CE, DePasquale CE, Mliner SB, Gunnar MR. Peer problems among postinstitutionalized, internationally adopted children: relations to hypocortisolism, parenting quality, and ADHD symptoms. *Child Dev*. 2019;90(3):e339–e355. https://doi.org/10.1111/cdev.12986.
31. Gretchenvan S, Martin R, Francesca M, B FT, M MI, Mike S. Transgenerational inheritance of behavioral and metabolic effects of paternal exposure to traumatic stress in early postnatal life: evidence in the 4th generation. *Environ Epigenetics*. 2018. https://doi.org/10.1093/eep/dvy023.
32. Sánchez MM, Ladd CO, Plotsky PM. Early adverse experience as a developmental risk factor for later psychopathology: evidence from rodent and primate models. *Dev Psychopathol*. 2001;13(3):419–449. https://doi.org/10.1017/S0954579401003029.
33. Mueller B, Bale T. Impact of prenatal stress on long term body weight is dependent on timing and maternal sensitivity. *Physiol Behav*. 2006:605–614. https://doi.org/10.1016/j.physbeh.2006.05.019.
34. Freide E, Weinstock M. The effects of prenatal exposure to predictable or unpredictable stress on early development in the rat. *Dev Psychobiol*. 1984;17(6):651–660. https://doi.org/10.1002/dev.420170607.
35. Monteiro S, Roque S, de Sá-Calçada D, Sousa N, Correia-Neves M, Cerqueira JJ. An efficient chronic unpredictable stress protocol to induce stress-related responses in C57BL/6 mice. *Front Psychiatry*. 2015;6. https://doi.org/10.3389/fpsyt.2015.00006.
36. Gillman MW, Rifas-Shiman S, Berkey CS, Field AE, Colditz GA. Maternal gestational diabetes, birth weight, and adolescent obesity. *Pediatrics*. 2003;111(3):e221–226. https://doi.org/10.1542/peds.111.3.e221.
37. Wang D, Levine JLS, Avila-Quintero V, Bloch M, Kaffman A. Systematic review and meta-analysis: effects of maternal separation on anxiety-like behavior in rodents. *Transl Psychiatry*. 2020;10(1). https://doi.org/10.1038/s41398-020-0856-0.
38. Huot RL, Gonzalez ME, Ladd CO, Thrivikraman KV, Plotsky PM. Foster litters prevent hypothalamic-pituitary-adrenal axis sensitization mediated by neonatal maternal separation. *Psychoneuroendocrinology*. 2004;29(2):279–289. https://doi.org/10.1016/S0306-4530(03)00028-3.
39. Zimmerberg B, Sageser KA. Comparison of two rodent models of maternal separation on juvenile social behavior. *Front Psychiatry*. 2011;2. https://doi.org/10.3389/fpsyt.2011.00039.
40. Strzelewicz AR, Ordoñes Sanchez E, Rondón-Ortiz AN, et al. Access to a high resource environment protects against accelerated maturation following early life stress: a translational animal model of high, medium and low security settings. *Horm Behav*. 2019;111:46–59. https://doi.org/10.1016/j.yhbeh.2019.01.003.
41. Slopen N, Fitzmaurice G, Williams DR, Gilman SE. Poverty, food insecurity, and the behavior for childhood internalizing and externalizing disorders. *J Am Acad Child Adolesc Psychiatry*. 2010;49(5):444–452. https://doi.org/10.1097/00004583-201005000-00005.
42. Hoeijmakers L, Lucassen PJ, Korosi A. The interplay of early-life stress, nutrition, and immune activation programs adult hippocampal structure and function. *Front Mol Neurosci*. 2015;7:1–16. https://doi.org/10.3389/fnmol.2014.00103.

43. Andersen SL. Trajectories of brain development: point of vulnerability or window of opportunity? *Neurosci Biobehav Rev.* 2003;27(1−2):3−18. https://doi.org/10.1016/S0149-7634(03)00005-8.
44. Keyes KM, Eaton NR, Krueger RF, et al. Childhood maltreatment and the structure of common psychiatric disorders. *Br J Psychiatry.* 2012;200(2):107−115. https://doi.org/10.1192/bjp.bp.111.093062.
45. Raineki C, Cortes MR, Belnoue L, Sullivan RM. Effects of early-life abuse differ across development: infant social behavior deficits are followed by adolescent depressive-like behaviors mediated by the amygdala. *J Neurosci.* 2012:7758−7765. https://doi.org/10.1523/JNEUROSCI.5843-11.2012.
46. Rincón-Cortés M, Sullivan RM. Emergence of social behavior deficit, blunted corticolimbic activity and adult depression-like behavior in a rodent model of maternal maltreatment. *Transl Psychiatry.* 2016;6(10):e930. https://doi.org/10.1038/tp.2016.205.
47. Zhang J, Abdallah CG, Chen Y, et al. Behavioral deficits, abnormal corticosterone, and reduced prefrontal metabolites of adolescent rats subject to early life stress. *Neurosci Lett.* 2013;545:132−137. https://doi.org/10.1016/j.neulet.2013.04.035.
48. Doherty TS, Blaze J, Keller SM, Roth TL. Phenotypic outcomes in adolescence and adulthood in the scarcity-adversity model of low nesting resources outside the home cage. *Dev Psychobiol.* 2017;59(6):703−714. https://doi.org/10.1002/dev.21547.
49. Naninck EFG, Hoeijmakers L, Kakava-Georgiadou N, et al. Chronic early life stress alters developmental and adult neurogenesis and impairs cognitive function in mice. *Hippocampus.* 2015;25(3):309−328. https://doi.org/10.1002/hipo.22374.
50. Dandi E, Kalamari A, Touloumi O, et al. Beneficial effects of environmental enrichment on behavior, stress reactivity and synaptophysin/BDNF expression in hippocampus following early life stress. *Int J Dev Neurosci.* 2018;67:19−32. https://doi.org/10.1016/j.ijdevneu.2018.03.003.
51. Benetti F, Mello PB, Bonini JS, Monteiro S, Cammarota M, Izquierdo I. Early postnatal maternal deprivation in rats induces memory deficits in adult life that can be reversed by donepezil and galantamine. *Int J Dev Neurosci.* 2009;27(1):59−64. https://doi.org/10.1016/j.ijdevneu.2008.09.200.
52. Brunson KL, Kramár E, Lin B, et al. Mechanisms of late-onset cognitive decline after early-life stress. *J Neurosci.* 2005;25(41):9328−9338. https://doi.org/10.1523/JNEUROSCI.2281-05.2005.
53. Molet J, Heins K, Zhuo X, et al. Fragmentation and high entropy of neonatal experience predict adolescent emotional outcome. *Transl Psychiatry.* 2016;6. https://doi.org/10.1038/tp.2015.200.
54. Oomen CA, Soeters H, Audureau N, et al. Severe early life stress hampers spatial learning and neurogenesis, but improves hippocampal synaptic plasticity and emotional learning under high-stress conditions in adulthood. *J Neurosci.* 2010:6635−6645. https://doi.org/10.1523/JNEUROSCI.0247-10.2010.
55. Ivy AS, Rex CS, Chen Y, et al. Hippocampal dysfunction and cognitive impairments provoked by chronic early-life stress involve excessive activation of CRH receptors. *J Neurosci.* 2010;30(39):13005−13015. https://doi.org/10.1523/JNEUROSCI.1784-10.2010.
56. Rice CJ, Sandman CA, Lenjavi MR, Baram TZ. A novel mouse model for acute and long-lasting consequences of early life stress. *Endocrinology.* 2008;149(10):4892−4900. https://doi.org/10.1210/en.2008-0633.

57. Kentrop J, Smid CR, Achterberg EJM, et al. Effects of maternal deprivation and complex housing on rat social behavior in adolescence and adulthood. *Front Behav Neurosci*. 2018;12. https://doi.org/10.3389/fnbeh.2018.00193.
58. Roth TL, Lubin FD, Funk AJ, Sweatt JD. Lasting epigenetic influence of early-life adversity on the BDNF gene. *Biol Psychiatry*. 2009;65(9):760–769. https://doi.org/10.1016/j.biopsych.2008.11.028.
59. Sullivan RM, Holman PJ. Transitions in sensitive period attachment learning in infancy: the role of corticosterone. *Neurosci Biobehav Rev*. 2010;34(6):835–844. https://doi.org/10.1016/j.neubiorev.2009.11.010.
60. Chen Y, Rex CS, Rice CJ, et al. Correlated memory defects and hippocampal dendritic spine loss after acute stress involve corticotropin-releasing hormone signaling. *Proc Natl Acad Sci USA*. 2010;107(29):13123–13128. https://doi.org/10.1073/pnas.1003825107.
61. Smith BL, Morano RL, Ulrich-Lai YM, Myers B, Solomon MB, Herman JP. Adolescent environmental enrichment prevents behavioral and physiological sequelae of adolescent chronic stress in female (but not male) rats. *Stress*. 2018;21(5):464–473. https://doi.org/10.1080/10253890.2017.1402883.
62. Yankelevitch-Yahav R, Franko M, Huly A, Doron R. The forced swim test as a model of depressive-like behavior. *JoVE*. 2015;(97):2015. https://doi.org/10.3791/52587.
63. Walf AA, Frye CA. The use of the elevated plus maze as an assay of anxiety-related behavior in rodents. *Nat Protoc*. 2007;2(2):322–328. https://doi.org/10.1038/nprot.2007.44.
64. Lueptow LM. Novel object recognition test for the investigation of learning and memory in mice. *JoVE*. 2017;2017(126). https://doi.org/10.3791/55718.
65. Molet J, Maras PM, Kinney-Lang E, et al. MRI uncovers disrupted hippocampal microstructure that underlies memory impairments after early-life adversity. *Hippocampus*. 2016;26(12):1618–1632. https://doi.org/10.1002/hipo.22661.
66. Farrell MR, Holland FH, Shansky RM, Brenhouse HC. Sex-specific effects of early life stress on social interaction and prefrontal cortex dendritic morphology in young rats. *Behav Brain Res*. 2016;310:119–125. https://doi.org/10.1016/j.bbr.2016.05.009.
67. Bondar NP, Lepeshko AA, Reshetnikov VV. Effects of early-life stress on social and anxiety-like behaviors in adult mice: sex-specific effects. *Behav Neurol*. 2018;2018. https://doi.org/10.1155/2018/1538931.
68. Nelson CA, Bos K, Gunnar MR, Sonuga-Barke EJS. V. The neurobiological toll of early human deprivation. *Monogr Soc Res Child Dev*. 2011;76(4):127–146. https://doi.org/10.1111/j.1540-5834.2011.00630.x.
69. Perroud N, Salzmann A, Prada P, et al. Response to psychotherapy in borderline personality disorder and methylation status of the BDNF gene. *Transl Psychiatry*. 2013;3. https://doi.org/10.1038/tp.2012.140.
70. Arp JM, ter Horst JP, Loi M, et al. Blocking glucocorticoid receptors at adolescent age prevents enhanced freezing between repeated cue-exposures after conditioned fear in adult mice raised under chronic early life stress. *Neurobiol Learn Mem*. 2016;133:30–38. https://doi.org/10.1016/j.nlm.2016.05.009.
71. van Ijzendoorn MH, Palacios J, Sonuga-Barke EJS, et al. I. Children in institutional care: delayed development and resilience. *Monogr Soc Res Child Dev*. 2011;76(4):8–30. https://doi.org/10.1111/j.1540-5834.2011.00626.x.
72. Kioukia-Fougia N, Antoniou K, Bekris S, Liapi C, Christofidis I, Papadopoulou-Daifoti Z. The effects of stress exposure on the hypothalamic-pituitary-adrenal axis,

thymus, thyroid hormones and glucose levels. *Prog Neuro Psychopharmacol Biol Psychiatry*. 2002;26(5):823–830. https://doi.org/10.1016/S0278-5846(01)00297-4.

73. Huda MN, Ahmad SM, Alam MJ, et al. Infant cortisol stress–response is associated with thymic function and vaccine response. *Stress*. 2019;22(1):36–43. https://doi.org/10.1080/10253890.2018.1484445.
74. Turan OM, Turan S, Funai EF, et al. Ultrasound measurement of fetal adrenal gland enlargement: an accurate predictor of preterm birth. *Am J Obstet Gynecol*. 2011; 204(4). https://doi.org/10.1016/j.ajog.2010.11.034, 311-311.e10.
75. Meyer PW, Müller LE, Zastrow A, et al. Heart rate variability in patients with post-traumatic stress disorder or borderline personality disorder: relationship to early life maltreatment. *J Neural Transm*. 2016;123(9):1107–1118. https://doi.org/10.1007/s00702-016-1584-8.
76. Hughes K, Bellis MA, Hardcastle KA, et al. The effect of multiple adverse childhood experiences on health: a systematic review and meta-analysis. *Lancet Public Health*. 2017;2(8):e356–e366. https://doi.org/10.1016/S2468-2667(17)30118-4.
77. Ganella DE, Allen NB, Simmons JG, et al. Early life stress alters pituitary growth during adolescence-A longitudinal study. *Psychoneuroendocrinology*. 2015;53:185–194. https://doi.org/10.1016/j.psyneuen.2015.01.005.
78. Moriceau S, Shionoya K, Jakubs K, Sullivan RM. Early-life stress disrupts attachment learning: the role of amygdala corticosterone, locus ceruleus corticotropin releasing hormone, and olfactory bulb norepinephrine. *J Neurosci*. 2009:15745–15755. https://doi.org/10.1523/JNEUROSCI.4106-09.2009.
79. Baram TZ, Yi S, Avishai-Eliner S, Schultz L. Developmental neurobiology of the stress response: multilevel regulation of corticotropin-releasing hormone function. *Ann N Y Acad Sci*. 1997;814:252–265. https://doi.org/10.1111/j.1749-6632.1997.tb46161.x.
80. Moussaoui N, Jacobs JP, Larauche M, et al. Chronic early-life stress in rat pups alters basal corticosterone, intestinal permeability, and fecal microbiota at weaning: influence of sex. *J Neurogastroenterol Motil*. 2017;23(1):135–143. https://doi.org/10.5056/jnm16105.
81. DePasquale CE, Raby KL, Hoye J, Dozier M. Parenting predicts strange situation cortisol reactivity among children adopted internationally. *Psychoneuroendocrinology*. 2018;89:86–91. https://doi.org/10.1016/j.psyneuen.2018.01.003.
82. Fisher PA, Gunnar MR, Dozier M, Bruce J, Pears KC. Effects of therapeutic interventions for foster children on behavioral problems, caregiver attachment, and stress regulatory neural systems. *Ann N Y Acad Sci*. 2006;1094:215–225. https://doi.org/10.1196/annals.1376.023.
83. Fisher PA, Van Ryzin MJ, Gunnar MR. Mitigating HPA axis dysregulation associated with placement changes in foster care. *Psychoneuroendocrinology*. 2011;36(4): 531–539. https://doi.org/10.1016/j.psyneuen.2010.08.007.
84. Yirmiya K, Motsan S, Zagoory-Sharon O, Feldman R. Human attachment triggers different social buffering mechanisms under high and low early life stress rearing. *Int J Psychophysiol*. 2020;152:72–80. https://doi.org/10.1016/j.ijpsycho.2020.04.001.
85. Kompier NF, Keysers C, Gazzola V, Lucassen PJ, Krugers HJ. Early life adversity and adult social behavior: focus on arginine vasopressin and oxytocin as potential mediators. *Front Behav Neurosci*. 2019;13. https://doi.org/10.3389/fnbeh.2019.00143.
86. Callaghan BL, Gee DG, Gabard-Durnam L, et al. Decreased amygdala reactivity to parent cues protects against anxiety following early adversity: an examination across

3 years. *Biol Psychiatry Cogn Neurosci Neuroimaging*. 2019;4(7):664–671. https://doi.org/10.1016/j.bpsc.2019.02.001.

87. Gee DG, Gabard-Durnam LJ, Flannery J, et al. Early developmental emergence of human amygdala-prefrontal connectivity after maternal deprivation. *Proc Natl Acad Sci USA*. 2013;110(39):15638–15643. https://doi.org/10.1073/pnas.1307893110.
88. Hair NL, Hanson JL, Wolfe BL, Pollak SD. Association of child poverty, brain development, and academic achievement. *JAMA Pediatr.* 2015;169(9):822–829. https://doi.org/10.1001/jamapediatrics.2015.1475.
89. Hanson JL, Nacewicz BM, Sutterer MJ, et al. Behavioral problems after early life stress: contributions of the hippocampus and amygdala. *Biol Psychiatry.* 2015;77(4):314–323. https://doi.org/10.1016/j.biopsych.2014.04.020.
90. Notaras M, van den Buuse M. Neurobiology of BDNF in fear memory, sensitivity to stress, and stress-related disorders. *Mol Psychiatry.* 2020. https://doi.org/10.1038/s41380-019-0639-2.
91. Blaze J, Roth TL. Caregiver maltreatment causes altered neuronal DNA methylation in female rodents. *Dev Psychopathol.* 2017;29(2):477–489. https://doi.org/10.1017/S0954579417000128.
92. Blaze J, Scheuing L, Roth TL. Differential methylation of genes in the medial prefrontal cortex of developing and adult rats following exposure to maltreatment or nurturing care during infancy. *Dev Neurosci.* 2013;35(4):306–316. https://doi.org/10.1159/000350716.
93. Doherty TS, Chajes JR, Reich L, Duffy HBD, Roth TL. Preventing epigenetic traces of caregiver maltreatment: a role for HDAC inhibition. *Int J Dev Neurosci.* 2019;78:178–184. https://doi.org/10.1016/j.ijdevneu.2019.05.002.
94. Hing B, Sathyaputri L, Potash JB. A comprehensive review of genetic and epigenetic mechanisms that regulate BDNF expression and function with relevance to major depressive disorder. *Am J Med Genet B Neuropsychiatr Genet.* 2018;177(2):143–167. https://doi.org/10.1002/ajmg.b.32616.
95. Hoye JR, Cheishvili D, Yarger HA, Roth TL, Szyf M, Dozier M. Preliminary indications that the attachment and biobehavioral catch-up intervention alters DNA methylation in maltreated children. *Dev Psychopathol.* 2019. https://doi.org/10.1017/S0954579419001421.
96. Naumova OY, Lee M, Koposov R, Szyf M, Dozier M, Grigorenko EL. Differential patterns of whole-genome DNA methylation in institutionalized children and children raised by their biological parents. *Dev Psychopathol.* 2012;24(1):143–155. https://doi.org/10.1017/S0954579411000605.
97. Doherty TS, Forster A, Roth TL. Global and gene-specific DNA methylation alterations in the adolescent amygdala and hippocampus in an animal model of caregiver maltreatment. *Behav Brain Res.* 2016;298:55–61. https://doi.org/10.1016/j.bbr.2015.05.028.
98. Coley EJL, Demaestri C, Ganguly P, et al. Cross-generational transmission of early life stress effects on HPA regulators and bdnf are mediated by sex, lineage, and upbringing. *Front Behav Neurosci.* 2019;13. https://doi.org/10.3389/fnbeh.2019.00101.
99. Barker V, Walker RM, Evans KL, Lawrie SM. Methylation of glucocorticoid receptor (NR3C1), BDNF and oxytocin receptor genes in association with childhood maltreatment in schizophrenia and schizoaffective disorder. *Schizophr Res.* 2020;216:529–531. https://doi.org/10.1016/j.schres.2019.11.050.

100. Unternaehrer E, Meyer AH, Burkhardt SCA, et al. Childhood maternal care is associated with DNA methylation of the genes for brain-derived neurotrophic factor (BDNF) and oxytocin receptor (OXTR) in peripheral blood cells in adult men and women. *Stress*. 2015;18(4):451–461. https://doi.org/10.3109/10253890.2015.1038992.
101. Thaler L, Gauvin L, Joober R, et al. Methylation of BDNF in women with bulimic eating syndromes: associations with childhood abuse and borderline personality disorder. *Prog Neuro Psychopharmacol Biol Psychiatry*. 2014;54:43–49. https://doi.org/10.1016/j.pnpbp.2014.04.010.
102. Pilkay SR, Combs-Orme T, Tylavsky F, Bush N, Smith AK. Maternal trauma and fear history predict BDNF methylation and gene expression in newborns. *PeerJ*. 2020; 2020(5). https://doi.org/10.7717/peerj.8858.
103. Guo J, Yang Y, Zheng X, et al. CpG methylation of brain-derived the neurotrophic factor gene promoter as a potent diagnostic and prognostic biomarker for post-traumatic stress disorder. *Int J Clin Exp Pathol*. 2018;11(10):5101–5109.
104. Masrour FF, Peeri M, Azarbayjani MA, Hosseini MJ. Voluntary exercise during adolescence mitigated negative the effects of maternal separation stress on the depressive-like behaviors of adult male rats: role of NMDA receptors. *Neurochem Res*. 2018;43(5): 1067–1074. https://doi.org/10.1007/s11064-018-2519-6.
105. Hötting K, Röder B. Beneficial effects of physical exercise on neuroplasticity and cognition. *Neurosci Biobehav Rev*. 2013;37(9):2243–2257. https://doi.org/10.1016/j.neubiorev.2013.04.005.
106. Abbink MR, Naninck EFG, Lucassen PJ, Korosi A. Early-life stress diminishes the increase in neurogenesis after exercise in adult female mice. *Hippocampus*. 2017;27(8): 839–844. https://doi.org/10.1002/hipo.22745.
107. Blake H. Physical activity and exercise in the treatment of depression. *Front Psychiatry*. 2012;3. https://doi.org/10.3389/fpsyt.2012.00106.
108. Heyman E, Gamelin FX, Goekint M, et al. Intense exercise increases circulating endocannabinoid and BDNF levels in humans-possible implications for reward and depression. *Psychoneuroendocrinology*. 2012;37(6):844–851. https://doi.org/10.1016/j.psyneuen.2011.09.017.
109. Monteiro BMM, Moreira FA, Massensini AR, Moraes MFD, Pereira GS. Enriched environment increases neurogenesis and improves social memory persistence in socially isolated adult mice. *Hippocampus*. 2014;24(2):239–248. https://doi.org/10.1002/hipo.22218.
110. do Prado CH, Narahari T, Holland FH, Lee HN, Murthy SK, Brenhouse HC. Effects of early adolescent environmental enrichment on cognitive dysfunction, prefrontal cortex development, and inflammatory cytokines after early life stress. *Dev Psychobiol*. 2016; 58(4):482–491. https://doi.org/10.1002/dev.21390.
111. Gunnar MR, DePasquale CE, Reid BM, Donzella B. Pubertal stress recalibration reverses the effects of early life stress in postinstitutionalized children. *Proc Natl Acad Sci USA*. 2019;116(48):23984–23988. https://doi.org/10.1073/pnas.1909699116.
112. Fisher P, Heidi EB, Chamberlain P. Early intervention foster care: a model for preventing risk in young children who have been maltreated. *Child Serv*. 2010:159–182. https://doi.org/10.1207/s15326918cs0203_3.
113. Dozier M, Peloso E, Lindhiem O, et al. Developing evidence-based interventions for foster children: an example of a randomized clinical trial with infants and toddlers. *J Soc Issues*. 2006;62(4):767–785. https://doi.org/10.1111/j.1540-4560.2006.00486.x.

114. Keller SM, Nowak A, Roth TL. Female pups receive more maltreatment from stressed dams. *Dev Psychobiol*. 2019;61(6):824–831. https://doi.org/10.1002/dev.21834.
115. McGonigle P, Ruggeri B. Animal models of human disease: challenges in enabling translation. *Biochem Pharmacol*. 2014;87(1):162–171. https://doi.org/10.1016/j.bcp.2013.08.006.
116. Commons KG, Cholanians AB, Babb JA, Ehlinger DG. The rodent forced swim test measures stress-coping strategy, not depression-like behavior. *ACS Chem Neurosci*. 2017;8(5):955–960. https://doi.org/10.1021/acschemneuro.7b00042.
117. Bracken MB. Why animal studies are often poor predictors of human reactions to exposure. *J R Soc Med*. 2009;102(3):120–122. https://doi.org/10.1258/jrsm.2008.08k033.
118. Pound P, Ebrahim S, Sandercock P, Bracken MB, Roberts I. Where is the evidence that animal research benefits humans? *Br Med J*. 2004;328(7438):514–517. https://doi.org/10.1136/bmj.328.7438.514.
119. Grossman AW, Churchill JD, McKinney BC, Kodish IM, Otte SL, Greenough WT. Experience effects on brain development: possible contributions to psychopathology. *J Child Psychol Psychiatry Allied Discip*. 2003;44(1):33–63. https://doi.org/10.1111/1469-7610.t01-1-00102.

SECTION

5

Pharmacological and nonpharmacological approaches to reprogram the developmental programming to improve the health outcomes

CHAPTER

The emerging field of epigenetic editing: implication for translational purposes for diseases with developmental origin

13

Minglu Wang[1,a], X. Shawn Liu[2]

[1]*Department of Physiology & Cellular Biophysics, Columbia University Medical Center, Columbia University, New York, NY, United States;* [2]*Department of Physiology & Cellular Biophysics, Columbia Stem Cell Initiative, Herbert Irving Comprehensive Cancer Center, Columbia University Medical Center, Columbia University, New York, NY, United States*

Summary of the epigenetic editing field

Epigenetic writers, erasers, and their small molecule inhibitors

DNA methylation is established by the de novo methyl transferases DNMT3a and DNMT3b and maintained across cell division by DNMT1, which preferentially recognizes hemimethylated DNA over nonmethylated DNA.[1] DNA methylation can be removed by ten−eleven translocation methylcytosine dioxygenases (TETs) including TET1, TET2, and TET3.[2] Histone modifications are catalyzed by different enzymes. Various histone acetyl transferases (HATs), and histone deacetylases (HDAC) catalyze or remove acetylation on lysine. Histone methyl transferases (HMTs) and demethylase catalyze or remove methylation on lysine, and the protein arginine methyltransferases (PRMT) catalyze arginine methylation on the histone tail. Small molecule inhibitors are chemical compounds screened from small molecule libraries that interfere with specific biological processes. Some small molecule inhibitors target epigenetic processes and are used in basic research as well as in therapy development. The targets of these inhibitors are usually writers or erasers of the epigenetic marks.

DNA hypomethylating agents such as DNA methyltransferase inhibitors (DNMTi) reduce DNA methylation and have been used in anticancer treatment.

[a]Current address: Genome Biology Unit, European Molecular Biology Laboratory (EMBL), Heidelberg, Germany; Collaboration for joint PhD degree between EMBL and Heidelberg University, Faculty of Biosciences, Heidelberg University, Heidelberg, Germany.

Perinatal and Developmental Epigenetics. https://doi.org/10.1016/B978-0-12-821785-6.00002-5

5-Azacytidine (5-azaC, or 5-aza) and 5-Aza-2′-deoxycytidine (5-azadC) are two representative DNMTis used in the clinical setting.[3] Small molecule inhibitors of histone modification are also used in cancer treatment, e.g., HDAC inhibitors. However, these small molecule inhibitors are not locus specific and can potentially lead to side effects. Thus, the dosage of epigenetic drugs based on small molecule inhibitors is limited, which might compromise their potency. Therefore, the development of specific epigenetic editing tools is important for translational purposes.

Targeted epigenetic editing

Targeted DNA methylation editing

Targeted DNA methylation editing can be achieved by fusing DNMT or TET with DNA-binding proteins such as zinc finger proteins (ZF), transcription activator-like effectors (TALEs), and catalytically dead Cas9 (dCas9).

Targeted gene activation by erasing DNA methylation

The catalytic domain of TET1 or TET2 can be targeted to erase DNA methylation. To increase the efficacy of demethylation, TET1 can be tethered with DNA repair proteins, since DNA repair follows the oxidization of 5-mC to ultimately remove DNA methylation.[4] In addition, 5-mC can be directly excised by a protein found in plants, repressor of silencing 1 (Ros1). dCas9-Ros1 was used for excising DNA methylation and gene reactivation.[5]

Targeted gene repression by inducing DNA methylation

DNMT3A and DNMT3B establish de novo DNA methylation in a replication-independent manner, and DNMT3A is more catalytically active than DNMT3B.[6] Thus, human DNMT3A is the most frequently used effector protein for induction of DNA methylation.[7] Since DNMT3A is a large molecule (130 kDa), only its catalytic domain is used as the effector. The catalytic domain of DNMT3A has been fused with ZF, TALE, and dCas9 in previous studies. Additionally, the potency of DNMT3A-based targeted DNA methylation can be amplified by fusing DNMT3A with DNMT3L,[8] which stimulates the activity of DNMT3A. Besides DNMT3A, other DNA methyltransferases, including DNMT1 and MIWI2,[9] can also be fused with DNA-binding proteins to mediate targeted DNA methylation, but are less frequently used than DNMT3A.

Targeted histone modification editing

Targeted gene activation by editing histone modification

The first generation of targeted gene activation was achieved through fusing DNA-binding proteins with protein factors associated with transcriptional activation, e.g., VP16 or p65. These factors were later demonstrated to recruit enzymes that catalyze active histone marks. VP16 is a herpes virus protein that functions as a transcriptional activator.[10] VP64 is a tetramer of VP16, constructed for more potent transcriptional activation.[11]

The mechanism of VP64-mediated activation was then found to be related to histone modifications, primarily acetylation of histone H3 at lysine 27 (H3K27Ac).[12] H3K27Ac is an active enhancer mark[13] and is required for enhancer activity.[14] Acetylation of lysine on the histone tail reduces the positive charge on the lysine side chain and thus loosens its interaction with negatively charged DNA, thus making the DNA more accessible to transcription factors.

Later, it was found that H3K27Ac is directly catalyzed by p300, as shown by genetic deletion[15] or pharmacologically inhibition of p300.[14] The discovery of enzymes that catalyze specific histone marks gave rise to the second generation of targeted transcriptional activation tools. As an example, p300 was also targeted to activate enhancers, and dCas9-p300 activates the distal enhancer that is irresponsive to editing by dCas9-VP64.[16]

P65, another transcriptional activator used in first-generation epigenetic editing, is an NF-kappa B transcription factor with strong transcriptional activation capability.[17] Targeting p65-TetR (tet repressor) to a sequence with TetO (tet operator) insertion is associated with acetylation of histone H3 at lysine 9 (H3K9ac), although the effect on transcriptional activation was less potent than VP16.[18]

In addition to acetylation of the lysine side chain, methylation of lysine residue 4 on histone H3 (H3K4me3) at transcription start sites (TSS) is also associated with transcriptional activation. PRDM9 is a histone methyltransferase that targets H3K4.[19] The effect of dCas9-or ZF-PRDM9 targeting to gene promoters on gene activation is moderate and much lower compared with dCas9- or ZF-VP64.[20] Interestingly, gene activation by ZF-PRDM9 is only persistent for nonhypermethylated genes.[20] This may be because DNA methylation excludes H3K4me3.

Targeted gene silencing by editing histone modification

The corresponding first-generation epigenetic editing tool for gene inactivation includes Kruppel-associated box (KRAB) and Sin3a, both of which recruit proteins that either catalyze repressive histone marks or remove active histone marks. KRAB is a protein domain conserved from yeast to human and is both necessary and sufficient to repress transcription in transgenic assays.[21] KRAB is targeted to promoters of genes for transcriptional repression.[22] KRAB recruits KAP-1, which serves as a scaffold to recruit SETDB1, a SET domain protein that catalyzes H3K9me,[23] and HDAC.

H3K9me is a marker for inactive promoters. H3K9me can also be catalyzed by SET domain–containing HMTs, G9a and SUV39H1. Targeting ZF-G9a or ZF-SUV39H1 to the promoters of endogenous genes is sufficient for gene repression and deposition of H3K9me, which spreads to 1 kb away from the ZF-binding site. The spreading of H3K9me is mediated by HP1. This suggests that H3K9me has a causative role in gene repression and is not merely a by-product.[24] Interestingly, recruiting HMT and HDAC at the same time can boost the repression effect,[24] which reflects natural KRAB function.

Like KRAB, Sin3a is a gene-silencing effector protein in epigenetic editing. Sin3a belongs to a transcriptional repression complex, which includes Sin3b, Mad1, and Max.[25] The Sin interaction domain (SID) of Mad1 has been fused with TALE for targeted repression and shows better repression capability than TALE-KRAB.[26] Directly fusing Sin3a with TALE leads to the reduction of H3K9Ac, a histone marker of active transcription.[27] Sin3a is associated with HDAC PRD3 and is able to reduce level of H3K4me3, a histone mark for active promoters. The latter is potentially achieved by inhibiting S-adenosylmethionine (SAM), a donor for methyl groups for epigenetic and protein posttranslational modification.[28]

Target noncoding RNA

Similar to messenger RNA regulation, the programming of noncoding (nc) RNA includes regulation of its expression and RNA editing that rescues mutations. Besides manipulating DNA methylation and histone modification, one can use antisense oligonucleotide (ASO) to reduce ncRNA expression and virus delivery to increase ncRNA expression. The CRISPR system can also be used for targeted RNA editing, and it is also possible to regulate the spatial location of the long ncRNA using CRISPR-DISPLAY.[29]

Among the aforementioned methods to regulate ncRNA, ASO is the only approach that has been clinically approved. ASO is a chemically modified DNA molecule that binds to RNA target to form a DNA–RNA complex, which induces degradation of the target RNA. ASO can target short or long ncRNA. The main limitation of this technology, considering its clinical application, is that its effects are transient. This is because the ASO is cleaved once it binds to its target.

Developmental disease that have the potential to be treated with epigenetic editing

Epigenetic editing can potentially be applied to treat developmental diseases such as imprinting disorders and X-linked disorders (Fig. 13.1).

Imprinting disorders

Angelman syndrome

Etiology

Angelman syndrome is a rare neurodegenerative disease that causes mental retardation, epileptic seizures, and gait ataxia. The gene associated with this disease is *UBE3A* on the chromosome 15, an imprinted gene expressed on the maternal allele. UBE3A is an ubiquitin ligase, an enzyme that adds ubiquitin molecules to proteins to target them for degradation. UBE3A also functions as a transcription coactivator.[30] UBE3A localizes to both the nuclei and dendrites of neurons.[30] UBE3A targets P27, P53, EPHEXIN-5, and ARC for proteasome-mediated degradation. P27 and P53 are important for the proliferation of neurons.[31] EPHEXIN-5 is important for

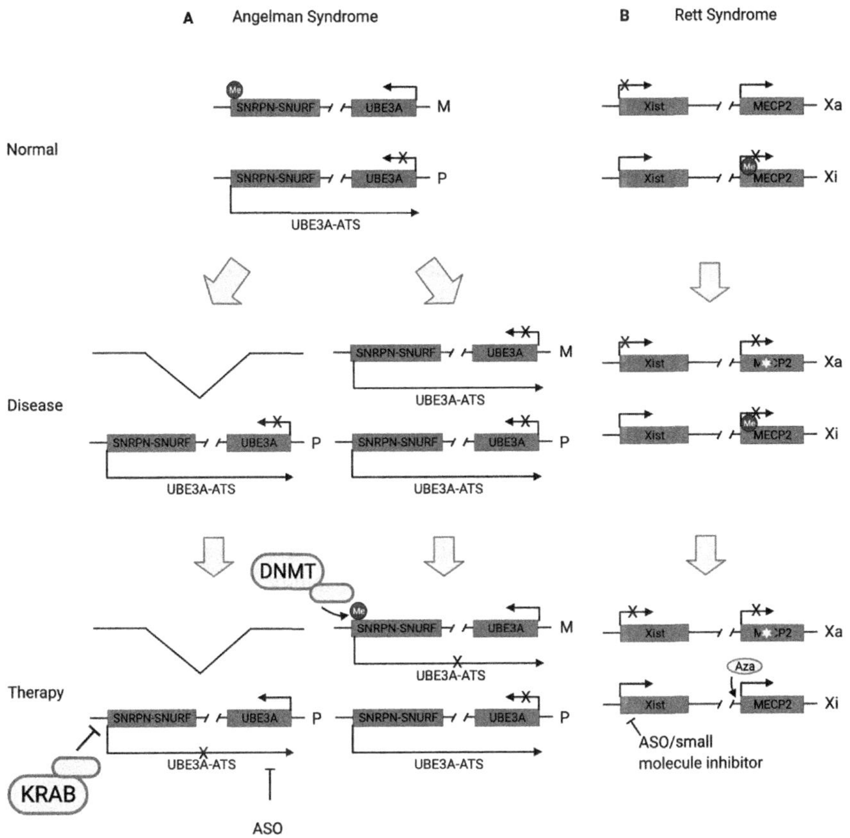

FIGURE 13.1 Summary of developmental diseases and epigenetic editing as a potential therapy.

(A) The Angelman syndrome caused by maternal deletion can be treated by reactivating the silenced paternal UBE3A by inhibiting the expression of paternal UBE3A-ATS through ASO or KRAB. The disease caused by maternal imprinting disorder can be rescued by restoring the methylation of the promoter of UBE3A-ATS through DNMT. M, maternal chromosome; P, paternal chromosome. UBE3A-ATS, antisense transcript of UBE3A. ASO, antisense oligonucleotide. KRAB and DNMT can be targeted to the loci through zinc finger (ZF) proteins or dCas9. (B) Rett syndrome is caused by heterozygous mutations in the *MECP2* gene, as illustrated by the star. Either using small molecule inhibitors to inhibit *XIST* transcription or the combination of ASO and Aza (DNMT inhibitor) to silence *XIST* and reactivate *MECP2* on the Xi is a potential therapeutic approach.

Created with BioRender.com.

synapse formation.[32] Elevated ARC increases α-amino-3-hydroxy-5-methyl-4-isoxazolepropionic acid (AMPA) receptor endocytosis and thus impairs synaptic transmission in excitatory neurons. In addition, as a transcriptional coactivator, UBE3A regulates the transcription of *CBLN1*[33] and *ARC*,[34] which are important in synaptic function. Thus, the neurons degenerate, and synapses are impaired in AS.

Normally, UBE3A antisense transcript (UBE3A-ATS) is expressed from the paternal allele as the promoter for maternal allele is hypermethylated. Interestingly, UBE3A only monoallelically expressed in neuron cells, because the UBE3A-ATS in nonneuron cells cannot elongate to *UBE3A* site and is subsequently terminated by a boundary enriched by multiple CCCTC-binding factor (CTCF) proteins.[35] Angelman syndrome is caused primarily by four genetic or epigenetic mechanism: (1) maternal deletion of a ~6 Mb region in 15q11.2–q13 encompassing the imprinting center and *UBE3A* gene, which accounts for more than 70% of patients and is phenotypically the most severe type; (2) mutation in the maternal copy of *UBE3A* accounts for 11% of patients[36]; (3) imprinting defects including deletion of the imprinting center, or the maternal methylation mosaicism of the exon 1 of *SNURF-SNRPN* (the promoter of *UBE3A-ATS*), which accounts for 6% of patients[37]; and (4) paternal uniparental disomy, which accounts for 3% of patients.[36]

Previous efforts to develop treatment

The concern of gene therapy for Angelman syndrome is that viral delivery of DNA or mRNA by virus can cause "gene overdose," and *UBE3A* duplication potentially contributes to autism.[38] As there is a silent allele of *UBE3A* on the paternal allele in Angelman syndrome patients, one attractive strategy is to reactivate this allele. This can be achieved by inhibiting the extension of the *UBE3A-ATS*. Drug screening using mouse cortical neurons found that topoisomerase I inhibitor significantly increased paternal *UBE3A* expression. However, this method lacks specificity since other genes can also be affected by topoisomerase I inhibitor. In addition, this drug has toxicity, making it difficult for clinical application.

Epigenetic editing to treat Angelman syndrome

Anti-UBE3A-ATS oligonucleotides can inhibit the extension of the *UBE3A-ATS* transcript in cultured mice neurons.[39] In cultured mice neurons, the ASO is tolerated and is effective in reducing *UBE3A-ATS* transcription and upregulating paternal *UBE3A* expression. When the ASO is administered once via intracerebroventricular injections, UBE3A expression in the cortex and spinal cord reaches 80% of the normal maternal allele for 4 months and then decreases to baseline. One month after the injection, performance in memory testing was improved. However, performance in marble burying, accelerating rotarod tests, and open field did not improve.

In a mouse neuroblastoma cell line, anchoring KRAB to the initiation site of the *UBE3A-ATS* reduced the transcription of *UBE3A-ATS*.[40] An HIV TAT domain for cell penetration was added to the ZF-KRAB construct. This allowed the fusion protein to cross the blood–brain barrier. The purified protein was delivered into the brain via peripheral injection, resulting in widespread distribution in the mouse brain. In the hippocampus and cerebellum, which contributes to the cognitive[41] and motor defects,[42] respectively, protein expression as measured by quantifying the immunohistochemistry signal in treated Angelman syndrome mice model achieved about 80% of the signal measured in wild-type control mice. One of the major limitations of this method is that the reactivation is transient, probably because they injected purified protein. The half-life of ZF-KRAB is 8–24 h, so it requires injection every other day.

In addition, targeting dCas9-DNMT3A to exon 1 of *SNURF-SNRPN* in single oocyte can restore the DNA methylation on the TSS of UBE3A-ATS in oocytes, which can be maintained after in vitro fertilization.[43] Embryo transfer experiments and further behavior testing in mice confirmed that the somatic tissue of the offspring had established the methylation of *SNURF-SNRPN*. This type of epigenetic editing will be suitable for treating the 6% of Angelman syndrome patients with imprinting defects.

Prader–Willi syndrome

Etiology

Prader–Willi syndrome (PWS) is a disease manifested as overeating with psychotic and endocrine disorders. The imprinting center of PWS (PWS-IC) controls the paternal epigenotypes of genes including *MAGEL2*, *NDN*, and *SNURF-SNRPN* in 15q11.2-q13. These genes normally express their paternal copy. Research has highlighted that the C/D box small nucleolar (sno)RNA cluster *SNORD116* in the intron of *SNURF-SNRPN* is responsible for the majority of the PWS symptoms.[44] Deletion of *SNORD116* caused hypothalamus defects and hyperphagia, resulting in obesity in a subset of mice.[45] *SNORD116* is suggested to modulate many neuronal gene transcripts, including those that regulate the circadian clock and energy expenditure.[46] The majority of PWS patients have 5.4 Mb or 6 Mb deletion on the paternal 15q11.2-q13 that include the PWS-related genes mentioned before.[47] Approximately 20%–30% of the patients have maternal uniparental disomy. About 1%–3% PWS patients have imprinting disorders.[48]

Previous efforts to develop treatment

Growth hormone therapy can increase the height and reduce the body mass of the patients. However, whether the cognitive and behavior defects can be ameliorated by growth hormone treatment is still under investigation.

To relieve additional PWS defects, epigenetic editing to reactivate the maternal copy of *SNORD116* might be promising. The maternal allele of *SNORD116* is silenced by a complex formed by SETDB1 and the zinc finger protein ZNF274. ZNF274 is relatively specific for this locus since it has fewer than 500 binding sites in the genome. Moreover, ZNF274 recruits SETDB1 for gene-silencing effect. H3K9me3 mediated by histone H3K9 methyltransferase SETDB1 is suggested to be the mechanism for silencing *SNORD116*.[49] However, knocking down ZNF274 with RNAi resulted in only a moderate increase of maternal *SNORD116* expression, less than 1/1000 of the paternal copy.[49] This may be because the remaining ZNF274 can still recruit enough SETDB1 for repression, or alternatively because there are other proteins that recruit SETDB1 to the maternal *SNORD116* locus. Knocking down of SETDB1 by short hairpin (sh)RNA decreases the repressive histone mark at the maternal *SNORD116* and partially restores the maternal SNORD116 level.[49]

H3K9me2 mediated by histone H3K9 methyltransferase G9a is another potential mechanism for silencing *SNORD116*.[50,51] Small molecule compounds that inhibit G9a were screened based on reactivation of *SNURF-SNRPN* in mice embryonic fibroblasts.[52] The inhibitors bind and block the catalytic domain of G9a, and activated *SNORD116* in human skin fibroblast derived from a PWS patient. In a PWS mouse model, this inhibitor reactivated the maternal *SNURF-SNRPN* to 50% of the paternal level and rescued perinatal lethality model at the postnatal stage after a single intraperitoneal injection. This inhibitor was tolerated and acted through the reduction of H3K9me2 and H3K9me3.[52] Importantly, this method is not specific since the H3K9 methylation was reduced not only at the *SNORD116*.

Potential therapy by specific epigenetic editing

Potentially, fusing H3K9 demethylase LSD1 with dCas9 may specifically reduce repressive histone marks to allow *SNORD116* reactivation. Of note, the development of therapeutic methods needs to be evaluated on appropriate animal models. However, the effects of the treatment on obesity and hyperphagia cannot be assessed in current animal models.[47]

X-linked intellectual disability

Rett syndrome

Etiology

X chromosome inactivation (XCI) is a mechanism for dosage compensation between males and females in mammal, during which one of the two X chromosomes of females is transcriptionally silenced. The XCI is mediated by the expression of *XIST*, an lncRNA from the future inactivated X chromosome (Xi). The expression of *XIST* in early embryogenesis is sufficient and necessary for silencing of Xi. *XIST* represses transcription in *cis* only on the X chromosome from which it transcribed and is required for both establishment and maintenance of XCI.[53]

Rett syndrome (RTT) is a disease that causes the loss of motor and language ability, and it is mainly observed in females. RTT is caused by loss-of-function mutations on the X-linked gene *MECP2*. MECP2 is a master protein in transcriptional regulation in neurons and is important for synaptic connection.[54]

Previous efforts to develop treatment

Current treatment for RTT is mainly symptomatic. The directions for developing future therapy include targeting MECP2 or its downstream molecules. Because MECP2 is a master protein in transcriptional regulation, targeting any single molecule downstream of MECP2 can only ameliorate an aspect of the disease. Gene therapy has major problem with gene overdose, since *MECP2* duplication has been linked to autism spectrum disorders.

There is a wild-type *MECP2* allele in most female RTT patients. Thus, reactivating the wild-type *MECP2* in Xi might be a promising treatment for RTT.

Importantly, reactivation in the adult stage can rescue the phenotype. This is demonstrated by experiments where a stop cassette was inserted into the *MECP2* gene in male mice and deleted in adult stage.[55]

Developing specific epigenetic editing therapy

Small molecule inhibitors of *XIST* have been used.[53] shRNA screening was used to identify factors required for maintaining XCI in a female mouse embryonic fibroblast cell line. Inhibiting PDPK1, a serine–threonine kinase and regulator of phosphatidylinositol 3-kinase (PI3K)/AKT signaling, reversibly reactivates *MECP2* in mouse cortical neurons and fibroblast cells derived from RTT patients. However, this study did not follow-up with MECP2 protein detection or neuron function examination. In a later study by the same group, two drugs were used, an inhibitor of ACVR1, a molecule involved in cell signaling, and an inhibitor of SGK1, a PDPK1 effector substrates in mice and induced RTT neuron derived from human-induced pluripotent stem cell (iPSC).[56] In RTT neurons, *MECP2 mRNA* levels increased to about 10% compared with the positive control, and the neuronal morphology becomes comparable with the positive control. However, there is no quantitative mRNA measurement nor further functional examination in living mice in this study.

An ASO against XIST, and 5-Aza, a DNMT1 inhibitor, have been used in combination for therapeutic effects in RTT neurons in vitro.[57] 5-Aza can cross the blood–brain barrier, so the delivery is relatively easy. The XIST ASO and 5-Aza had a synergistic effect achieving 2.2% of reactivation compared with the *MECP2* mRNA level in the active allele. Although the reactivation seems to be moderate, the phenotypical improvement was significant, suggesting partial restoration of MECP2 expression is effective.[57] In the future, single-cell techniques could be applied to investigate whether the 2.2% overall reactivation is a huge reactivation from a few cells, or a moderate reactivation from a large population of cells.

However, the major limitation of these studies is specificity. Other genes subject to XCI have been reactivated even to a greater extent compared with *MECP2*. In addition, the efficacy of improving *MECP2* transcription is suboptimal. Considering the XCI is a complicated process involving different epigenetic players, it might be promising to combine different editing strategies, e.g., DNA methylation and histone modification on the *MECP2* locus to tackle RTT in the future.[7]

Fragile X syndrome

Etiology

Fragile X syndrome (FXS) is characterized by intellectual disability and an autistic spectrum phenotype. FXS is caused by loss of fragile X mental retardation protein (FMRP) encoded by the *FMR1* gene during brain development. FMRP is an RNA-binding protein in neurons. FMRP regulates membrane protein in the synapse and maintains normal synaptic plasticity. In FXS, a CGG trinucleotide repeat (>200 times) expansion mutation at the 5′ untranslated region (UTR) of *FMR1* is followed by DNA hypermethylation, which results in heterochromatin formation at the *FMR1* promoter and subsequent silencing of *FMR1* expression in FXS. There is no cure for FXS so far.

Previous approach to develop treatment

There have been no effective treatment so far for FXS.[58] Since in FXS, the *FMR1* coding region is intact, epigenetically reactivating the *FMR1* gene is a promising approach. 5-azadC can irreversibly block DNMT. An initial test using 5-azadC is effective to reactivate mRNA but not translation,[59] since the multiple CGG repeats inhibit translation.[60] Combined treatment with 5-azadC and HDAC inhibitors causes a synergistic effect in reactivating *FMR1* transcription.[61] 5-azadC induces histone acetylation, and H3K4 methylation, and reduces H3K9 methylation at the *FMR1* locus.[62] However, the toxicity of 5-azadC has raised concern.[63] Moreover, the effects of 5-azadC treatment only last for 4 weeks.[64] This transient reactivation was resilenced by the recruitment of EZH2, a component of the polycomb repressive complex2 (PRC2) component, which catalyzes H3K27 trimethylation.[65] Compound 1a, an EZH2 inhibitor, was able to delay the resilencing, as the reactivation remains significant at 4 weeks posttreatment.[65] However, these previous efforts all utilized nonspecific inhibition of the epigenetic modifiers, which might lead to undesired side effects.[56]

Targeted epigenetic editing to treat FXS

Recently developed targeted DNA methylation editing tools[66] have been applied to treat FXS by demethylating the CGG repeats by dCas9-TET1. The heterochromatin status of the *FMR1* promoter was switched to an euchromatin state, as measured by anti-RNPII, H3K4me3, H3K27ac, HeK27me3, and H3K9me3 chromatin immunoprecipitation sequencing (ChIP-seq). The reactivation of *FMR1* can be observed in iPSC, iPSC-derived neurons (iN), and iN engrafted into the mouse brain. Finally, direct editing in postmitotic neurons in the mouse brain was able to reactivate *FMR1*, with FMR1 protein expression lasting for 1 month in the mouse brain.[67]

Insights for the application of epigenetic editing in the treatment of these developmental diseases

For any molecular therapy, there are several considerations including delivery, off-target effects, epigenetic efficacy, phenotypical potency, persistency, and immunogenic effects. Delivery and immunogenic effects are more general issues for gene editing, which have been reviewed in detail.[68–70] Thus, this chapter will cover off-target effects, efficacy of influencing transcription, phenotypical potency of the epigenetic editing, and persistency of the epigenetic modification (Fig. 13.2).

Off-target effects

Off-target is defined as the recruitment of DNA-binding protein—effector fusion to nontargeted sites in the genome.[71] We would like to use the DNA-binding protein

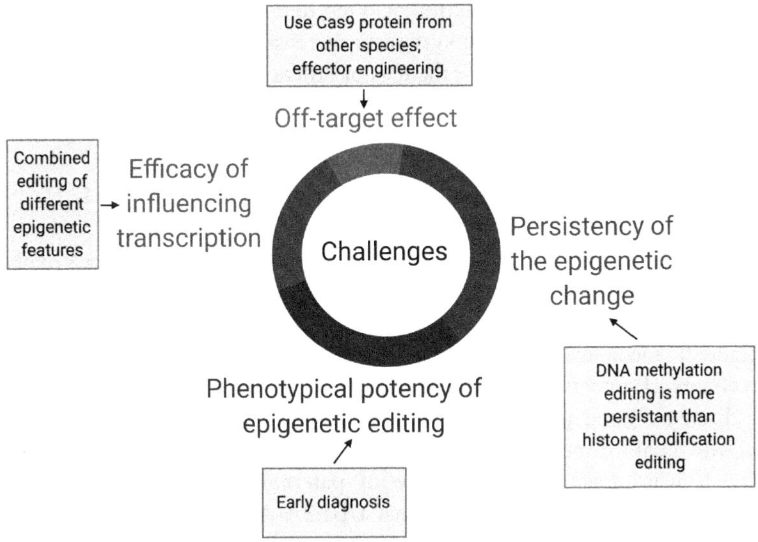

FIGURE 13.2 Summary of challenges of epigenetic editing for the treating developmental diseases.

This figure summarizes the challenges of epigenetic editing for treating developmental diseases and the potential solutions. The relative difficulty of each challenge is proportioned by the length of the colored band.

Created with BioRender.com.

dCas9 as an example to illustrate the off-target effect since it is most well characterized. One reason for off-target effect is that the choice of loci used for designing single guide (sg)RNA might be limited to the neighbor of the desired site. This limitation is due to the restriction by the protospacer-adjacent motif (PAM), a sequence required for target recognition by Cas9 (or dCas9). This restriction caused by PAM can be ameliorated by using dCas9 protein from another species with a different PAM sequence.[7] In addition, a PAMless editor, which is an engineered *Streptococcus pyogenes* Cas9 (spCas9), SpRY, recognizes NRN (R is A or G) and NYN (Y is C or T) PAMs instead of the NGG PAM for the original spCas9.[72] The design of this engineered variant was inspired by a high-throughput PAM determination assay that assesses the contribution of each amino acid residue in the PAM interacting domain of the SpCas9 protein.[72]

Another cause for an off-target is the effector fused with dCas9. This reason for off-target activity is specific to epigenetic editing. dCas9-DNMT3A without guide (g)RNA can cause global hypermethylation, potentially by the free diffusion of the catalytical domain of DNMT3A.[73] In an engineered cell line with DNMT3A–DNMT3L double knockout and DNMT1 knockdown, transfection with the catalytical domain of DNMT3A alone leads to 20% increase in global methylation.[73] One solution could be to mutate the amino acid in the catalytical domain of DNMT3A that binds to DNA.[74] Such a mutation also led to a reduction of methyl transferase activity; however, using a Suntag system to increase the number of DNMT3A proteins recruited can compensate for the loss of efficacy.[74] For other effectors, another potential solution might be engineering the epigenetic effector for DNA binding–inducible activation.[7]

Efficacy of influencing transcription

How effectively can epigenetic editing change transcription in diseases depends on several aspects including the epigenetic mechanism of transcriptional regulation of the disease-related gene, and the delivery methods used. This section will focus on the epigenetic mechanism, rather than the delivery methods as they are not the focus of this chapter. Epigenetic editing could be effective when the epigenetic regulation of the disease-related gene was straightforward. For example, in Angelman syndrome, the goal of epigenetic editing was to reactivate the silenced paternal *UBE3A*. It was clear that the silence of paternal *UBE3A* is mediated by the UBE3A-ATS. Thus, using ASOs against UBE3A-ATS was able to reactivate the paternal UBE3A protein expression to 90% of the normal maternal copy.[39] However, when the epigenetic mechanism is more complicated, as with the X chromosome inactivation of *MECP2* locus in Rett syndrome, epigenetic editing using XIST ASO and DNMT inhibitor only increased *MECP2* transcription to 2% of that observed with the active X chromosome. The challenge here is that XCI involves multiple layers of epigenetic modifications, including DNA methylation, histone modification, and chromatin 3D structure. Editing a single epigenetic mark might not be efficient for transcriptional regulation. For example, depositing histone methylation, a repressive histone mark, was not able to repress gene transcription.[75] Thus, the solution could be to use a strategy of combined editing with multiple modifications achieved by using orthogonal Cas proteins such as Cas9 and Cpf1, from different species. In addition, multiple zinc finger proteins could be delivered to the same cell for combined epigenetic editing.

Phenotypical potency of the epigenetic editing

The developmental diseases discussed in this chapter are related to brain development, which occurs in the prenatal stage. Prenatal defects are usually diagnosed when the phenotypical defects are manifested, much later than when the defective molecular process happens. Thus, whether the disease is reversible at a later stage is an important question for treating developmental diseases.

For RTT, in the groundbreaking attempt in this direction in a mouse model, the *Mecp2* gene was silenced by inserting a stop cassette into the gene and then reactivating it through cassette deletion, thereby rescuing the RTT phenotype.[76] This proof-of-concept work gave rise the field of X chromosome reactivation (XCR). Similarly, for FXS, since excessive mGluR-mediated signaling due to lack of FMRP is critical for disease development,[77] administration of mGluR inhibitor was effective in reversing the phenotype in young adult stage of FXS mice model, much later than the development of the disease.[78]

However, in AS, restoration of the paternal UBE3A in adult mice model by ASO targeting to UBE3A-ATS failed to rescue the majority of the neurocognitive defects of Angelman syndrome measured in behavior tests.[39] This was explained by later experiments that systematically examined the window of neurocognitive function impaired in Angelman syndrome. Although motor dysfunction can be rescued in adolescent, epilepsy, anxiety, and repetitive behavior can only be rescued in early development.[79] Thus, effective editing therapy should be combined with timely diagnosis.

Persistency of the epigenetic change

The persistency of the epigenetic change induced by the therapy is determined by two aspects, the epigenetic mechanism and the delivery method. Combining treatment with ZF-KRAB and ZF-DNMT3A-3L was able to achieve silencing that was resistant to transcriptional activators.[80] Similarly, recently published research shows that fusing KRAB and DNMT3A-3L to N and C terminal of dCas9 can achieve sustained epigenetic silencing cross-cell division and differentiation.[81] It is speculated that DNA methylation is the key for this persistency. In addition, certain histone modifications were also reported to be rigorous: HP1alpha can deposit H3K9me3 and induce the spread of heterochromatin state up to 10 kb, which can be inherited cross-cellular generations.[82] The basic speculation is that modified histone will recruit more histone modification enzymes so that this mark can be spread. However, this speculation has two additional requirements: the histone modification enzyme occasionally can lead to modification beyond nearby genes, and the cooperation between more than one edited nucleosome.[83]

The consideration of the delivery method is complicated. Since all the developmental diseases discussed in this chapter require brain delivery of the epigenetic editing machinery, crossing blood—brain barrier is required. For manipulation of DNA methylation or histone modification through DNA-binding protein—epigenetic effector fusion construct, purified protein with peptide that aids the crossing of blood—brain barrier was able to be delivered to the brain; however, its half-life is 8—24 h and therefore requires injection every the other day.[40] A single tail vein injection of AAV8 serotype that targets hepatocytes in the liver can lead to desired protein expression level for 6 months[84]; however, it is still difficult for AAV8 to cross the blood—brain barrier. Due to the transient nature of ASO-induced RNA degradation, this method of targeting noncoding RNA usually requires weekly injections.[85]

Future direction: manipulation of 3D chromatin topology to treat developmental diseases

Chromatin is packaged in nuclei as 3D topological structures. In megabase scale, the genome could be divided into A and B compartments based on high-throughput chromosome conformation capture (Hi-C) data.[86] Each compartment is defined by the preference of intracompartment interaction over intercompartment interaction. Compartment A displaces the nucleus interior, with higher transcriptional activities. Compartment B occupies the nucleus periphery, with lower transcriptional activities.[86] In megabase and submegabase scales, the chromatin folds into topologically associated domains (TADs).[87] TADs are believed to increase the frequency of the interaction between the enhancer and promoter of genes within the domain, thus contributing to transcriptional regulation. TADs are likely composed of an insulated neighborhood, at the scale of about 200 kb, where enhancers prefer to interact with promoters by forming loops within the insulated region.[88] The formation of chromatin 3D structure is mediated by protein factors. TADs and the insulated neighborhood are mainly mediated by cohesin and CTCF binding at the boundary.[87] In vertebrates, enhancer–promoter loops can be mediated by YY1,[89] or site-specific factors such as Ldb1.[90]

To dissect the causal relationship between chromatin topology and transcription, specific manipulation of chromatin topology is required. Manipulation of insulated neighborhoods has been discussed in detail elsewhere,[88] and this section will focus on manipulation of chromatin loops, specifically. First, site-specific looping factors can be anchored to manipulate loops. Deng et al. have used a ZF protein to target Ldb1 to the hemoglobin locus.[90] Second, others have used nonmammalian factors to induce looping. Morgan et al. fused dCas9 with a chemically induced proximity system, e.g., the plant S-(+)-abscisic acid (ABA) signaling pathway.[91] dCas9 can also be fused with proteins from an optogenetic system that dimerize under blue light.[92] Lastly, endogenous chromatin organizer proteins such as YY1[89] and CTCF[93] have been fused to dCas9 to induce loops. Manipulation of enhancer–promoter loops was capable of regulating the transcription of some, but not all, genes tested.[90,94]

Interestingly, growing evidence supports the role of 3D chromatin topology in the developmental diseases discussed in the previous sections of this chapter. For example, reactivation of wild-type genes in the X chromosome might be useful for treating X-linked developmental diseases.[56] Several studies have suggested that chromatin topology might play a role in X chromosome inactivation. For instance, many TADs were attenuated across the inactivated X chromosome but were found to correlate with clusters of genes that had escaped XCI in the mouse brain.[95,96] Cohesin was found to interact with *XIST* and was depleted from the inactive X chromosome in mice.[97] Thus, in the future, manipulation of chromatin topology would be a new and exciting direction to treat diseases such as Rett syndrome. In addition, CTCF is involved in Angelman syndrome. In healthy individuals, at the Angelman syndrome loci, the paternal *UBE3A* is repressed by UBE3A-ATS in the neurons.[35] However, in nonneuronal cells, UBE3A is biallelically expressed, because UBE3A-ATS cannot elongate to the paternal *UBE3A* and thus is terminated by CTCF blocking.[35] CTCF occupies the site between the TSS of *UBE3A-ATS*, and the *UBE3A* gene restricts the long-distance interaction on each side of the CTCF-binding site[35] Thus, restoring CTCF binding locally might restore the restriction and block the elongation of UBE3A-ATS in the paternal allele in neurons, thus reactivating the paternal *UBE3A* to potentially treat Angelman syndrome.

References

1. Okano M, Bell DW, Haber DA, Li E. DNA methyltransferases Dnmt3a and Dnmt3b are essential for de novo methylation and mammalian development. *Cell.* 1999;99(3): 247–257. https://doi.org/10.1016/S0092-8674(00)81656-6.
2. Tahiliani M, Koh KP, Shen Y, et al. Conversion of 5-methylcytosine to 5-hydroxymethylcytosine in mammalian DNA by MLL partner TET1. *Science (New York, NY).* 2009;324(5929):930–935. https://doi.org/10.1126/science.1170116.
3. Creusot F, Acs G, Christman JK. Inhibition of DNA methyltransferase and induction of Friend erythroleukemia cell differentiation by 5-azacytidine and 5-aza-2′-deoxycytidine. *J Biol Chem.* 1982;257(4):2041–2048.
4. Taghbalout A, Du M, Jillette N, et al. Enhanced CRISPR-based DNA demethylation by Casilio-ME-mediated RNA-guided coupling of methylcytosine oxidation and DNA repair pathways. *Nat Commun.* 2019;10(1). https://doi.org/10.1038/s41467-019-12339-7.
5. Devesa-Guerra I, Morales-Ruiz T, Pérez-Roldán J, et al. DNA methylation editing by CRISPR-guided excision of 5-methylcytosine. *J Mol Biol.* 2020;432(7):2204–2216. https://doi.org/10.1016/j.jmb.2020.02.007.
6. Challen GA, Sun D, Mayle A, et al. Dnmt3a and Dnmt3b have overlapping and distinct functions in hematopoietic stem cells. *Cell Stem Cell.* 2014;15(3):350–364. https://doi.org/10.1016/j.stem.2014.06.018.
7. Liu XS, Jaenisch R. Editing the epigenome to tackle brain disorders. *Trends Neurosci.* 2019;42(12):861–870. https://doi.org/10.1016/j.tins.2019.10.003.
8. Siddique AN, Nunna S, Rajavelu A, et al. Targeted methylation and gene silencing of VEGF-A in human cells by using a designed Dnmt3a-Dnmt3L single-chain fusion protein with increased DNA methylation activity. *J Mol Biol.* 2013;425(3):479–491. https://doi.org/10.1016/j.jmb.2012.11.038.
9. Kojima-Kita K, Kuramochi-Miyagawa S, Nagamori I, et al. MIWI2 as an effector of DNA methylation and gene silencing in embryonic male germ cells. *Cell Rep.* 2016; 16(11):2819–2828. https://doi.org/10.1016/j.celrep.2016.08.027.
10. Sadowski I, Ma J, Triezenberg S, Ptashne M. GAL4-VP16 is an unusually potent transcriptional activator. *Nature.* 1988;335(6190):563–564. https://doi.org/10.1038/335563a0.
11. Beerli RR, Segal DJ, Dreier B, Barbas CF. Toward controlling gene expression at will: specific regulation of the erbB-2/HER-2 promoter by using polydactyl zinc finger proteins constructed from modular building blocks. *Proc Natl Acad Sci USA.* 1998; 95(25):14628–14633. https://doi.org/10.1073/pnas.95.25.14628.
12. Memedula S, Belmont AS. Sequential recruitment of HAT and SWI/SNF components to condensed chromatin by VP16. *Curr Biol.* 2003;13(3):241–246. https://doi.org/10.1016/S0960-9822(03)00048-4.
13. Creyghton MP, Cheng AW, Welstead GG, et al. Histone H3K27ac separates active from poised enhancers and predicts developmental state. *Proc Natl Acad Sci USA.* 2010; 107(50):21931–21936. https://doi.org/10.1073/pnas.1016071107.

14. Raisner R, Kharbanda S, Jin L, et al. Enhancer activity requires CBP/P300 bromodomain-dependent histone H3K27 acetylation. *Cell Rep*. 2018;24(7): 1722–1729. https://doi.org/10.1016/j.celrep.2018.07.041.
15. Jin Q, Yu LR, Wang L, et al. Distinct roles of GCN5/PCAF-mediated H3K9ac and CBP/p300-mediated H3K18/27ac in nuclear receptor transactivation. *EMBO J*. 2011;30(2): 249–262. https://doi.org/10.1038/emboj.2010.318.
16. Hilton IB, D'Ippolito AM, Vockley CM, et al. Epigenome editing by a CRISPR-Cas9-based acetyltransferase activates genes from promoters and enhancers. *Nat Biotechnol*. 2015;33(5):510–517. https://doi.org/10.1038/nbt.3199.
17. Schmitz ML, Baeuerle PA. The p65 subunit is responsible for the strong transcription activating potential of NF-χB. *EMBO J*. 1991;10(12):3805–3817. https://doi.org/10.1002/j.1460-2075.1991.tb04950.x.
18. Bergmann JH, Jakubsche JN, Martins NM, et al. Epigenetic engineering: histone H3K9 acetylation is compatible with kinetochore structure and function. *J Cell Sci*. 2012; 125(Pt 2):411–421. https://doi.org/10.1242/jcs.090639.
19. Wu H, Mathioudakis N, Diagouraga B, et al. Molecular basis for the regulation of the H3K4 methyltransferase activity of PRDM9. *Cell Rep*. 2013;5(1):13–20. https://doi.org/10.1016/j.celrep.2013.08.035.
20. Cano-Rodriguez D, Gjaltema RAF, Jilderda LJ, et al. Writing of H3K4Me3 overcomes epigenetic silencing in a sustained but context-dependent manner. *Nat Commun*. 2016;7. https://doi.org/10.1038/ncomms12284.
21. Margolin JF, Friedman JR, Meyer WKH, Vissing H, Thiesen -HJ, Rauscher FJ. Kruppel-associated boxes are potent transcriptional repression domains. *Proc Natl Acad Sci USA*. 1994;91(10):4509–4513. https://doi.org/10.1073/pnas.91.10.4509.
22. Gilbert LA, Larson MH, Morsut L, et al. CRISPR-mediated modular RNA-guided regulation of transcription in eukaryotes. *Cell*. 2013;154(2):442–451. https://doi.org/10.1016/j.cell.2013.06.044.
23. Schultz DC, Ayyanathan K, Negorev D, Maul GG, Rauscher FJ. SETDB1: a novel KAP-1-associated histone H3, lysine 9-specific methyltransferase that contributes to HP1-mediated silencing of euchromatic genes by KRAB zinc-finger proteins. *Genes Dev*. 2002;16(8):919–932. https://doi.org/10.1101/gad.973302.
24. Snowden AW, Gregory PD, Case CC, Pabo CO. Gene-specific targeting of H3K9 methylation is sufficient for initiating repression in vivo. *Curr Biol*. 2002;12(24): 2159–2166. https://doi.org/10.1016/S0960-9822(02)01391-X.
25. Ayer DE, Laherty CD, Lawrence QA, Armstrong AP, Eisenman RN. Mad proteins contain a dominant transcription repression domain. *Mol Cell Biol*. 1996;16(10): 5772–5781. https://doi.org/10.1128/MCB.16.10.5772.
26. Cong L, Zhou R, Kuo YC, Cunniff M, Zhang F. Comprehensive interrogation of natural TALE DNA-binding modules and transcriptional repressor domains. *Nat Commun*. 2012;3. https://doi.org/10.1038/ncomms1962.
27. Konermann S, Brigham MD, Trevino AE, et al. Optical control of mammalian endogenous transcription and epigenetic states. *Nature*. 2013;500(7463):472–476. https://doi.org/10.1038/nature12466.
28. Liu M, Pile LA. The transcriptional corepressor SIN3 directly regulates genes involved in methionine catabolism and affects histone methylation, linking epigenetics and metabolism. *J Biol Chem*. 2017;292(5):1970–1976. https://doi.org/10.1074/jbc.M116.749754.

29. Shechner DM, Hacisuleyman E, Younger ST, Rinn JL. Multiplexable, locus-specific targeting of long RNAs with CRISPR-display. *Nat Methods*. 2015;12(7):664–670. https://doi.org/10.1038/nmeth.3433.
30. Bhattacharjee MB. The Angelman syndrome ubiquitin ligase localizes to the synapse and nucleus, and maternal deficiency results in abnormal dendritic spine morphology. *Hum Mol Genet*. 2008;17(1):111–118.
31. Gil-Perotin S, Haines JD, Kaur J, et al. Roles of p53 and p27(Kip1) in the regulation of neurogenesis in the murine adult subventricular zone. *Eur J Neurosci*. 2011;34(7): 1040–1052. https://doi.org/10.1111/j.1460-9568.2011.07836.x.
32. Margolis SS, Salogiannis J, Lipton DM, et al. EphB-mediated degradation of the RhoA GEF Ephexin5 relieves a developmental brake on excitatory synapse formation. *Cell*. 2010;143(3):442–455. https://doi.org/10.1016/j.cell.2010.09.038.
33. Krishnan V, Stoppel DC, Nong Y, et al. Autism gene Ube3a and seizures impair sociability by repressing VTA Cbln1. *Nature*. 2017;543(7646):507–512. https://doi.org/10.1038/nature21678.
34. Greer PL, Hanayama R, Bloodgood BL, et al. The Angelman syndrome protein Ube3A regulates synapse development by ubiquitinating arc. *Cell*. 2010;140(5):704–716. https://doi.org/10.1016/j.cell.2010.01.026.
35. Hsiao JS, Germain ND, Wilderman A, et al. A bipartite boundary element restricts UBE3A imprinting to mature neurons. *Proc Natl Acad Sci USA*. 2019;116(6): 2181–2186. https://doi.org/10.1073/pnas.1815279116.
36. Williams CA, Driscoll DJ, Dagli AI. Clinical and genetic aspects of Angelman syndrome. *Genet Med*. 2010;12(7):385–395. https://doi.org/10.1097/GIM.0b013e3181def138.
37. Buiting K, Groß S, Lich C, Gillessen-Kaesbach G, El-Maarri O, Horsthemke B. Epimutations in Prader-Willi and Angelman syndromes: a molecular study of 136 patients with an imprinting defect. *Am J Hum Genet*. 2003;72(3):571–577. https://doi.org/10.1086/367926.
38. Germain ND, Chen P-F, Plocik AM, et al. Gene expression analysis of human induced pluripotent stem cell-derived neurons carrying copy number variants of chromosome 15q11-q13.1. *Mol Autism*. 2014;5:44. https://doi.org/10.1186/2040-2392-5-44.
39. Meng L, Ward AJ, Chun S, Bennett CF, Beaudet AL, Rigo F. Towards a therapy for Angelman syndrome by targeting a long non-coding RNA. *Nature*. 2015;518(7539): 409–412. https://doi.org/10.1038/nature13975.
40. Bailus BJ, Pyles B, Mcalister MM, et al. Protein delivery of an artificial transcription factor restores widespread Ube3a expression in an Angelman syndrome mouse brain. *Mol Ther*. 2016;24(3):548–555. https://doi.org/10.1038/mt.2015.236.
41. Jiang Yh, Armstrong D, Albrecht U, et al. Mutation of the Angelman ubiquitin ligase in mice causes increased cytoplasmic p53 and deficits of contextual learning and long-term potentiation. *Neuron*. 1998;21(4):799–811. https://doi.org/10.1016/S0896-6273(00)80596-6.
42. Jay V, Becker LE, Chan FW, Perry TL. Puppet-like syndrome of Angelman: a pathologic and neurochemical study. *Neurology*. 1991;41(3):416–422. https://doi.org/10.1212/wnl.41.3.416.
43. Wei Y, Lang J, Zhang Q, et al. DNA methylation analysis and editing in single mammalian oocytes. *Proc Natl Acad Sci USA*. 2019;116(20):9883–9892. https://doi.org/10.1073/pnas.1817703116.

44. Gallagher RC, Pils B, Albalwi M, Francke U. Evidence for the role of PWCR1/HBII-85 C/D box small nucleolar RNAs in Prader-Willi syndrome. *Am J Hum Genet*. 2002;71(3): 669–678. https://doi.org/10.1086/342408.
45. Rodriguez JA, Zigman JM. Hypothalamic loss of Snord116 and Prader-Willi syndrome hyperphagia: the buck stops here? *J Clin Invest*. 2018;128(3):900–902. https://doi.org/10.1172/JCI99725.
46. Powell WT, Coulson RL, Crary FK, et al. A Prader-Willi locus lncRNA cloud modulates diurnal genes and energy expenditure. *Hum Mol Genet*. 2013;22(21):4318–4328. https://doi.org/10.1093/hmg/ddt281.
47. Cassidy SB, Schwartz S, Miller JL, Driscoll DJ. Prader-Willi syndrome. *Genet Med*. 2012;14(1):10–26. https://doi.org/10.1038/gim.0b013e31822bead0.
48. Kim SJ, Miller JL, Kuipers PJ, et al. Unique and atypical deletions in Prader-Willi syndrome reveal distinct phenotypes. *Eur J Hum Genet*. 2012;20(3):283–290. https://doi.org/10.1038/ejhg.2011.187.
49. Cruvinel E, Budinetz T, Germain N, Chamberlain S, Lalande M, Martins-Taylor K. Reactivation of maternal SNORD116 cluster via SETDB1 knockdown in Prader-Willi syndrome iPSCs. *Hum Mol Genet*. 2014;23(17):4674–4685. https://doi.org/10.1093/hmg/ddu187.
50. Xin Z, Allis CD, Wagstaff J. Parent-specific complementary patterns of histone H3 lysine 9 and H3 lysine 4 methylation at the Prader-Willi syndrome imprinting center. *Am J Hum Genet*. 2001;69(6):1389–1394. https://doi.org/10.1086/324469.
51. Xin Z, Tachibana M, Guggiari M, Heard E, Shinkai Y, Wagstaff J. Role of histone methyltransferase G9a in CpG methylation of the Prader-Willi syndrome imprinting center. *J Biol Chem*. 2003;278(17):14996–15000. https://doi.org/10.1074/jbc.M211753200.
52. Kim Y, Lee HM, Xiong Y, et al. Targeting the histone methyltransferase G9a activates imprinted genes and improves survival of a mouse model of Prader-Willi syndrome. *Nat Med*. 2017;23(2):213–222. https://doi.org/10.1038/nm.4257.
53. Bhatnagar S, Zhu X, Ou J, et al. Genetic and pharmacological reactivation of the mammalian inactive X chromosome. *Proc Natl Acad Sci USA*. 2014;111(35): 12591–12598. https://doi.org/10.1073/pnas.1413620111.
54. Amir RE, Van den Veyver IB, Wan M, Tran CQ, Francke U, Zoghbi HY. Rett syndrome is caused by mutations in X-linked MECP2, encoding methyl-CpG-binding protein 2. *Nat Genet*. 1999;23(2):185–188.
55. Giacometti E, Luikenhuis S, Beard C, Jaenisch R. Partial rescue of MeCP2 deficiency by postnatal activation of MeCP2. *Proc Natl Acad Sci USA*. 2007;104(6):1931–1936. https://doi.org/10.1073/pnas.0610593104.
56. Przanowski P, Wasko U, Zheng Z, et al. Pharmacological reactivation of inactive X-linked Mecp2 in cerebral cortical neurons of living mice. *Proc Natl Acad Sci USA*. 2018;115(31):7991–7996. https://doi.org/10.1073/pnas.1803792115.
57. Carrette LLG, Wang C-Y, Wei C, et al. A mixed modality approach towards Xi reactivation for Rett syndrome and other X-linked disorders. *Proc Natl Acad Sci USA*. 2018; 115(4):E668–E675. https://doi.org/10.1073/pnas.1715124115.
58. Muzar Z, Lozano R. Current research, diagnosis, and treatment of fragile X-associated tremor/ataxia syndrome. *Intractable and Rare Diseases Research*. 2014;3(4):101–109. https://doi.org/10.5582/irdr.2014.01029.
59. Chiurazzi P, Pomponi MG, Willemsen R, Oostra BA, Neri G. In vitro reactivation of the FMR1 gene involved in fragile X syndrome. *Hum Mol Genet*. 1998;7(1):109–113. https://doi.org/10.1093/hmg/7.1.109.

60. Feng Y, Zhang F, Lokey LK, et al. Translational suppression by trinucleotide repeat expansion at FMR1. *Science*. 1995;268(5211):731–734. https://doi.org/10.1126/science.7732383.
61. Chiurazzi P, Pomponi MG, Pietrobono R, Bakker CE, Neri G, Oostra BA. Synergistic effect of histone hyperacetylation and DNA demethylation in the reactivation of the FMR1 gene. *Hum Mol Genet*. 1999;8(12):2317–2323. https://doi.org/10.1093/hmg/8.12.2317.
62. Tabolacci E, Pietrobono R, Moscato U, Oostra BA, Chiurazzi P, Neri G. Differential epigenetic modifications in the FMR1 gene of the fragile X syndrome after reactivating pharmacological treatments. *Eur J Hum Genet*. 2005;13(5):641–648. https://doi.org/10.1038/sj.ejhg.5201393.
63. Tabolacci E, Chiurazzi P. Epigenetics, fragile X syndrome and transcriptional therapy. *Am J Med Genet*. 2013;161(11):2797–2808. https://doi.org/10.1002/ajmg.a.36264.
64. Pietrobono R, Pomponi MG, Tabolacci E, Oostra B, Chiurazzi P, Neri G. Quantitative analysis of DNA demethylation and transcriptional reactivation of the FMR1 gene in fragile X cells treated with 5-azadeoxycytidine. *Nucleic Acids Res*. 2002;30(14):3278–3285. https://doi.org/10.1093/nar/gkf434.
65. Kumari D, Usdin K. Sustained expression of FMR1 mRNA from reactivated fragile X syndrome alleles after treatment with small molecules that prevent trimethylation of H3K27. *Hum Mol Genet*. 2016;25(17):3689–3698. https://doi.org/10.1093/hmg/ddw215.
66. Liu XS, Wu H, Ji X, et al. Editing DNA methylation in the mammalian genome. *Cell*. 2016;167(1):233–247.e17. https://doi.org/10.1016/j.cell.2016.08.056.
67. Liu XS, Wu H, Krzisch M, et al. Rescue of fragile X syndrome neurons by DNA methylation editing of the FMR1 gene. *Cell*. 2018;172(5):979–992.e6. https://doi.org/10.1016/j.cell.2018.01.012.
68. Charlesworth CT, Deshpande PS, Dever DP, et al. Identification of preexisting adaptive immunity to Cas9 proteins in humans. *Nat Med*. 2019;25(2):249–254. https://doi.org/10.1038/s41591-018-0326-x.
69. Wagner DL, Amini L, Wendering DJ, et al. High prevalence of Streptococcus pyogenes Cas9-reactive T cells within the adult human population. *Nat Med*. 2019;25(2):242–248. https://doi.org/10.1038/s41591-018-0204-6.
70. Yin H, Kauffman KJ, Anderson DG. Delivery technologies for genome editing. *Nat Rev Drug Discov*. 2017;16(6):387–399. https://doi.org/10.1038/nrd.2016.280.
71. Doudna JA. The promise and challenge of therapeutic genome editing. *Nature*. 2020;578(7794):229–236. https://doi.org/10.1038/s41586-020-1978-5.
72. Walton RT, Christie KA, Whittaker MN, Kleinstiver BP. Unconstrained genome targeting with near-PAMless engineered CRISPR-Cas9 variants. *Science*. 2020;368(6488):290–296. https://doi.org/10.1126/science.aba8853.
73. Galonska C, Charlton J, Mattei AL, et al. Genome-wide tracking of dCas9-methyltransferase footprints. *Nat Commun*. 2018;9(1):597. https://doi.org/10.1038/s41467-017-02708-5.
74. Hofacker D, Broche J, Laistner L, Adam S, Bashtrykov P, Jeltsch A. Engineering of effector domains for targeted DNA methylation with reduced off-target effects. *Int J Mol Sci*. 2020;21(2). https://doi.org/10.3390/ijms21020502.
75. O'Geen H, Ren C, Nicolet CM, et al. DCas9-based epigenome editing suggests acquisition of histone methylation is not sufficient for target gene repression. *Nucleic Acids Res*. 2017;45(17):9901–9916. https://doi.org/10.1093/nar/gkx578.

76. Guy J, Gan J, Selfridge J, Cobb S, Bird A. Reversal of neurological defects in a mouse model of Rett syndrome. *Science*. 2007;315(5815):1143–1147. https://doi.org/10.1126/science.1138389.
77. Huber KM, Gallagher SM, Warren ST, Bear MF. Altered synaptic plasticity in a mouse model of fragile X mental retardation. *Proc Natl Acad Sci USA*. 2002;99(11):7746–7750. https://doi.org/10.1073/pnas.122205699.
78. Michalon A, Sidorov M, Ballard TM, et al. Chronic pharmacological mGlu5 inhibition corrects fragile X in adult mice. *Neuron*. 2012;74(1):49–56. https://doi.org/10.1016/j.neuron.2012.03.009.
79. Silva-Santos S, Van Woerden GM, Bruinsma CF, et al. Ube3a reinstatement identifies distinct developmental windows in a murine Angelman syndrome model. *J Clin Invest*. 2015;125(5):2069–2076. https://doi.org/10.1172/JCI80554.
80. Amabile A, Migliara A, Capasso P, et al. Inheritable silencing of endogenous genes by hit-and-run targeted epigenetic editing. *Cell*. 2016;167(1):219–232.e14. https://doi.org/10.1016/j.cell.2016.09.006.
81. Nuñez JK, Chen J, Pommier GC, et al. Genome-wide programmable transcriptional memory by CRISPR-based epigenome editing. *Cell*. 2021;184(9):2503–2519.e17. https://doi.org/10.1016/j.cell.2021.03.025.
82. Hathaway NA, Bell O, Hodges C, Miller EL, Neel DS, Crabtree GR. Dynamics and memory of heterochromatin in living cells. *Cell*. 2012;149(7):1447–1460. https://doi.org/10.1016/j.cell.2012.03.052.
83. Dodd IB, Micheelsen MA, Sneppen K, Thon G. Theoretical analysis of epigenetic cell memory by nucleosome modification. *Cell*. 2007;129(4):813–822. https://doi.org/10.1016/j.cell.2007.02.053.
84. Thakore PI, Kwon JB, Nelson CE, et al. RNA-guided transcriptional silencing in vivo with *S. aureus* CRISPR-Cas9 repressors. *Nat Commun*. 2018;9(1). https://doi.org/10.1038/s41467-018-04048-4.
85. Stein CA, Castanotto D. FDA-approved oligonucleotide therapies in 2017. *Mol Ther*. 2017;25(5):1069–1075. https://doi.org/10.1016/j.ymthe.2017.03.023.
86. Lieberman-Aiden E, van Berkum NL, Williams L, et al. Comprehensive mapping of long-range interactions reveals folding principles of the human genome. *Science (New York, NY)*. 2009;326(5950):289–293. https://doi.org/10.1126/science.1181369.
87. Dixon JR, Selvaraj S, Yue F, et al. Topological domains in mammalian genomes identified by analysis of chromatin interactions. *Nature*. 2012;485(7398):376–380. https://doi.org/10.1038/nature11082.
88. Hnisz D, Day DS, Young RA. Insulated neighborhoods: structural and functional units of mammalian gene control. *Cell*. 2016;167(5):1188–1200. https://doi.org/10.1016/j.cell.2016.10.024.
89. Weintraub AS, Li CH, Zamudio AV, et al. YY1 is a structural regulator of enhancer-promoter loops. *Cell*. 2017;171(7):1573–1588.e28. https://doi.org/10.1016/j.cell.2017.11.008.
90. Deng W, Lee J, Wang H, et al. Controlling long-range genomic interactions at a native locus by targeted tethering of a looping factor. *Cell*. 2012;149(6):1233–1244. https://doi.org/10.1016/j.cell.2012.03.051.
91. Morgan SL, Mariano NC, Bermudez A, et al. Manipulation of nuclear architecture through CRISPR-mediated chromosomal looping. *Nat Commun*. 2017;8:15993. https://doi.org/10.1038/ncomms15993.

92. Phillips-Cremins JE, Rege M, Valeri J, et al. LADL: light-activated dynamic looping for endogenous gene expression control. *Nat Methods*. 2019;16(7):633–639. https://doi.org/10.1038/s41592-019-0436-5.
93. Kubo N, Ishii H, Xiong X, et al. CTCF promotes long-range enhancer-promoter interactions and lineage-specific gene expression in mammalian cells. *bioRxiv*. 2020. https://doi.org/10.1101/2020.03.21.001693.
94. Kim JH, Rege M, Valeri J, et al. LADL: light-activated dynamic looping for endogenous gene expression control. *Nat Methods*. 2019;16(7):633–639. https://doi.org/10.1038/s41592-019-0436-5.
95. Deng X, Ma W, Ramani V, et al. Bipartite structure of the inactive mouse X chromosome. *Genome Biol*. 2015;16:152. https://doi.org/10.1186/s13059-015-0728-8.
96. da Rocha ST, Heard E. Novel players in X inactivation: insights into Xist-mediated gene silencing and chromosome conformation. *Nat Struct Mol Biol*. 2017;24(3):197–204. https://doi.org/10.1038/nsmb.3370.
97. Minajigi A, Froberg J, Wei C, et al. Chromosomes. A comprehensive Xist interactome reveals cohesin repulsion and an RNA-directed chromosome conformation. *Science (New York, NY)*. 2015;349(6245). https://doi.org/10.1126/science.aab2276.aab2276.

Index

'*Note:* Page numbers followed by "t" indicate tables and "b" indicate boxes.'

D

F

G

H

I

J

K

9780128217856